2nd International Conference on Communication, Image and Signal Processing (CCISP 2017)

Journal of Physics: Conference Series Volume 960

Shenzhen, China
17 – 19 November 2017

ISBN: 978-1-5108-5851-0
ISSN: 1742-6588

Printed from e-media with permission by:

Curran Associates, Inc.
57 Morehouse Lane
Red Hook, NY 12571

CURRAN ASSOCIATES INC.
proceedings
.com

Some format issues inherent in the e-media version may also appear in this print version.

Printed by Curran Associates, Inc. (2018)

For permission requests, please contact the Institute of Physics
at the address below.

Institute of Physics
Dirac House, Temple Back
Bristol BS1 6BE UK

Phone: 44 1 17 929 7481
Fax: 44 1 17 920 0979

techtracking@iop.org

Additional copies of this publication are available from:

Curran Associates, Inc.
57 Morehouse Lane
Red Hook, NY 12571 USA
Phone: 845-758-0400
Fax: 845-758-2633
Email: curran@proceedings.com
Web: www.proceedings.com

Table of contents

Volume 960

2017 2nd International Conference on Communication, Image and Signal Processing (CCISP 2017)

17–19 November 2017, Shenzhen, China

Accepted papers received: 19 January 2018
Published online: 12 February 2018

Preface

Papers

Communication Engineering

Signal Processing

PREFACE

CCISP 2017 was held in held in Shenzhen, China during Nov.17-19, 2017. CCISP 2017 was organized by Asia Pacific Institute of Science and Engineering (APISE). The conference provides a useful and wide platform both for display the latest research and for exchange of research results and thoughts in Communication, Image and Signal Processing. The participants of the conference were from almost every part of the world, with background of either academia or industry, even well-known enterprise. The success and prosperity of the conference is reflected high level of the papers received.

The proceedings are a compilation of the accepted papers and represent an interesting outcome of the conference. This book covers 3 chapters: Communication Engineering, Image Processing and Signal Processing.

We would like to acknowledge all of those who supported CCISP 2017. Each individual and institutional help were very important for the success of this conference. Especially we would like to thank the organizing committee for their valuable advices in the organization and helpful peer review of the papers.

We sincerely hope that CCISP 2017 will be a forum for excellent discussions that will put forward new ideas and promote collaborative researches. We are sure that the proceedings will serve as an important research source of references and the knowledge, which will lead to not only scientific and engineering progress but also other new products and processes.

Prof. Yucong Duan, Hainan University, China
Prof. Abdessamad Ben Hamza, Concordia University, Canada

COMMITTEES

Conference Chair

Prof. Yucong Duan, Hainan University, China

Publication Chair

Dr. Navid Asadi, University of Florida, USA

International Technical Committee

Prof. Abdessamad Ben Hamza, Concordia University, Canada

Prof. Pradip Sircar, Indian Institute of Technology, Kanpur

Dr. Ruksar Fatima, KBN College of Engineering, India

Dr. Taolue Chen, Middlesex University London, United Kingdom

Dr. Niranjana S., Manipal Institute of Technology (MIT), India

Dr. Numan Celik, Brunel University, United Kingdom

Dr. Michael Cree, University of Waikato, New Zealand

Dr. Giuseppe Serra, University of Udine, Italy

Dr. Zhang Yibo, University of Macau, Macau

Dr. Cheng Wang, University of Houston, USA

Dr. Lorenzo Baraldi, University of Modena and Reggio Emilia, Italy

Dr. Khalid Abbad, University of Sidi Mohammed Ben Abdellah, Morocco

Dr. Ayesha Choudhary, Jawaharlal Nehru University, India

Peer review statement

All papers published in this volume of *Journal of Physics: Conference Series* have been peer reviewed through processes administered by the proceedings Editors. Reviews were conducted by expert referees to the professional and scientific standards expected of a proceedings journal published by IOP Publishing.

The Esns-Based Traction Electric Quantity Prediction of High-Speed Train with Adaptive Model Inputs Temporal Scale

Kaixuan Chang[1,*], Xiaoxiao Shu[1], Yichao Zheng[1], Jing Tian[1] and Wei Li[1]

[1]State Grid Jiangxi Electric Power Research Institute, Nanchang 330096, China

*Corresponding author: kxchang@126.com

Abstract. The precise and reliable speed prediction provides a basis for predicting the traction electric quantity of high-speed train (HST). However, various existing mathematical speed control modes of HST may be insufficiently applicable because of the complex nonlinear and uncertain disturbance. To address this issue, in this paper, a speed predictive model of HST is established by taking the advantages of the efficient and robust prediction capability of echo state networks (ESNs) in dealing with nonlinear time-series. In consideration of the uncertainties of the temporal scale of the model inputs, an adaptive temporal scale selection criterion is set up to further enhance the model prediction accuracy. Then, based on the given traction characteristic curve of HST, the ESNs-based traction electric quantity prediction framework is established. Some experimental results illustrate that the proposed predictive framework can achieve a good performance in terms of predicting accuracy and reliability.

1. Introduction

The traction electric quantity consumed by high-speed train (HST) is mainly determined by the traction control strategy, which is closely related to the operation speed of HST. Besides, the precise speed predictive control technique of HST plays an essential role in realizing the energy-efficient railway transportation [1, 2]. Over the years, considerable research effort has been investigated on the model predictive control (MPC) of HST. Based on the Newtonian mechanical analysis, a nonlinear speed control model of HST was given in [3] and a predictive speed control model for HST was presented in [4], while an energy-saving control model of HST was proposed in [5]. Actually, in many cases, it is even impossible to obtain a suitable physically speed control model for HST due to the complex nonlinearity and the lack of knowledge of some critical parameters. On the other hand, as pointed out in [6], the physical phenomena that generate the temporal signals may occur at different speed. In other words, the input temporal scale of the predictive model of HST may be various in different operation states including traction, coasting and braking.

Echo state networks (ESNs) have the salient advantage of that the complex training task of classical RNN is reduced to a simple linear regression over readout weights while well reserving the memory capacities of RNNs. While ESNs showing a good modelling performance on time-series dataset, the temporal scale selection of the model input is out of the previous literature. Fortunately, the minimum description length (MDL) method has shown a great potential in analyzing real-world time-series data [6], and MDL-based criterions were developed for the semi-supervised classification in [7] as well as for model order selection in [8].

Thus, in this paper, ESNs is employed to deal with the input order uncertainties and approximately represent the nonlinear mapping of the speed and control force of HST, comparing with an adaptive selection method for the model input temporal scale. Subsequently, the MPC of HST is performed by using a PSO-based searching scheme. Finally, based on the predictive speed and control force, traction electric quantity prediction framework is built. The numerical experiments with real-world operation data of HST demonstrated the effectiveness and feasibility of the proposed predictive framework.

2. Preliminary and problem formulation

It is a fact that the properties of time-series appear in different ways depending on the temporal scale of observation, which is essential for modelling an unknown system accurately. In other words, as stated as the principle of automatic scale selection in [6], the selected temporal scale can be regarded as reflecting a characteristic length of the corresponding structure in the data under certain conditions.

Here, we take Mackey-Glass time-series for example, of which prediction model is

$$\begin{aligned} &x(t+\Delta t) = f(\mathbf{X}) \\ &\mathbf{X} = [x(t), x(t-\Delta t)\cdots, x(t-k\Delta t)], \Delta t \in \mathbb{Z}, k \in \mathbb{Z} \end{aligned} \tag{1}$$

where $\Delta t = 6$ and $k = 3$ donate the temporal scale and order of the model inputs, respectively. $\mathbf{X}$ donates an data observation point.

Similarly, for the speed prediction of HST, as a typical time-series prediction task, the data observation interval is determined by the temporal scale of model inputs, as shown in follows:

$$\begin{aligned} &\mathbf{x}(t) = f(\mathbf{z}(t-1), u(t)) \\ &v(t) = \xi(\mathbf{x}(t)) \end{aligned} \tag{2}$$

$$\mathbf{z}(t) = \{\mathbf{v}(t-md), \mathbf{u}(t-nd)\} \tag{3}$$

where $u(t)$, $\mathbf{z}(t-1)$, $\mathbf{x}(t)$ and $v(t)$ are the current system control input, model predictive input vector, output states and current speed of HST, respectively. $f(\cdot)$ donates the nonlinear mapping from the control force to the speed of HST. d is the data observation temporal scale, $m, n \in \mathbb{Z}$ are the orders of prediction input ($\mathbf{v}$) and output ($\mathbf{u}$), respectively, and here $m = n$.

Given m and d, as well as a sufficiently large number of training data, we can obtain the speed control error

$$J(t) = \frac{1}{M}\sum_{i=1}^{M} \xi(v_{d,i}(t), v_i(t)), \quad v_i(t) = f(m_i, d_i, \boldsymbol{\theta}_i, \mathbf{z}(t)) \tag{4}$$

where $\boldsymbol{\theta}$ donates the parameter vector of ESNs, M is the trial times at each iteration.

Based on the speed prediction model, the MPC of HST is to determine the current optimal control input $u(t)$ by solving the following optimization problem

$$\begin{aligned} &\underset{U(t)}{\arg\min} \ \frac{1}{T}\sum_{j=0}^{T-1} J(v(t+j), u(t+j)) \\ &s.t. \begin{cases} v(t+1) = f(\mathbf{z}(t), u(t), \boldsymbol{\theta}) \\ u(t+j) = g(cl_j), cl_j \in \{-1, 0, 1\} \\ U(t) = \{u(t), u(t+1), ..., u(t+K-1)\} \\ 0 < v(t) < V(t) \end{cases} \end{aligned} \tag{5}$$

where T is the control and prediction horizon, $g(\cdot)$ represents nonlinear mapping between the control force of HST and the operation states of HST control inputs cl_i, where "-1, 0, 1" donate the control states of braking, coasting and traction, respectively.

With the obtained optimal control strategy and the corresponding speed, we can predict the traction electric quantity of HST Q as follows

$$Q_j(t) = \psi(u(t+j), v(t+j)) \tag{6}$$

where ψ donate the traction characteristic relationships of HST.

3. The ESNs-based MPC of HST with adaptive temporal scale selection

In this section, the ESNs-based prediction model of HST is introduced companied with the adaptive temporal scale selection criterion. Then, the MPC of HST is performed with the PSO-based optimization strategy.

3.1 ESNs-based speed prediction model of HST

The typical ESNs can be defined by the state update and output equations, and for simplicity, we focus on the basic ESNs without output feedback, of which the architecture is shown in Figure 1 [9].

Based on Figure 1, the state update and output equation are shown as follows:

$$\mathbf{X}(t) = f(\mathbf{W}_{\mathbf{in}}^{T}\mathbf{z}(t) + \mathbf{W}\mathbf{X}(t-1) + \mathbf{B}) \tag{7}$$

$$v(t) = g(\mathbf{W}_{\mathbf{out}}^{T}\mathbf{X}(t)) \tag{8}$$

where $\mathbf{W}_{\mathbf{in}}^{T}$, $\mathbf{W}$ and $\mathbf{W}_{\mathbf{out}}^{T}$ are the weight matrices of input connection, internal feedback and output connection, respectively. $\mathbf{X}(t) = [x_1(t), ..., x_l(t)]^T$ is the internal activation states vector of ESNs, $\mathbf{B} = [b_1, ..., b_l]^T$ are biases. f and g are the activation functions of the internal states and outputs of ESNs, which generally are chosen as sigmoid function and identity function, respectively.

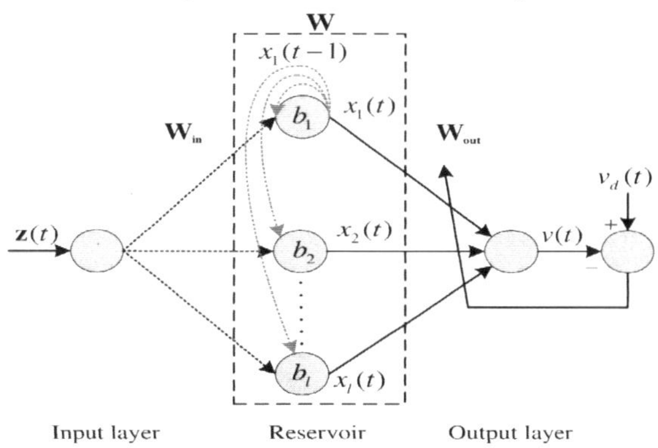

Figure 1. The ESNs-based speed prediction model of HST (each node has the connection like the one with b1; dashed arrows donate randomly generated and then fixed connection weights, and solid arrows donate trained connection weights)

3.2 Adaptive temporal scale selection

Motivated by the successful application of MDL in automatically discovering the intrinsic dimensionality of time-series [6], we apply the MDL-based criterion to adaptively choose the appropriate temporal scale of the model inputs. According to the idea that MDL method provides the same information as the minimum eigenvalue of the covariance matrix of observation datasets [8], we have the selected criterion for the temporal parameters defined as follows:

$$J_{m,d} = \frac{N}{2}\log(\lambda_m) + \frac{m+d}{2}\log N \tag{9}$$

where $J_{m,d}$ is the selected criterion, N is the number of observation data points, λ_m donates the minimal eigenvalue.

Then, as $\log(\cdot)$ is monotonically increasing, given a sufficiently large N, it can be derived out [9]

$$J_{m,d} = \sum_{i=1}^{c} \gamma_i \lambda_{m,i} \tag{10}$$

where c is the number of approximated linear subsets obtained by fuzzy clustering, and γ_i is the weighted value of the ith subset.

Finally, we obtained the selection criterion for temporal scale and input orders as follows:

$$[m,d]=\arg\min_{m,d}\left((\max(\frac{J(m,d)}{J(m-1,d)},\frac{J(m,d)}{J(m,d-1)}))\right) \tag{11}$$

3.3 Traction electric quantity prediction of HST with PSO-based optimization

PSO is a parallel and heuristic searching algorithm, which shows a good optimization performance in terms of efficiency and convergence rate. The update mechanism for the speed $v_j(k+1)$ and position $x_j(k+1)$ of particles is [10]:

$$\begin{aligned} & v_j(k+1)=\omega\cdot v_j(k)+\varphi 1\cdot r1\cdot(pb_j(k)-x_j(k))+\varphi 2\cdot r2\cdot(gb_j(k)-x_j(k)) \\ & x_j(k+1)=x_j(k)+v_j(k+1) \\ & \omega=\omega_{\max}-(\omega_{\max}-\omega_{\min})\cdot k/k_{\max} \end{aligned} \tag{12}$$

where j and k represent the particle and iteration times, $pb_j(n)$ and $gb_j(n)$ donate the current optimal position of a particle and the population, $\varphi 1$ and $\varphi 2$ are the acceleration coefficients of particles, $r1$ and $r2$ are random parameters valued in [0, 1], ω is the dynamic weighting factor.

For the MPC of HST, the root-mean-square error (e_v) and standard derivation (s_d) of the speed tracking accuracy are used as criterion indexes

$$e_v=\left(\frac{1}{M}\sum_{t=1}^{M}(v_d(t)-v(t))^2\right)^{1/2},\quad s_d=\left(\frac{1}{M}\sum_{t=1}^{M}(v_d(t)-\mu)^2\right)^{1/2} \tag{13}$$

where μ is the average of the predicted speed of M times running for each independent trial.

Then, based on the obtained predictive speed and the traction characteristic of HST[2], (6) can be rewrote as follows:

$$Q_j(t)=\begin{cases}\psi_1(u(t+j),v(t+j)), & 0<v(t+j)<50\\ \psi_2(u(t+j),v(t+j)), & 50\le v(t+j)<168\\ \psi_3(u(t+j),v(t+j)), & 168\le v(t+j)\le 350\end{cases} \tag{14}$$

4. Experimental studies

To evaluate the performance of the proposed prediction framework for the traction electric quantity of HST, two groups of simulation experiments (*a* and *b*) are carried out in this section based on the real-world operation data of HST which are collected from the driving cab of HST serviced on "Beijing-Shanghai high-speed railway" in China. Then, the ESNs-based speed prediction model of HST is trained with the structure of one input node, 60 internal nodes and one output node.

For the experiment *a*, the temporal scale was calculated according to (11) at the initial iteration and then fixed; on the contrary, for the experiment *b*, the temporal scale was updated by (11) at each predictive iteration. By applying the established prediction model and the indexes defined in (13), we performed the MPC of HST for 5 dependent trials, and each trial was running for M=50 times. Then, we obtained the evaluation results as shown in Table 1.

Table 1 Speed tracking error of the ENSs-based MPC of HST

Trails	1		2		3		4		5	
	a	b	a	b	a	b	a	B	a	b
e_v /e-3	5.2	4.8	5.1	4.6	4.9	4.5	5.3	5.0	5.2	4.9
S_d /e-4	2.0	1.6	1.7	1.7	1.8	1.5	1.9	1.7	1.7	1.4

where *a* and *b* represent the experimental results with fixed and adaptive temporal scale selection, respectively.

According to Table 1, we choose the simulation results of the third trial to demonstrate the speed tracking performance, as shown in Figure 2. Then, from Table 1 and Figure 2, we can see that the ESNs-based MPC of HST with both fixed and adaptive temporal scale could achieve a sound speed tracking performance. Furthermore, it can be observed from Table 1 that the MPC of HST with adaptive temporal scale selection (*b* in Table 1) obtain a better speed tracking performance in terms of accuracy and reliability than that of *a* in Table 1. Intuitively, as shown in Figure 2, it is found that the tracking speed in Figure 2b can follow the target speed of HST significantly more precisely than that in Figure 2a.

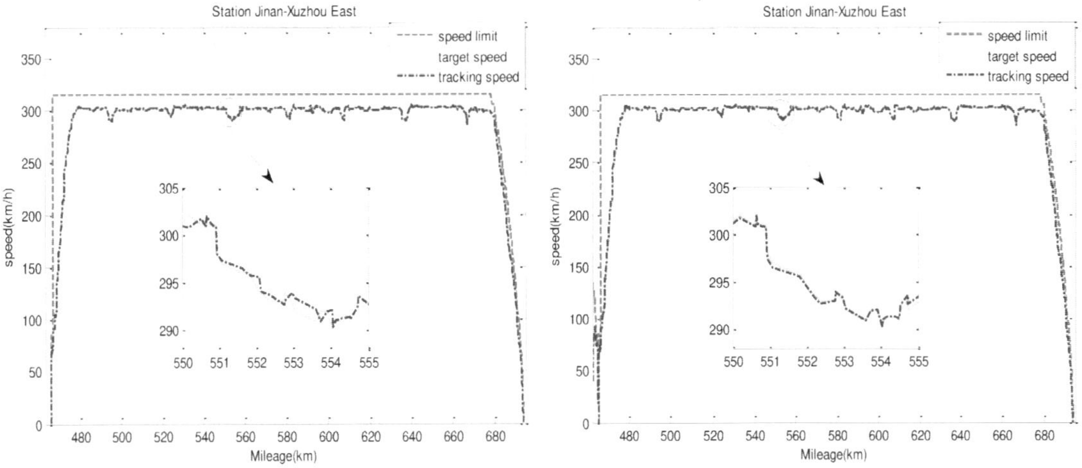

(a) With fixed temporal scale selection

(b) With the proposed adaptive temporal scale selection

Figure 2. The tracking speed of the ESNs-based MPC of HST

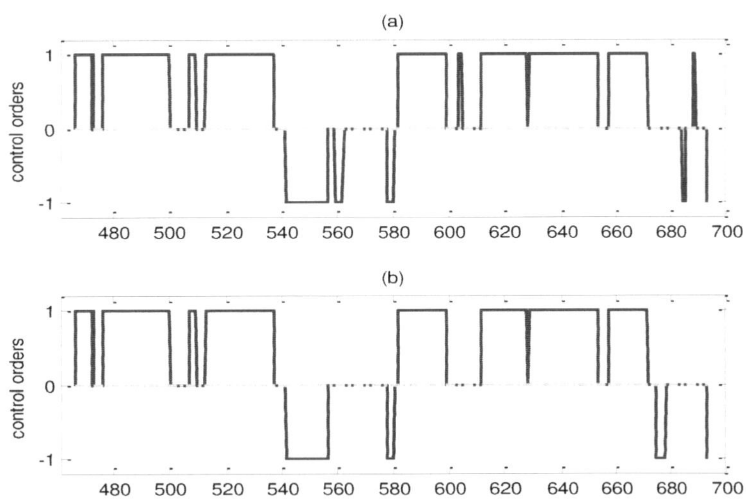

Figure 3. The predictive control orders of the ESNs-based MPC of HST (here (a) and (b) corresponding to that in Figure 2)

Then, according to equation (14), Figure 2 and Figure 3, we have the predicted traction electric quantity as shown in Figure 4.

The total traction electric quantities corresponding to (a) and (b) in Figure 4 are 6455Kwh and 6340Kwh, respectively.

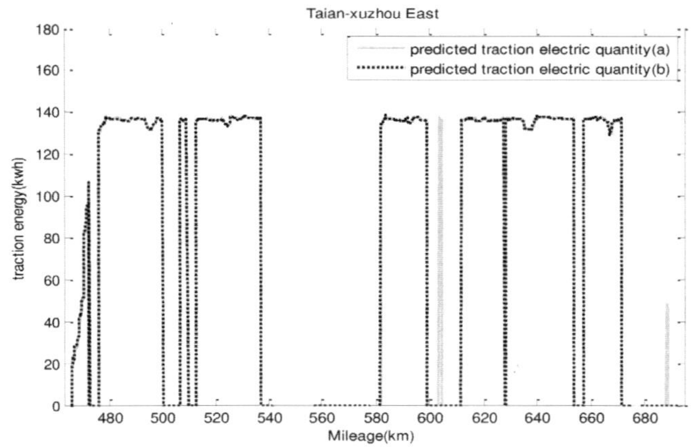

Figure 4. The predicted unit traction electric quantity of HST

5. Conclusion

The traction electric quantity prediction of HST is a kind of typical industrial process control problem, companying with the characteristics of complex nonlinearity and parameter uncertainties. Owing to the good training efficiency and memory capacities, ESNs show a promising potential to deal with this predictive control issue. Besides, the determination of temporal scale and order of the modeling inputs plays an important role in guaranteeing the predictive control performance. The further work about the proposed predictive control framework can be carried out from the aspects including efficient predictive control algorithms and online temporal scale selection criterions.

References

[1] Yang, H., Zhang, K.P., Wang, X., & Zhong, L.S. (2011). *Generalized multiple-model predictive control method of high-speed train*. Journal of the China Railway Society, 33(8), 80-87.

[2] Yang, H., Zhang, K.P., & Liu, H.E. (2016). *Online regulation of high speed train trajectory control based on ts fuzzy bilinear model*. IEEE Transactions on Intelligent Transportation Systems, 17(6), 1496-1508.

[3] Yang, G., Liu, M.G., Yu, L. (2013). *Nonlinear Predictive Control of Operation Process of High-speed Train*. Journal of the China Railway Society, 35(8): 16-21.

[4] Li, Z.Q., Yang, Z.C., Yang, H., Liu, J.M. (2015). *Dual Adaptive Generalized Predictive Control Method for High-speed Train*. China Railway Science, 36(6): 120-127.

[5] Liu, J.Q., Wei, Y.L., Hu, H. (2014). *Research on Optimization Control Method of Energy-saving Operation of High-speed Trains*. Journal of the China Railway Society, 36(10): 7-12.

[6] Lindeberg, T. (2016). *Temporal Scale Selection in Time-Causal Scale Space*. Journal of Mathematical Imaging and Vision, 1-45.

[7] Hu, B., Rakthanmanon, T., Hao, Y., Evans, S., Lonardi, S., & Keogh, E. (2011, December). *Discovering the intrinsic cardinality and dimensionality of time series using MDL*. In Data Mining (ICDM), 2011 IEEE 11th International Conference on (pp. 1086-1091). IEEE.

[8] Feil, B., Abonyi, J., & Szeifert, F. (2004). *Model order selection of nonlinear input–output models—a clustering based approach*. Journal of Process Control, 14(6), 593-602.

[9] Jaeger, H., & Haas, H. (2004). *Harnessing nonlinearity: Predicting chaotic systems and saving energy in wireless communication*. Science, 304(5667), 78-80.

[10] Clerc, M., & Kennedy, J. (2002). *The particle swarm-explosion, stability and convergence in a multidimensional complex space*. IEEE transactions on Evolutionary Computation, 6(1), 58-73.

An Efficient Offloading Scheme For MEC System Considering Delay and Energy Consumption

Yanhua Sun[1,2,3,a], Zhe Hao[1,2,3,b] and Yanhua Zhang[1,2,3]

[1]Beijing Advanced Innovation Center for Future Internet Technology, Beijing University of Technology, No 100, Pingleyuan, Beijing, China
[2]Beijing Laboratory of Advanced Information Networks, Beijing, China
[3]College of Information and Communications, Beijing University of Technology, No 100, Pingleyuan, Beijing, China

Email: [a]sunyanhua@bjut.edu.cn, [b]zoeahao@qq.com

Abstract. With the increasing numbers of mobile devices, mobile edge computing (MEC) which provides cloud computing capabilities proximate to mobile devices in 5G networks has been envisioned as a promising paradigm to enhance users experience. In this paper, we investigate a joint consideration of delay and energy consumption offloading scheme (JCDE) for MEC system in 5G heterogeneous networks. An optimization is formulated to minimize the delay as well as energy consumption of the offloading system, which the delay and energy consumption of transmitting and calculating tasks are taken into account. We adopt an iterative greedy algorithm to solve the optimization problem. Furthermore, simulations were carried out to validate the utility and effectiveness of our proposed scheme. The effect of parameter variations on the system is analysed as well. Numerical results demonstrate delay and energy efficiency promotion of our proposed scheme compared with another paper's scheme.

1. Introduction

In the foreseeable future 5G network, some applications require low latency computing but result in high energy consumption [1,2,3,4]. However, the computing capabilities and battery power of mobile devices (MD)are limited. These factors lead to conflicts between the applications and smart mobile devices, as well as, bring unprecedented challenges. Consequently, the idea of making the radio access network in close proximity to mobile devices that possess bandwidth and latency-critical data is becoming the consensus of the industry, and the MEC, which comes with the idea, has become the key technology to improve the users experience in 5G network [5]. MEC architecture allows MD to offload their computation-intensive tasks to MEC server via wireless cellular network [6].

It is well known that fog computing has been proposed to deploy computing resources closer to the terminals [7,8]. In recent years, studies on MEC architecture have become the hot spot gradually. In [9], a dynamic service migration problem in MEC were studied by authors. In [10], an analytical model for MEC was presented. In [11], a virtual network embedding problems were investigated, and network virtualization in the context of MEC networks were proposed by authors. In [12], an optimization problem to minimize the energy consumption of a MEC offloading system is researched by authors. In [13], a computation task scheduling policy of MEC offloading system is studied.

In this paper, a joint delay and energy efficient offloading scheme (JCDE) for multiuser MEC system is proposed, which takes into account both the latency demands and energy requirements of each MD in the system. Then, an optimization problem is formulated, in order to solve it, an iterative greedy algorithm is adopted. Numerical results show that proposed JCDE scenario achieves better performance compare with EECO scheme [12] and stochastic offloading scheme. Moreover, the impacts of different CPU capabilities of MEC server and MD on the offloading decisions are analyzed. Finally, the effects of MD's requirements on offloading assignments are investigated. The rest of the paper is organized as follows. The system model is introduced in Section2. The delay and energy efficient optimization problem is formulated in Section3. The JCDE is described in Section4. Then the numerical results are presented in Section5 and summary in Section6.

2. System model

As illustrated in figure 1, MD offload their tasks to MEC server through a 5G heterogeneous network. There is a Macro base station (MBS) equipped with a MEC server, besides there is a Small base station (SBS) whose service area is overlaid by that of the MBS. It is worth mentioning that MEC server has the ability to handle multiple tasks simultaneously.

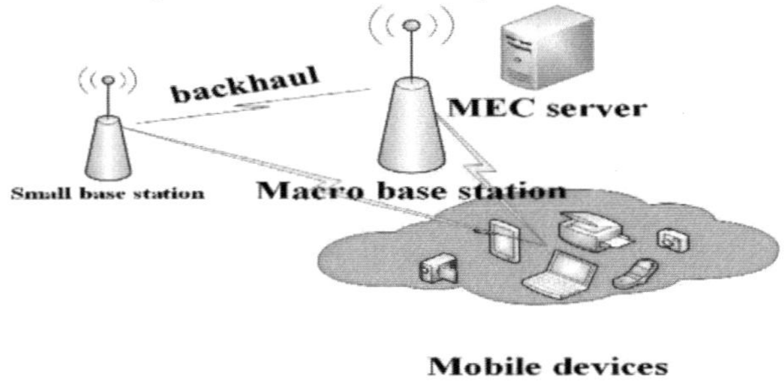

Figure 1. MEC system framework

We consider multiple MD which is denoted as $\mathcal{N} = \{1, 2, ..., N\}$. Each MD has a delay sensitive task or computation-intensive task. Each task can be described in three parameters as $T_i = \{I_i, O_i, c_i, t_i^{\max}\}$, $i \in \mathcal{N}$. I_i is the size of computational task (in bits), O_i is the size of output data which is computed by MEC server (in bits). c_i denotes the numbers of CPU cycles required for computing this task. For each MD i its task can either be executed locally on itself or accomplished on MEC server via offloading. $t_i^{\max}$ is the maximum latency requirement by the computational task.

3. Problem formulation

3.1. Key factors

In this system, some MD are running some applications that are sensitive to the delay and others that hope to save more energy. Considering different needs of different MD, we propose different weighting factors φ_i^T and φ_i^E for different MD in order to provide a better modeling flexibility and meet personalized user demands [8]. The former denotes latency factor and the latter denotes energy factor. $\varphi_i^T + \varphi_i^E = 1$ is satisfied. For MD can offload tasks to MEC server in two ways, $a_{i,j} = \{0,1\}$ is introduced as a task offloading decision indicator. Here $i \in \mathcal{N}$, when $a_{i,j} = 1$, the MD i selects mode j to process the task T_i. Otherwise, $a_{i,j} = 0$. $j = 1, 2, 3$ denotes the chosen modes. Specifically, $j = 1$ denotes the mode where the computational task T_i is executed locally, $j = 2$ corresponds to the mode where the task T_i is offloaded to MBS directly, $j = 3$ denotes the mode where the computational task is offloaded to MBS via the relay SBS. Thus, $a_{i,1} = 1$ represents that the MD i selects local computation. Similarly, $a_{i,2} = 1$ and $a_{i,3} = 1$ denote that MD i offload computational tasks to MEC server directly and via the relay SBS respectively.

3.2. *Local problem formulation*

Let F_i^L be the computational capability of MD i (in CPU cycles per second), different MD have different computational capabilities. Let δ_i^L be the energy consumption for one CPU cycle of MD. Thus, computational time of executing T_i locally is given as

$$t_{i,ex}^L = c_i / F_i^L \tag{1}$$

The energy consumption is calculated as

$$e_i^L = c_i \delta_i^L \tag{2}$$

Since there is no other time or energy consumption in the local execution, (1) represents the total delay of MD i accomplishing the task T_i locally. (2) represents the total energy consumption of MD i accomplishing T_i locally as well.

3.3. *Offloading problem formulation*

In our framework, a multi-user OFDMA in 5G heterogeneous networks is considered, the channels between the MD accessing the same base station are orthogonal to each other, therefore, there is only interference between the signals transmitted to the MBS and those to the SBS. The power consumption of the backhaul between the MBS and the SBS is neglected for this backhaul is shared with other communication infrastructures according to [12]. The backhaul transmission time is expressed as a proportional to the size of the data, and the coefficient is denoted by β.

When the MD i accesses the MBS, the uplink transmission rate can be given as

$$r_{i,u}^M = W \log_2\left(1 + \frac{P_i^M g_i^M}{I_i^S + n_0}\right) \tag{3}$$

Where W denotes the bandwidth (in Hz). P_i^M denotes the power of MD i transmitting data to MBS, it can be controlled by MBS according to power control mechanism [14]. g_i^M denotes the channel gain between the MD i and the MBS. I_i^S denotes the interference at the MBS which is caused by other MD's uplink transmission to the SBS. n_0 denotes the background noise power.

Similar to the above, the MD i accesses the SBS, the uplink transmission rate is given as

$$r_{i,u}^S = W \log_2\left(1 + \frac{P_i^S g_i^S}{I_i^M + n_0}\right) \tag{4}$$

Let F_i^S be the computational capability of MEC server and it is a constant for every task. Let δ_i^S be the energy cost of the MEC server for one unit CPU cycle, we assume $\delta_i^S < \delta_i^L$ on the basis of [12]. For MD i offloading T_i to MEC server directly, uplink transmission time is given as

$$t_{i,u}^M = I_i / r_{i,u}^M \tag{5}$$

Computing executing time is calculated as

$$t_{i,ex}^M = c_i / F_i^S \tag{6}$$

Downlink transmission time is calculated as

$$t_{i,d}^M = O_i / r_{i,d}^M \tag{7}$$

$r_{i,d}^M$ is the downlink transmission rate. For the size of output data O_i is much smaller than that of input data I_i [13]. $t_{i,d}^M$ is neglected ($t_{i,d}^M \approx 0$). Thus, (5) and (6) are the total time delay of MD i offloading T_i to MEC server through MBS directly.

Energy consumption of MD i offloading task to MBS directly includes two parts: uplink transmission energy, computational energy (ignoring downlink energy due to aforementioned reason similarly). They can be given respectively as

$$e_{i,u}^{M} = P_i^M I_i / r_{i,u}^{M} \tag{8}$$

$$e_{i,ex}^{M} = c_i \delta_i^S \tag{9}$$

According to the above, the total time and energy consumption of MD i offloading T_i to MEC server through MBS directly can be given as

$$t_i^M = t_{i,u}^M + t_{i,ex}^M \tag{10}$$

$$e_i^M = e_{i,u}^M + e_{i,ex}^M \tag{11}$$

Similarly, the total time consumption for the case where MD i selects to offload task through the SBS can be given as

$$t_i^S = t_{i,u}^S + t_i^{backhaul} + t_{i,ex}^S \tag{12}$$

Where $t_i^{backhaul}$ is the backhaul delay, calculated as

$$t_i^{backhaul} = I_i \beta \tag{13}$$

$t_{i,ex}^S = t_{i,ex}^M$ because computational execution time by MEC server for the same task is the same. Thus according to (5), (6), (8), (9), (10)and (11) the total duration of MD i selects offloading task through SS can be calculated as

$$t_i^S = I_i / r_{i,u}^S + I_i \beta + c_i / F_i^S \tag{14}$$

The total energy consumption can be calculated similarly as

$$e_i^S = P_i^S I_i / r_{i,u}^S + c_i \delta_i^S \tag{15}$$

Let introduce the task offloading decision indicator (3) and (4) can be rewritten as

$$r_{i,u}^M = W \log_2\left(1 + \frac{P_i^M g_i^M}{\sum_{l=1, l\neq i}^{N} a_{l,3} P_l^S g_l^M + n_0}\right) \tag{16}$$

$$r_{i,u}^S = W \log_2\left(1 + \frac{P_i^S g_i^S}{\sum_{l=1, l\neq i}^{N} a_{l,2} P_l^M g_l^S + n_0}\right) \tag{17}$$

3.4. System problem formulation

In summary, considering $a_{i,j}$, φ_i^T, φ_i^E, (1), (2), (5), (6), (8), (9), (10), (11), (12), (13), (14), (15), (16) and (17) the optimization problem is mathematically formulated to minimize the total time and energy consumption

$$\min \sum_{i=1}^{N} \{\varphi_i^T [a_{i,1} \frac{c_i}{F_i^L} + a_{i,2}(\frac{I_i}{W \log_2(1 + \frac{P_i^M g_i^M}{\sum_{l=1, l\neq i}^{N} a_{l,3} P_l^S g_l^M + n_0})} + \frac{c_i}{F_i^S})$$

$$+ a_{i,3}(\frac{I_i}{W \log_2(1 + \frac{P_i^S g_i^S}{\sum_{l=1, l\neq i}^{N} a_{l,2} P_l^M g_l^S + n_0})} + I_i \beta + \frac{c_i}{F_i^S})]$$

$$+ \varphi_i^E [a_{i,1} c_i \delta_i^L + a_{i,2}(\frac{P_i^M I_i}{W \log_2(1 + \frac{P_i^M g_i^M}{\sum_{l=1, l\neq i}^{N} a_{l,3} P_l^S g_l^M + n_0})} + c_i \delta_i^S) + a_{i,3}(\frac{P_i^S I_i}{W \log_2(1 + \frac{P_i^S g_i^S}{\sum_{l=1, l\neq i}^{N} a_{l,2} P_l^M g_l^S + n_0})} + c_i \delta_i^S)]\}$$

$$\begin{aligned}
&st. \\
&t_{i,ex}^{L} \le t_i^{\max}, t_i^{M} \le t_i^{\max}, t_i^{S} \le t_i^{\max} \\
&r_i^{M} \geqslant \frac{I_i}{t_i^{\max} - c_i / F_i^{s}}, r_i^{S} \geqslant \frac{I_i}{t_i^{\max} - ci / F_i^{S} - I_i \beta} \\
&a_{i,1} + a_{i,2} + a_{i,3} = 1 \\
&\varphi_i^{T} + \varphi_i^{E} = 1
\end{aligned} \tag{18}$$

4. JCDE algorithm

In this section, the details of JCDE scheme are given below:

1: **Initialization** of MD $i=\{1,2,...,N\}$, T_i, local mode $a_{i,1}$=0, MBS mode $a_{i,2}$=0, SBS mode $a_{i,3}$ =0, bandwidth, background noise, coefficient etc.
2: **repeat** for each MD in parallel (in a simulation, each MD is chosen as the first choice):
3: **while** three modes of devices $a_{i,1}+a_{i,2}+a_{i,3} \neq N$ **Do**
4: {choose a MD
5: compute the total consumption in three different modes
6: compare three results of this MD and choose the minimum consumption mode
7: choose another MD execute step4 under former MD's influence then go to 6}
8: **end while**
9: compare all results of parallel total consumptions
10: choose the minimum result and record its chosen mode $a_{i,1}$, $a_{i,2}$, $a_{i,3}$
11: **end**

5. Numerical results

In this section, we conduct simulations to validate the utility and effectiveness of proposed JCDE algorithm. Plenty of users are randomly distributed in a cell with a radius of 1000m, a MBS and a SBS are in this area. Bandwidth is 2MHz, the CPU capability of MEC server is {2,3,4,5} GHz/s, N= {10,20,30,40,50,60,70,80,90,100}, backhaul delay coefficient is 0.0001s/KB, delay constraint is 0.5s, CPU cycles required by computing a task are randomly distributed between 0.1 and 1.1GHz.

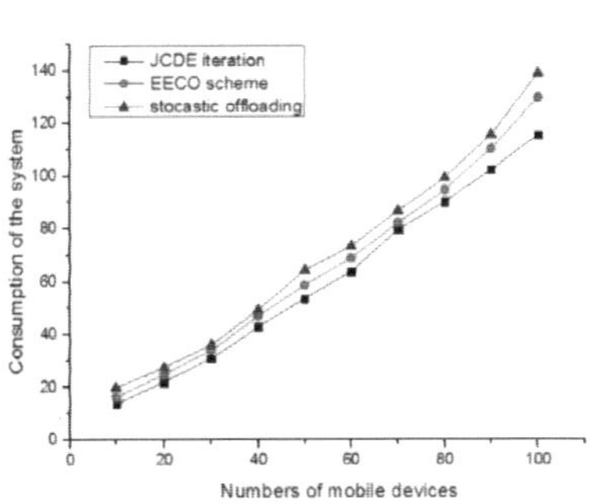

Figure 2. Consumption of system with different schemes

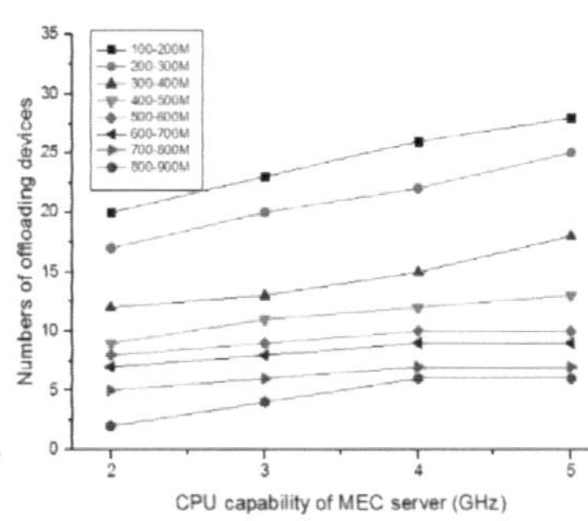

Figure 3. Influence of both CPU computing capabilities

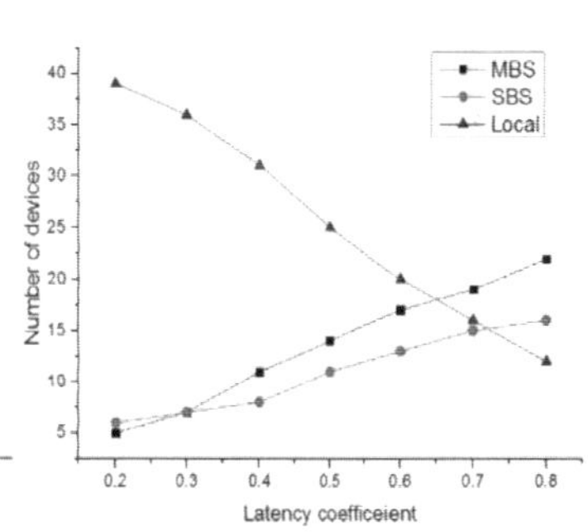

Figure 4. Influence of latency and energy coefficients

Figure 2 indicates the total consumption of the system with three different offloading schemes in terms of MD numbers. It is obvious that, regardless of MD numbers, consumptions of system in our JCDE scheme are always less than those in both EECO scheme [12] and stochastic offloading scheme. The simulation results validate the utility as well as effectiveness of our proposed scheme.

Figure 3 (N=50 in the simulations) shows that with the computational capability of MEC server CPU increasing and the computational capability of local CPU reducing, the number of MD which select to offload tasks is increases. Even though the computational capability of local CPU is strong, the computational capability MEC server CPU is weak, few MD still offload their tasks because they are close to the MBS and SBS and their tasks are delay-sensitive.

Figure 4 (the computational capability of MEC server CPU is 5GHz/s, N=50) shows latency and energy coefficient can affect offloading selections. It means the higher delay requirements of MD, the more assignments to offload, this is because the strong computing capability of MEC server. When the delay requirements of MD are low, most MD choose to perform computational tasks locally due to the large amount of energy consumed during offloading process.

6. Summary

In this paper, an optimization problem to minimize system consumption considering delay and energy for MEC system is formulated. To solve it, we design a JCDE algorithm and carry out simulations to validate the utility and effectiveness of proposed scheme. Numerical results demonstrate delay and energy efficiency promotion of our proposed scheme compared with EECO scheme and stochastic offloading scheme. In our future work we will study on NP-hard and more complex system model.

Acknowledgement

This work is jointly supported by the National Natural Science Foundation of China under Grant 61571021 and 61372089.

References

[1] Wang S and Dey S, Adaptive mobile cloud computing to enable rich mobile multimedia applications, *IEEE Trans. Multimedia,* vol 15, 4(2013), 870–883.
[2] http://www.ngmn.org/5g-white-paper.html.
[3] Shen J, Tan H,Wang J and Lee S, A novel routing protocol providing good transmission reliability in underwater sensor networks. *J. Internet Technol*, vol 16, 1(2015), 171-178.
[4] Pan Z, Zhang Y, Kwong S, Efficient motion and disparity estimation optimization for low complexity multiview video coding, *IEEE Trans. Broadcast*, vol 61, 2(2015), 166-176.
[5] 2016 Mobile Edge Computing– A Key Technology Towards 5G[EB/OL].
[6] ETSI 2014 *ETSI White Paper.*
[7] Zhanikeev M, A cloud visitation platform to facilitate cloud federation and fog computing, *computer,* vol 48, 5(2015), 80-83
[8] Sarkar S, Chatterjee S, Misra S, Assessment of the suitability of fog computing in the context of Internet of things, *IEEE Trans. Cloud Computing,* 99(2015).
[9] Wang S, Urgaonkar R, Zafer M, He T, Chan K, Leung K K, Dynamic service migration in mobile edge-clouds, *Proc. IFIP Netw.Conf,* 5(2015), 1–9
[10] Taleb T, Ksentini A, An analytical model for follow me cloud, *Proc. IEEE Global Commun Conf,* 12(2013), 1291–1296.
[11] Beck M T, Maier M, Mobile edge computing: Challenges for future virtual network embedding algorithms, *Proc. Adv. Eng. Comput.Appl.Sci*, 8(2014), 65–70.
[12] Zhang K, Mao Y M, Energy-efficient offloading for mobile edge computing in 5G heterogeneous networks, *IEEE Access*, vol 4(2016).
[13] Juan Liu, Yuyi Mao, Jun Zhang, and Khaled B. Letaief, Delay-optimal computation task scheduling for mobile-edge computing systems, *IEEE International Symposium on Information Theory,* 1451-1455
[14] Liu K W, Yang Z T, Xiong H X, Aplication Rsearch of Cmputers, vol 34(2017).

Analysis of Hardware Impairments on the Energy Harvesting Hybrid Relay Networks

K Guo[1], D Guo[1,*] and B Zhang[1]

[1]College of Communications Engineering, Army Engineering University, Nanjing, Jiangsu, 210007, China.

*Corresponding author: dxgguo@sina.cn

Abstract. In real communication systems, hardware impairments are often considered. When the system is suffering hardware impairments, the system performance will be worse. For this reason, in this paper, the impact of hardware impairments on the system performance for the energy harvesting hybrid relay networks is investigated. Especially, the closed-form expression of the outage performance and the optimal analysis of the instantaneous throughput for the system are derived, from the analysis we know that the impairments level leads great loss on the system performance. In addition, numerical results are derived to verify the correctness of our analysis.

1. Introduction

Now, an efficient solution to prolong the utility time of the wireless node is energy harvesting technology. In former works, if we want to change the lifetime of the wireless node, we must change battery or eliminated the cost of hard-wiring [1]-[5].

The wireless information and power transfer (SWIPT) technology was first proposed in [1] and [2]. This technology was proposed due to the wireless node can perform two functions, i.e. the node can harvest energy and decode the message at the same time. Until now, we know that it is more and more practical to analyse the energy harvesting technology, many literatures have analysed this technology [3]-[5]. Energy harvesting is needed for hybrid communication network duo to the limit the power of relay, and for some conditions, we could not change the battery of the relay, hence the hybrid relay networks is badly in need of energy harvesting technology.

In many former works, the authors just assumed the hardware in wireless communication node is perfect, which is not practical in relay system. Phase noise, I/Q imbalance and the other impairments factor will affect the hardware [6]-[8]. According to [9], a general hardware impairments (HIs) model was proposed to give an easy way to analyse HIs in real communication systems.

Hardware impairments is widely analysed in single-hop systems [10]-[12]. From the literatures, we know that once the network suffers (HIs), the network will have a fundamental capacity ceiling which is fixed value just the function of impairments level, in high rate systems this scene was seriously [13]. In [14], the authors analysed the impact of HIs on two-way relay networks. In [15], two selection schemes, namely, (opportunistic relay selection and partial relay selection scheme) was used in the networks to gain better system performance. Specially, the expressions for the system performance were derived. Recently, the system performance for the energy harvesting relay network and cognitive networks with hardware impairments has been analysed in [16]-[17], respectively.

In [18], a new selection scheme was used in the multi-relay networks in the presence of HIs and shadowed Rician (SR) channel. In [19], the impact of HIs on the satellite terrestrial networks was analysed, to its regret, the interference of the networks is ignored.

Only [19]-[20] researched effect of HIs on the hybrid relay networks, but the authors just analysed the HIs for regular hybrid communication system, neglecting the importance of energy harvesting on the system. The energy harvesting is a hot topic recently. Besides, taking the energy harvesting into consideration is very necessary for the hybrid networks. Hence, it is necessary for us to analyse the impact of HIs on energy harvesting hybrid relay networks. Based on these observations, the main innovations of the paper are concluded as, firstly, we investigate the impact of HIs on the energy harvesting hybrid relay networks. It is the first paper which analyse the effect of HIs on the energy harvesting hybrid relay networks. Secondly, we derive the expression of the outage performance of the system, as we know that the closed-from expression was first derived over hybrid networks. The paper suggests that both the fading channel parameters and HIs level affect the system performance. Finally, the optimal analysis for the instantaneous throughput is also derived, which give quick ways to estimate the impact of key parameters, i.e., the HIs level and the energy conversion efficiency on the throughput.

The structure of the rest paper is that, in Section II, the system model is introduced and the problem is formulated. In Section III, the signal-to-noise-and-distortion-rate (SNDR), the outage performance and the instantaneous throughput of the energy harvesting communication system was derived. In Section IV, the necessary simulation results are derived to verify the right of the theory results. In Section V, the conclusion is given.

2. The Illustration of the System and Channel Models

As illustrated in Figure 1, we analyse a dual-hop energy harvesting hybrid relay network with HIs. In the network, a single antenna source (S), a single antenna relay node (R) and a single antenna destination node (D) is considered. Each of them operates in the half-duplex mode, which is a scene suitable to the hybrid relay network. For the long distance, no direct link is assumed in this paper. T is the total time in which the information is transmitted from S to D and α is the factor of the total time when relay was harvested energy from S where $0 \le \alpha \le 1$. In the left time, $(1-\alpha)T$ is used for transmitting the information, due to R is a half-duplex relay, so $(1-\alpha)T/2$ is used for S to R to transmit the information and the other half $(1-\alpha)T/2$ is used for R to D to transmit the information.

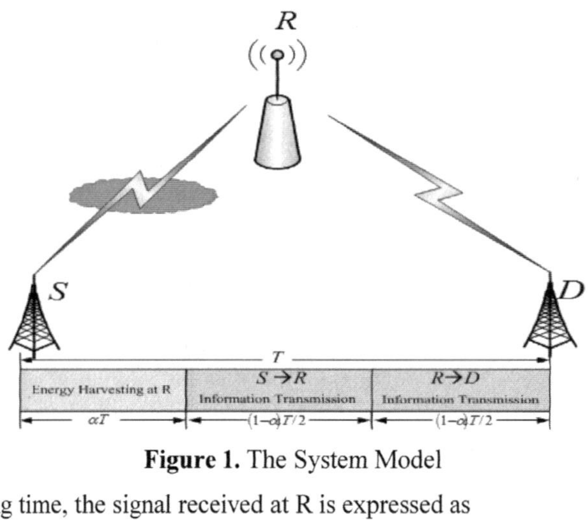

Figure 1. The System Model

In energy harvesting time, the signal received at R is expressed as

$$y_r^e = h_1 x_e + n_r^e \tag{1}$$

Where h_1 is the SR random variable (RV) for S$\rightarrow$R link, x_e the signal transmitted from S with power P, n_r^e the additive white Gaussian noise (AWGN), y_r^e the signal received by R that comes from energy harvested section with the power of E_R.

With the help of (1), in the harvesting period αT, the harvested energy, E_R is given by

$$E_R = \mu\alpha TP|h_1|^2 \tag{2}$$

Where $0 \le \mu \le 1$ is the energy efficiency.

During an information time, in the first half $(1-\alpha)T/2$, S broads the signal to R, the signal received at R is given by

$$y_r = h_1(x_s + \eta_{st}) + \eta_{rd} + n_r \tag{3}$$

Where x_s is the sent signal from S which has the power of P, η_{st} and η_{rd} are the distortion noise at S, R with the power $k_{st}^2 P$, $|h_1|^2 k_{rd}^2 P$, respectively. k_{st} and k_{rd} characterize the impairments level at S and R in the first half $(1-\alpha)T/2$. n_r is the AWGN at R with power $n_r \sim \mathcal{CN}(0, \delta_r^2)$, y_r is the received signal by R.

In the second half $(1-\alpha)T/2$, R forwards the signal received to D, hence its signal received at D is expressed as

$$y_d = h_2(y_r + \eta_{rt}) + \eta_{dd} + n_d \tag{4}$$

Where h_2 is the Rayleigh RV for R $\rightarrow$ D link, y_d is the signal received at D, η_{rt} and η_{dd} are the distortion noise at R and D respectively, with power $k_{rt}^2 P_r$ and $|h_2|^2 k_{dd}^2 P_r$, k_{rt} and k_{dd} characterize the level of impairments at R and D in the second half $(1-\alpha)T/2$. n_d is the AWGN at D with power $n_d \sim \mathcal{CN}(0, \delta_d^2)$, P_r is the power of y_r.

Through our work, it is assumed that the whole harvested energy is used for transmitting its signal, hence P_r is given by

$$P_r = \mu\alpha TP|h_1|^2\left[(1-\alpha)T/2\right]^{-1} = 2\varepsilon P|h_1|^2 \tag{5}$$

Where $\varepsilon = \mu\alpha(1-\alpha)^{-1}$, μ is the energy conversion efficiency.

3. The Performance of The System

In this section, we derived the end-to-end signal-to-noise-and-distortion-rate (SNDR), analytical expressions of the outage performance for our network. What's more, the optimal analysis of the instantaneous throughput is also got, in order to have a better performance, the optimal value α is also given.

3.1 The End-to-End SNDR of the Considered Network

With help of (3), the SNDR of the first and second half $(1-\alpha)T/2$ are given by, respectively.

$$\gamma_1 = |h_1|^2 P\left[\left(k_{st}^2 + k_{rd}^2\right)|h_1|^2 P + \delta_r^2\right]^{-1} \tag{6}$$

$$\gamma_2 = |h_2|^2 P_r\left[\left(k_{rt}^2 + k_{dd}^2\right)|h_2|^2 P_r + \delta_d^2\right]^{-1} \tag{7}$$

As AF protocol is used by R, hence the SNDR of the system is expressed as

$$\gamma_e = |h_1|^2|h_2|^2 PP_r\left[|h_1|^2|h_2|^2 PP_r\left(k_1^2 + k_2^2 + k_1^2k_2^2\right) + |h_1|^2 P\left(\delta_d^2 + k_1^2\delta_d^2\right) + |h_2|^2 P_r\left(\delta_r^2 + k_2^2\delta_r^2\right) + \delta_r^2\delta_d^2\right]^{-1} \tag{8}$$

Where $k_1^2 = k_{st}^2 + k_{rd}^2, k_2^2 = k_{rt}^2 + k_{dd}^2$.

Substituting (5) into (8) and after some simplifications, γ_e is given by

$$\gamma_e = 2\varepsilon\lambda_1^2\lambda_2\left(2\varepsilon\lambda_1^2\lambda_2 A + \lambda_1 B + 2\varepsilon\lambda_1\lambda_2 C + 1\right)^{-1} \tag{9}$$

Where $A = k_1^2 + k_2^2 + k_1^2 k_2^2$, $B = 1 + k_1^2$, $C = 1 + k_2^2$. In (9), we assume that $\delta_r^2 = \delta_d^2 = 1$, $\lambda_1 = P|h_1|^2 \delta_r^{-2}$ and $\lambda_2 = P|h_2|^2 \delta_d^{-2}$.

3.2 The Outage Performance

Before deriving the outage performance, we first discuss the channel fading for the S $\rightarrow$ R and R $\rightarrow$ D links. The channel model between S and R is formed as SR channel, for R both receive the signal of LOS and the signal from multi-path, the channel model between R and D is formed as Rayleigh fading which we just consider the multi-path effect.

From [21], the probability distribution function (PDF) for λ_1 is given by

$$f_{\lambda_1}(\lambda_1) = \alpha_1 \sum_{k_1=0}^{m_1-1} (1-m_1)_{k_1} (-\delta_1)^{k_1} (k_1!)^{-2} (\bar{\lambda}_1)^{-k_1-1} \lambda_1^{k_1} e^{-\Delta_1 \lambda_1} \tag{10}$$

Where $\Delta_1 = (\beta_1 - \delta_1)\bar{\lambda}_1^{-1}$, $\bar{\lambda}_1$ is the average signal-to-noise-ratio (SNR), $\alpha_1 \overset{\Delta}{=} \left(2b_1 m_1 (2b_1 m_i + \Omega_1)^{-1}\right)^{m_1} (2b_i)^{-1}$, $\beta_1 \overset{\Delta}{=} 0.5 \times b_1^{-1}$, $\delta_1 \overset{\Delta}{=} \Omega_1 \left[2b_1 (2b_1 m_1 + \Omega_1)\right]^{-1}$, ${}_1\Omega_1$, $2b_1$ and $m_1 \geq 0$ correspond to the average power of the LOS component, the average power of the multi-path component and the fading severity parameter ranging from 0 to ∞ , respectively, $(\cdot)_q$ is the Pochhammer symbol.

From above, we know that, the R $\rightarrow$ D channel fading we considered is Rayleigh fading, so the PDF of λ_2 is expressed as

$$f_{\lambda_2}(\lambda_2) = \bar{\lambda}_2^{-1} e^{-\frac{\lambda_2}{\bar{\lambda}_2}} \tag{11}$$

Where $\bar{\lambda}_2$ is the average SNR of the R $\rightarrow$ D link.

With the help of [9] and of (9), the outage probability is expressed as

$$\begin{aligned} P_{out}(x_0) = & \int_0^{\frac{Cx_0}{(1-Ax_0)}} \Pr\left\{2\lambda_2 \left(\varepsilon\lambda_1^2 - \varepsilon\lambda_1^2 Ax_0 - \varepsilon\lambda_1 Cx_0\right) \leq \lambda_1 Bx_0 + x_0\right\} f(\lambda_1) d\lambda_1 \\ & + \int_{\frac{Cx_0}{(1-Ax_0)}}^{\infty} \Pr\left(\lambda_2 \leq \lambda_1 Bx_0 + x_0 \left[2\left(\varepsilon\lambda_1^2 - \varepsilon\lambda_1^2 Ax_0 - \varepsilon\lambda_1 Cx\right)_0\right]^{-1}\right) f(\lambda_1) d\lambda_1 \end{aligned} \tag{12}$$

Substituting (11) and (12) into (13), by using [22] and after some messures, (12) is rewritten as

$$\begin{aligned} P_{out}(x_0) = & 1 - e^{-Cx_0\left[(1-Ax_0)\bar{\lambda}_2\right]^{-1}} + \alpha_1 \sum_{k_1=0}^{m_1-1} (1-m_1)_{k_1} (-\delta_1)^{k_1} (k_1!)^{-2} (\bar{\lambda}_1)^{-k_1-1} \\ & \times \int_{Cx_0(1-Ax_0)^{-1}}^{\infty} \left(1 - e^{-(\lambda_1 Bx_0 + x_0)\left[2\left(\varepsilon\lambda_1^2 - \varepsilon\lambda_1^2 Ax_0 - \varepsilon\lambda_1 Cx_0\right)\bar{\lambda}_2\right]^{-1}}\right) \lambda_1^{k_1} e^{-\Delta_1 \lambda_1} d\lambda_1 \end{aligned} \tag{13}$$

But try our best, (13) can not be solved to get a closed-from expression. To solve this hard question, we get the asymptotic mean of (13). From (9), we know that

$$2\varepsilon\lambda_1^2\lambda_2 \left(2\varepsilon\lambda_1^2\lambda_2 A + \lambda_1 B + 2\varepsilon\lambda_1\lambda_2 C + 1\right)^{-1} \leq 2\varepsilon\lambda_1\lambda_2 \left(2\varepsilon\lambda_1\lambda_2 A + 2\varepsilon\lambda_2 C + B\right)^{-1} \tag{14}$$

With the help of (12), (13) and (14), after ding necessary simplifications, the asymptotic expression of the outage performance is expressed (15), where $K_v(\cdot)$ represents the v th-order modified Bessel function of the second kind.

$$P_{out}(x_0) = \begin{cases} 1 - e^{-Cx_0\left[\varepsilon(1-Ax_0)\bar{\lambda}_2\right]^{-1}} + \alpha_1 \sum_{k_1=0}^{m_1-1} \sum_{v=0}^{k_1} \binom{k_1}{v} (1-m_1)_{k_1} (-\delta_1)^{k_1} (k_1!)^{-2} (\bar{\lambda}_1)^{-k_1-1} \left[Cx_0 (1-Ax_0)^{-1}\right]^{k_1-v} & \\ \times e^{-\Delta_1 Cx_0 (1-Ax_0)^{-1}} \left\{ v!\Delta_1^{-v-1} - 2\left[Bx_0 (2\varepsilon - 2\varepsilon Ax_0)^{-1} \bar{\lambda}_2^{-1} \Delta_1^{-1}\right]^{(v+1)/2} K_{v+1}\left(2\left[\frac{\Delta_1 Bx_0}{\bar{\lambda}_2 (2\varepsilon - 2\varepsilon Ax_0)}\right]^{1/2}\right)\right\}, & x_0 < \frac{1}{A} \\ 1, & x_0 \geq \frac{1}{A} \end{cases} \tag{15}$$

3.3 The Instantaneous Throughput Performance

We now analyse the instantaneous throughput, and try to get the optimal α for the system. With help of [5], the instantaneous throughput is given by

$$R(\alpha)=(1-\alpha)2^{-1}\log\left(1+2\varepsilon\lambda_1^2\lambda_2\left(2\varepsilon\lambda_1^2\lambda_2 A+\lambda_1 B+2\varepsilon\lambda_1\lambda_2 C+1\right)^{-1}\right) \tag{16}$$

The optimal α can be derived for solving this problem.

$$\alpha^* = \arg\max R(\alpha) \quad s.t.\ 0\le\alpha\le 1 \tag{17}$$

Given the fact that $R(\alpha)$ is a concave function of α, the optimal value α^* can be derived through the answer of this question $\frac{dR(\alpha)}{d\alpha}=0$. However, a closed-from answer of this problem is not avaliable. In order to solve this problem, a numerically evaluated value α^* is given.

Substituting $\varepsilon=\frac{\mu\alpha}{1-\alpha}$ into (16), (16) can be repressed as

$$R(\alpha)=2^{-1}(1-\alpha)\log\left(1+c_1\alpha\left(c_2\alpha+c_3\right)^{-1}\right) \tag{18}$$

Where $c_1=2\mu\lambda_1^2\lambda_2$, $c_2=2\mu\lambda_1^2\lambda_2A-\lambda_1B+2\mu\lambda_1\lambda_2C-1$ and $c_3=\lambda_1B+1$.

Substituting (18) into $\frac{dR(\alpha)}{d\alpha}=0$, we have the following expression

$$\frac{dR(\alpha)}{d\alpha}=-\frac{1}{2}\log\left(1+c_1\alpha\left(c_2\alpha+c_3\right)^{-1}\right)+c_1c_3(1-\alpha)\left\{2\left[\left(c_1+c_2\right)\alpha+c_3\right]\left(c_2\alpha+c_3\right)\right\}^{-1}=0$$
$$\Leftrightarrow c_1c_3(1-\alpha)\left\{\left[\left(c_1+c_2\right)\alpha+c_3\right]\left(c_2\alpha+c_3\right)\right\}^{-1}=\log\left(1+c_1\alpha\left(c_2\alpha+c_3\right)^{-1}\right) \tag{19}$$

The answer for (19) is the optimal value, but trying out best we can not solve this equation, hence in the numerical results we will give the numerical solution.

4. Numerical Results

In this section, some necessary simulations are given to show the right of the theory results. The simulation tool we used is MATLAB. In the following, we set the parameters $\mu=1$ for the energy conversion efficiency.

- case1: $k_1=k_2=0$;
- case2: $k_1=k_2=0.2$;
- case3: $k_1=k_2=0.3$;

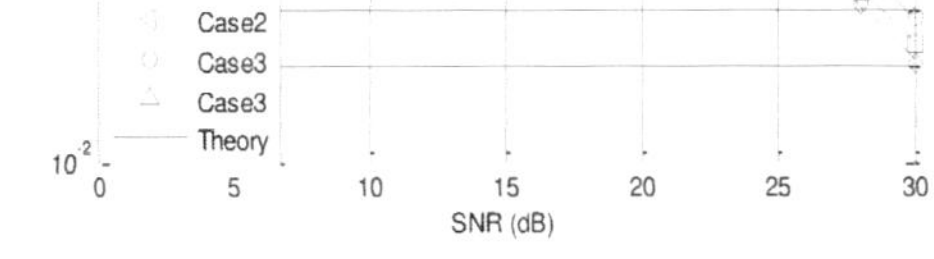

Figure 2. Outage probability of the system with ideal hardware and hardware impairments.

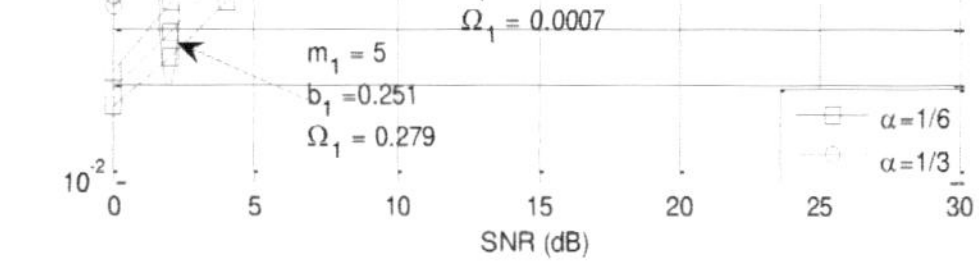

Figure 3. Outage probability of the system versus different x_0

In Figure 2, $\alpha=1/3$ is assumed. From Figure 2, we know that the outage performance will be significant reduced due to the self-interference on the network when the system has HIs Moreover, it also shows that the system has a worse performance when the channel is under heavy fading.

From Figure 3, we know that the system has an outage threshold when the system has HIs. The larger k_1,k_2 is, the lower outage threshold is. But if the system is under perfect hardware, there is no outage threshold, the outage probability will be 1 as the outage threshold increase to infinite. Figure 3 also depicts that, when α grows larger, the outage probability will become lower, which is easy to know, when α becomes larger, the power of relay becomes larger simultaneously.

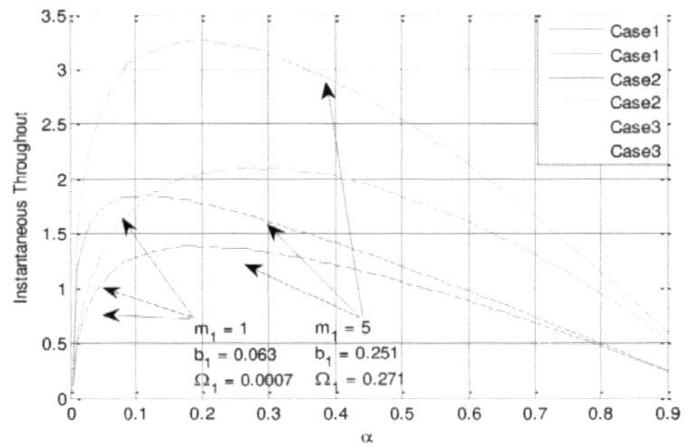

Figure 4. The instantaneous throughput versus different α .

Figure 4 illustrates the instantaneous throughput versus different α , from Figure 4, we know that α has an optimal value when the system has the largest instantaneous throughput value. It also depicts that the optimal α not only relates to k_1, k_2 but also relates to the channel fading. The heavy fading is, the lower α is; the larger k_1, k_2 are, the lower α is.

5. Conclusions

Summarized from our work, we studied the performance of the energy harvesting hybrid relay network with HIs. At first, we got the SNDR of the provided network. Secondly, the analytical expressions of the system performance which gave fast ways to estimate the effect of the HIs on the provided network. It is also found that the impairments level had great effects on the system performance. Moreover, the optimal analysis of the instantaneous throughput gave us the way to choose an optimal value α . We found it that the optimal value not only had the relationship with the impairments level but also with the channel fading parameters, which guided us how to balance the impairments level and the channel fading.

References

[1] L. R. Varshney, "Transporting information and energy simultaneously,"in Proc, IEEE Int'l Symp. Inf. Theory, Jul. 2008, pp. 1612-1616.

[2] R. Grover and A. Sahai, "Shannon meets Tesla: Wireless informationand power transfer," in IEEE Int'l Symp Inf. Theory, Jun. 2010, pp. 2362-2367.

[3] Y. Dong, J. Wang, B. Shim, and D. I. Kim, "Dearer: A distanceand-energy-aware routing with energy reservation for energy harvesting wireless sensor networks," IEEE Journal Sele. Areas., vol. 34, no. 12, pp. 3798-3813, Oct. 2016.

[4] L. P. Qian, G. Feng, and V. C. M. Leung, "Optimal transmission policies for relay communication networks with ambient energy harvesting relays," IEEE Journal Sele. Areas., vol. 34, no. 12, pp. 3754-3768, Oct. 2016.

[5] C. Zhong, H. A. Suraweera, G. Zheng, I. Krikidis, and Z. Zhang,"Wireless information and power transfer with full duplex relaying," IEEE Trans. Commun., vol. 62, no. 10, pp. 3447-3461, Sep. 2014.

[6] E. Costa and S. Pupolin, "m-QAM-OFDM system performance in the presence of a nonlinear amplifier and phase noise," IEEE Trans.Commun., vol. 50, no. 3, pp. 462-472, Mar. 2002.

[7] T. Schenk, "RF imperfections in high-Rate wireless systems: impact and digital compensation," Springer., pp. 2033-2043, Sep. 2008.

[8] C. Studer, M. Wenk, and A. Burg, "MIMO transmission with residual transmit-RF impairments," in Proc. 2010 ITG Work. Smart Ant., pp. 189-196.

[9] E. Bjornson, M. Matthaiou, and M.Debbah, "A new look at dual-hop relaying: performance limits with hardware impairments," IEEE Trans. Commun., vol. 61, no. 11, pp. 4512-4525, Nov. 2013.

[10] J. Qi, et. al., "Analysis and compensation of I/Q imbalance in amplify-and-forward cooperative systems," in Proc. 2012 IEEE Wireless Commun. Netw. Conf., pp. 215-220, Apr. 2012.
[11] E. Bjornson, P. Zetterberg, M. Bengtsson, and B. Ottersten, "Capacity limits and multiplexing gains of MIMO channels with transceiver impairments," IEEE Commun. Lett., vol. 17, no. 1, pp. 91-94, Jan. 2013.
[12] Z. Xinlin, M. Matthaiou, E. Bjornson, M. Coldrey, and M. Debbah, "On the MIMO capacity with residual transceiver hardware impairments," IEEE ICC 2014., Sydney, Nsw, Jun. 2014, pp. 5299-5305.
[13] Z. Xinlin, et. al., "Impact of residual transmit RF impairments on training-based MIMO systems," IEEE ICC 2014., Sydney, Nsw, Jun. 2014, pp. 4741-4746.
[14] M. Matthaiou, A. Papadogiannis, E. Bjornson, and M. ebbah, "Two-way relaying under the presence of relay transceiver hardware impairments," IEEE Commun. Lett., vol. 17, no. 6, pp. 1136-1139, Jun. 2013.
[15] T. T. Duy, et. al., "Procative relay selection with joint impact of hardware impairment and co-channel interference," IEEE Trans. Commun., vol. 63, no. 5, pp. 1594-1606, May. 2015.
[16] N. T. Do, et. al., "Performance analysis of multirelay RF energy harvesting cooperative networks with hardware impairments," IET Commun., vol. 10, no. 9, pp. 2551-2558, Dec. 2016.
[17] P. K. Sharma and P. K. Upadhyay, "Cognitive relaying with transceiver hardware impairments under interference constraints," IEEE Commun. Lett., vol. 20, no. 4, pp. 820-823, Apr. 2016.
[18] K. Guo, B. Zhang, Y. Huang, and D. Guo, "Outage Analysis of Multi- Relay Networks With Hardware Impairments Using SECps Scheduling Scheme in Shadowed- Rician Channel," IEEE ACCESS, vol. PP, no. 99, pp. 1-1, Mar. 2017.
[19] K. Guo, D. Guo, Y. Huang, X. Wang, and B. Zhang, "Performance analysis of a dual-hop satellite relay network with hardware impairments," 2016 WOCC., Chengdu, China, May, 2016, pp. 1-5.
[20] K. Guo, B. Zhang, Y. Huang, and D. Guo, "Performance analysis of two-way satellite terrestrial relay networks with hardware impairments," IEEE Wire. Commun. Lett., vol. PP, no. 99, pp. 1-1, May. 2017.
[21] N. I. Miridakis, D. D. Vergados, and A. Michalas, "Dual-hop communication over a satellite relay and shadowed Rician channels," IEEE Trans. Vehi. Tech., vol. 64, no. 9, pp. 4031-4040, Sep. 2015.
[22] I. S. Gradshteyn, I. M. Ryzhik, A. Jeffrey, and D. Zwillinger, "Table ofintegrals, series and products," 7thed.Amsterdam, Boston: Elsevier, 2007.

Orbit Codes over $M_2(\mathbb{F}_q)$ and Their Homogeneous Distance

E C T Corro[1,*], J M T Lampos[1], H S Palines[1] and V P Sison[1]
[1]Institute of Mathematical Science and Physics, University of the Philippines Los Baños, College, Laguna 4031, Philippines

*etcorro@up.edu.ph

Abstract. Let q be a power of a prime. The lattice of one-sided ideals of the finite unital non-commutative Frobenius ring $M_2(\mathbb{F}_q)$ of 2×2 matrices over the Galois field $\mathbb{F}_q$ is completely analyzed. It turns out that $M_2(\mathbb{F}_q)$ is a principal left semi-local ring in which each left ideal is generated by an idempotent element. The explicit forms of the non-trivial idempotents of $M_2(\mathbb{F}_q)$ are determined to give $q + 1$ proper non-trivial left maximal ideals each with q^2 elements. These are exactly the minimal left ideals as well. Using the structure of $M_2(\mathbb{F}_q)$ as a partial ordering of ideals, the generalized Möbius and Euler phi functions are applied to derive the explicit form of the homogeneous weight function on $M_2(\mathbb{F}_q)$. This weight depends on whether the element is the zero element, a zero divisor or a unit. A zero divisor gives the largest homogeneous weight. Moreover, orbit codes over $M_2(\mathbb{F}_q)$ are constructed via the action of the general linear group $GL(2,q)$ on $M_2(\mathbb{F}_q)$ by left translation. The orbit determined by a nonzero nonunit idempotent element of $M_2(\mathbb{F}_q)$ forms the nonzero elements of a minimal left ideal of $M_2(\mathbb{F}_q)$ which are all zero divisors. Consequently, it is shown that the minimum homogeneous distance of the orbit code generated by a nonzero nonunit idempotent element of $M_2(\mathbb{F}_q)$ approaches the Plotkin upper bound as the field size q becomes larger. Analogous results are obtained when the lattice of right ideals is considered and the action of $GL(2,q)$ on $M_2(\mathbb{F}_q)$ by right translation is used instead.

1.Introduction

In this paper, we consider the finite noncommutative Frobenius ring $M_2(\mathbb{F}_q)$ which admits a homogeneous weight. Frobenius rings are appropriate choices as alphabet for codes since the MacWilliams extension theorem is valid in such rings.

Codes over commutative rings are widely studied, however, only a few codes, such as the ones presented in [1] and [2], use noncommutative rings as alphabet. The construction of orbit codes based from the action of $GL(2,q)$ on $M_2(\mathbb{F}_q)$ is presented in this study. The concept of an orbit of a group action as a code is discussed in [3]. The orbit codes constructed in this study, which are determined by a suitable choice of nonzero nonunit idempotent element, are observed to approach the Plotkin bound in terms of the homogeneous distance.

This paper is organized as follows. In Section 2, definitions and concepts relevant to the study are presented. The results are elaborated in Section 3 which consists of four subsections. First, the lattice of left ideals of $M_2(\mathbb{F}_q)$ is derived in Section 3.1. Then the results in Section 3.1 are used to come up with an explicit form of the homogeneous weight of $M_2(\mathbb{F}_q)$ as shown in Section 3.2. Section 3.3

discusses the properties of the action of $GL(2,q)$ on $M_2(\mathbb{F}_q)$ by left translation which are used for the construction of orbit codes in Section 3.4.

2. Preliminaries

The concept of homogeneous weight was first introduced by Constantinescu and Heise in [4]. Homogeneous weight can be seen as a natural generalization of the Hamming weight on finite fields and the Lee weight on $\mathbb{Z}_4$.

A homogeneous weight on an arbitrary finite ring R with unity is defined as follows. Let $\mathfrak{R}$ be the set of real numbers and Rx be the principal (left) ideal generated by $x \in R$. A weight function $w : R \to \mathfrak{R}$ is called (left) *homogeneous* if the following conditions are satisfied.

(i) If $Rx = Ry$, then $w(x) = w(y)$ for all $x,y \in R$.

(ii) There exists a real number $\Gamma \geq 0$ such that $\sum_{y\in Rx} w(y) = \Gamma \cdot |Rx|$, for all $x \in R \setminus \{0\}$.

The definition for the right homogeneous weight follows the same analogy. The weight w is simply called a *homogeneous weight* if it satisfies both the left and right cases. The average value of w is denoted by the constant Γ. If $\Gamma = 1$, then w is a *normalized homogeneous weight*. The weight w on R^n is simply a natural extension from that on R, that is, $w(z) = \sum_{i=0}^{n-1} w(z_i)$ for $z = (z_0,z_1,...,z_{n-1}) \in R^n$. The homogeneous distance metric $\delta : R^n \times R^n \to \mathfrak{R}$ is defined by $\delta(x,y) = w(x - y)$ for $x,y \in R^n$.

The following theorem which was proven in [5], shows the existence and uniqueness of the homogeneous weight on a finite ring.

Theorem 2.1. *Let w be a weight defined on the finite ring R. The weight w is homogeneous if and only if there exists a positive real number c such that $w(x) = c\left(1 - \frac{\mu(0,Rx)}{|R^\times x|}\right)$ for all $x \in R$.*

The set $R^\times$ denotes the multiplicative group of units of R, and μ is the Möbius function on the poset of the principal left ideals of R given by (i.) $\mu(Rx,Rx) = 1$ for all $x \in R$, (ii.) $\mu(Ry,Rx) = 0$ if $Ry \nleq Rx$, and (iii.) $\sum_{Ry\leq Rz\leq Rx} \mu(Rz,Rx) = 0$ if $Ry \leq Rx$. Since the Möbius function, in this case, is defined on the poset of one-sided ideals of a finite ring, finding its values is not easy. Some related results involved deriving explicit formulas for the homogeneous weights of finite principal ideal rings [6], and the noncommutative Frobenius ring $M_2(\mathbb{F}_p)$ of 2×2 matrices over the prime field $\mathbb{F}_p$ [2]. In this paper, we come up with an explicit formula of the homogeneous weight of the noncommutative Frobenius ring $M_2(\mathbb{F}_q)$ where $\mathbb{F}_q$ is any finite field, and q is a power of a prime.

Furthermore, this paper also includes a construction of matrix codes in the form of orbit codes based on the action of the general linear group $GL(2,q)$ on $M_2(\mathbb{F}_q)$. In [3], a code C is an *orbit code* if it is an orbit of a group G on a metric set X. The minimum homogeneous distance $\delta_{\min}$ of C is defined to be $\delta_{\min} = \min\{\delta(x,y) | x,y \in C, x \neq y\}$. Applying the homogeneous distance on these orbit codes, it is observed that these codes are 'nearly' Plotkin optimal.

3. Results and Discussions

3.1. One-sided ideals of $M_2(\mathbb{F}_q)$

The ring $M_2(\mathbb{F}_q)$ is a left Artinian ring with unity whose (Jacobson) radical, $rad\ M_2(\mathbb{F}_q)$, the largest nilpotent ideal of $M_2(\mathbb{F}_q)$, is trivial, that is $rad\ M_2(\mathbb{F}_q) = \left\{\begin{pmatrix} 0 & 0 \\ 0 & 0 \end{pmatrix}\right\}$. It follows that each left ideal of $M_2(\mathbb{F}_q)$ is of the form $M_2(\mathbb{F}_q)e$, where e is an idempotent element of $M_2(\mathbb{F}_q)$. Hence, it is important to know the idempotent elements of $M_2(\mathbb{F}_q)$ to have an idea about the properties and explicit forms of the left ideals of $M_2(\mathbb{F}_q)$.

In [7], it was proven that given a commutative local ring R, for each $A \in M_2(\mathbb{F}_q)$, $A^2 = A$ if and only if $A = \begin{pmatrix} 0 & 0 \\ 0 & 0 \end{pmatrix}$ or $A = \begin{pmatrix} 1 & 0 \\ 0 & 1 \end{pmatrix}$ or $A = \begin{pmatrix} a & b \\ c & 1-a \end{pmatrix}$ with $bc = a - a^2$. This result applies to $M_2(\mathbb{F}_q)$ since $\mathbb{F}_q$ is a commutative local ring. Thus, we come up with the following corollary.

Corollary 3.1. *The matrix ring $M_2(\mathbb{F}_q)$ has $q^2 + q + 2$ idempotent elements.*

Particularly, each nonzero nonunit idempotent element of $M_2(\mathbb{F}_q)$ is of the form $\begin{pmatrix} 1 & k_1 \\ 0 & 0 \end{pmatrix}, \begin{pmatrix} 1 & 0 \\ k_2 & 0 \end{pmatrix}, \begin{pmatrix} 0 & k_3 \\ 0 & 1 \end{pmatrix}, \begin{pmatrix} 0 & 0 \\ k_4 & 1 \end{pmatrix}$ where $k_1, k_2, k_3, k_4 \in \mathbb{F}_q$, and $\begin{pmatrix} a & b \\ c & 1-a \end{pmatrix}$ where $a \in \mathbb{F}_q \setminus \{0,1\}$ and $b, c \in \mathbb{F}_q^{\times}$, the group of units of $\mathbb{F}_q$.

Using these idempotent elements and forming the left ideals $M_2(\mathbb{F}_q)e$ for each idempotent element $e \in M_2(\mathbb{F}_q)$, we come up the following theorem.

Theorem 3.2. *The matrix ring $M_2(\mathbb{F}_q)$ has $q + 1$ proper left ideals. Moreover, each proper left ideal is minimal with q^2 elements.*

The lattice of left ideals of $M_2(\mathbb{F}_q)$ is given by Figure 1

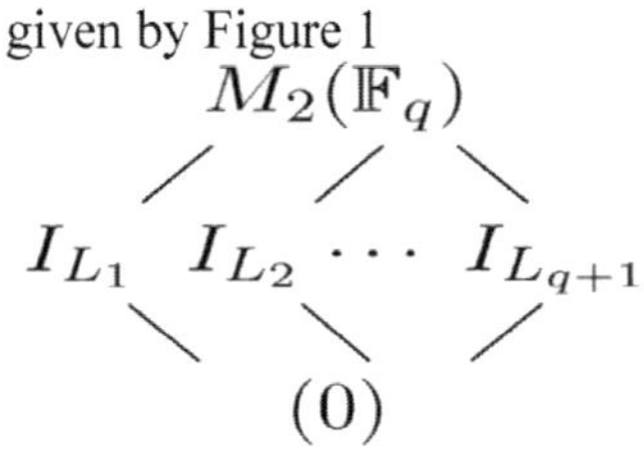

Figure. 1 Lattice of left ideals of $M_2(\mathbb{F}_q)$

where for each $i \in \{1,2,\ldots, q+1\}$, I_{L_i} denotes a distinct minimal left ideal of $M_2(\mathbb{F}_q)$.

Example 3.3. *The idempotent elements of $M_2(\mathbb{F}_2)$ are given by*

$$\begin{pmatrix} 0 & 0 \\ 0 & 0 \end{pmatrix}, \begin{pmatrix} 1 & 0 \\ 0 & 1 \end{pmatrix}, \begin{pmatrix} 1 & 0 \\ 0 & 0 \end{pmatrix}, \begin{pmatrix} 1 & 1 \\ 0 & 0 \end{pmatrix}, \begin{pmatrix} 1 & 0 \\ 1 & 0 \end{pmatrix}, \begin{pmatrix} 0 & 0 \\ 0 & 1 \end{pmatrix}, \begin{pmatrix} 0 & 1 \\ 0 & 1 \end{pmatrix}, \begin{pmatrix} 0 & 0 \\ 1 & 1 \end{pmatrix}.$$

Then the minimal left ideals of $M_2(\mathbb{F}_2)$ generated by $\begin{pmatrix} 1 & 0 \\ 0 & 0 \end{pmatrix}, \begin{pmatrix} 1 & 1 \\ 0 & 0 \end{pmatrix}$ and $\begin{pmatrix} 0 & 0 \\ 0 & 1 \end{pmatrix}$, respectively, are as follows:

$$\left\{\begin{pmatrix} 0 & 0 \\ 0 & 0 \end{pmatrix}, \begin{pmatrix} 1 & 0 \\ 0 & 0 \end{pmatrix}, \begin{pmatrix} 0 & 0 \\ 1 & 0 \end{pmatrix}, \begin{pmatrix} 1 & 0 \\ 1 & 0 \end{pmatrix}\right\},$$

$$\left\{\begin{pmatrix} 0 & 0 \\ 0 & 0 \end{pmatrix}, \begin{pmatrix} 1 & 1 \\ 0 & 0 \end{pmatrix}, \begin{pmatrix} 0 & 0 \\ 1 & 1 \end{pmatrix}, \begin{pmatrix} 1 & 1 \\ 1 & 1 \end{pmatrix}\right\},$$

$$\left\{\begin{pmatrix} 0 & 0 \\ 0 & 0 \end{pmatrix}, \begin{pmatrix} 0 & 0 \\ 0 & 1 \end{pmatrix}, \begin{pmatrix} 0 & 1 \\ 0 & 0 \end{pmatrix}, \begin{pmatrix} 0 & 1 \\ 0 & 1 \end{pmatrix}\right\}.$$

Note that the treatment for right ideals gives analogous results.

3.2. Homogeneous weight on $M_2(\mathbb{F}_q)$

In the paper [5] by Greferath and Schmidt, the homogeneous weight $w(x)$ of $x \in R$, where R is a finite ring with identity, is given by $w(x) = \Gamma[1 - \mu(0,Rx)/\varphi(x)]$, where μ is the Möbius function on the poset of the principal left ideals of R, and $\varphi(x)$ is the Euler phi function. Recall that for $x \in R$, $\varphi(x) = |\{y \in R : Ry = Rx\}|$. Applying our results in Section 3.1, we come up with the following results on the homogeneous weight of $M_2(\mathbb{F}_q)$. In the succeeding discussions, we let $R = M_2(\mathbb{F}_q)$ and 0 be the zero ideal of R.

Lemma 3.4. *If Z is the zero matrix, then $\mu(0,RZ) = 1$ and $\varphi(Z) = 1$.*
Proof:

$$\mu(0,RZ) = \mu(0,0) = 1 \text{ and } \varphi(Z) = |\{y \in R | Ry = 0\}| = |\{0\}| = 1. \quad (1)$$

Lemma 3.5. *If I_L is a proper left ideal of R, then $\mu(I_L,R) =- 1$.*
Proof:

Note that $\mu(I_L,R) + \mu(R,R) = 0$. Thus, $\mu(I_L,R) =- 1$ (2)

Lemma 3.6. *If $U \in GL(2,q)$, then $\mu(0,RU) = q$ and $\varphi(U) = q(q-1)(q^2-1)$.*
Proof:

Let $U \in GL(2,q)$. Then $\mu(0,RU) = \mu(0,R)$. By the definition of the Möbius function, $\mu(0,R) + \mu(I_1,R) + \cdots + \mu(I_{q+1},R) + \mu(R,R) = 0$, where $I_1,\cdots,I_{q+1}$ are minimal left ideals of R. Applying Lemma 3.5, we have $\mu(0,R) + (-1)(q+1) + \mu(R,R) = 0$ which implies that $\mu(0,RU) = \mu(0,R) = q$. On the other hand, $\varphi(U) = |\{y \in R{:}Ry = R\}| = |GL(2,q)| = q(q-1)(q^2-1)$. (3)

Lemma 3.7. *If B is a zero divisor of R, then $\mu(0,RB) =- 1$ and $\varphi(B) = q^2 - 1$.*
Proof:

Let $B \in R$ be a zero divisor. Thus, RB is a minimal left ideal of R. Hence, $\mu(0,RB) =- 1$. On the other hand, note that each minimal left ideal of $M_2(\mathbb{F}_q)$ has $q^2 - 1$ nonzero elements. Thus, $\varphi(B) = |\{y \in R{:}Ry = RB\}| = q^2 - 1$. (4)

With these results, we come up with an explicit formula of the homogeneous weight of $M_2(\mathbb{F}_q)$ as follows.

Theorem 3.8. *The homogeneous weight of $M_2(\mathbb{F}_q)$ is given by*

$$w_{\text{hom}}(x) = \begin{cases} 0 & \text{if } x = 0 \\ \Gamma\left(1 - \dfrac{1}{(q-1)(q^2-1)}\right) & \text{if } x \text{ is a unit} \\ \Gamma\left(\dfrac{q^2}{q^2-1}\right) & \text{if } x \text{ is a zero divisor} \end{cases}$$

Proof:

The proof follows immediately from Lemma 3.4 to Lemma 3.7. (5)

Moreover, it is worth noting that $M_2(\mathbb{F}_q)$ is partitioned into the set containing the zero element, the multiplicative group of units, and sets of zero divisors.

3.3. Action of $GL(2,q)$ on $M_2(\mathbb{F}_q)$

Consider the action of the general linear group $GL(2,q)$ on $M_2(\mathbb{F}_q)$ by left translation. Note that this action is an equivalence relation. Given $A \in M_2(\mathbb{F}_q)$, we denote the $\overline{A}$ as the orbit containing A. Then the orbits under this action are given as follows.

Let $Z,U \in M_2(\mathbb{F}_q)$ be the zero matrix and a unit, respectively. Then $\overline{Z} = \{Z\}$ and $\overline{U} = GL(2,q)$. Now, consider the idempotent elements $B_k = \begin{pmatrix} 1 & k \\ 0 & 0 \end{pmatrix}$ where $k \in \mathbb{F}_q$ and $B' = \begin{pmatrix} 0 & 0 \\ 0 & 1 \end{pmatrix}$ in $M_2(\mathbb{F}_q)$. Then for each $k \in \mathbb{F}_q$,

$$\overline{B_k} = \left\{\begin{pmatrix} u_{11} & u_{11}k \\ u_{21} & u_{21}k \end{pmatrix} \in M_2(\mathbb{F}_q){:}u_{11}, u_{21} \in \mathbb{F}_q \text{ not both } 0\right\}. \quad (6)$$

Here we have q distinct equivalence classes since we are considering the equivalence classes each containing one of B_k, $k \in \mathbb{F}_q$. Moreover, notice that for each $k \in \mathbb{F}_q$, B_k has $q^2 - 1$ elements. On the other hand,

$$\overline{B'} = \left\{\begin{pmatrix} 0 & u_{12} \\ 0 & u_{22} \end{pmatrix} \in M_2(\mathbb{F}_q) : u_{12}, u_{22} \in \mathbb{F}_q \text{ and } u_{12}, u_{22} \text{ not both } 0\right\}. \quad (7)$$

Similarly, $\overline{B'}$ has $q^2 - 1$ elements.

Example 3.9. *Consider the group action of $GL(2,2)$ on the matrix ring $M_2(\mathbb{F}_2)$ by left translation. Then, the following are the orbits of $M_2(\mathbb{F}_2)$ under this action. Let Z be the zero matrix in $M_2(\mathbb{F}_2)$. Then $\overline{Z} = \{Z\}$. Consider the identity element $I \in M_2(\mathbb{F}_2)$ which is a unit in $M_2(\mathbb{F}_2)$. So, $\overline{I} = GL(2,2)$. Now, consider the idempotent elements $B_0 = \begin{pmatrix} 1 & 0 \\ 0 & 0 \end{pmatrix}, B_1 = \begin{pmatrix} 1 & 1 \\ 0 & 0 \end{pmatrix}, B' = \begin{pmatrix} 0 & 0 \\ 0 & 1 \end{pmatrix} \in M_2(\mathbb{F}_2)$. Then $\overline{B_0} = \left\{\begin{pmatrix} 1 & 0 \\ 0 & 0 \end{pmatrix}, \begin{pmatrix} 0 & 0 \\ 1 & 0 \end{pmatrix}, \begin{pmatrix} 1 & 0 \\ 1 & 0 \end{pmatrix}\right\}$, $\overline{B_1} = \left\{\begin{pmatrix} 1 & 1 \\ 0 & 0 \end{pmatrix}, \begin{pmatrix} 0 & 0 \\ 1 & 1 \end{pmatrix}, \begin{pmatrix} 1 & 1 \\ 1 & 1 \end{pmatrix}\right\}$ and $\overline{B'} = \left\{\begin{pmatrix} 0 & 0 \\ 0 & 1 \end{pmatrix}, \begin{pmatrix} 0 & 1 \\ 0 & 0 \end{pmatrix}, \begin{pmatrix} 0 & 1 \\ 0 & 1 \end{pmatrix}\right\}$.*

3.4. Orbit codes over $M_2(\mathbb{F}_q)$

In this section, we construct nonlinear constant weight orbit codes over $M_2(\mathbb{F}_q)$. A (nonlinear) code C is a *constant weight* code provided every nonzero codeword has the same weight. In the following theorem, we use the explicit forms of the orbits of the group action of $GL(2,q)$ on $M_2(\mathbb{F}_q)$ by left translation to construct nonlinear constant weight orbit codes.

Theorem 3.10. *Let $B = \begin{pmatrix} 0 & 0 \\ 0 & 1 \end{pmatrix}$ and $B_k = \begin{pmatrix} 1 & k \\ 0 & 0 \end{pmatrix}$ where $k \in \mathbb{F}_q$. If $C = \overline{B} = \left\{\begin{pmatrix} 0 & a \\ 0 & b \end{pmatrix} : a,b \in \mathbb{F}_q\right\}$ or $C = \overline{B_k} = \left\{\begin{pmatrix} a & ak \\ b & bk \end{pmatrix} : a,b \in \mathbb{F}_q\right\}$, then C is a nonlinear constant weight orbit code over $M_2(\mathbb{F}_q)$ with minimum homogeneous distance $d_{\text{hom}} = \frac{q^2}{q^2-1}\Gamma$.*

Proof:

Each of the idempotent elements $B, B_k \in M_2(\mathbb{F}_q)$, where $k \in \mathbb{F}_q$, determines an orbit under the action whose elements are the nonzero elements of a minimal left ideal of $M_2(\mathbb{F}_q)$. Furthermore, note that each minimal left ideal of $M_2(\mathbb{F}_q)$ is an additive group whose elements are zero divisors and the zero element since it is a principal left ideal generated by a zero divisor. Thus, for each $k \in \mathbb{F}_q$, $\overline{B}$ and $\overline{B_k}$ are nonlinear constant weight orbit codes over $M_2(\mathbb{F}_q)$.

Now, we investigate the optimality of the code constructed in Theorem 3.10. As shown in Theorem 3.11, this generalized Plotkin bound is discussed in [8]. Given a block code C over the finite Frobenius ring R which is not necessarily linear, and the homogeneous weight w defined on C with average value $\Gamma \geq 0$, we take C as an (n,M,d) code of length n with M codewords and minimum distance d.

Theorem 3.11. *If C is an (n,M,d) code, then*

$$M(M-1)d \leq \textstyle\sum_{x,y \in C} w(x,y) \leq \Gamma n M^2. \quad (8)$$

It follows from this inequality that $d \leq \frac{\Gamma n M}{M-1}$. Hence, we come up with the following remark.

Remark 3.12. *Let C be an orbit code over $M_2(\mathbb{F}_q)$ as constructed in Theorem 3.10. Note that the code C is a $\left(1, q^2-1, \frac{q^2}{q^2-1}\right)$ code with respect to the normalized homogeneous weight. As the field size q becomes larger, the minimum normalized homogeneous distance of C approaches to the Plotkin bound. In this case, we say that C is 'nearly' Plotkin optimal.*

Example 3.13. *Let* C *be the orbit of* $A=\begin{pmatrix}1&1\\0&0\end{pmatrix}\in M_2(\mathbb{F}_2)$, *that is,* $C=\overline{A}=\left\{\begin{pmatrix}1&1\\0&0\end{pmatrix},\begin{pmatrix}0&0\\1&1\end{pmatrix},\begin{pmatrix}1&1\\1&1\end{pmatrix}\right\}$. *Observe that each element of* C *is a zero divisor of* $M_2(\mathbb{F}_2)$. *Furthermore, the nonzero elements of the minimal left ideal generated by* A *are the elements of* C. *Thus,* C *is a (nonlinear) constant weight code with a common normalized homogeneous weight of 4/3 and a minimum normalized homogeneous distance of 4/3.*

In Table 1, we illustrate how the minimum normalized homogeneous distance $d_{\min}$ of the code C over $M_2(\mathbb{F}_q)$ as constructed in Theorem 3.10, approaches the Plotkin bound d_P as we increase the field size q.

Table 1. Relative Values of the Homogeneous Distance and the Plotkin Bound.

q	$d_{\min}$	d_P	Relative Difference (%)
2	4/3	3/2	11.1111
3	9/8	8/7	1.5625
4	16/15	15/14	0.4444
8	64/63	63/62	0.0252
9	81/80	80/79	0.0156
16	256/255	255/254	0.0015

The relative difference in Table 1 is the value $\frac{d_P-d_{\min}}{d_P}\times 100$. It can be observed that as the field size q becomes larger, the rate of convergence of $d_{\min}$ to the Plotkin bound is reasonably fast. Practically, when $q\geq 16$, Plotkin optimality is safely achieved.

4.Summary and Conclusion

After determining the one-sided ideal structure of $M_2(\mathbb{F}_q)$, an explicit form of the homogeneous weight of $M_2(\mathbb{F}_q)$ was derived using the Möbius and Euler phi functions. Moreover, the action of the general linear group $GL(2,q)$ on $M_2(\mathbb{F}_q)$ by left translation provides a novel construction of nonlinear constant weight orbit codes over $M_2(\mathbb{F}_q)$ whose homogeneous distance approaches the Plotkin bound as the field size q becomes larger.

Acknowledgment

The first author acknowledges the support from the DOST-ASTHRD Program of the Philippine government.

References

[1] Oggier F and Solé P 2012 *IEEE Trans. Inform. Theory* **58** 734
[2] Falcunit D and Sison V 2014 *Proc. of the 2014 Int. Zürich Seminar on Communications, Zürich, Switzerland* 91
[3] Trautmann A L, Manganiello F, Braun M and Rosenthal J 2013 *IEEE Trans. Inform. Theory* **59** 7386
[4] Constantinescu I and Heise W 1997 *Probl. Inform. Transm.* **33** 208
[5] Greferath M and Schmidt S E 2000 *J. Combin. Theory* A **92** 17
[6] Fan Y and Liu H 2010 *Math. Ann. (Chinese)* A **31** 355
[7] Chen H, Yang X and Zhou Y 2006 *J. Algebra* **301** 280
[8] Greferath M and O'Sullivan M E 2004 *Discrete Math.* **289** 11

Performance of Coarse Timing Synchronization in Orthogonal Frequency Division Multiplexing System

BRUNO[1,*] and Zhai Xuping[1]

[1]Key laboratory of Specialty Fiber Optics and Optical Access Networks Joint International Research Laboratory of Specialty Fiber Optics and Advanced Communication, Shanghai Institute for Advanced Communication and Data Science, Shanghai University, Shanghai, China

*bruso167@gmai.com

Abstract. A novel symbol timing synchronization based on training symbols for OFDM system was designed and presented in this paper. The performance of the proposed method is tested in terms of the timing metric and mean square error (MSE), indoor no-noise, and Rayleigh fading channel, obtained by simulation. The over the air transmission of the proposed timing metric is evaluated by implementing a software defined radio based on the GNU Radio and universal software radio peripheral communication platforms. The simulation results show that the proposed timing metric indexed at the correct time point, which has smaller MSE than others methods, particularly at high signal-to-noise ratio. Therefore the proposed method works well, even in transmission through the air indoor laboratory environment.

1. Introduction

At present, orthogonal frequency division multiplexing (OFDM) has become the most popular modulation technique in wireless communication, owing to its high data rate transmission capability and high bandwidth efficiency. OFDM has been widely applied in practice, such as in digital audio broadcasting (DAB) and digital video broadcasting for terrestrial television (DVB-T) [1], [2], wireless local area network (WLANs), long-term evolution (LTE) [3], [4] and WiMAX. OFDM systems are considerably more sensitive to synchronization error due to the symbol timing and carrier frequency offset [5]-[7], which leads to the degradation of the system performance.

Different symbol timing offset estimation-based training symbols were proposed in literature. A well-known example is the Schmidl and Cox [8] method. This method uses a training symbol, which has identical halves in the time domain to estimate the symbol timing offset. However, its timing metric gives a plateau, which results in some uncertainty in determining the start of the OFDM frame. To avoid the ambiguity due to the timing metric, Minn et al. [9] modified the Schmidl and Cox method and proposed identical preambles in the time domain with the opposite signs. The Minn method performs well, but it still has a high mean square error in the inter-symbol interference (ISI) channel. To improve symbol timing synchronization, Park et al. [10] proposed another method, which causes an impulse-like timing response. This method is assumed to be a better estimator, but its timing metric has large sidelobes that lead to worse detection at the low SNR. Seung et al. [11] have designed a new time domain preamble to give a smaller MSE than the previous estimators, even in a multipath fading channel. To reduce the sidelobes in the Park estimator, A. M Khan et al. [12] modified the Park method and proposed a new preamble scheme that has a PN sequence of value 1 and -1, which performs better at low SNR.

The aim of this work is to design a new training symbol for coarse timing synchronization in OFDM receiver and compare it with three popular conventional existing estimators, which are the Schmidl and Cox, Minn and Park estimators. The proposed method is also used in real-time Transmission-based universal software radio peripheral (USRP) and GNU Radio as shown in Figure 3 to verify its performance by transmitting a text file from a laptop acting as a transmitter, through the air. At the receiver, the text file received by the antenna feeds into a PC to perform the synchronization task. Note that in this scenario, the complete OFDM system is needed, including channel coding and channel estimation.

2. System model

A general case of OFDM systems using the standard complex-valued baseband equivalent signal model can be expressed as

$$y(n)=\sum_{m=0}^{L-1}h(m)x(n-m) \tag{1}$$

where $h(\text{n})$ is the channel response, which has number of channels L. $x(\text{n})$ is the time domain signal, which can be expressed

$$x(n)=\frac{1}{\sqrt{N}}\sum_{k=0}^{N-1}c_k\exp\left(j\frac{2\pi nk}{N}\right) \tag{2}$$

where N is number of IFFT points and c_k are the complex information symbols. At the receiver, the timing offset is modelled as a delay in the received signal and the frequency offset as a phase distortion of the received data in the time domain. These two offsets and additive white Gaussian noise (AWGN) $w(\text{n})$ yield received signal:

$$r(\text{n})=y(n-\tau)\exp(\text{j}(2\pi\Delta fn+\phi))+\text{w}(\text{n}) \tag{3}$$

where τ is the integer unknown arrival time of a symbol, Δf is the frequency offset and ϕ is the initial phase. Therefore, the target of the timing synchronization is to estimate τ [13].

3. Conventional synchronization technique

3.1. Schmidl and Cox method

Schmidl and Cox proposed a training sequence that has identical halves in the time domain. Those identical halves can be generated by transmitting the zero on odd frequencies and PN sequence on the even frequencies. Then, by taking the IFFT, we obtain the halves in the time domain. Let N be the number of IFFT points in one OFDM symbol. Schmidl used the following preamble.

$$P_{SC}=[\text{A}_{\text{N/2}}\quad \text{A}_{N/2}] \tag{4}$$

where A represents the sample length $N/2$.

3.2. Minn method

In order to avoid uncertainty produced by timing metric of the Schmidl and Cox estimator, Minn proposed new training preamble is as follows

$$P_{Min}=[B_{N/4}\ \ B_{N/4}\ \ \text{-}B_{N/4}\ \ \text{-}B_{N/4}] \tag{5}$$

where B represents the sample length $N/4$.

3.3. Park method

In this method, the real value of PN sequence is transmitted on the even frequencies and zeros are transmitted on the odd frequencies and applied to the IFFT operation in the sequence. Park achieves an impulse-like timing response. The training symbol used in this method is given by

$$P_{Park} = [C_{N/4} \;\; D_{N/4} \;\; C^*_{N/4} \;\; D^*_{N/4}] \tag{6}$$

where C represents a sample length of $N/4$ generated by N number of IFFT of a PN sequence and C is symmetric to D. C^* and D^* are the conjugates of C and D respectively.

4. Proposed method

As is known, OFDM timing synchronization using preamble technique depends on the structure of the training symbol and definition of the timing metric. A sample of the proposed pattern preamble can be designed as follows:

$$P_{\mathrm{Pro}} = [\mathrm{E}_{N/2} \quad -\mathrm{E}_{N/2}] \tag{7}$$

where the real value E represents a sample of length $N/2$ generated by IFFT of a PN sequence.

This symbol pattern can be easily obtained by using the properties of the FFT. The training symbol can be produced by transmitting the complex values of the PN sequence on the odd frequencies and zeros on the even frequencies, as shown in following:

$$X = [0\; x_0 \;\; 0\; x_1 \ldots x_{N/2\text{-}1} \;\; 0 \ldots x^*_{N/2\text{-}1} \;\; 0\; x^*_1 \;\; 0\; x^*_0] \tag{8}$$

where $x_k = a + ib$; a, b are real values and x^*_k is the conjugate of x_k. Let N be the number of IFFT points in one OFDM symbol. Then, by applying the IFFT operation to expression (8) will produce the output signal in the time domain, as shown in (7). The timing metric for the proposed method can be written as follows:

$$P_{\mathrm{Pro}} = \frac{|P_4(\mathrm{d})|^4}{(R_4(\mathrm{d}))^4} \tag{9}$$

where the terms $P_4(\mathrm{d})$ and $R_4(\mathrm{d})$ can be expressed, respectively, as follows:

$$P_4(\mathrm{d}) = \sum_{n=0}^{N/2-1} r(\mathrm{n+d})\,\mathrm{r}(\mathrm{n+d}+\frac{N}{2}) \tag{10}$$

$$R_4(\mathrm{d}) = \sum_{n=0}^{N/2-1} |r(\mathrm{n+d})|^2 \tag{11}$$

and $r(\mathrm{n})$ is the received signal. Here, the 4th power is needed to reduce the subpeaks beside the main peak instead of the 2nd power (in Experimental). However, they yielded almost the same result in a Matlab simulation.

The timing metric defined in (9) creates a plateau, causing ambiguity in determining the correct symbol start position. To improve the proposed method, a new timing metric is defined. The key point of this improved method is to reduce the length of the plateau to achieve a shaped peak, which indicates the precise start of the FFT windows at the receiver. The plateau is a window with length equal to the cyclic prefix in the AWGN channel where the timing metric reaches a maximum; the start of the symbol can then be taken to be any point within the window. The length of the window could change in the frequency-selective channel. Therefore, to realize this concept, it is necessary to examine the output of the timing metric to find the exact length of the plateau. Let m be the length of the plateau and keep the preamble the same as that designed in (7) and (8); the only change here is the timing metric. The timing metric for the improved method can be defined as follows:

$$M_{ipro} = \frac{|P_5(\mathrm{d})|^4}{(R_5(\mathrm{d}))^4} \tag{12}$$

where the terms $P_5(\mathrm{d})$ and $R_5(\mathrm{d})$ can be written, respectively, as

$$P_5(\mathrm{d}) = \sum_{n=1}^{N/2+L} r(\mathrm{n+d})\,\mathrm{r}(\mathrm{n+d}+\frac{N}{2}) \tag{13}$$

$$R_5(d) = \sum_{n=1}^{N/2+L} |r(\mathrm{n+d})|^2 \tag{14}$$

Here, L = m-1 and $r(\mathrm{n})$ is the received signal.

5. Simulation results

This section investigates the performance of the proposed estimator by comparing it with three existing estimators in terms of timing metric and MSE using the indoor Rayleigh fading channel obtained by Matlab simulations. On the other hand, the performance of proposed method has been evaluated in terms of the timing metric in real-word transmission with the USRP and GNU Radio as the software defined radio (SDR) in an indoor laboratory environment; the OFDM parameters for this experiment setup are chosen as OFDM parameters for the IEEE 802.11a standard, as shown in Table 1.

Table 1. OFDM parameters for simulation and testbed.

OFDM Parameters	Simulations	Testbed
Number of FFT points	512	64
Number of subcarriers	256	52
Cyclic Prefix length	128	16
Number of taps	6	
Number of simulation runs	50000	
Centre frequency		2.412 GHz
Gain [dB]		38 dB
Distance between two USRP N210		3 m

5.1. Timing metric

Figure 1 shows the performance of the proposed algorithm with those of the Schmidl and Cox, Minn, and Park methods under the condition no noise and no channel distortion. The correct point is indexed 0 in the figure. The Schmidl and Cox method creates a plateau which causes ambiguity when estimating the correct time. The Minn method reduced the timing metric plateau, but when data is transmitted in the ISI fading channel, the adjacent samples to the metric peak have almost the same values as the main peak. This causes incorrect estimation and increases the MSE. The Park method achieves the impulse-like timing response. It is assumed to be the better estimator, but its timing metric has subpeaks which result in worse estimation in the fast fading channel. The proposed method achieved a shaped peak indexed at the correct timing point which indicates the precise starting point of the OFDM symbol.

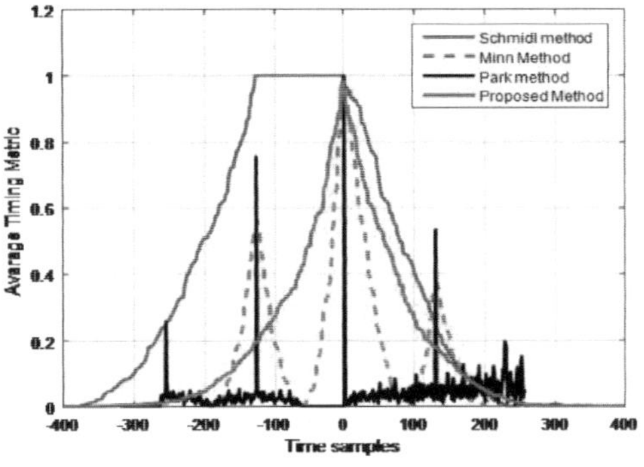

Figure 1. Comparison of timing metric estimators.

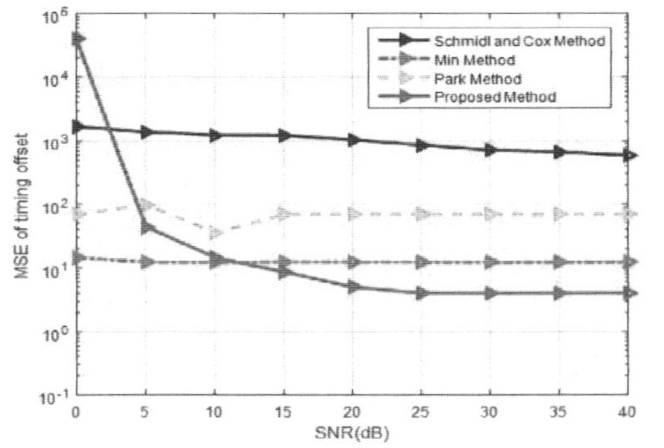

Figure 2. MSE of timing offset versus SNR.

5.2. Means square error(MSE)

Figure 2 shows the MSE of the estimators versus SNR in the Rayleigh fading channel. The results for the Schmidl and Cox method are the worst, and it has the greatest uncertainty in the calculated timing offset point owing to the plateau in the timing metric. The proposed method has a lower MSE than the Minn and Park methods. Therefore, the proposed method outperforms the Minn and Park methods.

5.3. Testbed description

This section describes the real-time testbed SDR implementation, based on the GNU Radio and USRP platforms, to investigate the performance of the proposed timing synchronization method. The experimental setup consists of one laptop, one PC, and two USRP N210 produced by Beijing Highmesh. Each USRP is equipped with one antenna and controlled by UHD software from the laptop. The daughter boards used for this experiment are RFX 2400, which could cover the frequency range between 2.3 – 2.9 GHz. Moreover, the canter frequency was set at 2.412 GHz.

As seen in Figure 3, during the transmission, the baseband signal from the host laptop is sent to the first USRP by an intermediate Gigabit Ethernet cable to ensure communication between the laptop and USRP device, where it is converted to an RF signal and transmitted through the air. At the receiver side, the analog signal from the air is captured by the antenna and gets processed in the second USRP board. Then, it is fed to the PC with GNU Radio installed to perform synchronization tasks and other signal processing.

Figure 3. Experimental setup in laboratory environment.

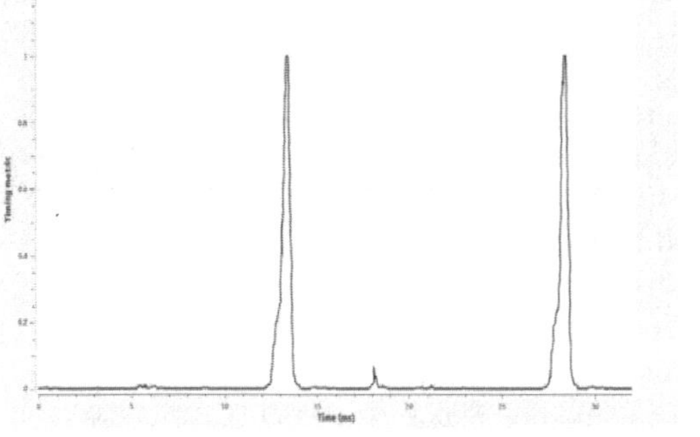

Figure 4. Example of experimentally obtained timing metric.

5.4. Testbed results

The coarse timing synchronization is implemented in block called synchronization. Its inputs are the samples from the USRP device in the form of a stream. As seen in in Figure 4, the first peak marks the start of the first OFDM frame; it feeds a fixed number of samples into the rest of the signal processing blocks and the second peak indicates another new frame that has been detected. If the second frame arrives shortly after the first one, it will be ignored.

6. Simulation results

In this paper, we presents a different symbol timing synchronization based on a preamble technique. The proposed timing offset estimator reduces the plateau that leads to the uncertainty estimation in the Schmidl timing offset estimation method. The simulation results show that the proposed method performs better than other methods in the indoor Rayleigh fading channel, particularly at high SNR. The proposed method can also detect the precise starting point of the FFT widows, even in real-time transmission through the air in an indoor laboratory environment.

Acknowledgments

This work was supported by the National Nature Science Foundation of china (61171085, 61401266), Innovation Program of Shanghai Municipal Education Commission (14ZZ096).

References

[1] BARUFF G, RUGINI L and BANELLI P 2014 Design and validation of a software defined radio testebed for dvb-t transmission *Radioengineering.* **23** 387-98
[2] Elsharief M, Zekry A and Abuelatta M 2014 Implementing a standard DVB-T system using MATLAB simulink *International Journal of Computer Applications.* **98** 27-32
[3] Ghosh A, Ratasuk R, Mondal B, Mangalvedhe N and Tomas T 2010 LTE-Advanced: Next-generation wireless broadband technology *IEEE Wireless Communications.* **17** 10-22
[4] Li P, Jia Y, Feng M, Ye C, Chen F and Fan H 2016 Real-time software defined radio platform for LTE-Advanced heterogeneous networks *Journal of Communications.* **11** 263-72
[5] Chang D, 2008 Effect and compensation of symbol timing offset in OFDM system with channel interpolation *IEEE Transactions on broadcasting.* **54** 761-70
[6] Mohseni S and Matin M A 2013 Study of the sensitivity of the ofdm wireless systems to the carrier frequency offset *International Journal of Distributed and Parallel Systems.* **4** 1-13
[7] Tripathi 2014 Effect of the carrier frequency offset on the wireless OFDM communications *International Journal Emerging Technology and Advanced Engineering.* **4** 307-10
[8] Schmidl T M and Cox D C 1997 Robust frequency and timing synchronization for OFDM *IEEE Transaction on communications.* **45** 1613-21
[9] Minn H, Bhargava V K and Letaief K B 2003 Robust timing and frequency synchronization for OFDM systems *IEEE Transaction on Wireless Communications.* **2** 822-39
[10] Park B, Cheon H, Kang C and Hong D 2003 A novel timing estimation method for OFDM systems *IEEE Communications Letters.* **7** 239-41
[11] Choi S D, Choi J M and Lee J H 2006 An initial timing offset estimation method for OFDM systems in Raleigh fading channel *Proceeding of IEEE vehicular Technology Conference.*
[12] Khan A M, Jeoti V, Zakariya M A and Ur Rehman M Z 2014 Robust symbol timing synchronization for OFDM systems using PN sequence *International Journal of Information and Electronics Engineering.* **4** 230-33
[13] Yi G, Gang L and Jianhua G 2008 A novel time and frequency synchronization scheme for OFDM systems *IEEE Transaction on Consumer Electronics.* **54** 321-25

Improved Fast Centralized Retransmission Scheme for High-Layer Functional Split in 5G Network

Sen XU[1,2,*], Meng HOU[1], Yu FU[3], Honglian Bian[2] and Cheng Gao[2]

[1]School of Information and Communication Engineering, Beijing University of Posts and Telecommunications, Beijing, 100876, China
[2]China Telecom Corporation Limited Beijing Research Institute, Beijing, 102209, P. R. China
[3]China Internet Network Information Centre, Beijing, China

*xusen.bri@chinatelecom.cn

Abstract. In order to satisfy the varied 5G critical requirements and the virtualization of the RAN hardware, a two-level architecture for 5G RAN has been studied in 3GPP 5G SI stage. The performance of the PDCP-RLC split option and intra-RLC split option, two mainly concerned options for high layer functional split, exist an ongoing debate. This paper firstly gives an overview of CU-DU split study work in 3GPP. By the comparison of implementation complexity, the standardization impact and system performance, our evaluation result shows the PDCP-RLC split Option outperforms the intra-RLC split option. Aiming to how to reduce the retransmission delay during the intra-CU inter-DU handover, the mainly drawback of PDCP-RLC split option, this paper proposes an improved fast centralized retransmission solution with a low implementation complexity. Finally, system level simulations show that the PDCP-RLC split option with the proposed scheme can significantly improve the UE's experience.

1. Introduction

Three typical families of typical applications including enhanced mobile broadband (eMBB), massive machine type communications (mMTC), and Ultra-Reliable and Low Latency Communications (URLLC) will be supported simultaneously in 5G system [1]. In order to satisfy these varied critical requirements, 5G RAN architecture has the need to be re-designed. 3GPP NR (New Radio) SI phase has been completed on April 2017[2]. Two typical architectures of 5G RAN have been approved to be specified in the ongoing 3GPP 5G WI stage, as illustrated in Figure 1.

One type of gNB is similar to 4G eNB with the full protocol stack supported at one base station node [3].This architecture is suitable for urban macro scenario and indoor hotspot scenario. To simplify the network and to drive down the operational cost (OPEX) of the network, another type of gNB adopts two-level RAN architecture. The upper layers of the NR radio stacks are centralized in Central Unit (CU) while the lower layers of NR functions are performed in Distributed Unit (DU). In order to exchange the control plane signalling and user plane data between CU and DU, a new interface named F1 interface has also been introduced. One DU can support multiple physical cells and belongs to single CU, which can provide multiple points for CP and UP respectively for the sake of scalability. The drive also involves the virtualization of user data handling, signalling and applications on commodity hardware as few as possible data centres, where hardware can be maintained inexpensively.

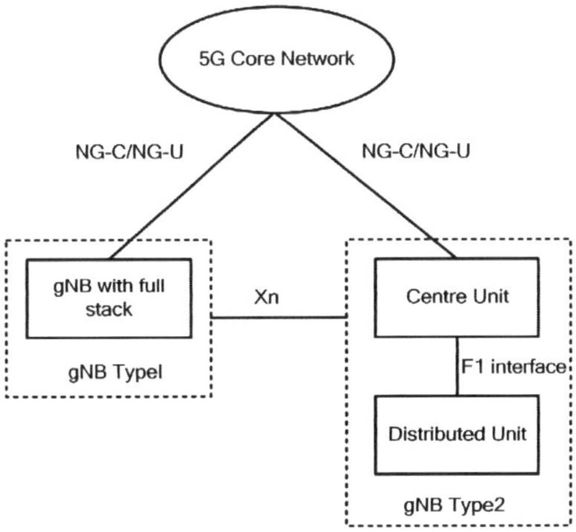

Figure 1 Illustration of two types of 5G RAN architecture

Currently 8 different functional splits between central and distributed units have been captured in [3], as illustrated in Figure 2.

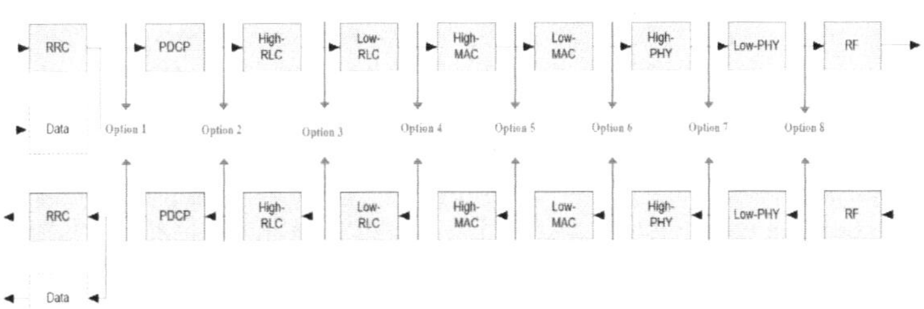

Figure 2 Function Split between central and distributed unit

The Option1~4 are defined as high layer split options while the Option5~8 are defined as low layer split options. 3GPP has decided that only one high layer split option will be specified in the 5G WI stage. And the Option2 and Option3 are mainly concerned options in 3GPP SI stage. As the sub-option of Option2, the Option2-1 is similar as 3C architecture in DC [4]. The characterization of Option2-1 is that RRC, PDCP are resided in the CU, while the RLC, MAC, physical layer and RF are resided in the DU. Due to the fundamental architecture for achieving a PDCP-RLC split have already been standardized for LTE Rel-12/13 Dual Connectivity (Option 3C), the Option2-1 is regarded as the most straightforward solution to 3GPP standardization. As the sub-option of Option3, the RLC sublayer is split into High RLC and Low RLC sublayers in the Option3-1. For ARQ related functions, such as RLC Acknowledge Mode operation are performed at the High RLC sublayer residing in the central unit, while the segmentation may be performed at the Low RLC sublayer residing in the distributed unit. The benefit of this architecture is to provide the centralization gains of ARQ in CU and use the end-to-end ARQ mechanism to recover the failure over transport network. Since the above options have its cons and pros, respectively. No consensus on the selection of higher-layer option had been reached until the completion of 3GPP NR SI stage.

The Option3-1 with some enhanced solutions is considered for a better performance with Option2-1 for the aspect of missing data retransmission and fastens recovery for DU change in the contributions [5] ~ [7]. An enhanced Option3-1 solution with Transport Link Control (TLC) function is proposed in the contribution [5], which supports RLC-like function for F1 interface. The Option 2-1 with TLC function has a better performance when compared with Option 3-1 in terms of recovering the lost packets. Meanwhile, the Option3-1 with an effective TCP mechanism can show better performance for fasten recovery and missing data retransmission in DU overload scenario [6]. And comparing to Option2-1, the Option3-1 can reduce processing and buffer requirements in DU when the DC change occurs [7].

Meanwhile, some contributions have put forward the opposite view with the performance of option2-1 outperforming Option3-1.With respect to Option3-1, the performance of Option2-1improves significantly while decreasing the CU-DU transport latency [8]. With lower transport capability requirement and less latency sensitive, the Option2-1 is a simpler choice for deployment. Moreover, due to ARQ function lies in DU, the RLC PDU retransmission delay can be avoided [9] [10].

Moreover, due to the non-ideal backhaul between CU and DU introducing the additional transmission delay for RLC retransmission, the enhanced solution for the fast-centralized retransmission of lost RLC PDUs is also needed for both split options, especially for Option2-1.

The remainder of this paper is organized as follows. In Section II, we firstly give the analysis for both options from the aspects such as implementation complexity and system performance. And an improved fast centralized retransmission is proposed in Section III. And the simulation results on the comparison of both Option2-1 and Option3-1 are given in Section IV. The conclusions are drawn in Section V.

2. The Evaluation on Two Options

From the operator's point view, the following evaluation criterions shall be taken into account:

- The better performance: The candidate option can provide a better system performance than the other.
- Easy Maintenance: the candidate option should be easy to configure a reasonable PDCP/RLC/MAC parameter to achieve the better performance.
- Less impact on NR standardization

2.1. The performance comparison

Since the ARQ function is resided in CU for Option3-1, the extra transport delay will be introduced in RLC retransmission. For the polling based RLC PDU retransmission without DU changes, this implies the RLC re-ordering parameters *t-Reordering* timer shall be increased by 4 times of one-way transport delay. And we take the Figure 3 as the example. For Option3-1, DU detects the instantaneous random error occurred or received the UE's assist report, it shall provide the error related information to CU for triggering retransmission, due to no ARQ function in DU. Subsequently the CU shall trigger RLC Polling Request to UE. And the RLC Polling Request message shall be transferred to DU entity. For Option2-1, it can omit the above two steps due to ARQ function resided in DU entity. And in step 3 and Step 4, DU transfer or trigger RLC Polling Request to UE and UE feedbacks the RLC Status Report to DU for indicates the lost RLC PDUs. For Option3-1, DU still need to transfer the RLC Status Report to CU and CU retransmits the RLC PDU to UE later. For Option2-1, the Step5 and Step6 can be omitted. Therefore, when one RLC PDU needs to be retransmitted, at least 4 times one-way transport delay will be introduced in ARQ RTT. Moreover, we must mention that the CU can quickly retransmit the lost PDU with 3C-like the flow control mechanism introduced in 3GPP Rel-12 for Option2-1. As the role of SeNB in DC, the DU shall provide the following information to CU via the X2-U interface [11]:

- reporting the highest PDCP PDU sequence number successfully delivered to the UE;
- the desired buffer size;
- the lost X2-U packets.

With the latest transmission information (e.g., the highest PDCP PDU successfully delivered to the UE and the lost X2-U packets), CU can identify which lost PDU needs to be retransmitted. In addition, the delay over air interface (e.g. about 8ms) can also be reduced. Furthermore, the above status information is piggyback with user date PDU, no explicit X2-C signalling is needed. Therefore, comparing to Option3-1, the delay for retransmission between DUs can be reduced sufficiently for Option2.

The Figure 4 gives simulation results on the PDCP throughput gain of AM bearer with t-Reordering timer of 30ms, 15ms, and 0ms. The three target RLC loss probabilities are considered as: 10^{-4}, 10^{-3}, and 10^{-2}. The t-Reordering timer is used by the receiving side of an AM RLC entity and receiving UM RLC entity in order to detect loss of RLC PDUs at lower layer. And we selected the t-Reordering timer of 40ms as the baseline for each target RLC loss probability. Considering the one-way delay of transport network is about several or ten milliseconds order of magnitude, the assumption on the extra 15ms and 30ms introduced in t-Reordering timer is reasonable. And we can observe that the UE throughput with the *t-Reordering* timer of 55ms decreases 0.8% ,6.8% and 16.4%, comparing to that with the t-Reordering timer of 0ms on the order of 0.01%,0.1%, and1% RLC loss probability.

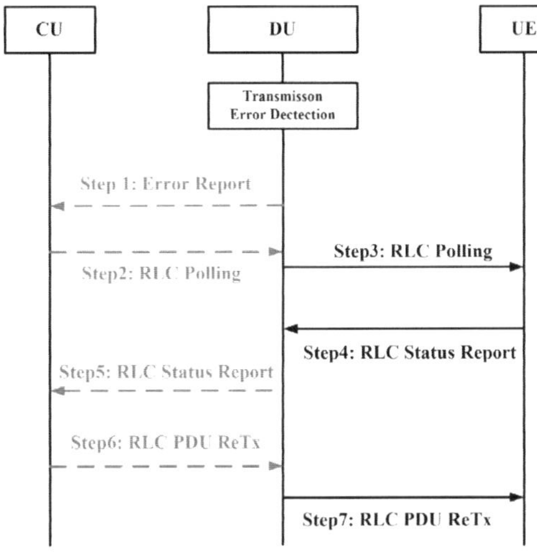

Figure 3 polling based RLC PDU Retransmission without DU changes

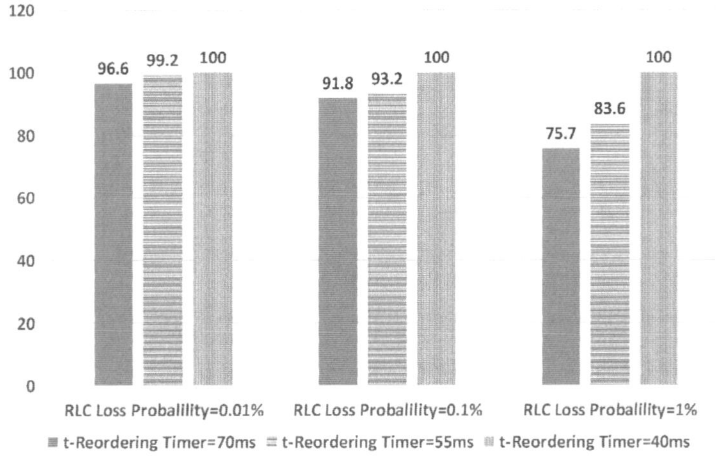

Figure 4 UE performance with variable RLC ARQ RTT

2.2. Parameters maintenance and optimization

For Option3-1, how to optimize RLC parameters (e.g., t-Reordering in H-RLC) for ARQ to meet various latency requirements and the performance changes of transport network is a big challenge for operators. And we still take the *t-Reordering* timer for example. This timer can affect the UE's experience as the above section described. And a reasonable timer value shall consider the follows factors:

- varying transport network latency
- varying queuing latency at the DU
- the delay over the air interface

If the DU queue and the delay of transport network are stable, a fixed value for *t-Reordering* timer can be configured to CU. However, in the real network the DU queue delay and the transport network delay may vary significantly over time, the operators can hardly configure a reasonable value for *t-Reordering* timer. If this timer is set too short, then the unnecessary retransmissions will be triggered, and network resources will also be wasted. If the timer is too long, UE will wait too much time before triggering retransmissions in case RLC-PDUs have been lost. Therefore, UE's performance will be decreased with unreasonable parameters for Option3-1. Moreover, in order to satisfy the very short slot length for URLLC, the joint optimization for HARQ and ARQ operation is needed. Through fast adaptation of ARQ, the ARQ and HARQ function located in the same entity (e.g., DU) can effectively improve UE's performance.

2.3. The impact on NR standardization

From the standardization point of view, the Option3-1 is more complex than Option2-1. Due to the signalling procedure and UP procedure have been specified in the DC topic of 3GPP Rel-12 and onwards, Option2-1 can enjoy this fruit to complete the standardization job as soon as possible. However, since the UP architecture need to be re-designed and specified, the complexity of the standardization work for Option3-1 will be higher than that for Option2-1. And from the implementation point of view, we can also conclude that the complexity of Option2-1 is less than Option3-1. Firstly, the interface and related procedures between PDCP and RLC are clearly defined in 3GPP related specifications. For CU and DU with different vendors, the overall system can be expected to work well. However, for Option3-1some new kinds of RLC internal information exchange and interface need to be defined. Considering the 3GPP specification doesn't restrict the implementation way for any protocol stack layers, the different thoughts of implantation way for RLC internal may be exist among vendors. And this may lead to many problems for CU and DU with different vendors. Moreover, another issue is that whether one set or two sets of protocol stack is required to separately support intra-gNB and inter-gNB. As the illustration in Figure 5,it is obvious that two sets of RLC protocol stack are needed for option 3-1, one is working for DU, and another is working for CU/DU separation. While for option 2-1, only one set of protocol stack is enough for both DC and CU/DU separation.

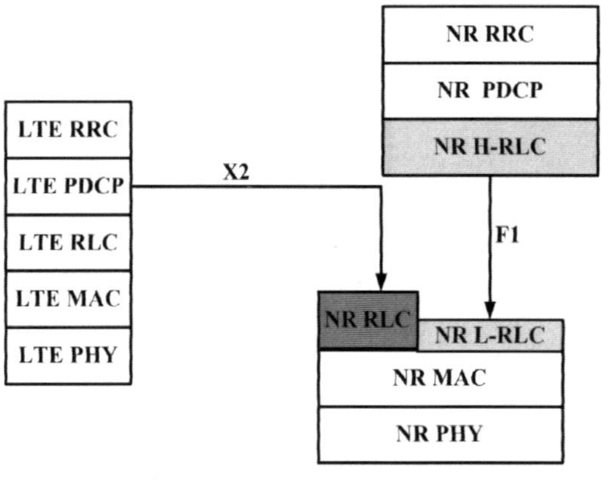

Figure 5 Two Sets of RLC Protocol Stack for DU in Option3-1

Based on the above analysis, we can conclude that the Option2-1 outperforms the Option3-1 from the aspect of UE's experience, the complexity of standardization and implantation, and network parameters optimization.

3. The improved fast centralized retransmission scheme

In order to reduce the interrupt delay, how to support the fast retransmission of lost PDCP PDUs during the intra-CU inter-DU handover scenario is regarded as the most difficult challenge for Option2-1. In this section, we firstly give the normal procedure without any improved solution for PDU retransmission in DU changes scenario, then an improved fast retransmission solution is proposed. As a contrast, the delay of the normal procedure for Option3-1 is also analyzed in this section.

3.1. The Basic Solution for Option2-1 and Option3-1

In this solution, after the handover procedure is completed, the UE triggers a PDCP status report to CU via target DU, which has been supported as PDCP data recovery. By decoding the PDCP status report, CU retransmit the lost PDCP PDUs to UE via target DU. As shown in Figure 6, the round trip transport delay will be suffered for the retransmission in this procedure.

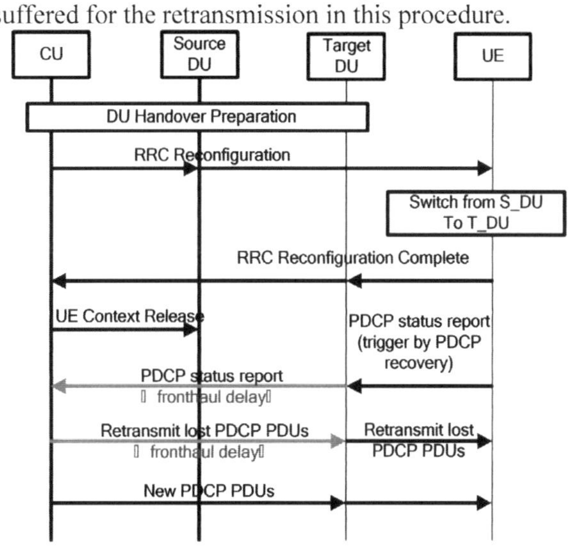

Figure 6 Basic Solution for Option2-1

For Option3-1, the RLC status report is triggered by RLC status report polling. The polling period delay and the transport delay will be suffered before the RLC status report is triggered. As shown in Figure 7, it will lead to additional delay for the lost RLC PDU retransmission.

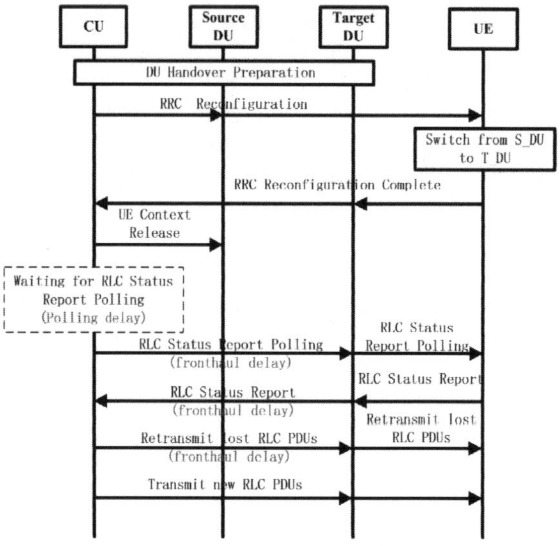

Figure 7 Basic Solution for Option3-1

3.2. The improved solution for PDU retransmission

The key point for this solution is to allow the target DU to buffer a copy of the PDCP PDUs in advance which are transmitting via source DU during the handover procedure. As the illustration in Figure 8, the target DU could buffer a copy of the PDCP PDUs in advance which are transmitting via source DU during the handover procedure. After the handover procedure is completed, UE triggers a PDCP status report to target DU. From the implementation perspective, target DU could be able to derive the SN of the lost PDCP PDUs based on the PDCP status report. Then target DU will retransmit the lost PDUs which have already been buffered in target DU directly and the bi-direction transport delay can be avoided. This solution can further reduce the latency time of lost RLC PDUs recovery.

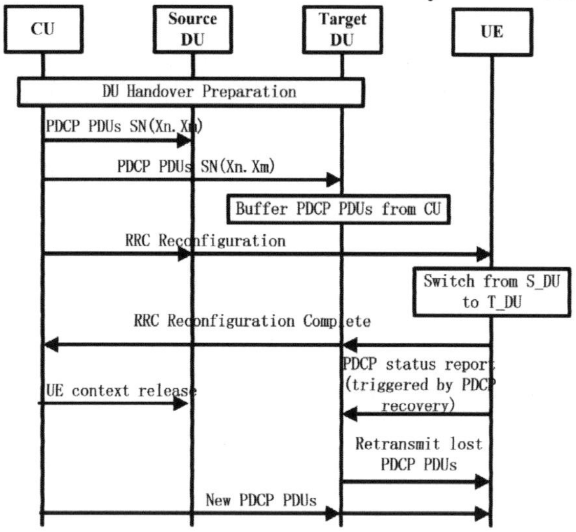

Figure 8 Improved Solution for lost PDCP PDUs retransmission procedure

The following table summarizes the delay for the lost data PDUs retransmission for the above mentioned solutions for both Option2-1 and Option3-1. As shown in Table 1, the delay performance of the ordinary retransmission procedure for Option 2-1 is better than that in Option 3-1. With the improved solution, the retransmission delay can be further reduced for Option2-1. Besides, the signalling procedure can also be further simplified.

Table 1 Summary of Lost PDU Retransmission Delay for Three Solutions

	Fronthaul delay[ms]	Polling period caused delay [ms]
Option2-1	2 messages transmission	0
Option3-1	3 messages transmission	Y
Proposed Solution	0	0

4. System Simulation for the improved Option2-1

In the section, the System level simulation for CU/DU split is performed to evaluate the performance of different options on in section IV. The traffic model in this simulation adopts FTP model. We consider two types of Download File Sizes in this simulation: 100M bits and 1G bits. The one-way transport delay is set to 15ms. Since the most of appliances are configured with AM mode for RLC in the real network, it is reasonable to set AM mode for RLC in this simulation. Moreover, the HARQ MCS selection is on the target of BLER=10%, the residual RLC BLER should be under 1% after HARQ retransmission processing. We also set RLC residual BLER 0.1% 0.5% 1% 2% in order to simulate both the normal and worse channel situation over air interface.

The Figure 9 shows the TCP throughput efficiency for both Option 2-1 and Option3-1, compared with efficiency simulated in the condition 0% residual RLC BLER TCP throughput. The capability of physical throughput for single user is assumed to 100Mbps.

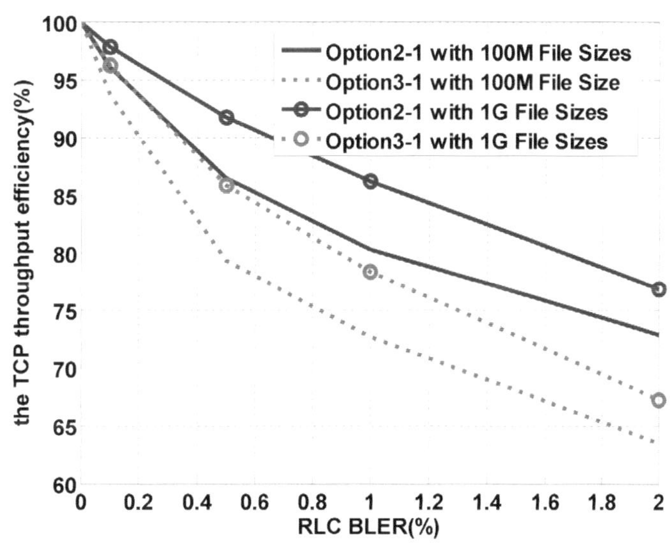

Figure 9 the TCP throughput efficiency for both Option 2-1 and Option3-1

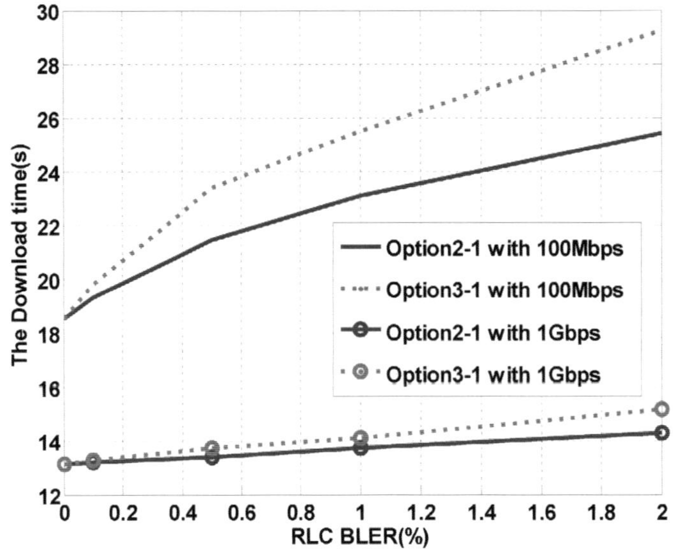

Figure 10 The performance comparison with 100M bits File Sizes

We can observe that the TCP throughput efficiency with Option2-1 can get the performance gain 3%, 7%, 8%, and 9%, comparing to Option3-1 for 100M file sizes of the order of 0.1%, 0.5%, 1%, and 2% RLC residual BLER. And for 1G bits file sizes, we can also observe that the TCP throughput efficiency with Option2-1 can get the performance gain 1.2%, 6%, 8%, and 9%, comparing to Option3-1 of the order of 0.1%, 0.5%, 1%, and 2% RLC residual BLER. Since the ARQ is located in CU, a round-trip transport delay is introduced into the RLC Status Report and RLC retransmission. With the effective TCP slow start mechanism, the delay of RLC retransmission will lead to some negative impact on the throughput. Therefore, the Option2-1 performance is obviously better than Option3-1 for short time TCP services.

In Figure 10 and 11, FTP download time is selected to comparison the performance of Option2-1 and Option3-1. Two types of user physical throughput for single UE are considered in this simulation: 100Mbps and 1Gbps. The download file size is assumed to 100M bits in Figure 10 and 1G bits in Figure 11.

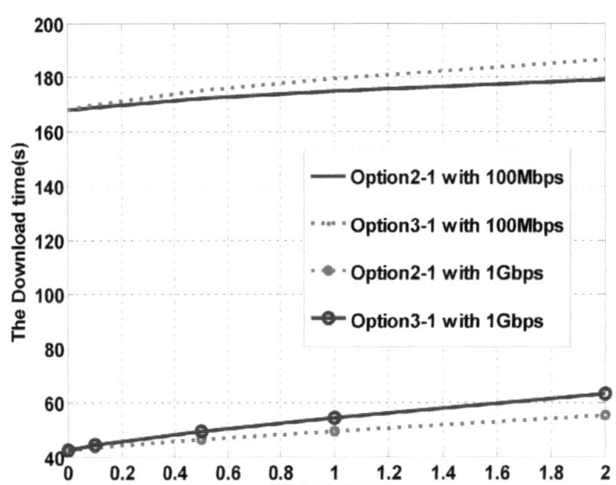

Figure 11 The performance comparison with 1G bits File Sizes

We can observe that the download time with Option2-1 can reduce 0.5seconds, 2seconds, 2.5seconds, and 4 seconds, compare to Option3-1 for 100M bits File Sizes with 100Mbps physical throughput of single user on the order of 0.1%, 0.5%, 1%, and 2% RLC residual BLER. And for 1G bits file sizes, we can also observe that the download time with Option2-1 can reduce 0.7seconds, 3.2seconds, 5seconds, and 8 seconds, comparing to Option3-1 for 100M bits File Sizes with 1Gbps physical throughput of single user on the order of 0.1%, 0.5%, 1%, and 2% RLC residual BLER.

Based on the above simulation results, we can conclude that the Option2-1 has the better performance than Option3-1.

5. Conclusion

In April 2017, the first phase of 3GPP 5G SI has been completed. To simplify the network and to drive down the OPEX of the network, CU/DU functional split has been considered as an important feature for 5G. This paper firstly summarized the standardization work of CU/DU functional split, and then provided our analysis on the comparison of two high layer functional split options from the aspect of UE's experience, the complexity of standardization and implantation, and network parameters optimization. Finally, the Option2-1 has been verified with the higher TCP throughput efficiency and shorter the download time for the fixed file sizes, comparing with the Option3-1. Then this work has proposed an improved solution to achieve the fast retransmission of lost PDCP PDUs in the intra-CU inter-DU handover scenario. As a result, we propose to adopt Option2-1 with the proposed solutions to improve user's experience.

References

[1] 3GPP TR 38.913 V14.0.0,Study on Scenarios and Requirements for Next Generation Access Technologies, October 2016

[2] 3GPP Draft_MeetingReport_RAN_75_170309_eom,3GPP TSG RAN Meeting #75 Dubrovnik,Croatia,March,2017

[3] 3GPP TR 38.801 V14.0.0, Study on new radio access technology: Radio access architecture and interfaces, March 2017

[4] 3GPP TS 36.300 V14.2.0, Evolved Universal Terrestrial Radio Access (E-UTRA)and Evolved Universal Terrestrial Radio Access Network(E-UTRAN);Overall description ,March 2017

[5] IAESI, Thales, Fairspectrum, VTT, Option 2 split with performance and reliable CU-DU connection ,R3-170973,3GPP TSG-RAN WG3 Meeting#95bis,Spokane, USA, April, 2017,

[6] CATT, Further Clarification of CU-DU Split Option 3-1 ,R3-170381,3GPP TSG RAN WG3 Meeting #95 Athens, Greece, February 2017

[7] Nokia, Alcatel-Lucent Shanghai Bell, AT&T, Conclusion on higher layer split option ,R3-170419,3GPP TSG-RAN WG3 Meeting #95 Athens, Greece, February, 2017

[8] ZTE, Clarification on CU-DU split options ,R3-162852, 3GPP TSG RAN WG3 Meeting#94 Reno, USA, November 2016

[9] Ericsson, Network deployments based on option 2, R3-170683, 3GPP TSG-RAN WG3 Meeting #95 Athens, Greece, February, 2017.

[10] Ericsson, Performance evaluation of option 3 ,R3-170685,3GPP TSG-RAN WG3 Meeting #95 Athens, Greece, February,2017

[11] 3GPP TS 36.425 V12.1.0, X2 interface user plane protocol ,Macrh,2015

A Distributed Compressive Sensing Scheme for Event Capture in Wireless Visual Sensor Networks

Meng Hou[1,2], **Sen Xu**[1], **Weiling Wu**[1] **and Fei Lin**[2,*]

[1] SICE, Beijing University of Posts and Telecommunications, Beijing, 100876, China
[2] SEEA, Qilu University of Technology, Jinan, 250353, China

*linfei@qlu.edu.cn

Abstract. Image signals which acquired by wireless visual sensor network can be used for specific event capture. This event capture is realized by image processing at the sink node. A distributed compressive sensing scheme is used for the transmission of these image signals from the camera nodes to the sink node. A measurement and joint reconstruction algorithm for these image signals are proposed in this paper. Make advantage of spatial correlation between images within a sensing area, the cluster head node which as the image decoder can accurately co-reconstruct these image signals. The subjective visual quality and the reconstruction error rate are used for the evaluation of reconstructed image quality. Simulation results show that the joint reconstruction algorithm achieves higher image quality at the same image compressive rate than the independent reconstruction algorithm.

1. Introduction

1.1. Wireless visual sensor networks

Wireless Visual Sensor Networks (WVSNs) is a sensor network composed of a large number of visual sensor nodes to monitor a certain area. It captures and monitors certain event in the field of view by means of image acquisition and processing, and transmits event information via wireless communication.

In general wireless sensor networks, scalar data such as temperature, humidity, illuminance and pressure, or speed and other vector data have relatively simple data types and small amount of data. The raw data can be transmitted directly, and finally the event is captured by the sink node, for example, the sink node can capture such event as "temperature exceeding the threshold" through temperature data.. Image data contain richer information, the use of a variety of image processing algorithms (such as pattern recognition, target tracking, behavior recognition, etc.), can achieve the capture of a lot of certain events [1]. For example, the use of target tracking algorithm can achieve the event capture of "the emergence of a human target" in the field of view.

However, the amount of data in the image is too large, in the wireless visual sensor network, if the event capture is completed by the sink node, the bandwidth of the wireless communication network cannot meet the demand. In addition, the power consumption of data processing is much lower than the power consumption of wireless transmission of data [2], therefore, WVSNs should use the underlying node to complete the event capture as much as possible and transmit event information in the communication network rather than the raw image data, thus avoiding a large amount of raw data being transmitted in the network.

In the video monitoring system, each camera independently and continuously transmits video data stream to the video server, and the higher resolution of the video is, the better it will be [3]. Different from the video monitoring system, WVSNs takes the event collecting rather than capturing clear video (or images) as the primary purpose. After the underlying nodes capture events, the WVSNs transmit event information directly to the network rather than the video stream. Therefore, the ability to flexibly deploy the wireless communication mode can meet the requirements of the transmission bandwidth.

1.2. Distributed compressive sensing

As a new signal processing method, CS (Compressive Sensing) can reconstruct the original signal at a signal receiving end with a small number of measurements. The requirements of compressive-sensing method for the signal are quite sparse [4-5]. The image signal can be sparsely represented, such as by domain transformation. For example, smooth images are sparse in Fourier basis, while piecewise smooth images are sparse on wavelet basis. Thus, for N-pixel images, it can be represented by K sparsity [6].

D. Baron et al. proposed a distributed compressive sensing based on the theory of compressive sensing, a compressive sensing which extends the single signal to the signal ensemble.

The DCS theory is based on the joint sparsity concept of the signal ensemble by using the correlation of the signals. In a typical DCS scenario, if the sensor's measurement signal is sparse and relevant under a certain basis, each sensor encodes the measurement signal through an incoherent bases (eg, a random matrix), and transmits the small amount of encoded data to the decoding side. Under appropriate conditions, the decoder can accurately reconstruct the measurement signal of each sensor [7].

DCS can further reduce the demand for communication bandwidth. The image signals collected by VSNs nodes deployed in the field of view usually have spatial correlation, so DCS provides an efficient image processing method for VSNs for capture events [8-9].

2. Scenario model

As shown in Figure 1, a scenario model for WVSNs can be represented as two parallel planes: Field of View (FoV, Field of View) Plane π, Sensor Node Deployment Plane π_1. In order to capture Event E in π, it is necessary to monitor the event area (EA, Event Area). The image acquisition area of the node of $S_i(i = 1,2,3,4)$ in π_1 can cover EA after splicing, then $S_i(i = 1,2,3,4)$ and the cluster head C_1 together constitute a cluster of capturing Event E.

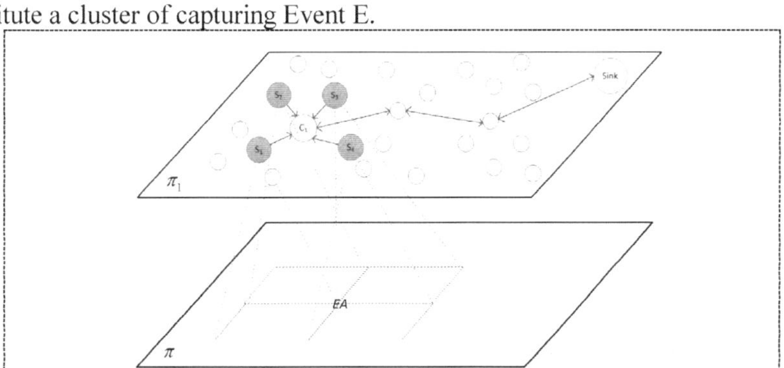

Figure 1. A scenario model for WVSNs.

The sensor node, cluster head, and sink make up a cluster tree network. The sensor node collects the original image as the encoding end of compressive sensing and generates the measurement and transmits it to the cluster head. As the compressive-sensing decoding terminal, the cluster head is used to reconstruct the original signal and capture the event through the image processing algorithm. And then the captured event is transmitted to the Sink node data through the multi-hop routing.

3. A distributed compressive sensing algorithm for event capture

Step 1: The sink node calculates the Event Area (EA) based on the coverage area required for the capture event, and selects the sensors node of the camera within the EA. These nodes form clusters and select head node. The multi-hop routing of head node and sink nodes is established to form cluster tree type of WVSNs.

Step 2: The head node generates random seed and measurement length (Default measurement length is used for the first time) and broadcasts within the cluster. The sensor node generates the measurement matrix after receiving the random seed and the measurement length. The sensor node measures (encodes) the acquired image signals and then transmits the measurement to the cluster head.

Step 3: The cluster head generates the joint measurement matrix according to the random seed, the length of the measurement and the number of nodes, and calculates the sparse base matrix according to the node acquisition signal length and the number of nodes to obtain the joint mapping matrix. The measurement of each node received are combined according to the EA image acquisition area splicing sequence, and the OMP algorithm is used to reconstruct (decode) the measured signal ensemble.

Step 4: The cluster head uses the image processing algorithm to capture the events in the reconstructed EA spliced image signal, and passes the event data to the sink node through the multi-hop route.

Step 5: The sink node increases or decreases the measurement length based on whether the capture event succeeds or not. After receiving the instruction, the cluster head regenerates the random seed and the new measurement for the next measurement. Assuming that the image quality threshold required for successful capture event is q_0, the reconstructed image quality $q \geqslant q_0$ is the condition for terminating the sink node and cluster head scheduling.

4. Measurement of image signal and joint reconstruction algorithm

Assuming that the image obtained in the acquisition event area EA is represented by I, the pixel is represented as $I \times I$. As shown in Fig. 1, it is stitched by the image $B\_B_i(i = 0,1,\cdots,N-1)$ collected by the sensor node $S_i(i = 0,1, \cdots,N-1)$ in EA. Assuming that the image $B_i(i = 0,1,\cdots,N-1)$ is in the same size; the pixels are $B \times B$. After the distributed sensing begins, the node receives the random seed and the measurement length indicated by the cluster head to generate the measurement matrix $\phi_i \in R^{M\times B}(i = 0,1,\cdots,N-1)$, wherein, M is the measurement length.

S_i uses ϕ_i to perform a random measurement . The measurements $y_i(i = 0,1,\cdots,N-1)$ are as follows:

$$Y = \begin{Bmatrix} y_0 \\ y_1 \\ \vdots \\ y_{N-1} \end{Bmatrix} = \begin{Bmatrix} \phi_0 B_0 \\ \phi_1 B_1 \\ \vdots \\ \phi_{N-1} B_{N-1} \end{Bmatrix} = \begin{Bmatrix} \phi_0 & & & \\ & \phi_1 & & \\ & & \ddots & \\ & & & \phi_{N-1} \end{Bmatrix} \begin{Bmatrix} B_0 \\ B_1 \\ \vdots \\ B_{N-1} \end{Bmatrix} \tag{1}$$

4.1. Independent reconstruction

S_i sends y_i to the cluster head. When the independent reconstruction algorithm is used, the cluster head receives y_i and respectively generates the mapping matrix θ_i according to the random seed, the signal length and the measurement length, then:

$$\theta_i = \phi_i \varphi_i \tag{2}$$

Wherein, $\phi_i \in R^{M\times B}$ is the measurement matrix, $\phi_i \in R^{B\times B}$ is a sparse base matrix. And then the optimization problem is solved as:

$$\theta_i = \phi_i \varphi_i \min \|s_i\|_0 \quad s.t. \quad y_i = \theta_i s_i \tag{3}$$

Wherein, s_i is the sparse coefficient of B_i in the base of $\phi_i \in R^{B \times B}$, $\widetilde{B}_i = \phi_i s_i$, then the recovery image $\widehat{B}_i$ is obtained and then the recovered $\hat{I}$ is obtained through the image splicing.

4.2. Joint reconstruction

S_i sends y_i to the cluster head. When the joint reconstruction algorithm is used, the cluster head receives the measurement of all nodes and generates the joint measurement matrix according to the random seed, the number of nodes, the length of the node signal and the length of the measurement [10].

$$\tilde{\Phi} = \begin{Bmatrix} \phi_0 & & & \\ & \phi_1 & & \\ & & \ddots & \\ & & & \phi_{\frac{I}{B}-1} \end{Bmatrix}, \tilde{\Phi} \in R^{\tilde{M} \times I} \tag{4}$$

Wherein, $\widetilde{M}$ is the length of the joint measurement, then $\widetilde{M} = (I/B) \times M$. And then the joint mapping matrix is obtained:

$$\Theta = \tilde{\Phi}\tilde{\Psi} \tag{5}$$

Wherein, $\tilde{\Psi} \in R^{I \times I}$ is the joint sparse base matrix; And then the optimization problem is solved:

$$\min \|s\|_0 \quad s.t. \quad Y = \Theta s \tag{6}$$

Wherein, s is the sparse coefficient of I in the base of $\tilde{\Psi} \in R^{I \times I}$, $\tilde{I} = \tilde{\Psi}s$, then the recovery image $\hat{I}$ is obtained.

5. Simulation results and analysis

The method of computer simulation is used to verify the scheme proposed in this paper. The image obtained by collecting the event area EA by 256-order gray-scale image is used to transform the sparse image signal by using DWT [11]. The stochastic measurement matrix is used for signal mapping; and the OMP algorithm is used to solve the optimization problem of the joint reconstruction [12].

5.1. Evaluation index

The reconstruction error is selected as the evaluation index of the reconstructed image quality to reflect proximity between the reconstructed image $\hat{\mathbf{I}}$ and the original image I. The reconstruction error rate is calculated as follows:

$$rer = \frac{\|\hat{I} - I\|_2}{\|I\|_2} \tag{7}$$

Namely, the ratio of the difference between the pixel values $\hat{\mathbf{I}}$ and I and the square root of the I pixel value.

5.2. Quality comparison of reconstructed images between independent reconstruction and joint reconstruction

Figure 2 is a local satellite image of a city; the image name is "innoway" with 256 * 256 pixels, 256-order gray-scale image to analog EA image I. When there are four nodes in EA, I is stitched by the images collected by the four sensor nodes, and the image collected by each camera sensor node is 128 * 128 pixels.

If the measurement M = 64, then the original image B_i with 128 * 128 pixels progressively projects through the mapping matrix, then 64 * 128 measurement matrix is obtained. At this point, the image compressive rate is:

$$r = \frac{M \times B}{B \times B} = \frac{64 \times 128}{128 \times 128} = \frac{1}{2} \tag{8}$$

Independent reconstruction (IR) method is to receive the measurement matrix to reconstruct the original image $B_i$and then is spliced into $\tilde{I}$.

Joint reconstruction (JR) method is to receive the measurement matrix for splicing, and then directly obtain the reconstructed EA image$\tilde{I}$ through the joint mapping matrix reconstruction.

Figure 3 is the reconstructed EA image $\tilde{I}$ of the independent reconstruction (left) and joint reconstruction (right) respectively, when the measurement is the same. As can be seen from the Figure, when M = 48, IR cannot distinguish things in the figure, and JR can roughly distinguish between streets and buildings; when M = 64, JR can distinguish between streets and buildings, IR can only roughly distinguish between streets and buildings.

Figure 2. The original image I of the “innoway”.

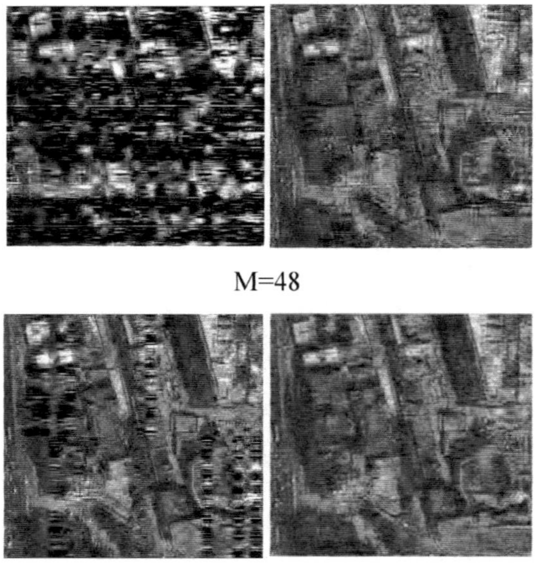

M=48

M=64

Figure 3. The reconstructed image $\tilde{I}$ of the “innoway”, left for independent reconstruction, right for the joint reconstruction.

Figure 4 is the reconstruction error curve of independent reconstruction and joint reconstruction, it can be seen that in the same measurement, the error of joint reconstruction is lower than that of the independent reconstruction. When the measurement is close to 128, the compression rate is close to 1, and the reconstruction error is close to but less than 0, and then a platform is as follows.

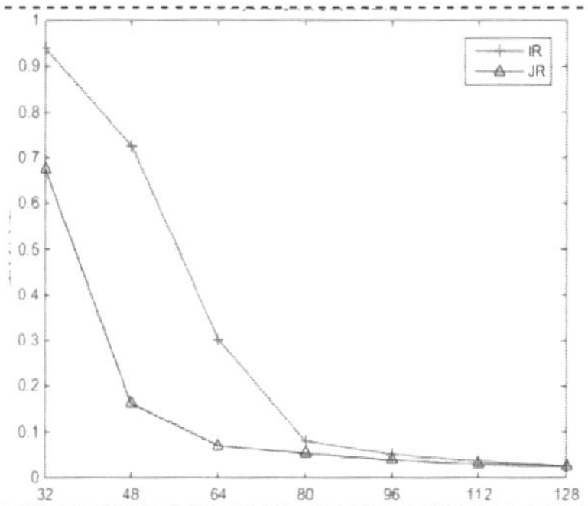

Figure 4. Comparison of reconstruction errors for independent reconstruction and joint reconstruction.

5.3. The influence of the number of nodes in EA on the reconstructed image quality

We use the image "lujiazui" as the EA image I with 512 * 512 pixels and 256 –order gray-scale image. The image $B_i(i = 0,1,\cdots,N-1)$ acquired by each node is 64*64 pixels, then N=64. If the measurement is M = 32, the compression rate is 1/2, the cluster head uses the joint reconstruction algorithm to restore the original image. There are three networking schemes:

a) Each four nodes make up a cluster, each cluster covers 1/16 area, with a total of 16 clusters;

b) Each 16 nodes make up a cluster, each cluster covers 1/4 area, with a total of 4 clusters;

c) 64 nodes form a cluster covering the entire EA area.

Because the compression rate of each node is the same, the data volume of the three networking schemes is the same, but the transmission method of the cluster head to the sink node is different. As can be seen from Fig. 5, the quality of the whole EA area image $\tilde{I}$ is better than the upper left corner 1/4 region, which is better than the 1/16 region of the upper left corner. It is shown that the more the number of nodes in the cluster is, the higher the quality of the cluster head reconstruction image is in the case of collecting the same amount of EA image I data.

Figure 5. Cluster head reconstruction image of scheme a), b) and c)

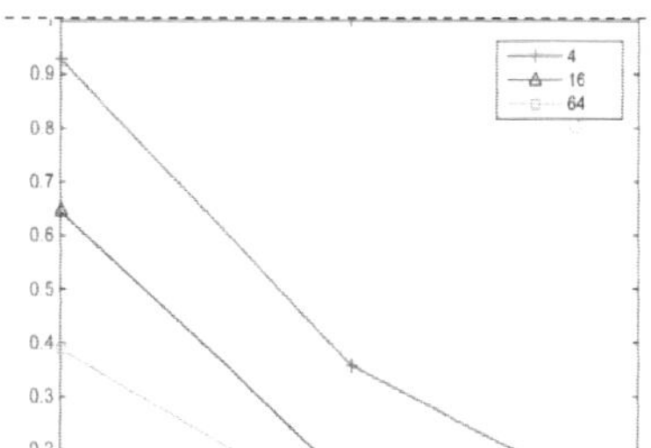

Figure 6. Comparison of the reconstruction errors for the different number of nodes.

Figure 6 is the reconstruction error curve with different numbers of nodes; we can see that in the case of the same measurement, the more the number of nodes, the lower the reconstruction error.

6. Conclusions

DCS is one of the distributed source codings (DSC) [13]. The sensor node acquires the original image signal by compression sensing coding, thereby reducing the amount of data. In the same EA, when the number of nodes increases, the spatial correlation of the image signal collected by the node increases. And the reconstruction error decreases when the compression rate is constant. In this paper, a DCS scheme is proposed in this paper, through analysis and simulation, it is verified that the scheme can reconstruct the original image with lower reconstruction error.

Acknowledgments

This paper is partially supported by the High School Science and technology program of Shandong Province (No. J14LN04) and Postgraduate Education Innovation Program of Shandong Province (No. SDYY16032).

References

[1] Shen H, Bai G. Routing in wireless multimedia sensor networks: A survey and challenges ahead[J]. Journal of Network & Computer Applications, 2016, 71:30-49.
[2] Charfi, Y., N. Wakamiya and M. Murata, Challenging issues in visual sensor networks. Wireless Communications, IEEE, 2009. 16(2): p. 44-49.
[3] Soro S, Heinzelman W. A Survey of Visual Sensor Networks[J]. Advances in Multimedia, 2009, 2009(1687-5680):21.
[4] Tropp J A, Gilbert A C. Signal Recovery From Random Measurements Via Orthogonal Matching Pursuit[J]. IEEE Transactions on Information Theory, 2007, 53(12):4655-4666.
[5] CandèandCand Transactions on Information Theory, 2007, 53(12):4655-4666.55-4666.atchinmeasurements[J]. Communications on Pure & Applied Mathematics, 2005, 59(8):1207-1223.
[6] Jiang H, Matthews K, Wilford P, et al. Video coding using compressive measurements: US, US 8929456 B2[P]. 2015.
[7] Prades-Nebot J, Ma Y, Huang T. Distributed Video Coding using Compressive Sampling[C]// Picture Coding Symposium. IEEE, 2009:1-4.
[8] D. Baron, M. F. Duarte, S. Sarvotham, M. B. Wakin, and R. G. Baraniuk, raniuk, . Baraniuk, iuk, raniuk, araniuk, B. Wakin, and R. G. Baraniuk, Picture Coding Symposium. In, Control, and Computing, (Monticello, IL), Sept. 2005.
[9] Tseng H Y, Shen Y C, Wu J L. Distributed video coding with compressive measurements[C]// International Conference on Multimedea 2011, Scottsdale, Az, Usa, November 28 - December. DBLP, 2011:1273-1276.
[10] Baron D, Wakin M B, Duarte M F, et al. Distributed compressed sensing[J]. Preprint, 2005, 22(10):2729 - 2732.
[11] Dong H, Zhuang B, Su F, et al. A novel distributed compressive video sensing based on hybrid sparse basis[C]// Visual Communications and Image Processing Conference. IEEE, 2015:320-323.
[12] Hayashi K, Nagahara M, Tanaka T. A User's Guide to Compressed Sensing for Communications Systems[J]. Ieice Transactions on Communications, 2013, E96-B(3):685-712.
[13] Alnuaimi, M., F. Sallabi and K. Shuaib. A survey of Wireless Multimedia Sensor Networks challenges and solutions. in Innovations in Information Technology (IIT), 2011 International Conference on. 2011.

Development of Na Adaptive Filter to Estimate the Percentage of Body Fat Based on Anthropometric Measures

Naydson Emmerson S P do Lago[1, *], Allan Kardec Barros[1], Nilviane Pires S Sousa[3], Carlos Magno S Junior[1], Guilherme Oliveira[1], Camila Guimares Polisel[4] and Ewaldo Eder Carvalho Santana[2]

[1] Department of Eletrical Engineering, Federal University of Maranhão, São Luís-MA, Brazil
[2] Department of Mathematics and Informatic, State University of Maranhão, São Luís MA, Brazil
[3] Center for Biological Sciences and Health, Federal University of Maranhão, São Luís-MA, Brazil
[4] Center for Biological and Health Sciences, Federal University of Mato Grosso do Sul, Brazil

[*]naydson.14@hotmail.com; naydsoemmeson@gmail.com

Abstract. This study aims to develop an algorithm of an adaptive filter to determine the percentage of body fat based on the use of anthropometric indicators in adolescents. Measurements such as body mass, height and waist circumference were collected for a better analysis. The development of this filter was based on the Wiener filter, used to produce an estimate of a random process. The Wiener filter minimizes the mean square error between the estimated random process and the desired process. The LMS algorithm was also studied for the development of the filter because it is important due to its simplicity and facility of computation. Excellent results were obtained with the filter developed, being these results analyzed and compared with the data collected.

1. Introduction

With the increase of obesity and sedentary lifestyles, the prevalences of diabetes and cardiovascular diseases are also rapidly increasing [1]. Obesity can be defined as excessive accumulation of body fat. It is about a complex public health problem that in the last decades has had its prevalence increased worldwide [2]. The period of highest risk for the incidence of obesity is the transition between adolescence and early adulthood, in both sexes and in various ethnic groups, as well as being a factor of important risk for cardiovascular diseases [3]. The Body Mass Index (BMI) is one of the most commonly used anthropometric indicators for being safe, easy to apply and inexpensive [4]. Waist circumference (WC) is an indicator of abdominal adiposity, which is often associated with the same obesity-related risk factors [5]. Although there is no consensus on the best anthropometric predictor of blood pressure levels high in the child and adolescent population, the importance of using simpler, more practical and cheaper cardiovascular risk. [6]. According to the Ending Childhood Obesity report from the World Health Organization (WHO) (2016) [7], the number of overweight children in low-middle income countries more than doubled between 1990 and 2014, going from 7,5 to 15.5 million.

There are several methods used to evaluate body composition, especially body fat, in adults and adolescents. Among the indirect methods, hydrostatic weighing and dual energy X-ray absorptiometry (DEXA) stand out, however, they are more difficult methods to apply in clinical and epidemiological researchs and as a screening method in public health due to the high cost and the need for a qualified technical team to carry out the measures. Among the double indirect methods, anthropometry is considered a simple, fast and low-cost method that can be applied in a large number of individuals. Many anthropometric indicators have been proposed to diagnose health risks by taking into consideration the increase of body fat. The most widely used is still the body mass index (BMI), which, however, has some limitations [8]. With a tendency to persist in adult life, it is necessary to establish the best indicator of distribution of body fat that allows early identification of adolescents to establish interventions and improve future cardiovascular health [9].

Therefore, it is of great importance to construct techniques for tracking and / or screening of the percentage of body fat, based on the analysis of low-cost and non-invasive indicators, such as anthropometric parameters (weight, height, waist circumference, among others) that are one of the best health indicators in the juvenile population [10], being universally applicable, with good acceptance of the population and proposed by the World Health Organization. Adaptive filters are interesting in cases where some of the characteristics of the application are not known. Thus, it can be said that adaptive filters are best suited for applications where the input signal conditions or the system parameters vary slowly and the filter is capable of self-tuning to compensate such changes. The classic algorithm LMS (Least Mean Squares) is one of the most known techniques for the development of an adaptive filter. The LMS uses a stochastic gradient algorithm to adapt the load of a filter. This adaptation consists in the continuous (and adaptive) adjustment of the values of the filter coefficients, taking as a measure the minimization of the error in the mean of the squares. Therefore, the objective is to implement an adaptive filter, based on the Wiener technique, that can infer the percentage of body fat of adolescents by noninvasive measures.

2. Methods

2.1 Sampling

The sample was obtained from the school population of 16 schools in the public school network of Sao Luis, Maranhao, Brazil. The schools were chosen at random. The sample consisted of 471 students, being 329 female and 142 male, aged 10 to 19 years. The following exclusion criteria were adopted: Female adolescents who were pregnant, breastfeeding or using contraceptives; Physical incapacity that would preclude or compromise anthropometric measures; Non-agreement of the responsible ones or of the participants; Absence in data collection. The present study is approved by the research ethics committee of the University Hospital of the Federal University of Maranhão, according to opinion 251/11.

2.2 Data collection

The admeasurement of all variables was performed by trained researchers using standardized techniques and calibrated equipment. All measurements were made in duplicate and the mean was considered in the data analysis. The following variables were collected: weight, height, waist circumference and percentage of body fat that was obtained through the Tetrapolar Bioimpedance apparatus (Sanny, Brazil). From the information obtained by the collected variables, a matrix X, of order 4 x 471 was generated, with the values of x1, x2, x3 and x4, respectively, weight, height, waist circumference and percentage of body fat that was obtained. The matrix description can be seen as follows:

$$\boldsymbol{X} = \begin{bmatrix} x_1(1),\cdots,\ x_1(n),\cdots,x_1(N) \\ x_2(1),\cdots,\ x_2(n),\cdots,x_2(N) \\ x_3(1),\cdots,\ x_3(n),\cdots,x_3(N) \\ x_4(1),\cdots,\ x_4(n),\cdots,x_4(N) \end{bmatrix} \tag{1}$$

Where $1 \leq n \leq N$.

The measurement of all information was collected by trained individuals to use calibrated equipment.

2.3 Wiener filter

The Wiener filter is considered one of the best linear least squares filters that can be used for prediction, estimation, interpolation, signal and noise filtering, etc. To design them, an appropriate prior knowledge of the statistical properties of the input signal is required. The problem is that this knowledge usually cannot be obtained. The objective of the Wiener filtering is to determine the impulse response of length N, having u(n) as input and given that the output of the filter is y (n). Then, we get the error signal e(n) from subtraction of the desired signal d(n) by y(n) as shown in the following equation:

$$e(n) = d(n) - y(n) \tag{2}$$

In order to minimize the error signal $e(n)$, we must choose the coefficients of the filter with the purpose to match $y(n)$ to $d(n)$. Thus, the function to be minimized as mean square error, is given by the following expression:

$$J = E\,[e(n)^2 = E[e(n)e^*(n)] \tag{3}$$

Where J is the cost function to be minimized.

To minimize the cost function J is given a vector w = [w0, w1,..., wN-1]T satisfying equation (3).

$$y(n) = w^H u(n) \tag{4}$$

Then we have from equation 4, the error signal expressed as:

$$e(n) = d(n) - w^H u(n) \tag{5}$$

Substituting equation 5 into equation 3 we obtain:

$$J = E[\,|d(n)|^2] - p^H w - w^H p + w^H R w \tag{6}$$

Where R is the input autocorrelation matrix, p is the vector composed of the cross correlation between the desired response and the input components. Thus, we can describe in the following matrix form:

$$\boldsymbol{R} = \begin{bmatrix} R_x[0] & R_x[1] & \cdots & R_x[N-1] \\ R_x[1] & R_x[0] & \cdots & R_x[N-2] \\ R_x[2] & R_x[1] & \cdots & R_x[N-3] \\ & & \vdots & \\ R_x[N-1] & R_x[N-2] & \cdots & R_x[0] \end{bmatrix} \tag{7}$$

$$\boldsymbol{p} = \begin{bmatrix} \boldsymbol{p}[0] \\ \boldsymbol{p}[1] \\ \boldsymbol{p}[2] \\ \vdots \\ \boldsymbol{p}[N-1] \end{bmatrix} \tag{8}$$

To find the coefficients of the filter w, we calculate the gradient of J, deriving and equating the equation to zero.

$$\Delta J = \frac{\partial j}{\partial w} = 2p - 2Rw \tag{9}$$

Therefore, obtaining the following result:

$$Rw_0 = p \tag{10}$$

$$w_0 = R^{-1}p \tag{11}$$

Being w0 the vector of optimal coefficients for the Wiener filter. From equation 6, we can determine the minimum mean quadratic error as follows:

$$J_{min} = E\,[|d(n)|^2] - p^H w_0 - w_0{}^H p + w_0{}^H R w_0 \tag{12}$$

Substituting w0 by equation 11 we will get:

$$J_{min} = E\,[|d(n)|^2] - p^H R^{-1}p - (R^{-1}p)^H p + (R^{-1}p)^H RR^{-1}p$$

$$= \sigma_d^2 - p^H R^{-1} p \tag{13}$$

Where $e_k = y_k - \mathrm{wt\,k}\ X_k$.

2.4 Correlation coefficient

It is often necessary to study the relationship between two or more variables. If the study deals with only two variables, we have a correlation and a simple regression. If we involve more than two variables, we have multiple correlation and regression. Regression and correlation deal only with the linear relationship between two variables. The correlation analysis provides a number that summarizes the degree of linear relationship between the two variables. That is, the correlation coefficient is adequate to evaluate only the linear relationship. The two variables may be perfectly related, but if it is not in a linear form, the value of the coefficient can be zero or close to zero. The correlation coefficient can vary between **-1** and **+1**, with a coefficient of +1, indicating a perfect positive linear correlation. A correlation coefficient of -1 indicates negative perfect linear correlation.. The most common is that the coefficient is in the interval between these two values. A correlation coefficient of 0 (zero) means that there is no linear relation between the two variables.

3. Results

When executing the developed algorithm, with the matrix *X* created in item *"C"*, we define a new matrix *Z* that will be the input of an adaptive filter whose output will be a vector *y1* of dimension *1x471*. From vector y1, a new matrix *Z2* is generated, and it contains the time delay determined by another algorithm that assists in the execution of this one. This delay identifies at which sampling position there is the best correlation index between the input matrix and the desired result. We can see in figure 1 below, the result of the delay and the correlation coefficient in relation to the matrix input of X:

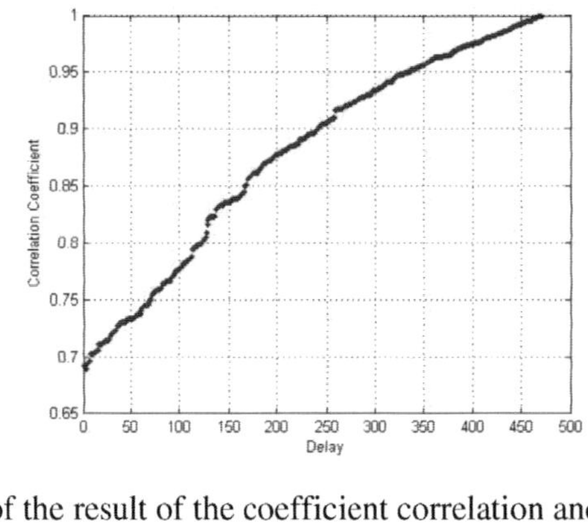

Figure 1 - Graph of the result of the coefficient correlation and the delay determined by the matrix X.

We can verify through figure 1 that the X axis is the input matrix of the sampling giving its maximum value of 471, and the Y axis being the result of the correlation coefficient in approaching the value +1. This shows, therefore, that the correlation result reached its expected maximum value. To make sure that the correlation coefficient was within the characteristics of its basic assumption, we generated the graph between the percentage of body fat and the adaptive measure obtained by the filter. Figure 2 below shows the generated graph:

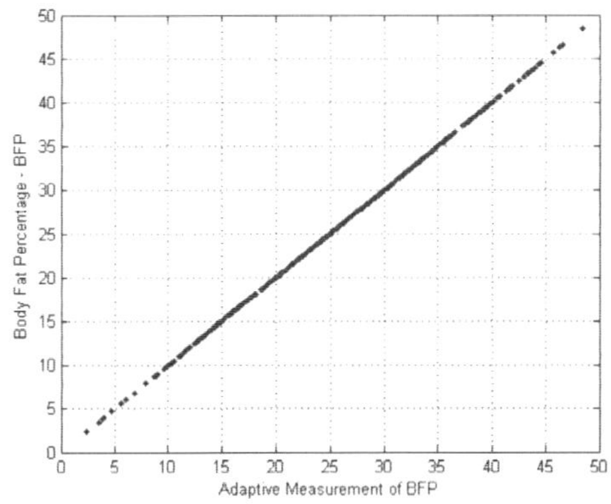

Figure 2 - Relationship between fat percentage and Measurement obtained.

We see that the relationship between the two variables meets the basic assumption of correlation coefficient, being them linear variables and perfectly related. With the matrices created and knowing the determination of the delay, and that the correlation coefficient is within the expected characteristics, we obtained the following result for our sample of the 471 students:

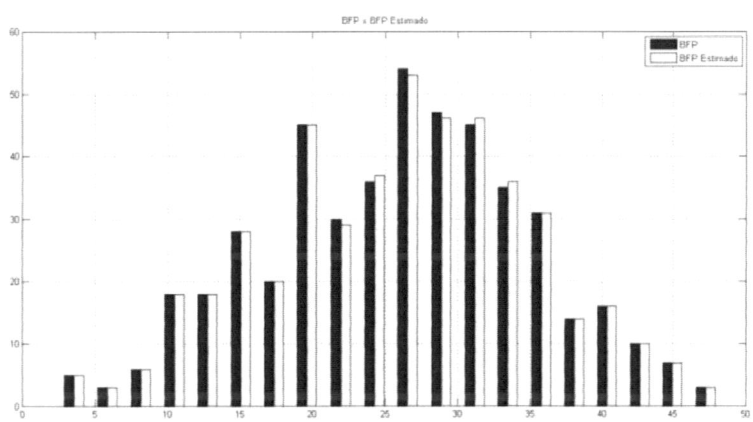

Figure 3 - Relation of the percentage of calculated body fat and that stimulated by the adaptive filter.

We can see in figure 3 the comparison between the information of Body Fat Percentage that had already been calculated - in blue color - and the result of the Body Fat Percentage stimulated by the developed filter - in Yellow color - that there were few divergences

4. Conclusion

The Wiener Adaptive Filter developed in this study proved to be efficient when executed and in showing the correlation result of coefficients of the defined matrices. The prevalence of overweight and obesity has increased significantly in recent decades, and is associated with the development of cardiovascular diseases and metabolic disorders, such as diabetes and hypertension, both in adults and children and adolescents. The anthropometric indicators, which are weight, height and waist circumference, are the most used indicators in clinical practice and in epidemiological studies to identify / evaluate overweight / obese individuals with a high risk of developing cardiovascular problems. Hence, the algorithm developed in this work may aid in the screening of individuals with high body fat percentage and thus, help in the prevention of comorbidities associated with overweight. The use of the filter developed in the present study by health professionals will be of great assistance since it will allow the analysis of the patient's body fat percentage through low cost, reproducible and accessible indicators. This will allow the screening of overweight and obese patients serving as a method of prevention for possible comorbidities associated with being overweight, since the filter will allow the early intervention of these adolescents.

References

[1] Chung H.I., Park S., Park J.M., Yoo E.G. (2016).Waist-to-Height Ratio as an Index for Cardiometabolic Risk in Adolescents: Results from the 1998-2008 KNHANES. Yonsei Med J. 57: 658-663.

[2] FILHO, M.S.P., et al. Cardiovascular, metabolic, and inflammatory risk factors and their relationship with obesity in children and adolescents: clinical and therapeutic aspects. Boletim científico de Pediatria, vol. 2, n.2, 2013.

[3] FRIGNANI, RR; PASSOS, MA; FERRARI, GL et al. Reference curves of the body fat index in adolescents and their association with anthropometric variables. J Pediatr. 91:248-255, 2015. http://10.1016/j.jped.2014.07.009.

[4] Mota J.F., Rinaldi A.E.M., Pereira A.F., Orsatti F.L., Burini R.C. (2011). Anthropometricindicators as risk markers for metabolic abnormalities. Ciências e Saúde Coletiva, 16 (9):3901-3908.doi:http://dx.doi.org/10.1590/S1413-81232011001000026

[5] Sousa, N. P. S., e Silva, A. C. S., Sousa, C. M., & Barros, A. K. Anthropometric Measures in the Prediction of High Fat Percentage in Female Adolescents.

[6] Oliveira A.V., Costa A.C.P.J., Pascoal L.V. et al. (2014). Correlação entre indicadores antropométricos e pressão arterial de adolescentes. Texto Contexto Enferm. 23(4):995-1003.

[7] OMS (Organização Mundial da Saúde). Ending Childhood Obesity (ECHO). Geneva: OMS, 2016.

[8] Pelegrini A, Silva DAS, Silva JMFL et al. Anthropometric indicators of obesity in the prediction of high body fat in adolescents. Rev Paul Pediatr. 2015;33:56-62. http://dx.doi.org/10.1016/j.rpped.2014.06.007

[9] PEREIRA P.F.; De FARIA F.R.; De FARIA E.R., et al. Indicadores antropométricos para identificar síndrome metabólica e fenótipo cintura hipertrigliceridêmica: uma comparação entre as três fases da adolescência. Rev Paul Pediatr., 33(2):194-20,2015.

[10] SOUZA, M.H.N. et al. Avaliação do estado nutricional e da saúde de crianças e adolescentes na prática assistencial do enfermeiro. Cogitare Enferm. 18(1): 29-35, 2013.

An Accurate Transmitting Power Control Method in Wireless Communication Transceivers

Naikang Zhang[1,*], Zhiping Wen[1], Xunping Hou[1] and Bo Bi[1]

[1]Beijing Microelectronics Technology Institute, Beijing, China

E-mail: *znk097@foxmail.com

Abstract. Power control circuits are widely used in transceivers aiming at stabilizing the transmitted signal power to a specified value, thereby reducing power consumption and interference to other frequency bands. In order to overcome the shortcomings of traditional modes of power control, this paper proposes an accurate signal power detection method by multiplexing the receiver and realizes transmitting power control in the digital domain. The simulation results show that this novel digital power control approach has advantages of small delay, high precision and simplified design procedure. The proposed method is applicable to transceivers working at large frequency dynamic range, and has good engineering practicability.

1. Introduction

Wireless communication systems such as radar, mobile phones, base stations, walkie-talkies, bluetooth devices, etc., consist of both receiver and transmitter to interact with outside world. For a wireless communication transceiver, it is necessary to control the transmit signal power for the following reasons: the instability of transmission circuit and environmental effects give rise to unstable transmitted signals without power control; wireless cellular network is able to achieve precise set of cell size and better convergence performance with power control; power dissipation can be reduced and interference to other frequency bands can be minimized with appropriate power control [1], et cetera.

In order to realize transmission power control, we need to detect the power size of transmitted signal for the first step. [2] employs a logarithmic amplifier to detect the signal from the power amplifier (PA) output terminal and convert the power into voltage that contains voltage ripple due to the non-constant amplitude envelope of input signal, and compares that voltage with ramp control signal. The comparison result is transferred as PA's control voltage. However, the logarithmic amplifier is only suitable for signal with small variation in peak-to-average ratio (PAR), and the operation frequency range is relatively narrow. This method is more applicable to dedicated power measurement integrated circuits.

Analog Devices has developed the True Power Detection series of RMS-OC integrated circuits [3] in 2001. Its performance is hardly susceptible to environmental temperature, and has a high linearity characteristic. The shortcoming of the device is large power consumption as the supply voltage is 2.7 to 5.5V, and the working current reaches as large as 2mA.

Some researchers propose to reuse the receiver in the transceiver while transmitting so as to achieve power control. To address the influence lead by gain variation on the receiver path, the signal strength detection circuit is placed close to the antenna. However, if the received signal is relatively

IOP Conf. Series: Journal of Physics: Conf. Series **960** (2018) 012009 doi:10.1088/1742-6596/960/1/012009

small compared to the receive path noise, the circuit will be difficult to achieve accurate detection and be susceptible to interference from nearby channels [4].

Most of the existing researches are based on analog measurement, although analog methods can achieve continuous measurement, their operation bandwidth is substantially low, and the analog circuits are generally tricky to design. Furthermore, the use of low-pass filter for average operation calls for very low cut-off frequency to ensure the stability of the result, meaning that the measurement speed is much slow. In view of this fact, this paper proposes a digital method to detect and control transmit signal power by multiplexing the receiver while it is idle. Operationally, the output signal of transmitter is coupled to the receiver input, and after a series of processing, it can be transformed into digital value representative of present attenuation. The power control word deduced by the attenuation value will be sent back to the attenuator on transmit path. Through this method, we can realize power control to transmitted signals with less extra circuit in a transceiver while ensuring performance.

2. System structure of power control

The power control system structure in a transceiver proposed in this paper is shown with reference to figure 1. Data to be transmitted is firstly provided to a digital-to-analog converter (DAC), the output of which is fed into a modulator. The modulated radio frequency signal is attenuated by an attenuator, and the output of the attenuator is provided to a power amplifier (PA) via a balun. Then the signal is amplified and transferred into an output filter. After filtered, the signal is passed to duplexer and isolator unit, which can alternatively allow an external signal at the antenna to be fed into a receiver circuit.

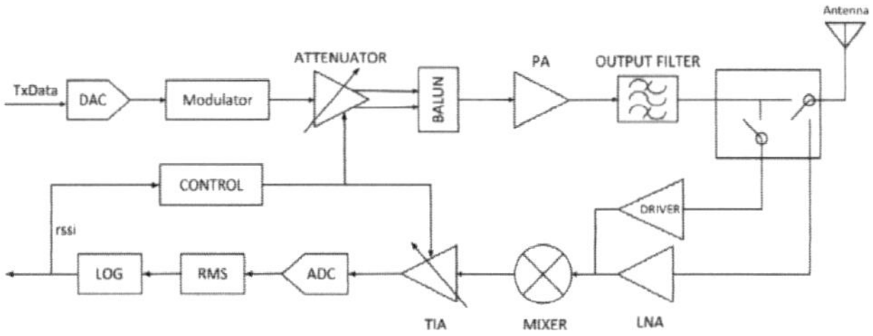

Figure 1. System Structure of Transmit Power Control.

The method offered in this paper is available in TDD mode only because it reuses the inactive receiver to perform power detection during transmitting. Since the transmitted signal is at high level with respect to received signal from outside, it is multiplexed into the receiver chain after the LNA. In other words, the signal coupled at the output port is applied via a downconverter input driver to the downconverter of receiving path. The output of the downconverter is amplified by a trans-impedance amplifier (TIA), the output of which is received by a low-pass filter, which attenuates blocking signals outside the channel. And the filter feeds an output to analog-to-digital converter (ADC). The digital output of ADC is fed into RMS module which develops a value indicative of received signal strength. The value is utilized to infer present attenuation value. And the attenuation value is passed to the control unit, which compares it with desired attenuation word from software and produces power control command to modules both at transmitting path and receiving path.

3. Theoretical design of power control system

3.1. Operating principle of signal strength detector

Power measurement of ac signals can be converted to measurement of true effective value of the signal voltage, which is generally defined as RMS. And the true RMS value of a signal is equal to the amount of direct current required to produce the same heat on the same resistive load.

Traditional method of detecting the effective value of voltage is achieved by the diode square law detection circuit or thermoelectric couple heating device. Both methods have many disadvantages, bringing about inaccurate measurement, poor linearity and thermal sensitivity, especially when detecting small radio frequency signals. The characteristic range of the diode square law is very narrow, and the detection performance becomes poor when the signal is too small or too large, resulting in large measurement error. Thermoelectric couple heating device is complicated, fragile and easy to be damaged at oversized signal. With the development of analog integrated circuits, people have found new way to detect signal voltage [5]. The valid values are defined by the following formula

$$V_{RMS} = \sqrt{\frac{1}{T}\int_0^T v^2 dt} \tag{1}$$

Nonetheless, there still exists some drawbacks of the analog method. Analog circuits are usually tricky to design and has poor flexibility, and their operation bandwidth are relatively narrow, and the filters used to perform averaging operation tend to bring about large time delay. Due to the drawbacks of analog measurement mentioned above, we measure the RMS value with a digital method.

The amplitude of AC signal is always changing, but when we subdivide the time into infinite amount of extremely tiny time Δt, the amplitude can be considered unchanged during that small time, we define that value as $e(i)$. Hence, we can get

$$\sum_{i=1}^{n} [e(i)^2 \Delta t] = E_{rms}^2 t \tag{2}$$

where n follows the formula below

$$n = \frac{t}{\Delta t} \tag{3}$$

Therefore, the power of a signal can be written as

$$E_{rms}^2 = \frac{1}{n}\sum_{i=1}^{n} [e(i)^2] \tag{4}$$

That is,

$$E_{rms} = \sqrt{\frac{1}{n}\sum_{i=1}^{n} [e(i)^2]} \tag{5}$$

According to above formula, firstly we need to discretize the analog signal. That operation can be realized through ADC sampling. When the sampling rate is much higher than the frequency of the measured signal, in other words, the time interval between the two sampling results of the ADC is very short, we can approximate that the amplitude of the measured signal does not change during Δt. Then we take advantage of (5) to calculate the RMS. For periodic signals, we can use one or more cycles to calculate, whilst for signals with no significant period, we can specify some time to perform the sampling and calculation.

For the purpose of realizing linear relationship between signal power and attenuation control word, we need to convert the detected power value to a logarithm scale, so the output of RMS module is coupled to a LOG circuit. We define the output of the LOG circuit as "rssi" (abbreviation of "received signal strength indicator").

3.2. Transmit power control method

3.2.1. Calibration work before power control. It is necessary to calibrate the circuits before performing regular power control. There are two purposes for calibration, one of which is to calculate the error between actual transmit power and ideal power, denoted by Trans_cal, and the other aim is to obtain the rssi value when the transmit signal power is maximum, defined as $rssi_{max\_cal}$.

Resultant from channel delays, imbalances, and other signal impairments occurring within the transmitter chain, signals are often impaired while propagated. In other word, when the attenuator in front of the power amplifier receives a certain attenuation command, it is likely that the signal is not transmitted at expected power. We need to figure out the error between the transmitted signal and the ideal signal before nominal power control. The calibration can be performed by transmitting a single tone signal with known power and gain, and measuring the output signal before antenna, and we can acquire the error as with the formula shows below,

$$trans\_cal = P\_meas - P\_des \tag{6}$$

where P_des is the desired power and P_meas is the measured power. For the purpose of facilitating subsequent operations, the power value used thereof is converted to logarithm scale, meaning its unit is dBm, and trans_cal is expressed in dB.

It should be noted that besides the non-ideal factors existing in transmitting path, there are also some unpredictable deviations in detection path resulting in inaccuracy of the rssi value we conclude finally, like gain error, phase mismatch, temperature effect and interference from other frequency band, et cetera. To address the effects mentioned before, it may be desirable to perform another calibration.

In order to prevent the ADC in power control loop from clipping and raise the SNR of the detector path, we need to make convenient changes of amplifier gain on the detector path. Accordingly, we have to subtract the gain along the path from the estimated power value so as to obtain real signal power at the antenna. The gain error between theoretical value and the actual one, together with other non-ideal factors on detector path described hereinbefore will give rise to inaccuracy of the final rssi value, so we can develop a calibration test to store the all of the errors. In light of the power control method which will be covered in detail later, we take present power value minus maximum power to get present attenuation value, followed by comparing present attenuation value with desired attenuation value requested, the result of which can be utilized to deduce a new attenuation control word. Since we need maximum power value to be subtracted, we can store it as well as non-ideal factors of detector path in a register during calibration. Operationally, we transmit a signal with maximum power by set attenuation word to 0, and obtain a rssi value representative of the transmitted signal power plus amplifier gain of detector path at the output of the log circuit, we subtract the trans_cal calculated above from that rssi value to eliminate the impact of the transmit path. The calculation is performed as shown in (7),

$$\begin{aligned} rssi_{\max\_cal} &= rssi\_cal - trans\_cal \\ &= rssi\_cal - P\_meas + P\_des \end{aligned} \tag{7}$$

where rssi_cal is the rssi value read during calibration, and the calculation result rssimax_cal will be stored in a register. As long as a new rssi is deduced in the power control loop during transmitting, the attenuation at that time can be simply obtained in accordance with the following formula,

$$pre\_atten = rssi - rssi_{\max\_cal} - \Delta g \tag{8}$$

where pre_atten stands for present attenuation, rssi indicates the measured signal power, and Δg represents the change of gain on detector path. It can be observed that the effects of non-ideal factors through detector path are removed by subtract operation, and the gain error is significantly shrunk.

3.2.2. Iterative algorithm of power control. The iterative method of generating attenuation control word can be expressed by (9),

$$\begin{aligned} atten(n) &= atten(n-1) + err\_atten \\ &= atten(n-1) + pre\_atten + des\_atten \end{aligned} \tag{9}$$

where atten(n) stands for the attenuation control word for transmit burst number n, atten(n-1) represents the previous control word, and des_atten is the desired amount of attenuation requested. It should be noted that des_atten is a positive value, while pre_atten deduced in accordance with (8) is negative, though both of the two indicate the degree of attenuation. In view of this fact, we add up the two together to acquire the error between their absolute value, and the error is added to current attenuation control word so as to get a new one. For example, if the attenuation detected is -10dB, whereas the attenuation requested is -20dB, i.e., that the des_atten received by the system is 20dB, which means the sum is 10dB, therefore, the current attenuation word is added to 10dB to acquire new attenuation word. The advantage of this method is the capacity to track the amount of attenuation requested. Whenever a new des_atten appears, the error changes accordingly, and atten(n) will be updated in time. Therefore, once the power control loop settles down, subsequent signals will be transmitted with correct power regardless of how much attenuation command changes.

For a further description, since there is no previous rssi calculated for the first burst of transmission, the control word atten_1 should be set specially. We determine herein atten_1 as des_atten minus trans_cal, and utilize a mux module to select between the preset control word and the iterative one according to transmit clock.

Figure 2 gives a more detailed illustration of the control module in figure 1, and illustrates the operation process of power control as stated hereinbefore.

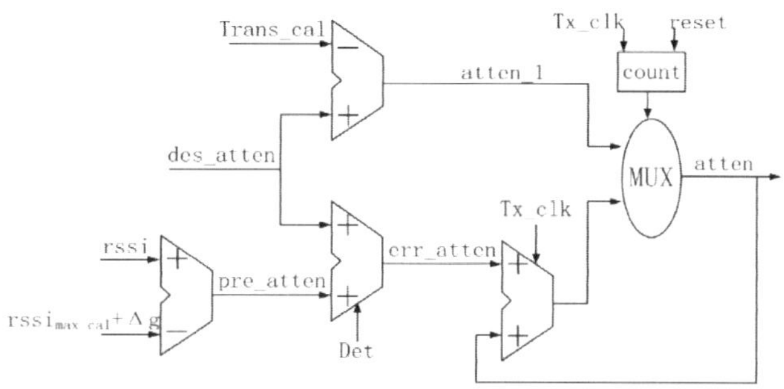

Figure 2. Structure Diagram of the Control Module.

Therein Tx_clk enables the transmitter at each burst, so atten output from the control module is latched on the rising edge of Tx_clk. Det is an internal signal which goes high on the rising edge of Tx_clk and enables the power control circuit, and falls down to latch the pre_atten, so its high level should be maintained long enough to ensure an accurate and stable power detection. The mux module selects atten_1 to be the control word after the reset signal is pulled high, and selects closed control loop result on the falling edge of the first Tx_clk burst after reset.

3.2.3. Gain adjustment of the power detection path. As noted above, signals output from transmitter may have fluctuation in their power caused by attenuation changes, strength differences caused by different modulation ways, and offset through the transmit path, et cetera. If we detect the power with receive path without consideration of gain compensation, the SNR of detected signals may float and there may be peak clipping during ADC works. To ensure the accuracy of detected signal power, we add 10dB to the gain of TIA in receive path at every 10dB increment of desired attenuation. The initial gain is set and stored in a circuit register upon power up [1]. Through this method, the ADC can work normally and the SNR can be maintained over 20dB.

4. Simulation result of the proposed power control method

The method put forward in this paper can be verified simply through building a transceiver module with MATLAB/Simulink tool kits. During simulation, the modulated signal ready to be transmitted is a single tone constant waveform at a frequency of 0.8GHz, and the attenuation requested is set to 0dB, 30dB, 20dB and 10dB successively, the output signal waveform in time domain is illustrated by figure 3 and figure 4 shows the transition of signal amplitude during the first burst, we can see that it takes some time on the order of 6.6 usec for the transmitter to settle down. Figure 5 and figure 6 describe subsequent signal waveform transitions when attenuation control word increases and decreases respectively. It can be observed that transmitter can achieve flexible and accurate switching in power with little transition time after the first burst, which indicates the advantage of (9).

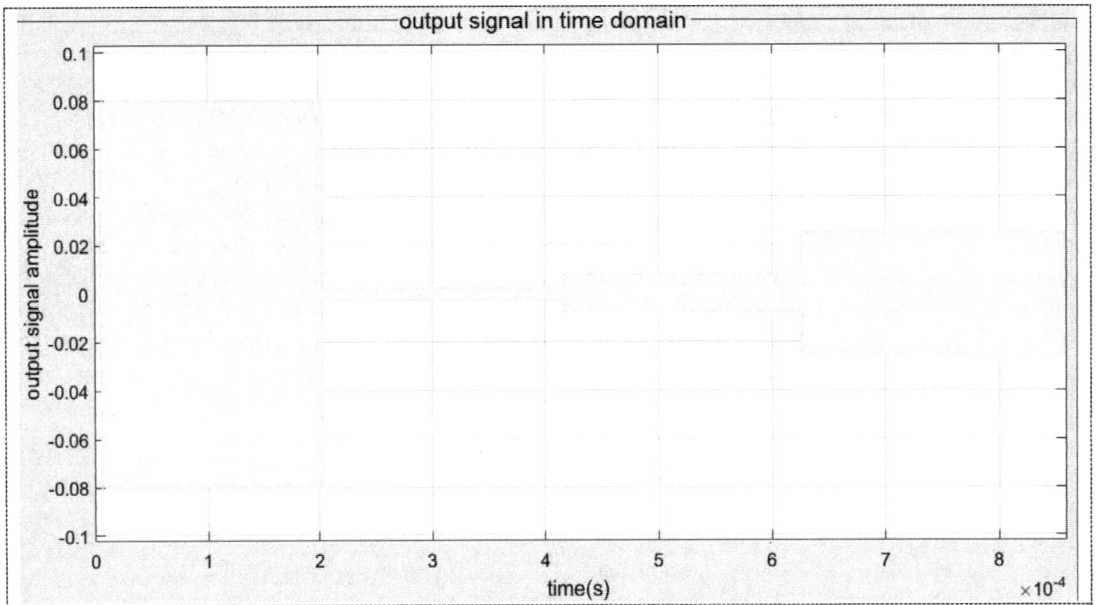

Figure 3. Transmitted Signal Waveform in Time Domain.

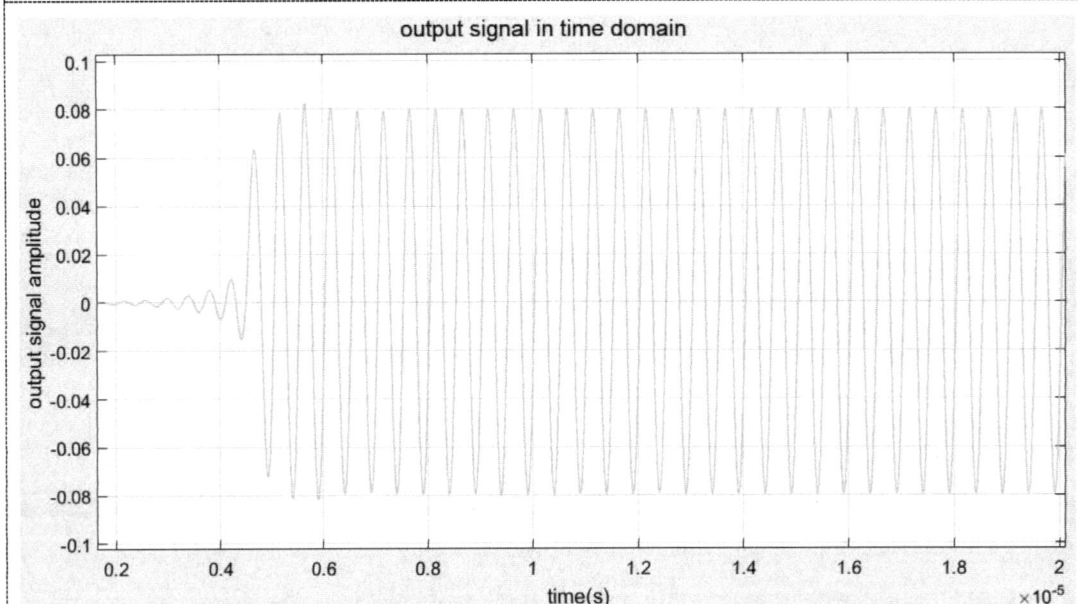

Figure 4. Transition of Signal Amplitude During the First Burst.

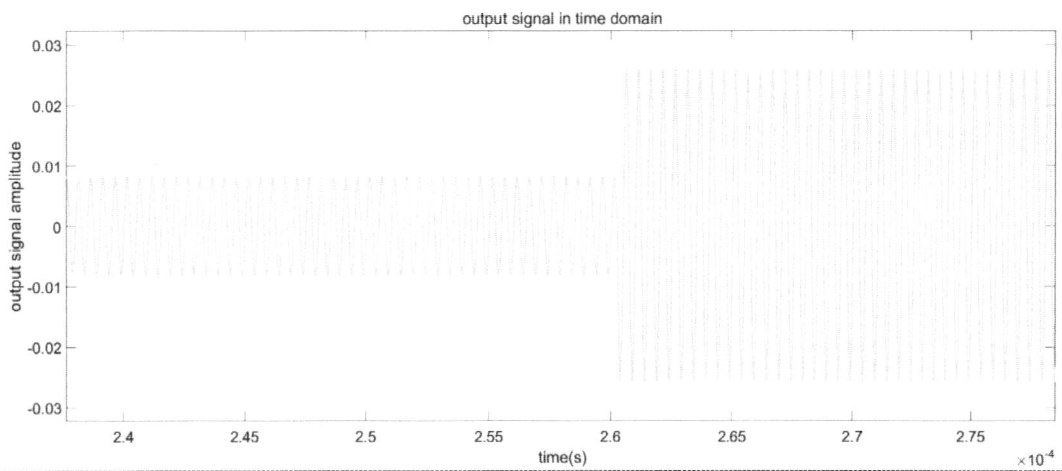

Figure 5. Transition for Decreased Attenuation Word.

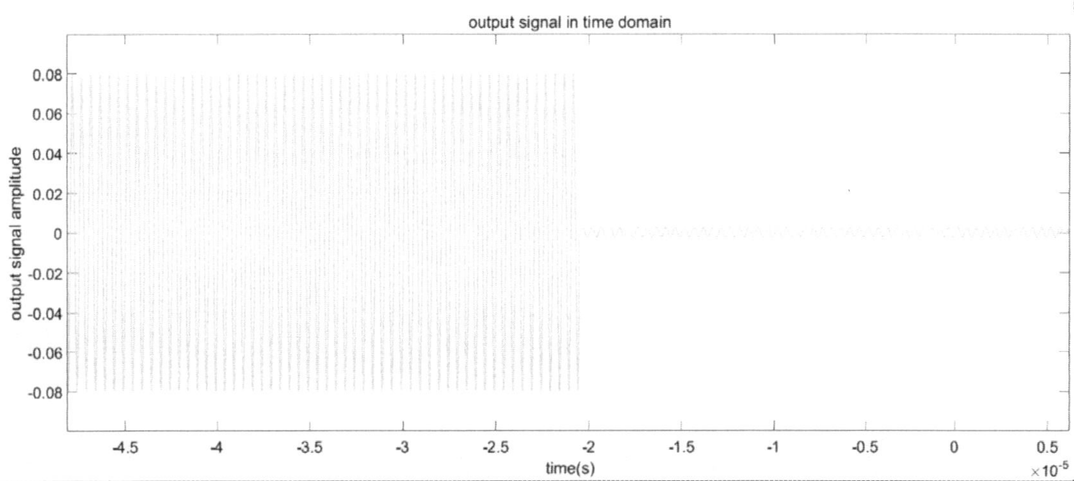

Figure 6. Transition for Increased Attenuation Word.

5. Conclusion

This paper points out the shortcomings of traditional power measurement methods, and puts forward power measurement method in the digital domain. It proposes to multiplex the receive path as signal detector in a transceiver while the receiver is idle. In addition, this paper uses subtraction and iteration to obtain the attenuation control word, which successfully eliminates the error through signal detection path, and keeps track of changes of attenuation command sent by the software. In summary, the proposed power control method introduces less additional circuit in a transceiver, and achieves power control to transmitted signals with the advantage of large frequency range, small delay and high precision. This new approach simplifies traditional circuit design and can be widely used in wireless communication transceivers.

References

[1] Masoud Zargari, and David K. Su, *et al.*, "A 5-GHz CMOS Transceiver for IEEE 802.11a Wireless LAN Systems"[J]: IEEE Journal of Solid-State Circuits, 2002, (37): 1688-1694.

[2] Robert J. McMorrow, Eamon Nash. "Edge Power Detector/ Controller": U.S., Patent 7260367 B2[P]. 2007-08-21.

[3] Liu Ye, Li Gang. "LF to 2.5 GHz TruPwr Detector AD8361" [J]: Foreign electronic components, 2000, (12): 5-7.
[4] Paul J. Husted, William J. McFarland. "Method and System for Noise Floor Calibration and Receive Signal Strength Detection": U.S., Patent US7245893[P]. 2007-07-17.
[5] Yulin Zhang, Xiaohong Zheng. "AD8361—0.1~2.5GHz True Power RMS Response Device and Introduction for Use" [J]: Application of Electronic Technology, 2001, (3): 73-74.

UMDR: Multi-Path Routing Protocol for Underwater Ad Hoc Networks with Directional Antenna

Jianmin Yang[1, 2], Songzuo Liu[1, 2, *], Qipei Liu[1, 2] and Gang Qiao[1, 2]

[1]Acoustic Science and Technology Laboratory, Harbin Engineering University, Harbin 150001, China;

[2]College of Underwater Acoustic Engineering, Harbin Engineering University, Harbin 150001, China

Email: [*]liusongzuo@hrbeu.edu.cn

Abstract. This paper presents a new routing scheme for underwater ad hoc networks based on directional antennas. Ad hoc networks with directional antennas have become a hot research topic because of space reuse may increase networks capacity. At present, researchers have applied traditional self-organizing routing protocols (such as DSR, AODV) [1] [2] on this type of networks, and the routing scheme is based on the shortest path metric. However, such routing schemes often suffer from long transmission delays and frequent link fragmentation along the intermediate nodes of the selected route. This is caused by a unique feature of directional transmission, often called as "deafness". In this paper, we take a different approach to explore the advantages of space reuse through multipath routing. This paper introduces the validity of the conventional routing scheme in underwater ad hoc networks with directional antennas, and presents a special design of multipath routing algorithm for directional transmission. The experimental results show a significant performance improvement in throughput and latency.

1. Introduction

In underwater ad hoc network, network nodes are typically equipped with omni-directional antennas. The omni-directional antenna is equally transmitting and receiving sound signals in all directions. Due to the nature of the broadcast, the conventional MAC protocol for ad hoc networks specifies that all nodes within the transmission range must suspend their possible transmission when a node is transmitting until the channel becomes idle again. These MAC schemes have a certain capacity limit for the network. With the recent progress in the research of "smart" or "adaptive" antennas, the directional transmission technology of underwater ad hoc networks has become possible. The directional antenna divides the omni-directional transmission area into a fixed number of sectors, as shown in Figure 1. The transmission of a sector does not affect the signal propagation of other sectors. Thus, the space area previously occupied by an omni-directional transmission now can be shared by several directional transmissions. This feature is called "space reuse". While directional antennas have the potential to increase capacity and throughput of the network, they also cause new problems such as increased routing discovery overhead, complex MAC and routing protocols. For example, in directional transmission, the broadcasts require the same packet to be transmitted in all antenna elements in order. This may result in long media access delays and different receive times at different neighbor nodes. The routing specification also becomes more complex because each node needs to know not only the next hop, but also the antenna elements of the next hop. This means the networks would have longer routing messages and larger routing tables.

In this work, we try to study the impacts of directed transmission on the underwater ad hoc network routing scheme. In our study, a typical directional MAC (DiMAC) protocol [3] was used. Three routing schemes are proposed. First, we apply traditional dynamic source routing (DSR) protocols directly to DiMAC. Second, we introduce some necessary modifications to make the DSR more suitable for directional transmission, and the resulting protocol is called DDSR. Finally, we propose a new underwater multi-path directional antenna self-organizing route (UMDR) protocol, which could effectively utilize the advantages of spatial reuse and improve the routing performance.

2. Relate works

Although the application of directional antenna in wireless networks has been widely studied, the research in underwater ad hoc network is relatively limited, which is mainly limited by the media access control protocol and channel difference. Some studies of wireless networks routing schemes using directional antennas are reported in [3] [4] [5]. In [4], the authors use directional antennas to improve the efficiency of on-demand routing protocols in mobile ad hoc networks. The idea is to use directed replay during route rediscovery. In the traditional route rediscovery process, routing request is flooded the entire network once. However, in most cases, this flooded is unnecessary because the previous direction of the target node is known. In this paper, the author assumes that each node knows its direction for other nodes. When the transmission is corrupted, the rediscovery process sends routing requests only in the previous direction of the destination. In this way, you can reduce the overall routing overhead. In [5], directional antennas are used to improve routing performance in both cases. One is due to mobility and dynamic network division of the situation, the other is in the middle of the node caused by the movement of the route repair process. The proposed method takes advantage of the important characteristics of the directional antenna -longer transmission distance. In [3], a simple medium access control (MAC) protocol called DiMAC is proposed, and the DSR routing protocol is evaluated based on DiMAC. Some improvements have also been introduced to improve DSR performance for directional transmission. Because of the unresolved "deafnes" problem, the authors conclude that the advantages of using directional antennas in self-organizing networks cannot be guaranteed, and in some cases it is better to use omni-directional antennas.

3. Antenna and MAC models

The antenna model in our study consists of N beam patterns, as shown in Figure 1. We assume that the main lobe of each beam has a conical radiation pattern of 2π / N radians. Antennas can work in both modes: omnidirectional and Directional. In directional mode, only one beam with G_d gain can be used at a time, and in the omnidirectional mode, the signal is received with the gain ofG_o. G_d is inversely proportional to the number of beams used. The transmission distance of the antenna is proportional to the transmission gain. Thus, the directional transmitter can achieve a longer distance than the omni-directional transmitter at the same transmission power.

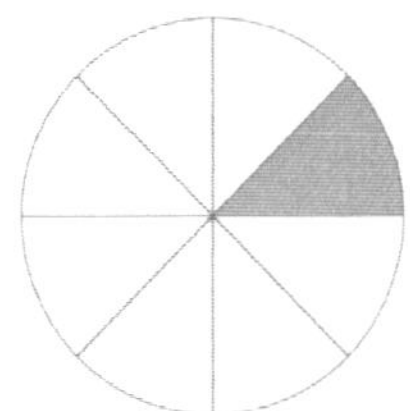

Figure 1. Directional Antenna

In our study, we used the DiMAC protocol as the MAC protocol. DiMAC is based on the IEEE 802.11 Distributed Coordination Function (DCF) and also uses RTS and CTS for channel reservation. In DiMAC, RTS / CTS packets are sent and received by directional antennas via a single specified antenna element. According to DiMAC protocol, three modules have been implemented based on 802.11 DCF:

(1).The channel reservation mechanism for each antenna element. This includes the "send" and "receive" timers, and the NAV state for indicating the channel condition of the component and the packet sent / received.

(2). Broadcast sweep function. The routing discovery phase requires broadcast. Some scans on some antenna elements may fail because these components may experience busy channels. There is a trade-off between retry attempts and routing discovery latencies. In our work, if the channel is busy, each element performs only one retry. Also, when the neighbor node receives the broadcast packet, it will not reply immediately because the sender may be sent through other antenna elements. The receiver must wait for a while to allow the sender to complete the scan.

(3). Neighbor table. For single-hop transfer, the sender must know not only the next hop node, but also the antenna element of the next hop. To do this, we must create and maintain neighbor tables to update and find information related to each antenna element.

Deafness [6] is a unique and critical issue for directional MAC protocols such as DiMAC. As shown in Figure 2, when node A is communicating with node B through one of its antenna elements, it cannot hear (or cannot respond to) any signal from other elements. The problem is that DiMAC uses directional transmission to perform RTS / CTS switching between node A and node B. Therefore, any other node (for example, node C, node D or node E) is located in a different direction and will not hear the RTS / CTS channel reserved. If node C, node D or node E attempts to communicate with node A during data transmission with node B, no reply will be received before the timeout. After multiple retransmissions, node C, node D or node E will deduce that node A has been removed and any route of node A is broken. Obviously, this feature will affect MAC performance and routing performance.

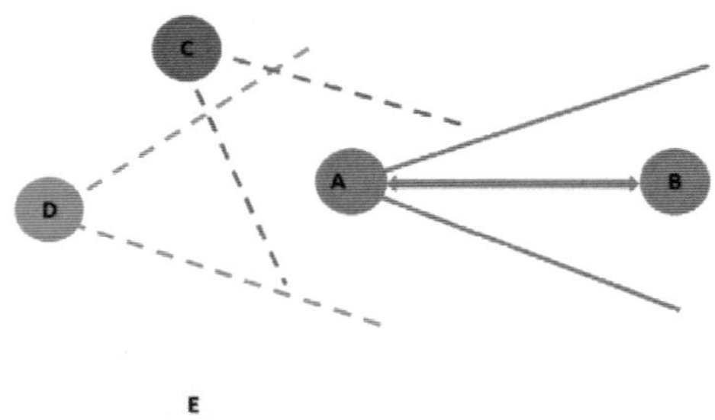

Figure 2. Node "Deafness"

4. Routing protocol description

Routing protocols in underwater ad hoc networks can usually be divided into two categories: reactive and active protocols. Most of the underwater ad hoc network routing protocols are reactive protocols. If the packet needs to be sent and no route is available to the destination, the reactive protocol performs route discovery. The reactive protocol achieves lower routing overhead at the expense of additional routing discovery cost.

Most of the existing routing protocols used for ad hoc networks select a single route for packet transmission based on the minimum number of hops. In this paper, we consider these traditional routing schemes to be ineffective for ad hoc networks with directional antennas. The main problem is the middle node deafness. More specifically, it is assumed that the data flow **F** is routed through the antenna element e of the active intermediate node **N**. If node **N** frequently performs data exchange on other antenna elements, stream **F** will soon be subjected to link break element e due to excessive deafness of node **N**. Traditional routing protocols will initiate expensive route rediscovery and data retransmission processes, although node **N** may become available for **F** soon. Deafness of all active nodes significantly increases the dynamics of the instantaneous topology of self-organizing networks. Its effect is even greater than mobility since switching from one antenna element to another may be very fast, and the duration of each session of the data exchange is very unpredictable. This poses a major challenge to the routing design of ad hoc networks with directional antennas.

We propose a reactive source routing protocol for underwater ad hoc networks using directional antennas. It is called underwater multi-path directional antenna self-organizing route (UMDR). In UMDR, each node maintains a routing table that lists the paths from the sender to each possible destination. Each node updates the routing table based on the overheard packet. A notable feature of UMDR is that the routing table records multiple routes to each destination, so when a route encounters a busy channel, it can immediately select an alternate route. The source node places the entire path in the packet header, and the intermediate node forwards the packet according to the path specified in the header. When there is a packet to be sent and there is no available route to the destination, the route discovery process is initiated.

During the route discovery process, the sender uses the sweeping mechanism to broadcast the routing request message. When a neighbor node receives a route request, it searches through its own routing table. If there is no available route to the destination, it will immediately rebroadcast the routing request over all its antenna elements, except that the antenna element which the message is received. If it has an available route, it will send a routing reply message to the sender after a brief delay, which allows the sender to complete the scan. During packet forwarding, each node passes the message to the next hop based on the route specified in the packet header. If the first attempt fails, UMDR will try two more times. If all of these attempts fail, the UMDR assumes that the neighbor node is busy and immediately uses alternate routes. If the alternate route is not available or the cost of the hop count is much higher than the original route, the node will continue to use the original route until it assumes that the link is actually corrupted. The idea here is to forward the packet to the next hop as soon as possible. Due to the possible spatial multiplexing provided by the directional antenna, the chances of establishing alternative routes are much higher than those of the omnidirectional transmission environment. Multipath routing minimizes the delay per hop, effectively reducing overall end-to-end latency. Records the number of attempts for each next hop, and when it exceeds a certain threshold, the UMDR assumes that the neighbor node may have moved to another location. The scanning function of the neighbor table is executed to locate the neighbor node in the adjacent antenna element. If the scanning process fails, the broken link is assumed and the node performs route update.

The forwarding process requires an alternative route with the cost similar to the cost of the original route. This depends on the number of alternate routes stored in the intermediate node. In order to increase multi-pathing knowledge, the intermediate node can forward the old routing request if it comes from a different sender, or has a shorter length. This will definitely produce more routing packets, which increases the routing overhead. However, the advantage of this approach is that the destination and intermediate nodes can learn more disjoint routing. One of the characteristics of using directional antennas is the broadcast delay caused by broadcasting. Therefore, the first received routing request may not represent the shortest route [3]. In UMDR, the target node delays its routing reply in a short period of time, which may be equal to the scan delay. It can also send alternative routes at a reasonable routing cost.

As the directional antenna transmission distance is longer, so the network links are not easy to be destroyed. At the same topology change rate, the node may take a long time to move out of the neighbor's range. However, in directed self-organizing networks, nodes may frequently move to adjacent antenna elements of the same node, especially when there are many antenna elements. This requires a switching mechanism. The handover function is implemented in the MAC protocol, where the location of the antenna elements of the neighboring nodes is periodically updated in the "neighbor table". For routing protocols, the handover and the actual link disconnection should be handled separately.

The routing tables and neighbor tables in the mobile node should be kept up-to-date. Each entry in these tables has a timer. Paused entries will be deleted in time. The antenna element entries in the neighbor table are updated when the switch occurs. When the link is disconnected, a "routing error" packet is generated. Any node that listens for this routing error packet will update its routing table based on the error message.

5. Simulation results and analysis

We check the performance of UMDR with the help of NS3. The simulation scenario is that of a 5 km by 5 km area, a total of 50 nodes are simulated. At the beginning, each node has a random initial position, and then it will move to the random destination at a randomly selected speed (evenly distributed between 0 and 10 m/s). When it reaches the destination, it will select another destination after a pause. In all scenarios, a send buffer with a size of 50 packets is maintained to store packets waiting for a route. The source node and the destination node are randomly selected from the network. All traffic contributes to network traffic, and its packet size is set to 512 bytes. 40 packets are used to test the overall performance.. For fair comparison, simulate the same traffic and mobility scenarios for different protocols. Each data point is calculated as the average of 20 runs of different mobility scenarios.

Two metrics are used to test performance: delivery proportion and end-to-end packet delay.

(1). Delivery proportion (DP) is the ratio of the number of received packets to the number of packets generated by the source node.

(2).End-to-end packet delay based only on successfully forwarded packets. It includes routing delay, queuing delay for each intermediate node, contention delay for MAC layer, and transmission delay per hop.

There is a trade-off when selecting the number of antenna elements. Longer transmission distances can be achieved with more antenna elements with narrow beamwidths, but at longer scan delays, higher control overhead and more switching operations. The simulation is based on an eight-element antenna system.

The performance curves in Figures 3 (a) and 3 (b) represent the combination of three routes and MAC protocols: DSR over 802.11 DCF, DSR over DiMAC (DDSR), and UMDR over DiMAC.

Figure 3 (a) shows that DDSR and UMDR achieve higher packet delivery rate than DSR. There are two reasons. One is that the longer transmission range for directional transmission reduces the number of hops for the selected route. And also because the transmission distance is longer, routing errors and route rediscovery become less frequent. The routing overhead is reduced accordingly. Another reason is that the directional antenna reuses the space between the antenna elements, thereby increasing the channel capacity. UMDR uses multi-paths to further reduce the number of rediscover routes, and then increase the time fraction of sending packets to improve performance. UMDR also distinguishes between switching, busy and link disconnected, and performs different actions accordingly. This method effectively reduces the number of route discovery.

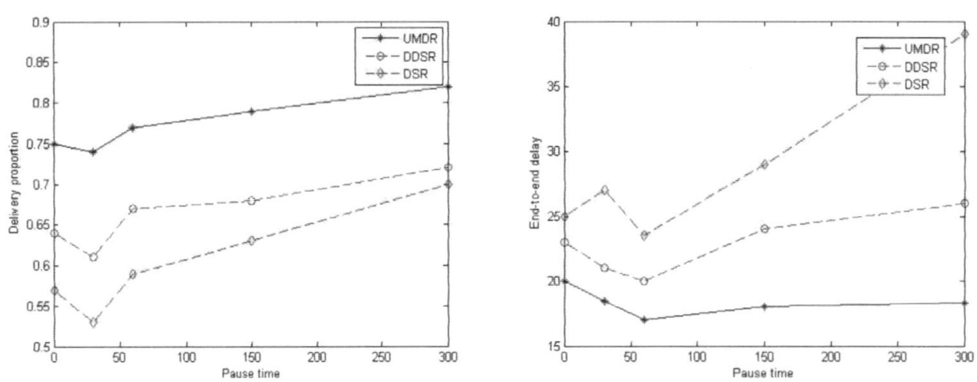

(a) Delivery Proportion (b) End-to-End Delay (sec.)

Figure 3. Performance comparisons of three routing and MAC protocol combinations

Figure 3 (b) shows the average end-to-end delay of successfully delivered packets. The delay of DDSR and UMDR is shorter than DSR, but the improvement is limited. Several features of UMDR increase the end-to-end delay of packets. First, the scanning operations used in DDSR and UMDR will increase the broadcast time. Second, the destination has a delay after the routing request arrives. UMDR has better latency because at the intermediate node it does not wait for too many retries, but instead switches to alternate routes.

6. Conclusion

In this paper, underwater multi-path directional antenna self-organizing routing protocol (UMDR) is proposed. Some problems related to the effective routing in the directional antenna self-organizing network are solved, and the modification of the multi-path route of the directional antenna is introduced. Performance improvements show that directional antennas can actually improve the performance of underwater ad hoc networks through appropriate routing and MAC protocols.

References

[1] D.B. Johnson and D.A. Maltz J 1996 *Mobile Computing* vol. 353 (Kluwer Academic)
[2] C.E. Perkins and E.M. Royer P 1999 Proc. Workshop Mobile Computing Systems and Applications
[3] R. Choudhury and N. Vaidya P 2003 Proc. 8th Conference on Personal and Wireless Communication
[4] A. Nasipuri and J. Mandava P 2000 Proc.IEEE International Conference on Computer Communication and Networks
[5] A. Saha and D. Johnson Master thesis 2003 Energy Saving and Parition Bridging using Directional Antennas in mobile ad hoc networks
[6] R. Choudhury, X. Yang, R. Ramanathan and N. Vaidya P 2002 Proc. of the ACM MOBICOM

Resonance Properties of Class I and Class II Neurons Differentially Modulated by Channel Noise

Lei Wang[1]

[1]Neuroscience and Intelligent Media Institute, Communication University of China

E-mail: *wanglei_nc@163.com

Abstract. Resonance properties of two different neuron types (Class I and Class II) induced by channel noise are investigated in this study. It is found that for Class I neuron, spiking activity is enhanced when certain noise intensity is presented, especially under weak current stimuli -- a typical phenomenon of stochastic resonance (SR); while for Class II neuron, in addition to perform the SR, certain noise intensity would inhibit neuronal activity under some current stimuli -- a typical phenomenon of inverse stochastic resonance (ISR). Moreover, we show that only sodium channel noise or potassium channel noise variation can achieve the similar phenomena. Consequently, the model results suggest that channel noise may exert differential roles in modulating the resonance properties of Class I and Class II neurons.

1. Introduction

Stochastic resonance (SR) is a typical phenomenon observed when certain noise inputs can enhance the weak-signal processing ability of systems or devices [1]. During the past decades, SR has been widely found in neural systems. It was revealed that additive current noise with certain intensity can help single neurons [2-3] and neuronal networks detect subthreshold signals [4-8].

In addition to SR, inverse stochastic resonance (ISR) has recently been found theoretically and experimentally in some neural systems. Generally, ISR refers to the occurrence of a minimum in firing rate which occurs for certain noise intensity [9-10]. For instance, Tuckwell and Jost [11-13] have suggested that current noise with weak intensity would inhibit the generation of repetitive spiking of neurons. Moreover, certain intensities of conductance noise [14] and color noise [15] have also been shown to inhibit rhythmic neuronal activity.

Besides the above noise sources, channel noise intrinsically exist in biological neurons has been well recognized to play significant roles in modulating neuronal activities [16-17]. The relationship between channel noise and ISR has also been investigated, e.g. [18-19]. However, these analyses were primarily on the classical or standard Hodgkin-Huxley (HH) neuron, which belongs to Class II according to previous categories [20-21], whether other classes of neurons can also demonstrate the channel noise induced ISR is still not completely clear. In view of this, this paper aims to investigate how intrinsic channel noise would modulate the resonance properties of Class I neuron, and compare to the results of Class II neuron.

2. Model and methods

Two ionic models were introduced to represent Class I and Class II neurons respectively. One model is referred from Traub and Miles [22], and the other model is referred from the classical HH neuron [23]. Neuronal membrane dynamics are described by the equations below:

$$C_m \frac{dV_m}{dt} = -g_{Na} m^3 h(V_m - V_{Na}) - g_K n^4 (V_m - V_K) - g_L (V_m - V_L) + I \tag{1}$$

$$\frac{dx}{dt} = \alpha_x (1-x) - \beta_x x + \sigma \xi_x (t) \quad (x = m, n, h) \tag{2}$$

where V_m is the membrane potential, C_m is the membrane capacitance, I is the current stimulus, g_{Na}, g_K and g_L are the maximal conductance of sodium, potassium and leakage, respectively. V_{Na}, V_K and V_L are the reversal potentials, respectively; m and h represent the activation and inactivation gating variables for sodium channel, and n represents the activation gating variable for potassium channel.

Specific parameters for Class I neuron are: $C_m = 0.2\ \mu F/cm^2$, $g_{Na} = 20\ mS/cm^2$, $g_K = 6\ mS/cm^2$, $g_L = 0.01\ mS/cm^2$, $V_{Na} = 50\ mV$, $V_K = -90\ mV$, $V_L = -65\ mV$

$$\alpha_m = 0.32(V_m + 50)/(1 - \exp(-(V_m + 50)/4)) \tag{3}$$

$$\beta_m = -0.28(V_m + 23)/(1 - \exp((V_m + 23)/5)) \tag{4}$$

$$\alpha_h = 0.128 \exp(-(V_m + 46)/18) \tag{5}$$

$$\beta_h = 4/(1 + \exp(-(V_m + 23)/5)) \tag{6}$$

$$\alpha_n = 0.032(V_m + 48)/(1 - \exp(-(V_m + 48)/5)) \tag{7}$$

$$\beta_n = 0.5 \exp(-(V_m + 53)/40) \tag{8}$$

Specific parameters for Class II neuron are: $C_m = 1\ \mu F/cm^2$, $g_{Na} = 120\ mS/cm^2$, $g_K = 36\ mS/cm^2$, $g_L = 0.3\ mS/cm^2$, $V_{Na} = 50\ mV$, $V_K = -77\ mV$, $V_L = -54.4\ mV$.

$$\alpha_m = 0.1(V_m + 40)/(1 - \exp(-(V_m + 40)/10)) \tag{9}$$

$$\beta_m = 4 \exp(-(V_m + 65)/18) \tag{10}$$

$$\alpha_h = 0.07 \exp(-(V_m + 65)/20) \tag{11}$$

$$\beta_h = 1/(1 + \exp(-(V_m + 35)/10)) \tag{12}$$

$$\alpha_n = 0.01(V_m + 55)/(1 - \exp(-(V_m + 55)/10)) \tag{13}$$

$$\beta_n = 0.125 \exp(-(V_m + 65)/80) \tag{14}$$

The effect of channel noise is employed by adding a noise term on the variation of gating variables (see ξx in equation (2)). Here σ denotes the relative noise intensity, larger value of σ means stronger effect of channel noise. Similar to previous reports, in which the noise input applied to neurons was usually adopted the Gaussian white noise [2,14,24], ξx in equation (2) is also generated by Gaussian processes with zero mean and the following convariance functions [25].

$$\langle \xi_x(t_1) \xi_x(t_2) \rangle = \frac{\alpha_x (1-x) + \beta_x x}{N_{Na}} \delta(t_1 - t_2) \quad (x = m, h) \tag{15}$$

$$\langle \xi_x(t_1) \xi_x(t_2) \rangle = \frac{\alpha_x (1-x) + \beta_x x}{N_K} \delta(t_1 - t_2) \quad (x = n) \tag{16}$$

where δ is the Dirac delta function, N_{Na} and N_K denote the total number of sodium and potassium channels respectively on a given membrane area. We choose $N_{Na} = \rho_{Na}\ S$ and $N_K = \rho_K\ S$, here ρ_{Na} and ρ_K are the channel densities of sodium and potassium, respectively. ρ_{Na} =10 μm^{-2} and ρ_K = 3 μm^{-2} for Class I neuron; ρ_{Na} =60 μm^{-2} and ρ_K = 18 μm^{-2} for Class II neuron. S means the total membrane area, and its value is 10 μm^2.

Simulations were performed by adopting the MATLAB software, and the fourth-order Runge-Kutta algorithm is employed to calculate the membrane potentials of each neuron in equation (1-2) with time step of 0.01 ms.

3. Results

Basic frequency-stimulus (*f-I*) curves of Class I and Class II neurons are demonstrated in Figure. 1, in which it is clear that Class I neuron fires slowly in response to weak stimuli, and displays a continuous *f-I* curve; while Class II neuron exhibits a discontinuous *f-I* curve for their inability to produce spikes below certain threshold intensities. These results are much similar to previous descriptions [20-21].

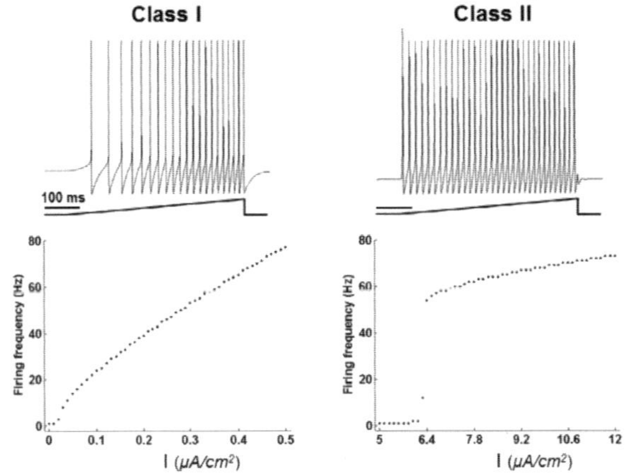

Figure 1. Frequency-stimulus (*f-I*) curves of Class I and Class II neurons.

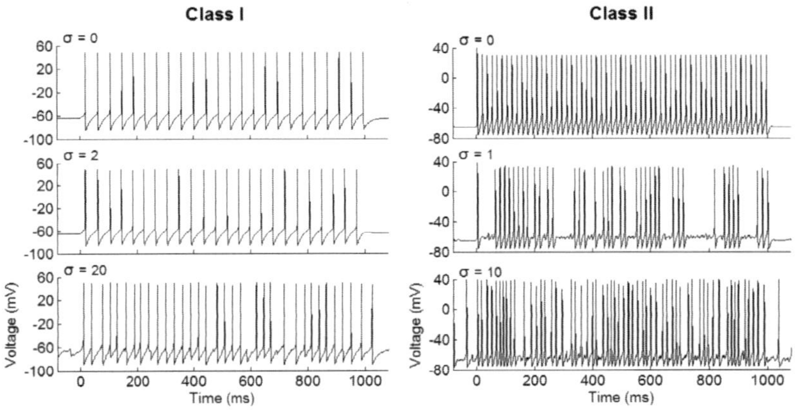

Figure 2. Spiking activities of Class I and Class II neurons under different noise intensities.

Noise-induced spiking behaviors of Class I and Class II neurons are shown in Figure. 2. It is apparent that Class I neuron is much robust to noise input, with little variation in the spiking activities, even though the noise intensity is large ($\sigma = 20$). On the contrary, Class II neuron is much sensitive to noise input, showing rather large variations in the spiking activities, even though the noise intensity is small ($\sigma = 1$).

It should be pointed out that due to the different ion dynamics in each model neurons, we use different ranges of channel noise, to obtain their overall variation trends.

Figure. 3 demonstrates the resonance properties of Class I and Class II neurons. For Class I neuron, certain noise intensity would significantly enhance its responsiveness to subthreshold stimulus (I = 0.01 $\mu A/cm^2$), and this increase trend persists even under relative larger suprathreshold stimulus (I = 0.05/0.40 $\mu A/cm^2$). For Class II neuron, certain noise intensity would also enhance its excitability to subthreshold stimulus (I = 6.0 $\mu A/cm^2$); while under stimulus near the excitation threshold, certain noise intensity would obviously inhibit its responsiveness (I = 6.5 $\mu A/cm^2$). Moreover, under larger suprathreshold stimulus (I = 10 $\mu A/cm^2$), noise input has little influence on neuronal activities. This group of results suggests that certain channel noise can induce the typical SR in Class I neuron; while for Class II neuron, channel noise can not only induce the typical SR, but also induce the ISR under some current injections.

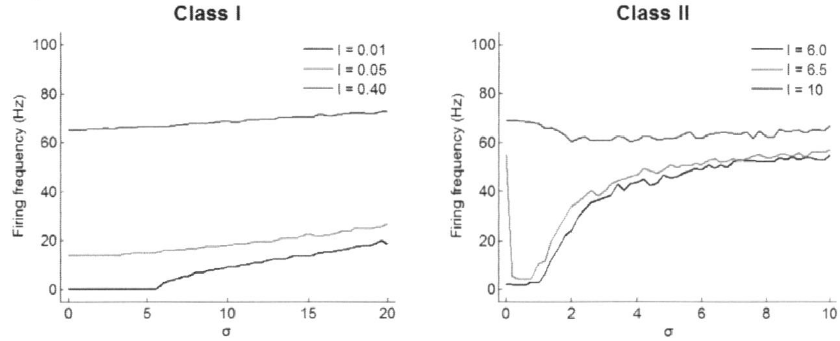

Figure 3. Channel noise induced SR (left) and ISR (right) in Class I and Class II neurons.

Contour diagrams which show the two-parameter dependence of resonance behaviors of Class I and Class II neurons are illustrated in Figure. 4. It is evident that excitability of Class I neuron is enhanced with the increase of noise intensity, while excitability of Class II neuron is inhibited by noise input under some current stimuli.

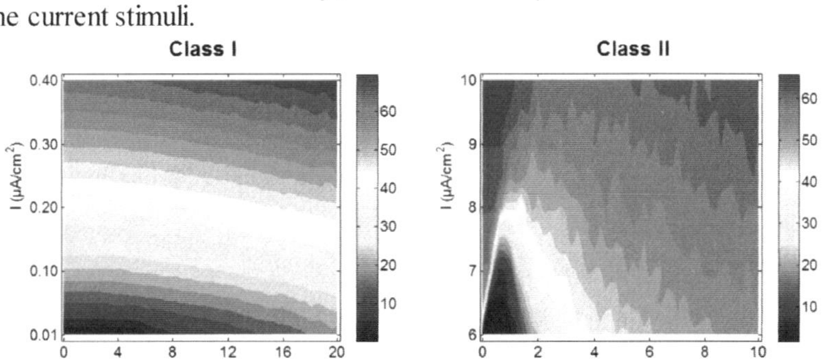

Figure 4. Two-parameter dependence of resonance behaviors in Class I and Class II neurons. Contour plots showing the SR (left) and ISR (right) are dependent on the noise intensity and stimulus amplitude.

As in the two neuron models we analyzed, there are two main kinds of ion channels: inward sodium channel and outward potassium channel. To explore how these two different ion channels contribute to the modulation of neuronal resonance properties, we separately vary noise intensity in one ion channel, and keep the other unchanged.

Model results presented in Figure. 5(A&C) show that sodium channel noise and potassium channel noise play opposite roles in modulating the resonance behaviors of Class I neuron. Specifically, increase of sodium channel noise would enhance neuronal activities, similar to the results in Figure. 4 (left); while increase of potassium channel noise would suppress neuronal activities.

Results for Class II neuron are demonstrated in Figure. 5(B&D), it is clear that increase of sodium channel noise and potassium channel noise both can induce similar variations of resonance behaviors with the results in Figure. 4 (right).

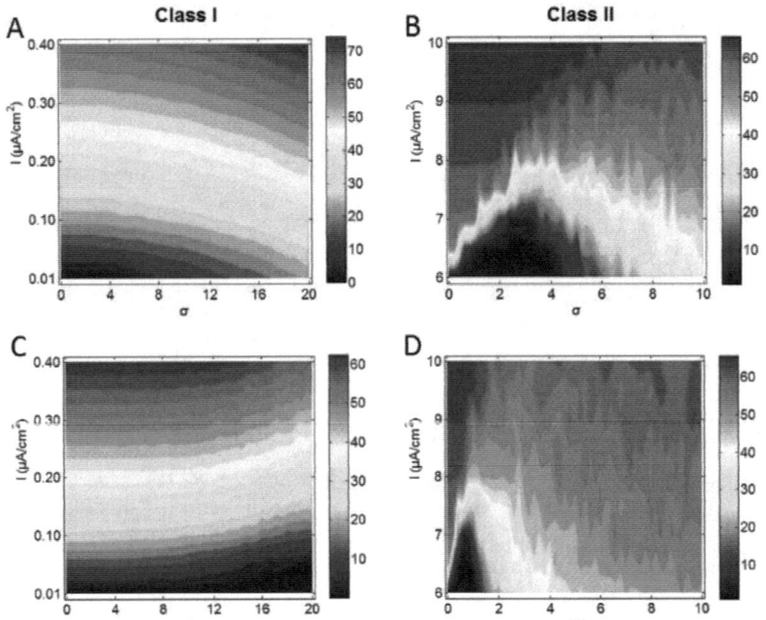

Figure 5. Two-parameter dependence of resonance behaviors in Class I and Class II neurons. Contour plots showing the SR (A&C) and ISR (B&D) are dependent on the noise intensity and stimulus amplitude, in which A&B represent sodium channel noise, while C&D represent potassium channel noise.

4. Conclusions

In summary, we have performed a numerical investigation on the effect of channel noise in modulating the neuronal resonance behaviors. By employing two different types of neuron (Class I and Class II) as bases, we show that spiking activity of Class I neuron is enhanced when channel noise with certain intensity is added, and this enhancement is particularly evident under weak current stimuli; spiking activity of Class II neuron is also enhanced by channel noise under a certain range of current stimuli, but for current stimuli near the excitation threshold, neuronal spiking activity is inhibited. In addition, we further demonstrate that variations of only sodium channel noise or potassium channel noise can achieve the similar phenomena. It should be mentioned that we have only investigated the influence of channel noise on the variation of resonance properties in two single model neurons, and did not consider any synaptic connections with other neurons. However, in the brain, neurons are organized in modules or networks to function, thus future investigations can analyze potential differences between networks composed of Class I neurons and Class II neurons.

References

[1] McDonnell M D and Abbott D 2009 PLoS Comput. Biol. 5 e1000348
[2] Jin X and Liang P J 2006 Chin. Phys. Lett. 23 500
[3] Danziger Z and Grill W M 2015 J. Comput. Neurosci. 38 53
[4] Ozer M, Perc M and Uzuntarla M 2009 Phys. Lett. A 373 964
[5] Yu H T, Guo X M, Wang J, Liu C, Deng B and Wei X L 2015 Commun. Nonlinear Sci. Numer. Simulat. 29 346
[6] Wang Q Y, Perc M, Duan Z S and Chen G R 2010 Int. J. Mod. Phys. B 24 1201
[7] Sun X J, Perc M, Lu Q S and Kurths J 2008 Chaos 18 023102
[8] Gu H G, Jia B, Li Y Y and Chen G R 2013 Physica A 392 1361
[9] Tuckwell H C and Jost J 2012 Physica A 391 5311
[10] Paydarfar D, Forger D B and Clay J R 2006 J. Neurophysiol. 96 3338

[11] Tuckwell H C and Jost J 2010 PLoS Comput. Biol. 6 e1000794
[12] Tuckwell H C and Jost J 2011 J. Comput. Neurosci. 30 361
[13] Tuckwell H C and Jost J 2009 Phys. Rev. E 80 031907
[14] Gutkin B S, Jost J and Tuckwell H C 2009 Naturwissenschaften 96 1091
[15] Guo D 2011 Cogn. Neurodyn. 5 293
[16] Wang L, Zhang P M, Liang P J and Qiu Y H 2013 Chin. Phys. Lett. 30 070506
[17] Wang B Y and Gong Y B 2015 Chin. Phys. B 24 118702
[18] Schmerl B A and McDonnell M D 2013 Phys. Rev. E 88 052711
[19] Uzuntarla M Cressman J R, Ozer M and Barreto E 2013 Phys. Rev. E 88 042712
[20] Hodgkin A L 1948 J. Physiol. 107 165
[21] Izhikevich E M 2007 Dynamical Systems in Neuroscience: the Geometry of Excitability and Bursting (MIT Press)
[22] Traub R D and Miles R 1991 Neuronal Networks of the Hippocampus (Cambridge University Press)
[23] Hodgkin A L and Huxley A F 1952 J. Physiol. 117 500
[24] Gan C B, Perc M and Wang Q Y 2010 Chin. Phys. B 19 040508
[25] Ozer M, Uzuntarla M, Perc M and Graham L J 2009 J. Theor. Biol. 261 83.

A Heuristic Algorithm for Automatic Classification of Ring River

Chengming Li[1], Yong Yin[1], Zhiwei Hao[1] and Xiaoli Liu[1,*]

[1] Chinese Academy of Surveying and Mapping, No28 Lianhuachi Xi Road, Haidian district, Beijing, China

Email: *83391860@qq.com

Abstract. In map generalization, a technical difficulty and key to expressing water elements is river classification, which has high requirements for accuracy and calculation speed. In the study of river classification, the current hierarchical algorithm is not quite efficient and lacks classification of ring rivers. Based on this, this paper establishes the attributes of river classification, its geometric rules and ring rules, and classifies different ring types for the existing in river system. When calculating river's longest path, we use the heuristic algorithm for iterative calculation to automatically conduct river classification. The experimental results show that this algorithm can achieve good results for the ring river system and improve the calculation efficiency and accuracy of river system classification.

1. Introduction

In the era of computer graphing, GIS has become increasingly mature, powerful, and more widely used. In river classification, we no longer have to select and view each part of the river. Instead, we only need to rely on automatic mapping for river classification based on the river's attributes, geometric features and other information. Compared to manual classification of river, automatic classification not only relieves people from tedious work, but also improves the accuracy of river classification.

At present, there have been several studies of automatic river classification. Hehai Wu[1] establishes a river tree structure for classification and believes that the river's main stem has the maximum length, so we can find the maximum length of the river to determine the main stem. However, in determining the main stem, merely focusing on the maximum length while ignoring river's flow and characteristics will lead to inaccurate classification. Paiva et al. [2] propose to use angle information for river classification based on the geological knowledge that at the confluence of the river, the flow of the upper and lower part should almost be on a straight line. Nevertheless, when classifying river only based on angles, there will be large inaccuracy in classification if there is angle deviation at the node of the river. Qingsheng Guo et al.[3] believe that the main stem is often the longest river, and it also meets the 180° assumption at the node. Thus, they propose to combine the aforementioned two methods of length and angle to classify the main stem and tributary of the river, but they fail to consider the attribute information and the practical significance of the river. Wei Wu et al. [4] use the flow, layer, name, and other attributes information in river classification, which strengthens the rationality of the river classification. However, classification based on angles only will lead to some errors. In the above research, the rings in the river system are not treated, and the efficiency in calculating the longest path of the river is relatively low.

In this paper, we use the attribute information and geometric rules to classify the river system. We divide the ring types into single input and single output rings and multi input and multi output ones, and treat each type of rings with methods of the shortest path and the angle-priority rule. Meanwhile, this paper addresses the problem of low efficiency in searching for the longest path in geometric rules. By applying the heuristic algorithm, we reduce the number of searches and thereby improving the efficiency.

2. Rules for river classification

In the existing research results, the basic methods for river classification can be summarized as follows [5-9]:

First, reorganize information of the river system, construct the river tree, and search a river by setting an outflow point from the river as a starting node. Look for the main stem and set it as the first-order stream, then remove it from the river system and continue searching from the tributary node of the main stem, setting it as the second-order stream. Iterate this process and perform "top down" classification until all parts of the river has been searched.

In the searching process, how to accurately determine the main stem is essential, which determines the final result of river classification [10-11]. In determining the main stem of the river, the combination of the attribute information with geometric features is adopted. The attribute information of the river includes: river name, river type, river flow, etc. Geometric features include: the angle of the river, the length of the river, and whether there exists a river ring. According to the above information, the rules for river classification are presented as follows:

2.1. Rule of attribute information

This is to establish a number of rules to determine the main stem from the river's attribute information. The rules are as follows (priority from top to bottom):

(1) The river flow constraint principle: if the flow of one stream is inconsistent with the that of another stream, they cannot constitute the same river.

(2) The river type constraint principle: streams of the same type preferentially constitute the same river.

(3) The river name constraint principle: streams with the same name preferentially constitute the same river.

2.2. Rule of geometric features

If there is no attribute information of the river or the information is not enough to determine the main stem of the river, we need to determine based on the shape of the river to. The geometric features mainly include the angle and the length of the river. The angle of the river is the angle between the current stream and the next stream, giving priority to the straight stream. The length of the river is to determine whether the next stream is on the longest path of the current stream, giving priority to those on the longest path. The rules are as follows (priority from top to bottom):

(1)Length priority principle: If the longest path of the stream is much longer than that of the other ones, then give priority to streams on the longest path to constitute the same river. As shown in Figure 1, according to this principle, we should choose the one with the longest path as the main stem of the river.

(2) Angle priority principle. If the longest path of the stream is similar to the length of the candidate stream, but the angle of the candidate stream is much better than the angle on that of the longest path, then the candidate stream should be chosen as the main stem. As shown in Figure 2, according to this principle, we should choose the one with a better angle as the main stem of the river.

Figure 1. Length priority principle

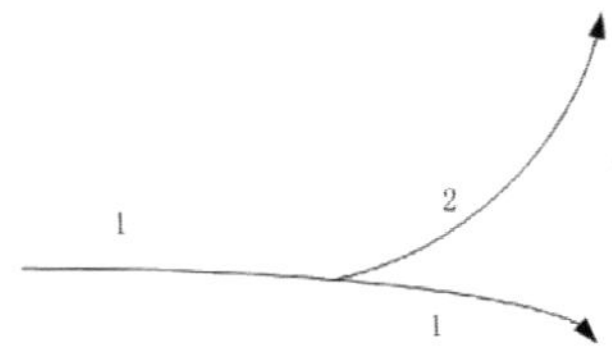

Figure 2. Angle priority principle

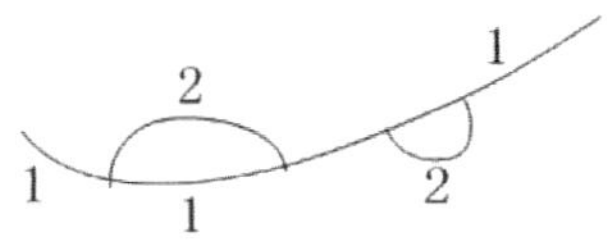

Figure 3. Single input and single output ring

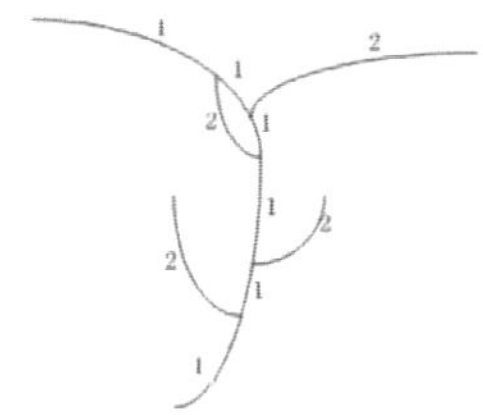

Figure 4. Multi input and multi output ring

2.3. Rule of rings

The basic principles above can be used in river classification for those without rings. However, when rings exist in the river system, the effect of the above rules in river classification is poor. Therefore, this paper has a detailed classification of rings to address this issue by applying different methods to different river rings.

(1) Single input single output ring. A single input single output ring with only one tributary and one distributary. Single input single output rings are the most common in river classification. Since the path of the bulging part of the ring is often longer than those of the straight ones, when applying the previous principles to rings, the stream with longer path will be selected, which will lead to classification error. Therefore, we need to study the ingle input single output ring in the river. After finding it, by selecting the stream with the shortest path from the tributary to distributary or the stream with the best angle of the arc at the entrance, you can greatly reduce the classification error for single input and single output rings. As shown in Figure 3, the stream with the shortest path in the single input and single output is selected as the main stem.

(2) Multi input and multi output ring. As the name suggests, multi input and multi output ring is the river ring that has multiple tributaries or multiple distributaries. The streams of this type of rings flow to a number of streams, as shown in Figure 4. In this case, we can follow the angle priority principle, i.e., between the tributary stream and the ring, we should select the one with better angles as the main stem. This will avoid most of classification errors in the more into the ring out of the grading error. As shown in Figure 4, for the multi input and multi output ring, we should select the stream with better angle as the main stem of the river.

3. Automatic river classification

Automatic river classification first requires building a river tree. We need to find a distributary point as a starting node to search. And then through the main stem identification, we can determine the order of the stream, and classify each stream through iterative processing. Main stem identification requires the use of river attribute information and geometric features, as introduced above. In the process of classification, the attribute information has higher priority than the geometric features, i.e., we should first classify based on the attribute information. In the absence of attribute information or when the attribute information is not sufficient for classification, then geometric features will be used.

Figure 5 is a river system. N1 is the distributary point, and the attribute information of the river system is listed in Table 1:

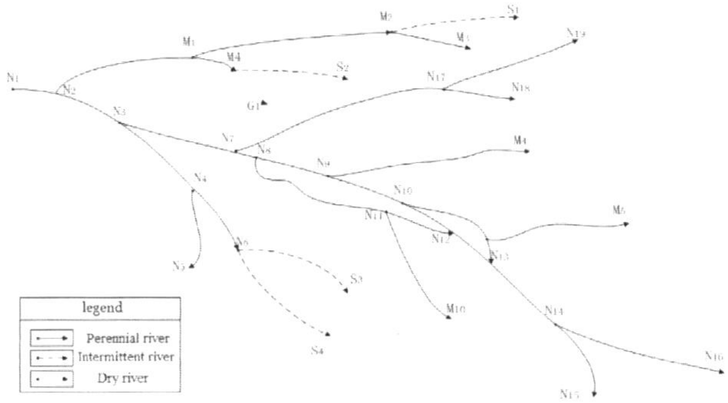

Figure 5. River network

Table 1. Attribute information of river network

Stream Section Identifier	River Type	River Name
(N1, N2) , (N2, N3) , (N3, N4) , (N4, N5) , (N4, N6) , (N3, N7) , (N7, N8) , (N8, N9) , (N9, N10) , (N10, N12) , (N8, N11) , (N11, N12) , (N10, N13) , (N12, N13) , (N13, N14) , (N14, N15) , (N14, N16)	Perennial river	Banqiao River
(N7, N17) , (N17, N18) , (N17, N19)	Perennial river	Sanya River
(N2, M1) , (M1, M4) , (M1, M2) , (M2, M3) , (N9, M4) , (N13, M5) , (N11, M10)	Perennial river	/
(M2, S1) , (M4, S2) , (N6, S3) , (N6, S4)	Intermittent river	/
(M4, G1)	Dry river	/

3.1. Determine main stem with attribute information

According to the rule of attribute information above, we can use the attribute information to determine the main stem of the river. In Figure 6, M2 (M2, S1) is a seasonal stream. Since the former stream is of different type, choose (M2, M3) as the main stem. Another example is when determining the main stem of the next order at N2. Because the two candidate streams are of the same type as (N1, N2), we use name information to determine the main stem. (N2, M1) has the same as (N1, N2), and (N2, N3) has the same name, so (N2, N3) is the main stem. At the N3 node, (N3, N7) and (N3, N4) are two candidate streams, both have the same types and same names. Therefore, it is necessary to determine the main stem using the geometric characteristics.

3.2. Determine main stem with geometric features

The rule of geometric features combines length information with angle information to determine the main stem. When using length information, we need to find the longest path of the stream. This can be considered as finding the longest path of the directed graph. The problem of finding the longest path of the river tree[12] is described as follows:

IOP Conf. Series: Journal of Physics: Conf. Series **960** (2018) 012012 doi:10.1088/1742-6596/960/1/012012

Let G=(V, A) be a fully connected graph, and V={0,1,2,...,n} be a set of nodes. is the geographical coordinate of node n. A={(i,j) i,j ϵV } is a set of arcs, where Node 0 is the set (X_n, Y_n) arting node. The goal is to find the longest path starting from the starting node.

In the problem of finding the longest path of the stream, river information is usually stored in a spatial database, which not only stores the attribute data of geographical elements, but also the spatial data. The location data of these geographic features could give heuristics in searching for the path. By introducing the heuristic information into the searching process, we could avoid a large number of invalid searches and improve the searching efficiency.

The heuristic algorithm for calculating the longest path can be described as below:

(1)Determine the queue of target node to be searched. Enter all nodes in the graph of degree 1, i.e., all the leaf nodes. Calculate the distance between all leaf nodes and the starting point, write as D(n), and solve for Max{D(n)}. Calculate D(i) , the distance between leaf node i and the starting node, and include this leaf node i in the target queue T if D(i)/Max{D(n)}≥k, kϵ(0,1], where k is the threshold value. The smaller the k, the more lenient the selection of target nodes, and thus the more nodes to be searched, and the longer the search will last, but the accuracy would be higher.

(2) Search the river map. Using the idea of heuristic search, f(n) is the valuation function at node n. We use f(n)=g(n)+h(n) to evaluate, where the cost function g(n) represents the length of the arc from the current node m to the extended node n, i.e., $Length_{mn}$. The evaluation function h(n) represents the Euclidian distance from the extended node n to the target node t, i.e., $Dist_{nt} = \sqrt{(X_n^2 - X_t^2)^2 - (Y_n^2 - Y_t^2)^2}$. Through calculating min{f(N1),f(N2),f(N3)...}, we can select the next node n. See if there are other attempted paths at the current node m. If yes, include them into the incomplete search node queue S.

(3) Set node n as the current node and continue searching by repeating Step 2 until meeting a leaf node. Check whether this leaf node is in the target node queue T to be searched T. If it is in T, delete this node from T.

(4) If the target node is empty, then end searching. If it is not empty, set the current node as the ending node of the incomplete search node queue S, and go back to Step 2 to continue searching.

For example, in Figure 7, (N3, N7) and (N3, N4) are two candidate streams at node N3, and both have the same attribute information, so geometric features are needed to determine the main stem. The heuristic algorithm is used to calculate the longest path from N3 through N7 and from N3 through N4, and then compare to select the main stem.

In Figure 6, first we calculate the longest path from N3 through N7. After screening, the target node set is {N15, N16}, and the set of incomplete search node is empty. The current starting node is N7, which has two expandable nodes N17 and N8, and the target node is N16. We can calculate f(N7,N17)=g(N7,N17)+h(N7,N17)=$Length_{N7N17}$+$Dist_{N17N16}$,f(N7,N8)=g(N7,N8)+h(N8,N17)=$Length_{N7 N8}$+$Dist_{N8N17}$, and choose min{f(N7,N17), f(N7,N8)}. Because f(N7,N8) is smaller, we choose N8 as the next node. At the same time, we include node N7 into the set of incomplete search nodes, and the search continues until reaching the target node N16. Then delete N16 from the target node set, and include node N14 into the set of incomplete search nodes. The next target node is N15. Take the newly added node N14 from the incomplete search set, and continue searching until reaching N15. Delete N15 from the set of target nodes, and set becomes empty, so the program ends.

According to the above algorithm, the longest paths starting from N3 through N7 and starting from N3 through N4 are MaxLengthN3N7 and MaxLengthN3N4, respectively. We get $MaxLength_{N3N7}$/$MaxLength_{N3N4}$≥k, k>1. That is, the longest path starting from N3 through N7 is much longer than that from N3 through N4, so we select (N3, N7) as the main stem.

There are rings at nodes N8 and N10, and most are multi input multi output rings. Thus, we should choose streams with better angles as the main stems. Therefore, (N8, N9), (N9, N10), (N9, N9), (N9, N9), (N9, N9), (N9, N9), (N9, N9), (N9, N9), (N9, N9), (N9, N9) N10, N12), (N12, N13) are selected as the main stems.

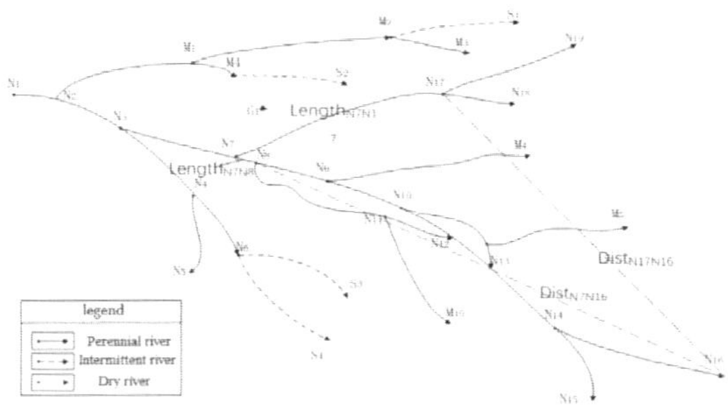

Figure 6. Heuristic algorithm to select the main stem

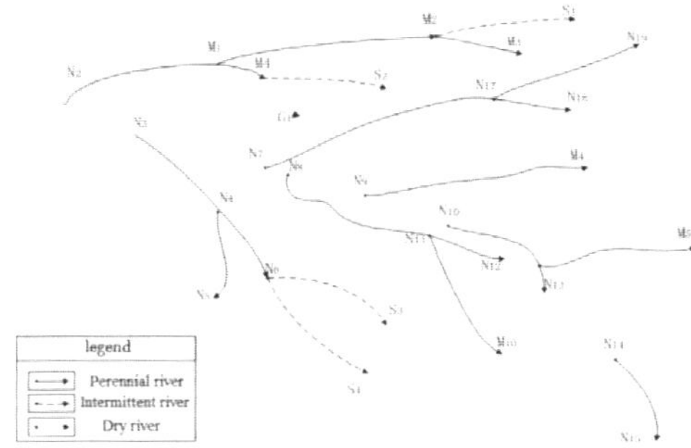

Figure 7. Result of one iteration

3.3. Automatic river classification

Searching results show that the streams from node N1 to node N16 are first-order streams. Then remove the first-order streams from the river system and iterate. Figure 7 shows the results after one iteration.

Continue searching all nodes on the main stem until all streams have been searched and classified. Figure 8 shows the final classification results.

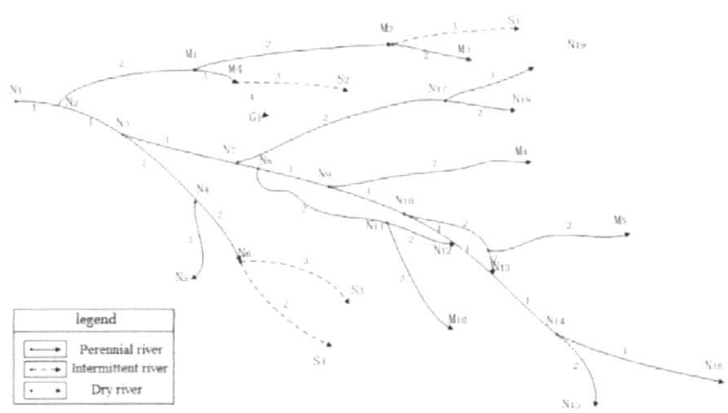

Figure 8. Classification results

4. Experiment and analysis

The experimental platform is based on the WJ-III stepless map workstation. The development environment is based on Microsoft Visual Studio 2010 and adopts C ++ programming language. The experiment environment is a personal computer with Windows 7 system (64-bit), which has Intel Core i7-6700 (3.40GHz) CPU, and 4GB RAM. The data for experiment is the monitored geographical information of the water system. We conduct the river classification experiment. The information of the water system includes river flow, river type, and river name.

The experiment is conducted using the water system data in two regions. We compare the results with and without optimization using the heuristic algorithm.

Table 2. Comparison of length algorithm before and after optimization

Comparison Content	Region 1	Region 2
Total number of arcs in river	1093	13012
Complete ergodic classification time (second)	1.984	15.05
Heuristic classification time (second)	1.881	12.71
Increase rate in time (%)	5.3	18.4

Unoptimized ergodic number of arcs	3189	42532
Optimized ergodic number of arcs	1961	35521
Reduction ratio in ergodic number (%)	62.6	19.7
Difference between heuristic and ergodic classification (difference ratio)	0（0%）	0（0%）

Table2 shows the experimental results. It shows that in river classification in different regions, the optimized algorithm can reduce the ergodic number of arcs and time in different degrees, but the classification results are the same, and the efficiency is improved. Time reduction is more obvious for a larger total number of arcs. The analysis of accuracy and time of the heuristic algorithm are as follows:

(1) Algorithm accuracy: The difference between the heuristic algorithm and the original algorithm lies only in the first step of selecting the ending nodes of the river, which might filter out the correct results. But in the experiment, after the classification of attribute information, the streams using this algorithm are often some smaller tributaries, which are usually short and straight, so the correct ending nodes are very rarely screened out.

(2) Algorithm time: In the unoptimized case, the algorithm is ergodic, which searches each arc to find the longest path. The optimized algorithm will first find the candidate ending nodes that meet the criteria, and then search for these ending nodes through heuristic information, which reduces the number of searches and thus shortens the searching time of the algorithm. In the best case of optimization, when searching for the longest path, there is only one target node that forms the longest path from the starting node, and each node selected by the heuristic algorithm on this path. In this case, the total number of executions of this algorithm is the number of nodes on the longest path. In the worst case, we need to search all the arcs. It is the same as when not optimized. In general, the time of the optimized algorithm is between the best and the worst case.

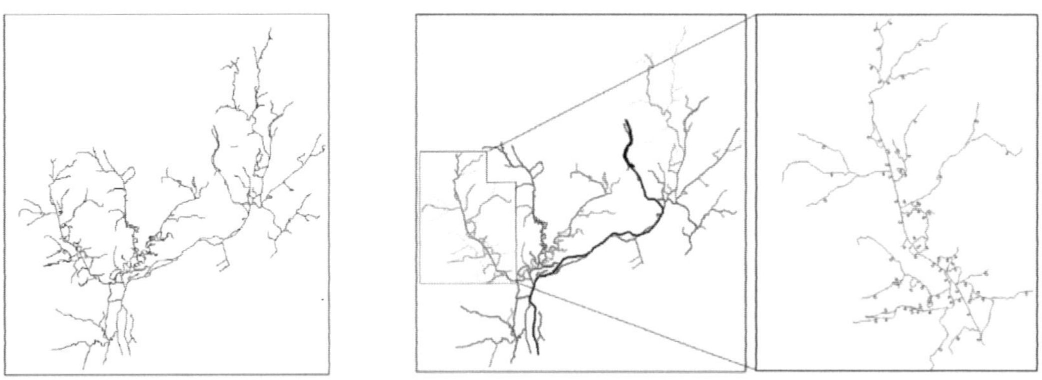

(a) Original river system (b)River classification results

Figure 9. Classification comparison results

Figure 9(a) is the original river system of one of the two experimental regions. The classification results are distinguished by color and thickness, the darker colors and thicker lines represent the higher river hierarchy. From the classification results in Figure 9(b), we obtain good classification results. The results by optimized algorithm are accurate and well-structured, especially for complex river system, such rings and multiple nodes in the river. The optimized classification effect meets the mapping requirements, and beats the existing algorithm.

5. Conclusion

This paper improves the algorithm of river classification. As for accuracy, the heuristic algorithm combines the attribute information with geometric characteristics, and reduces classification error in complex river systems (especially rivers with rings) through treating different types of rings. As for efficiency, it improves the algorithm of finding the longest path of the river system, which reduces the number of searches. The validity and rationality of this method are supported by experiments. In the process of river classification, we can include more geometric features of the river, such as the density of the river and the number of left tributaries and right tributaries of the river, which can further improve the accuracy of river classification.

Acknowledgments

This research was supported by the National Surveying and Mapping Projects (A1713) and the National SciTech Support Plan (2015BAJ06B01).

References

[1] Wu H. Structured Approach to Implementing Automatic Cartographic Generalization. 1996. *Journal of Wuhan Technical University of Surveying & Mapping*, **21**(3), pp 277-285.

[2] Paiva J, Egenhofer MJ, Frank A. Spatial Reasoning About Flow Directions: Towards an Ontology for River Networks. 1992. International Archives of Photogrammetry and Remote Sensing, 24(B3), pp 318-324.

[3] Guo Q, Huang Y. Automatic reasoning on main streams of tree river networks. 2008. Geomatics & Information Science of Wuhan University, 33(9), pp 978 – 981.

[4] Wei WU, Chengming LI, Yin Y, et al. An automatic plotting algorithm of the gradual change of river based on directed topology. 2016. Science of Surveying & Mapping, 41(12), pp 89 – 93

[5] Zhang Y. Structured Design of Dendritic River Networks Based on Graph[J]. Editorial Board of Geomatics & Information Science of Wuhan University, 2004, 29(6):537-536.

[6] Zhang Q N, Quan H. Construction and Application of River-Tree. 2005. Acta Scientiarum Naturalium Universitatis Sunyatseni, 44(6), pp 101 – 104.

[7] Ai Z. The river net structural model and its construction in geography information system. 1995. Surveying & Mapping of Sichuan, 2, pp 75-79.

[8] Guo Q S. Analysing the Characters of the Networks of Rivers and Structuralizing the Tree-like Network of Rivers Automatically.1999. Surveying & Mapping of Geology & Mineral Resources, 4, pp7–9.

[9] Zhang Q N. Line Generalization Based on Dynamic Segmentation.2004. Acta Scientiarum Naturalium Universitatis Sunyatseni, 43(2), pp 104-107.

[10] Zixing Ai, Niu R F, Hehai Wu, et al. The analyzing method of river network based on dynamic tree topology.2006. Science of Surveying & Mapping, 31(3), pp 80-81.

[11] Xiao T, Fang W U, Qi H, et al. A Multi-criteria Decision Model for Identifying Master River and Its Application in River System Construction.2005. Acta Geodaetica Et Cartographic Sinica, 34(2):154-160.

[12] Wang J X. Algorithms for Longest Path:A Survey.2009. Computer Science, 36(12), pp 1 – 4.

Anti-jamming Technology in Small Satellite Communication

Zixiang Jia[1]

[1]School of Electrical and Information Engineering, Tianjin University, Tianjin 300000, China

Email: 1165910129@qq.com

Abstract. Small satellite communication has an increasingly important position among the wireless communications due to the advantages of low cost and high technology. However, in view of the case that its relay station stays outside the earth, its uplink may face interference from malicious signal frequently. Here this paper classified enumerates existing interferences, and proposes channel signals as main interference by comparison. Based on a basic digital communication process, then this paper discusses the possible anti - jamming techniques that commonly be realized at all stages in diverse processes, and comes to the conclusion that regarding the spread spectrum technology and antenna anti-jamming technology as fundamental direction of future development. This work provides possible thought for the design of new small satellite communication system with the coexistence of multi - technologies. This basic popular science can be consulted for people interested in small satellite communication.

1. Introduction

Classical mechanics and other modern physics provided academic support for the embryonic form of cross-continental long distance communication. In 1945, AC Clarke purposed an ideal that put the relay station outside the earth, which eventually led to the birth of worldwide satellite communications technology [1]. People immediately foresaw the coming of satellite era in synchronous orbit. Through the adjustment of orbit height and frequency band, the new communication based on on-board relay station could give global radio coverage [2]. In 1954, American designed the first passive satellites. Regarding the moon as natural satellite, United States Navy (USN) sent telegrams successfully. However, due to the relative change of the position between earth and moon, it is not reliable to continually use moon as relay station [3]. As a result, the opposing superpowers started to design artificial satellite. In 1957, the former Soviet Union launched the first active satellite named Sputnik I [4]. During 60s to 80s, synchronous satellites such as Early bird and INTELSAT brought a wider range of coverage area and further distance of communication [4]. Thus the study of satellites is gradually going from military to civilian. To the 21st century, mobile communication system application to low orbit widened the civil market with valuable frequency band and business. The development of satellite communication is less than a hundred years, in which the anti-jamming technique plays an important role [5].

At present, satellites have a variety of different functions. They can be classified according to the frequency band as well as the use. As the classification of the height of orbit, the satellite can be divided into Geosynchronous Earth Orbit (GEO), Medium Earth Orbit (MEO) and Low earth orbit (LEO)[2]. Among them, the coverage of satellite located in medium and low orbit are not as wide as high orbit's. So, it is necessary to connect multiple satellites with entire constellation to achieve the complete coverage of time and space domain around the whole world [3]. However, at that time, the cost of a large satellite that was heavier than one ton, which was at least 100 million dollars each. This could be typically evidenced by famous Alphasat from Inmarsat, 350 million dollars per satellite. Therefore, considering the cost of manufacturing and launch, small satellites defined by less quality than 900KG draw the attention of scientist [4]. Iridium system is the first introduction of mobile communication system composed by 66 small satellites [4]. The entire cost of this system only needs more than one billion dollars. The cost of some small constellations is almost as same as single satellite synchronization. This has a great attraction for medium-sized enterprise and Institute lacking of research funds. As a result, the development of small satellite systems has gradually become the mainstream.

Prior to the application of satellite broadband to personal communication, the early small satellites were limited to the background of the era [3]. Military use resulted in the importance of anti-jamming capability of satellites. This has become an important evaluation standard for communication system in addition to cost savings, simplified construction and shortened transmission time [4]. Both the countermeasure of tactical interference and the confidentiality of communication signals are important propositions at that time. The small satellite communication didn't come across the globalization for civil use until 21st century. At that time, Anti-jamming technique in small satellite communication was still the focus of this area. The focus could be divided into electromagnetic interference, natural interference and a variety of other factors [5].

Thereby this paper is going to discuss anti-jamming technique in small satellite communication. The first part will introduce some possible interference in small satellite communication. The second part gives detailed study and analysis of diverse anti-jamming. And the last part proposes suitable anti-jamming program in different process of small satellite communication.

2. Possible interference

Even if the relay station from small satellite communication is located outside the earth, it is still influenced by interference from external factors, which is a force majeure. Thus the research on anti-jamming technology has been an important subject of small satellite communication.

Depending on the sources, the possible interference in the communication process could be classified in two parts. The first is natural interference from environment and physical phenomenon. The second is man-made interference from faulty operations and electronic equipment [6]. Both of these are likely to cause typical electromagnetic interference. They will make significant varying degrees of impact on the uplink and downlink signals [7].

2.1 Natural interference

The natural interference could be mainly regarded as harsh climate and special astronomical phenomena, causing passive effect on the radio waves in the space. Some natural phenomena, including rainstorm, blizzard and thunder, will cause physical damage to the ground base station. The climatic noise is the main reason for the impact of satellite communication. This can be typically evidenced in rain attenuation, an energetic attenuation of the radio signal caused by rainfall. That means the radio waves will be absorbed and scattered by dense water molecules when through the atmosphere. At the same time, the ground receiving system will be affected by rain noise [8]. Above the band commonly used in satellite communications, the Ku-band is believed to be most seriously affected due to the similar size of wavelength compared with raindrops [8]. Another instance is sun outage. When time passes the spring equinox or the autumnal equinox, the orbital positions of the earth, satellite and sun overlap into a straight line. The downlink signal from ground station will become practically unusable resulting from severe noise created by solar radiation [9]. Thus sun outage interference cannot be overcome with current technical means; there won't be more detailed analysis.

IOP Conf. Series: Journal of Physics: Conf. Series **960** (2018) 012013 doi:10.1088/1742-6596/960/1/012013

2.2 Man-made interference

Man-made interference in satellite communication mainly refers to mistaken spatial layout and channel allocation. In view of the fact that there is totally hundreds of in-orbit satellites, plus disused space garbage (satellite debris), indubitably the outer space will be more crowded in the future. Particularly, satellites sharing geosynchronous orbits will be subject to severe adjacent satellite interference [7]. Thus small satellites are mainly located in the low orbit with unified connection controlled by the same link. The influence for small satellite communication is not as serious as synchronous satellite's. What should be in consideration is adjacent channel interference, which is caused by partial overlap of the different carrier frequencies [7]. At present, the small satellite system based on mobile business has been popular in most parts of the world. In view of the number of users has far exceeded expectations, the impact of adjacent channel interference is worthy of attentions.

2.3 Electromagnetic interference

Electromagnetic interference is the focus of small satellite anti-jamming technology to overcome. The transponder of small satellite is different from synchronous satellites'. Synchronous satellites use geosynchronous. In order to allocate frequency resources reasonably, the data is almost transparent. The downstream receiver thus can be easily disturbed by illegal signals [6]. Although small satellite communication is not bothered by this issue, other factors like ground microwave, radar, radio and others are still possible to damage terrestrial base stations during the process of long distance communication. It is essential to prepare for possible malicious attacks from third parties. Both active interference from the jammer that spontaneously generates electromagnetic energy, and passive interference through the reflector to weaken the effective radio waves, is the range that needs to be considered.

Directly against these possible interferences, the next section will introduce and illustrate some potential solutions in detail, with the analysis of specific anti-jamming method.

3. Small satellite communication system

Similar to synchronous satellite system, small satellite systems are always divided in two parts, the ground station and the on-board system.

Assume a digital satellite communication system as an example.

When a baseband signal is transmitted by the source, it will experience the source coding, the channel coding stage, and modulation respectively. Then the modulation signal will be sent to satellites by antenna using upstream channel. This weak signal is received by the satellite communication antenna. After that, it comes to the amplification, frequency conversion and power amplification in the communication transponder. Finally the amplified on-board radio wave will be re-sent to the ground station by the satellite communication antenna. This will achieve long distance communication between two ground stations or multiple ground stations.

Meanwhile, satellite communications have the most complex wireless channels as well as the complexity of their relay stations located outside the stars. As a result there may be more threats, and the various processes of entire communication systems have their own anti-jamming means correspondingly.

The first that needs to consider is the natural phenomenon of interference. However, rain attenuation, sun outage and other natural interference cannot be resolved by recent existing techniques. What the ground station could do is just making forecasts for advancing notice as well as timely repairing faulty equipment to reduce losses.

The second problem is the power limitation of satellite communications. Although cannot equal synchronous satellites, low-orbit system still need a large amount of power to launch its satellites to the corresponding orbital height. Furthermore, the on-orbit satellites can only rely on limited means to survive such as solar energy, while the application of multiple access technology generated intermodulation interference. Under comprehensive consideration the utilization of frequency band is not high in satellite communication, which means it can be easily interfered by other carrier signals.

The third is the interference from the illegal signals from the third party, which is also the main problem among satellite system.

In general, the interference to be considered is mainly noise and false signals. This paper thus makes a decision about interference that may exist in the ground station and on-board environment respectively.

4. Anti-jamming in entire process

One of the focuses of the development of modern satellite communications is to study the noise in the wireless communication process to improve the effectiveness and reliability of transmission.

4.1 Modulation techniques

In the channel coding stage before modulation, such as Pulse Code Modulation (PCM), it is possible to increase the interference tolerance by changing the encoding method, or add extra partial code in the information code group according to certain rules. So that it can check and correct potential error according to the corresponding rule of verifier [7]. To more detail, the cost of satellite communication is enormous, which is also one-side communication. This system is mainly suitable for Forward Error Correction (FEC), which belongs to the application without the requirement of receiver to reverse the request to retransmit the data. That means the transmitting station will send the data attached by a certain redundant error correction code sent together, which based on the receiver to inspect its error [7].

4.2 Wave filtering

It needs higher cost and technical requirements to interfere with the downlink signals transmitted by satellites. So the active interference, using illegal Radio-frequency (RF) signals to attack the uplink signal transmitted by ground station, is the main method [6]. It thus becomes the focus of anti-jamming technology. To deal with signals that may partially or completely coincide with the modulation signal, which will causes serious confusion, filtering will be the most effective method.

In modulation process, this paper lifts an uplink working on the Ku-band as an example. Its frequency band is about 12 to 18GHz. By comparison the modulation wave occupies the narrow bandwidth relatively. And for the possible narrowband interference, the encoded discrete signal is Fourier transformed to eliminate the small frequency point of the interference in the frequency domain model. Then relevant procedure reduces the attenuation of the modulated signal and returns it to the time domain [5]. Furthermore, in the more complex model there is also the broadband interference at same time. In this situation, the combination of frequency domain and space-time filtering is more suitable [5].

4.3 Spread spectrum technique

The third step is to use the spread spectrum technique (SS) to further process the modulation signal before the ground station transmitter transmits the modulation signal. In this way, the system could occupy much more bandwidth than the bandwidth of the modulation signal. And the receiver uses corresponding spreading code to recover the data. The first method is direct sequence spread spectrum (DS) [7]. The transmitter sends extended series of spectrum, so that the unit band power becomes smaller, which means the signal power spectral density becomes low. Then the entire communication can be hided in the background of channel noise and thermal noise. So that the signal will be submerged in the noise, and be hard to be found by third parties. At the receiving end, the same spreading code sequence is used for dispreading. And the existing noises also dispread in this process as well as widened, and then through an intermediate-frequency filter. Thus its interference power will be greatly reduced. The second method is Frequency-Hopping Spread Spectrum (FHSS) [7]. The transmitter uses a certain code sequence to select the Multi-Frequency Shift Keying (MFSK), and constantly change the transmission frequency in a wide range. The number of these random frequencies can reach to thousands or even million. Real modulation signal is therefore difficult to be intercepted by a third party which aims to use the illegal signal with corresponding frequency for interference.

4.4 *Smart antenna technology*
The fourth step, through the antenna to send uplink signals. This process also has abundant anti-jamming technology. One typical method is the adaptive antenna technology [7]. The antenna automatically control and optimize the array. When the system is subject to interference, the nulling of antenna pattern automatically point to the direction of the interference source, achieving interference suppression. Another method is Multi-Beam Antenna (MBA), which generally has 19 emission source beams [7]. This system can detect the location of the interference source according to the sensor. Then it controls the relative amplitude and phase of each beam through network beams forming. As a result, the gain of the antenna in the direction of interference vanishes.

4.5 On-board anti-jamming technology
The fifth step, the signal reaches the satellite. Transparent transponder is one of the most vulnerable satellite communication system links. Because of data transparency, it is easily interfered by third parties. This process corresponds to a variety of on-board anti-jamming technology, for which the development is relatively slow. Using the satellite switching (SS) as an example, which also called the exchange of multi-beam satellites. Firstly, the system uses time division multiplexing (TDM) before the circuit exchange [10]. Taking advantages of time interval of sampling pulse makes it possible to transmit multiple baseband modulation signals in the single channel. This process will also use a large number of invalid bands from direct sequence. Another method is Code Division Multiple Access (CDMA) [10]. That is using the orthogonal address code to send the modulation signal to achieve the beam forming. Then, dispread and spread the user unit on the satellite to achieve connectivity between different channels through exchanging. As a result, the efficiency and convenience of the entire satellite communication system will be greatly improved. At the same time, this MBA is more controllable in the face of interference, for example, giving up the interference beam to reduce the overall system loss.

4.6 Others
In the sixth step, the relay station transmits the downlink signal to the ground receiving station. This process mainly uses the C-band. The frequency band is about 4 to 8GHz, lower than Ku-band. The on-board modulation mainly faces broadband interference. Compared to narrowband, space-time filtering is more appropriate [6].

The seventh step, the signal will be disposed under dispreading, demodulation and decoding to obtain the corresponding original signal. This process is mainly based on the feedback and destination of many anti-jamming technologies from previous process. At the same time, the ground station to accept the signal usually works on the S-band. It should pay attention to the ground microwave interference and radar interference [11]. They may cause the increase of Bit Error Rate (BER) for useful signals. Receiver can adjust the elevation of receiving antenna though measuring the interference signal to reduce the bit error rate, or directly using electromagnetic shielding means [11].

The above is based on a specific satellite communication digital transmission process, from which this paper enumerates the anti-jamming technologies in various stages that can be used.

5. Conclusion

The previous parts of the classification introduce the development of small satellites, the possible interference in the communication process, and the specific analysis of the existing major anti-jamming technology in a complete communication process.

Satellite communication is one of the most important means of wireless communication. Based on the idea of controlling cost and optimal performance, the construction of constellation made by multi-beam satellites is undoubtedly the main direction of satellite development.

Among them, the noise treatment, while increasing the signal-to-noise ratio (SNR) to approach the Shannon limit at the same time, is the common research theme of all communication methods. Therefore, anti-jamming technology will continue to be a long-term mission of small satellite communications in the future.

For most of the wireless communications, including satellite communications, spread spectrum technology and antenna anti-jamming technology are the main anti-jamming means. It is because the user base of satellite communication is larger. Spread spectrum, which has no connection with the location of interference, has more advantages than others. Although satellite communication witnesses later development of on-board anti-jamming technology, the achievement of the functions of the ground station on the star as well as the multi-user signal exchange have greatly improved the communicating efficiency and fault tolerance rate.

In addition, in the future, it is appropriate to focus on the coexistence of a variety of anti-jamming technologies of communication system, requiring the ability of on-board signal processing technology, smart antenna technology, spread spectrum technology, optimizing the modulation, to ensure that the interference for different types can be disposed by corresponding confrontation technologies to resist and control.

References

[1] Clarke A C 1966 Extra-terrestrial relays: can rocket stations give world-wide radio coverage *Progress in Astronautics & Rocketry* **19** p 3-6
[2] Zhang G 1996 Study on the orbit height of satellite mobile communication system JOURNAL-CHINA INSTITUTE OF COMMUNICATIONS 17 p 111-114
[3] Orii T, Ono T, Aoyama J I, Todome K and Shibayama Y 2002 Development of versatile small satellite Acta Astronautica 50(9) p 557-567
[4] Zhang X G 2000 The status, characteristics and development direction of small satellites Telecommunications Information 5 p 27-29
[5] Chen Q 2011 Analysis on anti-jamming technology for satellite navigation receiver Radio Engineering
[6] Han D 2013 Anti-jamming of uplink signals and typical interferences in satellite earth stations Journal of Shijiazhuang Vocational Technology Institute
[7] Gu C Y, Chen X F and Yi K 2004 Prospect of anti-jamming techniques in the satellite communications Systems Engineering & Electronics
[8] Gao X J 2009 Simulation research on interference from nature in ku-band satellite Communication Northeastern University
[9] Barewald J R 1981 Calculating earth station sun outage times Satellite Communications 5 p 57-62
[10] Xia Z, Liang J, Yuan X G and Sun J 2007 The application of On-board Shifting technologies in anti-jamming communication satellites Microcomputer Information 36 054
[11] Zhao L 1998 Interference in digital communication satellite from ground microwave Communication and computing technology

A Software Defined Radio Based Airplane Communication Navigation Simulation System

L He[1], H T Zhong[1,*] and D Song[1]

[1]School of Aeronautics, Northwestern Polytechnical University, Xi'an, China

[*]714663885@qq.com

Abstract. Radio communication and navigation system plays important role in ensuring the safety of civil airplane in flight. Function and performance should be tested before these systems are installed on-board. Conventionally, a set of transmitter and receiver are needed for each system, thus all the equipment occupy a lot of space and are high cost. In this paper, software defined radio technology is applied to design a common hardware communication and navigation ground simulation system, which can host multiple airplane systems with different operating frequency, such as HF, VHF, VOR, ILS, ADF, etc. We use a broadband analog frontend hardware platform, universal software radio peripheral (USRP), to transmit/receive signal of different frequency band. Software is compiled by LabVIEW on computer, which interfaces with USRP through Ethernet, and is responsible for communication and navigation signal processing and system control. An integrated testing system is established to perform functional test and performance verification of the simulation signal, which demonstrate the feasibility of our design. The system is a low-cost and common hardware platform for multiple airplane systems, which provide helpful reference for integrated avionics design.

1. Introduction

Radio communication and navigation system plays an important role in ensuring the safety of civil aircrafts in flight. The function and performance of airplane systems need to be tested before they are installed on the airplane. Conventionally, a set of transmitter and receiver should be designed on ground to communicate with the systems on-board [1-4]. Normally a set of hardware was designed for each system, which took a lot of space and time consuming.

Software defined radio (SDR) is a new technology in the field of radio engineering, which uses as much as software to replace the work of hardware [5]. Normally a common hardware platform is used for multiple systems and different signal processing software is developed for different system. Also software can be divided into common module and specialized one and the common module can be called by different systems. SDR make it possible to integrate many systems and at the same time reduce the weight and size.

Many researches have been performed in the field of signal simulation. FPGA or DSP techniques are used for avionics, radar and GPS system [6-8]. In this paper, a set of communication/navigation simulation system is designed based on SDR to give a way of avionics testing and verification.

2. Basic principles of airplanes communication and navigation system

This section describes the basic principles of communication and navigation systems of civil airplane [2, 8].

2.1. Communication system

Airplane communication includes HF and VHF system. VHF is used between flight crew and ground crew during take-off/landing or flying over control airspace. HF is a long range communication system used for contact with ground stations and the other airplane. The modulation method used by HF is AM or SSB, the working frequency is 2M to 29.999 MHz with 1kHz gap. VHF uses AM or FM modulation, where AM frequency band is 117.975M to 137MHz with 8.33kHz gap, FM frequency band are 108M to 173.975MHz and 225M to 399.975MHz.

2.2. Automatic direction finder

The automatic direction finder (ADF) determines the relative bearing to conventional broadcast stations or beacons through calculating the direction of the receiving radio. ADF works in 190k~1750kHz to communicate with civil radio station and NDB station with different identification signal, which is composed of two English character Morse code transmitted 20~30 characters per minute. The signal transmitted through horizontal, longitudinal and vertical antenna are

$$\begin{aligned} V_1 &= A\cos\theta\cos\omega t \\ V_2 &= A\sin\theta\cos\omega t \\ V_3 &= B\sin\omega t \end{aligned} \tag{1}$$

Where A and B is the amplitude, θ is the relative bearing, ω is the angular frequency of the carrier.

After a series of processing, including balance modulation, mixing with local oscillator, frequency detection and LPF, we can obtain the bearing angle

$$\theta' = \arctan\left(\frac{\sin\theta}{\cos\theta}\right) \tag{2}$$

3. Airplane communication navigation simulation system design

3.1. System architecture based on SDR

Normally the architecture of SDR includes two parts: front end and back end. The front end comprises analog front end, A/D, D/A, and digital front end. The back end is baseband signal processing. According to airplane communication and navigation system, we give the architecture of the simulation system, shown in Figure 1.

A wideband antenna receives all the signal of different systems. The avionics software is composed of multiple functions for the communication and navigation systems. Through interactive menu and control to the hardware interface, the current status, working frequency, data wave and configuration parameter are displayed to the user.

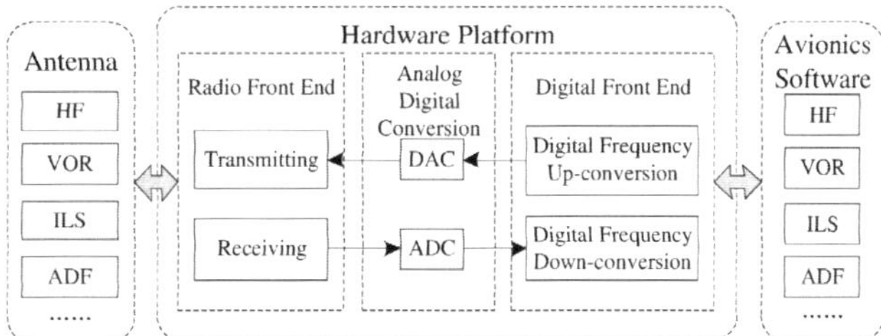

Figure 1. Airplane communication system architecture based on SDR.

3.2. Hardware configuration

The hardware we used for our system is shown in Figure 2. Voice is collected by the sound card of the computer and is input to the computer to be modulated and processed by the avionics software. Then it is fed to USRP N210 through Ethernet. After D/A transform and frequency conversion it is transmitted. The receiving process is reverse.

The USRP transceiver is a kind of universal software defined radio peripheral developed by NI company and popular used by radio system simulation and design. The principle is shown in Figure 3.

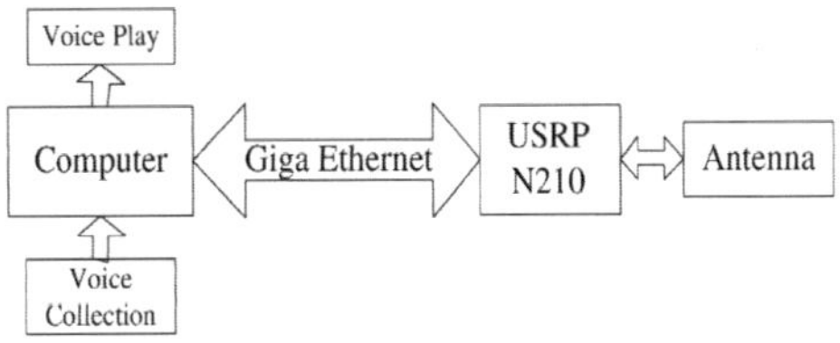

Figure 2. Hardware configuration.

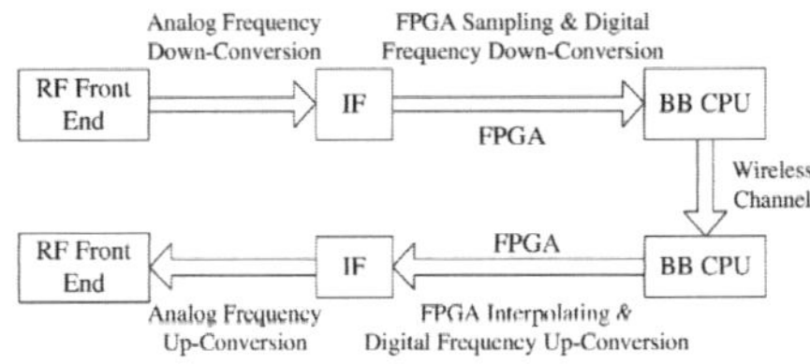

Figure 3. Principle of USRP N210.

USRP N210 mainly includes FPGA motherboard, radio board and analog digital transform. FPGA is used for baseband digital signal processing. The FPGA motherboard is connected to a computer through Giga Ethernet. All the digital wave signal processing, e.g., modulation and demodulation, are performed on the CPU of the computer. All the universal high speed processing, e.g., frequency up/down conversion, interpolating and sampling, are executed in FPGA. The radio board can be selected for different radio systems.

3.3. Software architecture

Signal processing software design is the key point of our simulation system. For each communication and navigation system, there is a transmitting/receiving and signal calculating method.

Through detail analysis of the principle of each system, we can extract some modules that used by several system as common module. These common modules are: carrier generation, mixer, amplifier, LPF, BPF, CIC, pulse identification, random pulse generation, etc. And the special software modules of different systems are shown in Table 1.

Table 1. Software modules of diffrent system.

System	Software Module
HF	SSB mod/demod, Hilbert transform,
VHF	FM mod/demod,
VOR	Signal generation, AM/FM mod/demod, Morse code generation, Bearing angle calculation
ADF	Signal generation, I/Q signal selection, AM mod/demod, Bearing angle calculation
ILS	Signal generation, DC estimation, Morse code generation, AM mod/demod, DDM calculation, Warning module

3.4. Software design

There is several application software that USRP N210 can support, i.e., GNU Radio, Visual C++, Matlab/Simulink, LabVIEW and etc. By comparing them we selected LabVIEW because it can be used on our Windows operation system, is easy to developed and has friendly man-machine interface. Especially, it includes many toolkits for radio signal processing, which make it suitable for our systems. The combination of USRP hardware and the software give a popular used prototype platform.

The parameter of some function module should be configured for designated signal before transmitting and receiving. The main parameters are IP address, IQ sampling rate, antenna and gain, et ac. A voice collection and playback program is design based on the sound card of the computer to verify it proper functioning. And a test programs are needed for the navigation systems to verify that the signals we generate are correct.

For airplane communication system, the software includes AM, SSB and FM transmitting and receiving program. For navigation system, based on the principles mentioned in part two, there needs to design specific program for each system.

As an example, we give a flow chart of ADF receiving program, shown in Figure 4. Figure 5 gives an example of a receiving program compiled in LabVIEW.

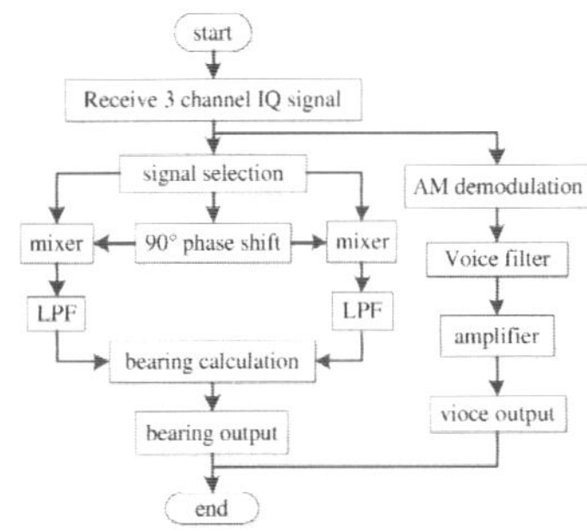

Figure 4. Signal receiving flow chart of ADF sytem.

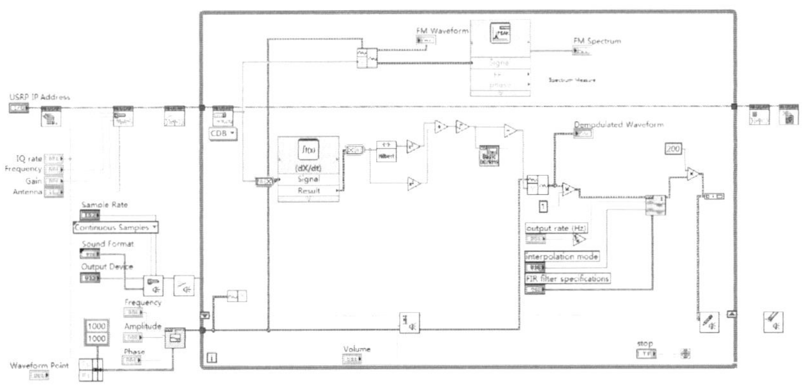

Figure 5. An example of the compiled LabVIEW program.

4. Simulation test

The test platform composes two computers, two USRP N210 and antennas.

Two computers work as transmitter and receiver. The transmitter collects voice signal or generate specified signal for the communication and navigation system. After modulation and digital processing, they are fed to USRP through Ethernet. USRP transform the signal to radio band and transmit them through the antenna. The receiving process is vice versa.

As mentioned above, before the communication and navigation system simulation tests, first we perform voice collection/playback and navigation signal test, as shown in Figure 6.

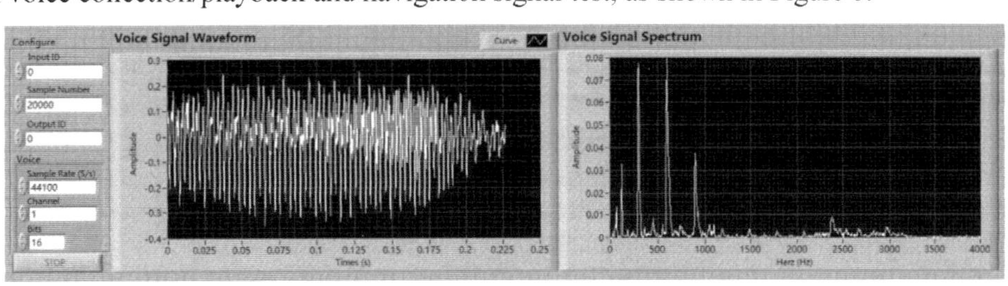

Figure 6. Signal test and anaysis.

Figure 7 is an example of a receiving program. The left part of the panel is the parameter configuration area and the right is the received wave form, frequency spectrum.

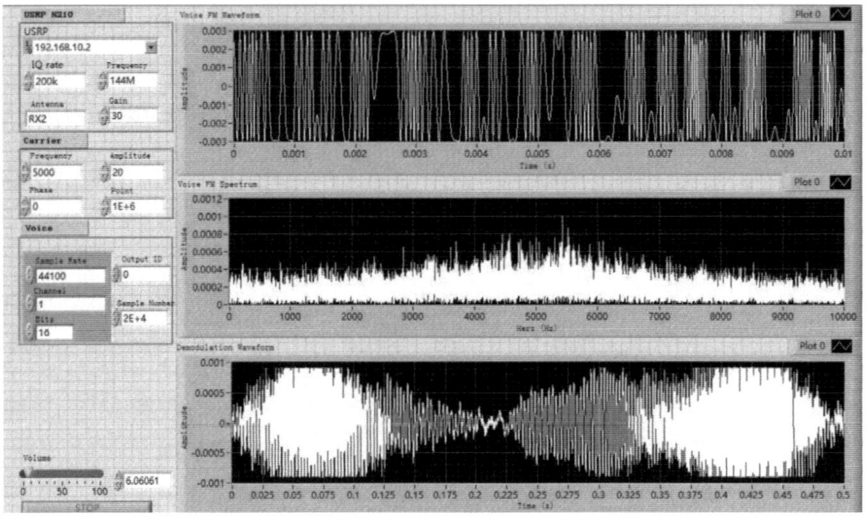

Figure 7. Signal receiving pannel.

Measure should be taken to verify the correctness of the design. For communication systems, we can listen to the received voice and compare the quality with the collected one. For the navigation systems, the received angle or deviation can be compared with the pre-defined parameter of the transmitted signal. An example is given in Figure 8 to show the error of ADF bearing signal. It shows that the error is between ±0.01°, which satisfies the design requirement.

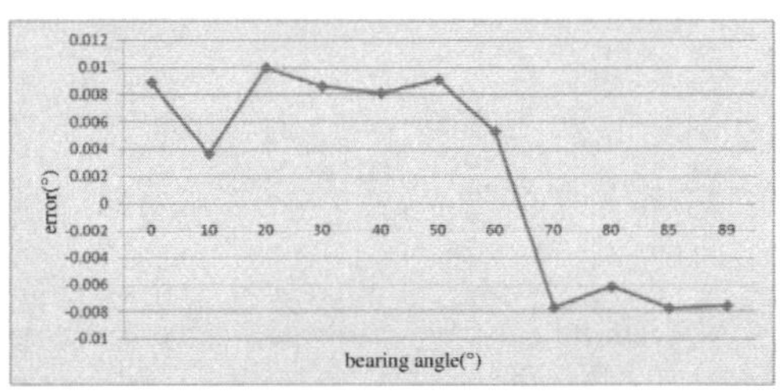

Figure 8. ADF bearing error.

5. Conclusion

This paper illustrates a communication/navigation simulation system designed on the bases of SDR hardware USRP and LabVIEW. It can generate signal of various airborne avionics and gives a way to test them. This simulation system is low-cost, light and easy to design. And more systems can be added to it according to one's needs of the system design.

References

[1] Shaw G R and Lopato J Software Defined Radio as a solution to testing RF Avionics 2014 *Autotestcon IEEE* 290-92.

[2] Li Jun. Aeronautical radio signal simulation technology. Master Degree Thesis, Northwestern Polytechnical university, 2015.

[3] Xie F P, Li Y, Liu X Y and Cheng Y F Design and Implementation of Echo Simulator for Airborne Wind-shear Radar 2012 Computer Measurement & Control 1039-1042.

[4] Leosrisook C, Pongpheaw K and Horapong K An Implementation of Portable ILS Localizer Signal Receiver using SDR 2014 ISCIT 131-135.

[5] Lou C Y, Xu J L and Yang X N Software-Defined Radio: Principles and Practice Electronic Industry Press 2014.

IOP Conf. Series: Journal of Physics: Conf. Series **960** (2018) 012014 doi:10.1088/1742-6596/960/1/012014

[6] Cheng Y Q, Xu K, Hu Y F, Pillai P, Baddoo J, Smith A, etal.. Technology demonstrator of a novel software defined radio-based aeronautical communications system 2014 IET Sci. Meas. Technol. 8 370–79.
[7] Akopian D and Soghoyan A, A Labview-Based Fast Prototyping Software Defined GPS Receiver Platform 2013 IEEE Global Conference on Signal & Information Processing 1230-33.
[8] Han Bin. Software design of universal radio navigation system for general aircraft. Master Degree Thesis, Northwestern Polytechnical university, 2015.

Performance Analysis of Ivshmem for High-Performance Computing in Virtual Machines

Pavle Ivanovic[1] and Harald Richter[1]

[1]Clausthal University of Technology

Email: *pavle.ivanovic@tu-clausthal.de; hri@tu-clausthal.de

Abstract. High-Performance computing (HPC) is rarely accomplished via virtual machines (VMs). In this paper, we present a remake of ivshmem which can change this. Ivshmem was a shared memory (SHM) between virtual machines on the same server, with SHM-access synchronization included, until about 5 years ago when newer versions of Linux and its virtualization library libvirt evolved. We restored that SHM-access synchronization feature because it is indispensable for HPC and made ivshmem runnable with contemporary versions of Linux, libvirt, KVM, QEMU and especially MPICH, which is an implementation of MPI - the standard HPC communication library. Additionally, MPICH was transparently modified by us to get ivshmem included, resulting in a three to ten times performance improvement compared to TCP/IP. Furthermore, we have transparently replaced MPI_PUT, a single-side MPICH communication mechanism, by an own MPI_PUT wrapper. As a result, our ivshmem even surpasses non-virtualized SHM data transfers for block lengths greater than 512 KBytes, showing the benefits of virtualization. All improvements were possible without using SR-IOV.

1. Introduction

High-Performance-Computing (HPC) is traditionally performed on a supercomputer or on a parallel computer. Virtual machines (VMs) or clouds instead are only scarcely used for HPC, because it is the experience of the respective user community that they are not efficient enough for codes based on the Message Passing Interface (MPI) [1], which is the standard for HPC. The reason for that behaviour is that communicating MPI processes are slowed down by the TCP/IP data transmission that is needed for inter-VM and inter-server communication in a virtualized data centre or in a cloud. As a consequence, a parallel computer or supercomputer can typically not be replaced by a local or distributed TCP/IP system. Also, in cases where a single HPC server would be sufficient, a shared-memory communication, which would be much faster than TCP/IP, is not possible because the VMs are isolated from each other. On the other hand, virtual resources and clouds can be obtained in the Teraflops and Terabytes range for a few dozen Euros only, for example from auctions in the Amazon Web Service (AWS), which makes HPC in VMs and clouds very appealing. In this paper, we focus on the performance analysis of communicating MPI processes in two VMs on the same server with MPICH [2] as library. This library is used by us for inter-VM block transfers, via both TCP/IP and ivshmem [3], which is an inter-VM shared memory. The data transfer speeds are compared to those between two non-virtualized MPI processes in host OS, which are using standard SHM shared memory. It could be shown that for larger block sizes the performance is the same for both, virtualized and non-virtualized MPI SHM inter-process communication, thus proving the efficiency of ivshmem. For larger block sizes, ivshmem is even better. The rest of the paper is organized as follows: in chapter

2, an overview on inter-VM communication technologies is given. In chapter 3, our adaption of ivshmem to MPICH is explained from a functional and implementation point of view. Performance results for MPI block transfers between VMs are given and discussed in chapter 4.

2. State-of-the-art in inter-vm communication

There are a several technologies publicly available for inter-VM communication inside of the same server, which are: 1.) TCP/IP-based Internet communication, 2.) Virtio-net [4], 3.) vHost-net [5], 4.) virtio-vsocks, 5.) ivshmem and 6.) MVAPICH2-virt [6]. The first five methods are more efficient implementations of TCP/IP-based communication, while the latter two are based on shared memory, which is superior with respect to speed. In this paper, we focus on standard TCP/IP and ivshmem, because TCP/IP is used for inter-VM communication by most MPI implementations, although it provides for the slowest performance. We compare TCP/IP with our ivshmem adaptation, because in contrast to MVAPICH2-virt, it features SHM access synchronization which is indispensable for HPC.

2.1. TCP/IP-based Inter-VM Communication on the Same Server

In standard TCP/IP configuration, a sending MPI process in the guest OS of an emulated PC (source VM) transmits data to a receiving MPI process in another emulated PC (target VM) by calling the send () function of the Berkeley sockets in the source VM TCP/IP protocol-stack. The sender provides the IP address and port number of the receiver as call parameter for send. In case of full hardware virtualization by software, the call results in a guest-OS device driver-call for an emulated Ethernet card, which is intercepted by that QEMU [7] that has launched the source VM. Subsequently, this QEMU hands-over the TCP/IP packet to the single KVM in the host OS kernel which injects the packet into the Internet by means of the host-OS Ethernet device-driver and the real Ethernet card. This happens if no virtual switch (vSwitch) is engaged. Subsequently, the packet returns from the Internet, because the receiver is in the same server. The real Ethernet card accepts its own packet and gives it via its device driver to KVM which forwards it to the target QEMU. Meanwhile the receiving MPI process has called the recv() function in its TCP/IP-protocol stack, which results in a target-VM device-driver call that is intercepted by KVM. Subsequently, the receiver QEMU hands-over the received packet to the target-VM Ethernet device-driver which delivers it to the receiving MPI process. Additionally, source and target QEMUs have to emulate two Ethernet cards (vNICs) as virtual PCI devices for their VMs. The described actions incur a lot of overhead which is bad for HPC. This can be reduced by hardware accelerators for virtualization such as SR-IOV [8], but never fully eliminated. Overhead stems also from the TCP/IP protocol stack which would not be needed, because only a transmission inside of the same server is performed and not a worldwide communication.

2.2. Virtio-net

The KVM hypervisor has two interfaces: the first is used by the QEMUs it cooperates with. The second API is called Virtio that provides for "paravirtualization". This is a more efficient IO virtualization method than the so-called full software emulation. Besides being an API, Virtio is also a library of paravirtualized device drivers for the guest OS. For example, Virtio-net is the paravirtualized device-driver for a virtual Ethernet card (vNIC). Internet-based inter-VM communication can profit from Virtio-net if a Berkeley send call results in calling Virtio-net. This driver is aware that it is executed in a VM, and it is therefore actively cooperating with its QEMU. With Virtio-net, KVM does not need to intercept guest-OS device-driver accesses to emulated PC devices, because they are not performed. Instead, the parameters for virtio calls are directly forward to QEMU. Technically, the Berkeley send call is not mapped onto a vNIC in guest OS, but via QEMU onto a Virtio-net send queue ("virtqueue") in host OS. Virtqueues are much simpler and thus faster than vNICs, because they are only buffers. The rest in Virtio-net happens as described in section 2.1. Please note also that the paravirtualized guest-OS device-driver is called front-end driver, while the modified host-OS Ethernet-driver is termed back-end driver. The original host-OS back-end driver cannot be used in Virtio-net because its input is a Linux data-structure called sk_buf, while the front-end driver outputs so-called virtqueue entries. Both, the front-end and the modified back-end driver, are contained in the Virtio library. The first disadvantage of Virtio is that TCP/IP is still engaged. The second is that the virtqueue entries have to be handled by two QEMUs each, one for the source, the other for the target VM. Finally, for every data frame sent by the real Ethernet card, each QEMU has to make a system call to KVM, which is a time-consuming procedure, because it requires a full process context switch, with all MMU page table entries reloaded.

2.3. vHost-net

VHost-net implements a paravirtualized guest-OS device-driver for an Ethernet card in a more efficient way than Virtio-net does. This is accomplished by outsourcing parts of QEMU's job for managing virtqueues in the user space of host OS to a vHost driver in the host OS kernel. QEMU is still needed for the control plane of the communication, but the data plane, which moves data from the guest OS driver to the real Ethernet card, is completely implemented by the vHost driver. This is possible because the host OS kernel can directly access the virtqueue entries that are output from the paravirtualized guest-OS device-driver, since the VM is a host OS process. This skips QEMU as intermediate stage for every packet. Therefore, Internet-based inter-VM communication can benefit from vHost-net, although TCP/IP is still engaged.

2.4. Virtio-vsock Server

Virtio-vsock is from the user's point of view a simple Linux socket. However, it appears as the same socket in two different operating systems, which is remarkable. One end of this socket is in guest OS, the other in host OS. This is, of course, not directly possible, but requires the support of that QEMU that is responsible for the guest OS. Virtio-vsock thus provides for a communication between guest OS and host OS. Two Virtio-vsocks connected in a row allow for an inter-VM communication with the host OS as intermediate station. The two big advantages of Virtio-vsock are that TCP/IP is not engaged, and that the emulation of two vNICs is also not needed. It allows also inter-VM communication for VMs without vNICs. Virtio-vsock is a newer QEMU feature, but, as every socket, it implies data buffering. Therefore, ivshmem is superior to it with respect to latency.

2.5. Ivshmem

Ivshmem is a virtual PCI device in a guest OS which is emulated by KVM/QEMU. It establishes a Linux/POSIX shared memory (SHM) between the VM and its host OS. Ivshmem enables zero-copy VM-to-Host communication and vice versa, which is very efficient with respect to bandwidth and latency, because no internal data buffer exists. Ivshmem can also be used for inter-VM communication with the host OS SHM as intermediate step. Ivshmem is implemented by mapping its virtual PCI device memory to the host OS SHM. This is possible, because the memory is emulated by QEMU as a data structure inside of itself, and because multiple QEMUs can communicate with each other in host OS, as depicted in figure 1.

2.6. MVAPICH2-virt

Among other features, MVAPICH2-virt implements ivshmem. However, it does not provide for any access-synchronization for it, because, in contrast to our ivshmem remake, it does not make usage of the ivshmem-server. If mutual exclusion in SHM access is needed, then using ivshmem in MVAPICH2-virt means engaging TCP/IP, together with a monitoring protocol in order to emulate that feature. This is still beneficial compared to pure TCP/IP data exchange, but inferior to ivshmem with blocking read synchronization.

3. Our adaption of ivshmem to mpich

Ivshmem was supported for a few years by libvirt and QEMU as a virtual PCI device that allowed for shared memory (SHM) between guests and host. However, this shared memory existed without synchronization for SHM read/write accesses, which was insufficient for most applications. Therefore, a complementary synchronisation system, called Shared-Memory-Server (SMS) was created later, which became part of QEMU as its ivshmem-server. This server could send interrupts to VMs on the same machine, thus allowing unblocking of an application in a target VM, which was waiting for data from an application in a source VM. Mutual blocking and unblocking was used for SHM access synchronization without consuming CPU cycles (no spinlock), which is indispensable for HPC. However, this valuable feature could only be implemented if both applications were made as the user parts of Linux user-IO device-drivers (uio). Additionally, it was the task of the user to write, by means of code examples of both, the kernel and user part of the uio driver and to integrate his application into this driver. This was a not applicable for HPC with MPI, because MPI processes are never drivers. The reason for ivshmem being limited to uio device drivers was that the one and only possible spinlock-free synchronization between VMs in Linux is possible by means of a blocking read in the user part of uio. As soon as an interrupt arrives at a VM, this user part is called as an interrupt service routine and the blocking read at the target VM is unblocked. Unfortunately, with newer Linux versions, unblocking did not work anymore, with the consequence that ivshmem was again nearly useless because of missing SHM-access synchronization. Although a patch was suggested by the creator of ivshmem, the situation was still difficult, because meanwhile libvirt has also developed further and did not cooperate well any more with ivshmem and the ivshmem-server in QEMU. Finally, no proper documentation existed for ivshmem, which complicated the situation. Our contribution is to patch codes for both, the kernel and the user part of uio device drivers for ivshmem and to find proper version and configuration matches for Linux, QEMU and libvirt so that everything works again. We have also created a documentation and delivered an example usage of ivshmem for HPC by creating a performance test-code based on the integration of ivshmem into MPICH, together with a new SHM communication and synchronization channel inside of MPICH, thereby replacing the inefficient TCP/IP inter-VM communication. In particular, we achieved ivshmem to run for the first time with Ubuntu version 14.04, QEMU version 2.9.50, libvirt version 2.0 and MPICH version 3.2.

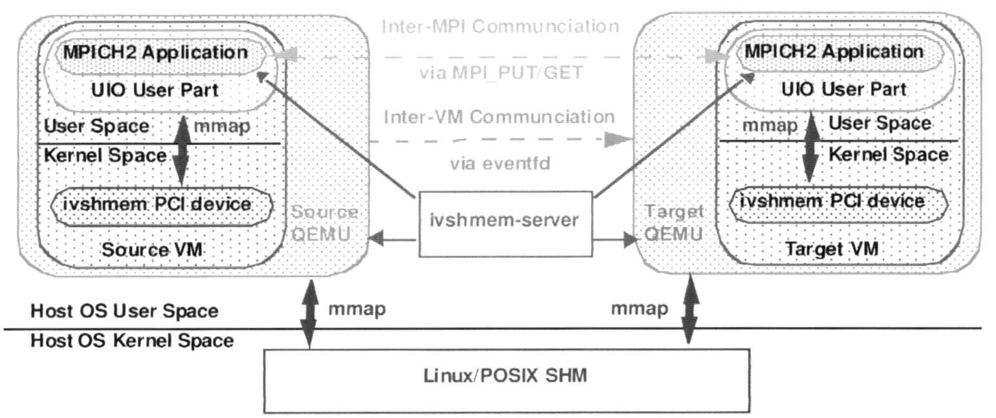

Figure 1. Architecture of ivshmem. The address space of the MPI guest-OS application is first "mmapped" to ivshmem virtual PCI-memory and then mapped to Linux/POSIX SHM.

The state that is reached by our work is as follows: we know and documented publicly how to integrate MPICH-based HPC processes to be the user part of uio device drivers: before running the HPC processes, a uio-ivshmem driver patched by us must be loaded into the kernel, together with a proper XML configuration file that supports the so-called doorbell function of ivshmem. Then, it is possible again to use it with blocking read as 5 years ago. In detail: two inter-process communication (IPC) mechanisms are needed inside of the ivshmem-server: 1.) an IPC between the two QEMU instances that are responsible for the source and target VMs. 2.) An IPC between the ivshmem-server and the HPC applications. From the viewpoint of host OS, all codes are user-mode host-OS processes, which simplifies IPC. For the inter-QEMU IPC, Linux eventfds are used. For the IPC between the ivshmem-server and the HPC applications, sockets are engaged. As soon as a QEMU instance receives an eventfd from another QEMU, it raises an interrupt in its VM with the help of KVM. Additionally, we have engaged Message-Signalled-Interrupts (MSI) in the ivshmem virtual PCI device, thereby providing for the user a numeric value as a user parameter. This requires a proper XML-input configuration-file for libvirt. Finally, in order to integrate ivshmem into MPICH, a custom MPICH was developed that replaces the standard MPI one-sided MPI_Put() call by a wrapper of the same name. If MPI_PUT is called by a HPC code, then our wrapper is executed, which makes all MPICH changes fully transparent to the user and its HPC code. The patched user parts of the uio drivers have two important tasks: 1.) they are mapping ivshmem PCI registers into the address space of the user application, so that the modified MPICH can access them. 2.) They are sending and receiving interrupts to the VMs which in turn call the MPICH application processes. The blocking read () is performed at the target VM on a device file. The PCI-device register-mapping into guest-OS user-space is accomplished by mmap() and ensures a simple SHM access from the HPC code. There are registers and sets in the ivshmem PCI device as follows: 1.) a PCI-BAR0 base-address register which points to a register set, including the so-called doorbell register. 2) A PCI BAR2 register, that points to the SHM. Unblocking a target HPC application process is simply accomplished by writing in our MPICH code an appropriate value to the ivshmem doorbell register, which is accessible by the application and by MPICH.

4. Performance measurements

Performance measurements were accomplished in the same server with varying block sizes and by means of the following three scenarios: a) an inter-VM communication with MPI_PUT for one-sided data exchange and with MPI_WIN_LOCK for passive remote memory access (RMA). This uses the standard MPICH Nemesis-sock channel and results in standard TCP/IP communication. b) An inter-VM communication with MPI_PUT and MPI_WIN_LOCK via ivshmem and SHM synchronization. This required the integration of ivshmem in MPICH. c) A host-OS communication with the standard MPICH via the SHM Nemesis channel. The third scenario was used as a reference for a) and b). Results for the elapsed times are depicted in figure 2, and for bandwidths in figure 3.

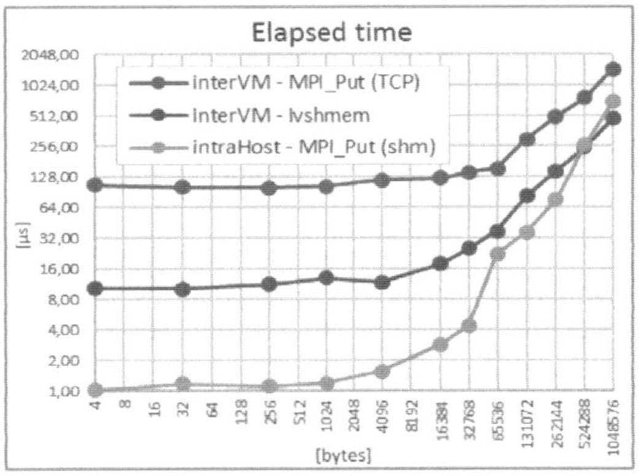

Figure 2. Measurements of elapsed time for various message sizes

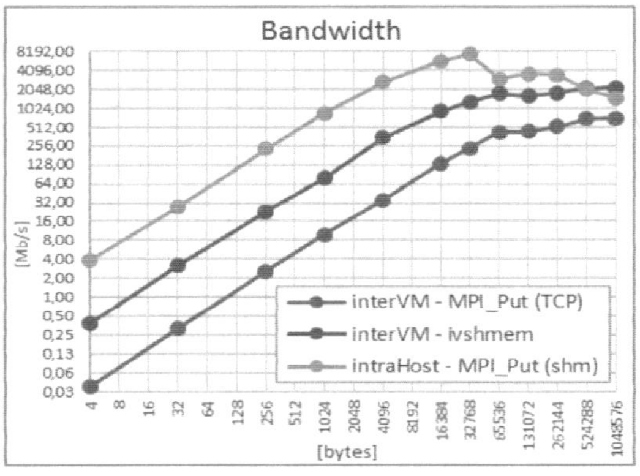

Figure 3. Results of calculated bandwidth for various message sizes

The elapsed time is calculated at the sender by dividing by two the time difference between the first byte sent and the reception of a notification from the receiver about transaction completion, with SHM synchronisation included. Bandwidth is calculated by dividing the elapsed time by the respective message sizes. Based on these measurements, we could observe four phases in figure 2 and figure 3: I.) Phase 1 for 0 to 4 kB block lengths for the transferred message: in this phase, we can see that the SHM synchronisation time is dominating the data transfer time, because the elapsed time does not increase with an increased message size. Furthermore, the amortization costs for SHM synchronisation are decreasing with increased message size. Finally, ivshmem surpasses TCP/IP by a factor of up to ten in both, bandwidth and elapsed time, respectively. II.) Phase 2 for 4-64 kB block lengths: in this phase, we can see that the elapsed time increases with the message size, which is visible at the ivshmem curve. The maximum reachable bandwidth is approximately 7.4 GB/s at a block length of 32 KB. Also for TCP/IP, a maximum value for its linear slope is reached. The reason for the latter is that only one TCP/IP packet is needed for all message sizes. Furthermore, the curve for the elapsed time is sharply changing its slope angle. III.) Phase 3 for 64-256 kB block lengths: in this phase, we can see that a transmission saturation is reached at about 256 KB block length with data rates of 2,1 and 0,25 GBs/s. Here, ivshmem surpasses TCP/IP by a factor of four. IV.) Phase 4 for block lengths 512kB: we can see a declining reference curve, while ivshmem becomes the fastest communication. This is a remarkable result, because it means that virtual communication can even be faster than real communication for very large block sizes. The reason for that surprising behaviour is that ivshmem does not make any time-consuming system-call, in contrast to host-OS applications which are coupled by SHM. Ivshmem operates fully in user space because of mmap and the blocking-read feature, which is in the user part and thus in the user space of uio, although the SHM is physically located in host-OS system-space. This is a highly efficient method and because of that, ivshmem is second to none in HPC. As a result, the performance difference between TCP and ivshmem in phase IV is a factor of three with respect to bandwidth, although TCP implementations are highly efficient as well for large message sizes. Finally, it must be mentioned that the block length for the peak ivshmem bandwidth is also influenced by the chosen synchronization and communication type. For example, with active RMA, MPI Start-Post-Wait-Complete and point-to-point MPI_Send, the maximum bandwidth is obtained at a message size of 256kB, resulting in an ivshmem bandwidth benefit four times higher than TCP/IP.

5. Conclusions

We have fixed non-spinlock access synchronization in ivshmem shared memory, which is useful for communication between virtual machines in the same server. Ivshmem is not TCP/IP based, but can replace it, thus delivering a much better performance. Additionally, we made ivshmem runnable with contemporary versions of Linux, QEMU, libvirt and MPICH. Furthermore, an up-to-date documentation for it was provided, including "How-Tos". Finally, MPICH was adapted by us to our ivshmem remake, and we demonstrated its usefulness by an example HPC code. The MPICH adaptation transparently replaces the MPI one-sided MPI_Put() call, which is frequently used in HPC shared-memory codes. Finally, we have shown by performance measurements, that ivshmem is superior for all message lengths to that MPICH communication type, which is based on a TCP/IP channel. Without our MPICH adaptation, it cannot be avoided that TCP/IP will be used by MPICH for inter-VM communication on the same server, which is useless for HPC. The performance benefit with ivshmem is between factors of 3-10 compared to TCP/IP. For large messages (≥ 512kB), ivshmem even surpasses non-virtualized host-OS Linux/POSIX SHM performance. This means that inter-VM communication can be faster than non-virtualized SHM data transfers for some cases.

References

[1] Gropp W, Lusk E and Thakur R 2014 *Using MPI: Portable Parallel Programming with the Message-Passing Interface, Third Edition* (MIT Press)
[2] Amer A et al. 2015 MPICH User's Guide, Version 3.2 (Chicago: Argonne National Laboratory)
[3] MacDonell C 2011 Shared-Memory Optimizations for Virtual Machines (Edmonton: University of Alberta)
[4] Tsirkin M, Huck C and Moll P 2013 Virtual I/O Device (VIRTIO) Version 1.0 (OASIS)
[5] Lan S 2016 vHost User Multiqueue Configuration and Use (Intel)
[6] Panda D 2017 MVAPICH2-Virt 2.2 (NOWLAB, Columbus: The Ohio State University)
[7] Weil S 2017 QEMU version 2.10.0 User Documentation (QEMU, www.qemu.org)
[8] Kutch P, Johnson B 2017 SR-IOV for NFV Solutions. Practical Considerations and Thoughts. (Intel, Networking Division)

Research on Collaborative Technology in Distributed Virtual Reality System

ZhenJiang Lei[1], JiJie Huang[2], Zhao Li[1], Lei Wang[1], JiSheng Cui[1] and Zhi Tang[1]

[1] State Grid Liaoning Electric Power Co., Ltd. Jinzhou power supply company
[2] Beijing Kedong Electric Control System Co. Ltd., Beijing, China

Email:*chxhjjco-p@163.com; leizhenjiang@sina.com,

Abstract. Distributed virtual reality technology applied to the joint training simulation needs the CSCW (Computer Supported Cooperative Work) terminal multicast technology to display and the HLA (high-level architecture) technology to ensure the temporal and spatial consistency of the simulation, in order to achieve collaborative display and collaborative computing. In this paper, the CSCW's terminal multicast technology has been used to modify and expand the implementation framework of HLA. During the simulation initialization period ,this paper has used the HLA statement and object management service interface to establish and manage the CSCW network topology, and used the HLA data filtering mechanism for each federal member to establish the corresponding Mesh tree. During the simulation running period, this paper has added a new thread for the RTI and the CSCW real-time multicast interactive technology into the RTI, so that the RTI can also use the window message mechanism to notify the application update the display screen. Through many applications of submerged simulation training in substation under the operation of large power grid, it is shown that this paper has achieved satisfactory training effect on the collaborative technology used in distributed virtual reality simulation.

1. Introduction

In order to improve the immersion sense, distributed virtual reality technology applied to the joint training simulation not only needs real-time uploading target motion data and uses CSCW (Computer Supported Cooperative Work) terminal multicast technology to display, but also takes into account the engineering simulation (HLA / RTI) technology to ensure the temporal and spatial consistency of the simulation. Because HLA has been widely used in joint training, distributed virtual reality development is required using of CSCW's technology to develop real-time RTIs under the overall architecture of HLA and enabling the RTI to support both collaborative computing and collaborative display.

Real-time RTI development technology in the field of high-performance simulation has been made a lot of research, for example, using shared memory technology[1] to improve the real-time communication RTI performance or building an infiniband network and using direct memory access (RDMA, Remote Direct Memory Access) communication mechanism[2] to optimize the RTI. These studies replace the underlying communication code of RTI using shared memory or memory mapping, but the joint training simulation is done in a TCP / IP-based LAN environment, which requires extending the underlying communication code of RTI, instead of replacing it.

Based on the virtual reality technology, this paper discusses how to combine CSCW and HLA / RTI technology to improve the fidelity of multi-person collaborative training simulation in 3D virtual scene.

2. Virtual reality application background

Based on the virtual reality of the transformer equipment operation and maintenance simulation training system application architecture shown in Figure 1, including substation equipment, virtual reality (VR) helmet and the grid computing server. Each substation equipment, the grid computing program, trainer with a VR helmet and the teacher constitutes a federated member of HLA [3], There are many visualization model components of electrical equipment in a substation, and to achieve immersive real-time rendering of virtual reality, at least to ensure that the graphics refresh rate of not less than 60 frames / sec, so in order to achieve real-time display of three-dimensional scenes need to use real-time communication technology.

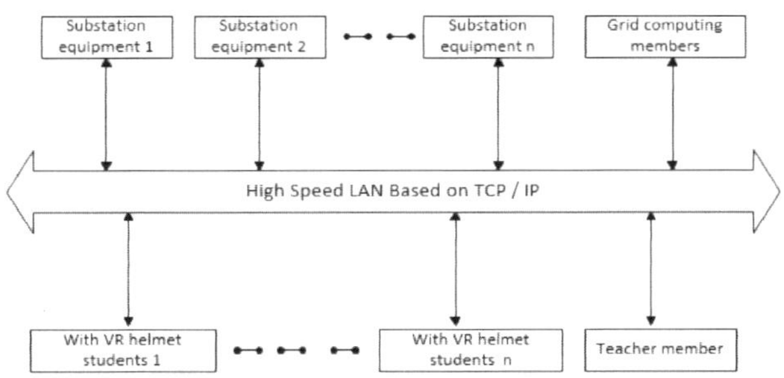

Figure 1. VR-based operation and maintenance training structure of substation equipment

The current high-speed Ethernet up to 1GbE or even 10GbE, need to reduce the delay from the target to choose, and the delay related to the two issues are transmission and serialization, but in the delay time serialization time only a small part. The huge delay of the connection layer is mainly introduced by the interface and the system at each end of the connection. For a particular package, it is 135 microseconds through a 1GbE connection, and it can only fall to 75 microseconds through a 10GbE connection, and most of the other costs are from the device in both ends, not from the connection itself.

In this paper, a 1GbE network card and network cable are used to build high-speed Ethernet, and the speed of the two-way transmission between the two nodes is 75 microseconds, which is higher than the speed measured in the literature[2] with high-speed RTI to update the object properties. So it can use CSCW multicast technology to real-time communication.

3. CSCW real-time communication mode

CSCW's three elements are cooperation, coordination and communication, CSCW cooperation mode from time and space, can be divided into synchronous mode, distributed synchronization mode, asynchronous mode and distributed asynchronous mode[4][5]. Among them, the synchronous mode system is used for the local group (LAN) between the face-to-face real-time collaborative operating system, which is required to achieve "You see that I see [6] [7]. CSCW in the collaborative operation will inevitably encounter multiple users to operate on the same object, so how to ensure that these operations are necessary to orderly implementation is necessary, at home and abroad for collaborative editing, collaborative design, web conferencing and other collaboration Concurrent control on the work system has been studied. The concurrency control method of the shared object is divided into the following typical concurrency control method [8]: token control method, locking method, operation information serialization, concurrency control based on operation conversion.

CSCW's communication method adopts the terminal system multicast mode. The terminal system multicast is a practical multicast system developed by Carnegie Mellon University (CMU), including member management and packet copy. This transition from router-supported multicast to multicast supported by the end system can solve most of the problems of IP multicast (transmission unreliability). The multicast tree is the core structure of the terminal system multicast. Its creation is divided into two steps: first initialize a mesh (Mesh), so that each node in Mesh is connected; then create a spanning tree on Mesh. Spanning tree to the corresponding node on Mesh as the root node.

HLA uses the LBTS algorithm to achieve the federal members of the time and space consistency, to ensure the correctness of the simulation logic, but at the expense of the real-time simulation. Because the realization of LBTS algorithm can only use a centralized star network topology. Obviously, the star library topology, which does not communicate with each other in the RTI library components, increases traffic and reduces the rate of information exchange among members. The CSCW terminal multicast mode is introduced for this purpose, and the above coordination problem faced by CSCW can be solved by the time management of HLA. The CSCW multicast communication used in the training simulation only completes the unsupervised and restricted message sending and receiving of HLA, Extended RTI communication as shown in Figure 2:

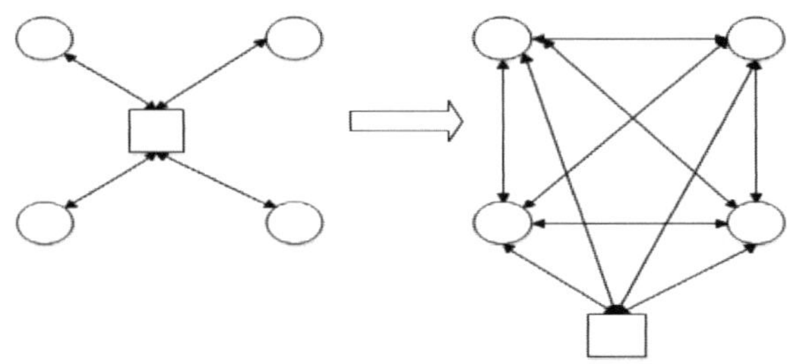

Figure 2. Terminal multicast extension RTI communication architecture

Where the box represents the service process of the RTI, and the circle represents the libRTI thread called by the federated member. The left star topology in the graph preserves the function of all six categories of HLA / RTI services, while the mesh communication topology (Mesh) between the four new federated members on the right shows the terminal multicast function of CSCW Federated members interact with each other's three-dimensional drawing information, which does not affect the temporal and spatial consistency of the simulation event, but only to achieve the co-simulation system visual "you see that I see", used to improve the virtual reality environment immersive fidelity.

4. RTI and CSCW communication topology

The open source CERTI is modified to extend the libRTI thread using the HLA service supported by RTI which adds the CSCW terminal multicast function. The libRTI library of the CERTI contains two interface of the RTI member ambassador and RTI callback ambassador. The application uses the RTI callback ambassador interface to define their own function, which can be queried and executed by calling the RTI member ambassador interface tick or by responding the message sent by the libRTI thread. The tick interface implementation includes the realization of time management that meets the space-time consistency of the callback event and implements the corresponding callback interface using the LBTS algorithm. The transmission of real-time data of interest in this article is executed by responding to a message from the libRTI thread.

The first step is to establish a dynamic terminal multicast network, including the establishment and deletion of member multicast links, which can be implemented through the subscription and distribution functions of RTI's declaration management and object management. When a federate member joins the federation, it declares the subscription and publish of the cosimNetCard object class, which contains the two attributes of cardIP and cardPort (Attribute cardPate reliable timestamp) in the FOM. Then each member of the union calls the registerObjectInstance interface, which second object name parameter uses the string of "FedName: cardIP: cardPort" in which the FedName indicates the member name, the cardIP indicates the member network IP address, and the cardPort indicates the member network port.

Each federate member judges whether the object class of the object is cosimNetHandle in the RTI callback interface of the object, and if its second parameter is a string of "FedName: cardIP: cardPort" in the parsing call parameter, then this federate member can monitor real-time data communication to another federate member by using a link tag which is created when this federate member initiates the connection using the network address and port. The link tag is a [FedName, cardIP, cardPort, linkID] quaternion which is joined the local queue (gradually forming Mesh connectivity map), and its linkID or FedName can be used as a keyword to find this quaternion. When another federate member named as FedName calls DeleteObjectInstance, the callback interface of this federate member, which first parameter uses the above "FedName: cardIP: cardPort" string, can eliminate the corresponding quaternion in the queue indexed by the FedName.

When a federate member sends real-time data, the data is packetized and sent to each linkID in the quaternion queue (which can also filter out the actual recipients to form a Mesh spanning tree according to the application). In addition to monitoring the network events from the RTI service and the interface calls from the local RTI federation side, the network port for real-time data communication is monitored, as shown in Figure 3 below. When the real-time communication data is monitored, the libRTI thread sends a Windows message to the display window, informs the display window to receive the message and updates the display.

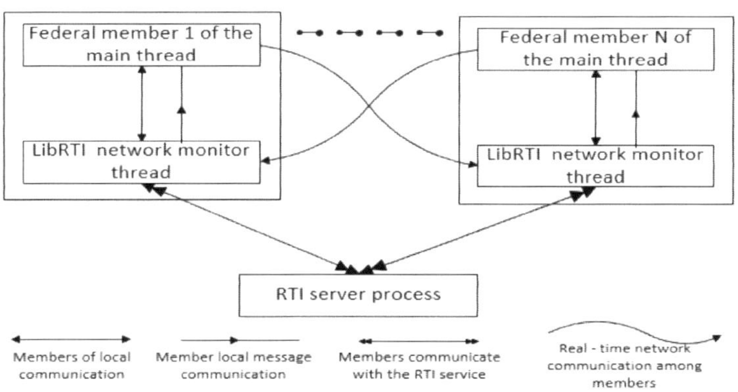

Figure 3. The terminal multicast extends the communication channel of the RTI component

Federate members have established a networked communication link according to the application requirements to build real-time terminal multicast queue, and a Mesh spanning tree to filter data according to meet the conditions of the members in order to reduce traffic and improve execution efficiency.

5. Constructing an experiment using the terminal multicast

Based on the virtual reality technology, interactive training includes three-dimensional projection, three-dimensional arc curtain, image virtual reality helmet, action capture, simulation workstations and network switches, using virtual reality helmet and three-dimensional arc curtain projection two immersive stereoscopic display technology.

The experiment supports multiple people collaborative work to complete a task. Multiple trainees can be different roles to log into the training scene, team work together to complete a multi-person participation in the task, as shown in Figure 4. When the trainees operate, all the federate members, including the large grid simulation program, are still moving forward in a fixed simulation step, and the propulsion strategy is both limited and controlled; both the federates and the faculty members are listening to the libRTI Threads which uses the windows message mechanism to send message of the movement and operation, and is responsible for updating the display screen in the display screen.

6. Conclusion

The introduction of virtual reality technology in distributed interactive simulation based on HLA / RTI can increase the visualization effect of simulation, but with the fidelity requirement of user experience, the realism of RTI network transmission is put forward higher requirements. It is necessary to divide the transmitted data into the virtual scene interactive data and the data supporting the simulation of time and space consistency, which can be realized with the original RTI of the six categories of services. The best method to exchange scene interactive data is the directly transferring between federate members. In this paper, CSCW's terminal multicast mechanism is introduced in the federate component libRTI of the open source CERTI to realize the direct transmission of the motion and operation data between the trainers, and the real-time demand of the scene data interaction has been meet. From the last two years for the operation of large-scale simulation training applications, it is shown that this paper has achieved satisfactory training effect on the collaborative technology used in distributed virtual reality simulation.

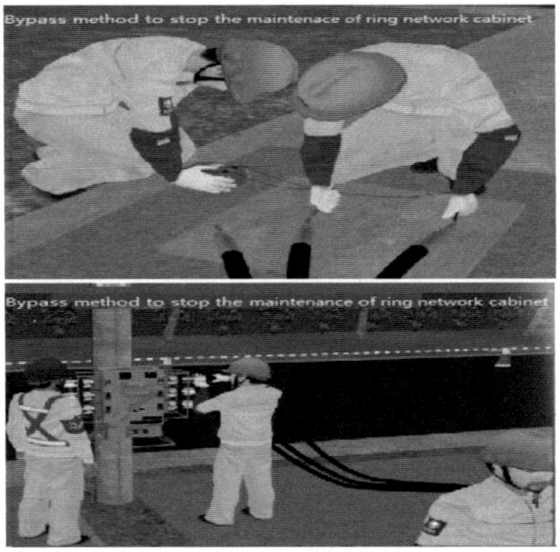

Figure 4. Simulation Training of Transformer Operation Based on Extended

References

[1] Zhang zhihui, Li bohu1,Chai xudong, Huang jijie and Hou baocun 2014 HP-HLA/RTI Prototype Oriented on Shared Memory Environmen.*Journal of System Simulation* **26L2** 315-22.

[2] Xing chi,Li bohu 2016 Research on RTI Communication Mechanism on Infiniband Network Architecture ACTA ELECTRONICA SINICA 44L2 327-33.

[3] Li weiqing, Wu huizhong, Lin changnian 2006 Research on Virtual Environment of Substation Training Simulator Journal of System Simulation 18S1 123-126.

[4] Penichet V.M.R., Marin I. and Gallud J.A 2007 J.A Classification Method for CSCW Systems Electronic Notes in Theoretical Computer Science 168L8 237-47.

[5] Zhao Jianmin and Long Xiaochun 2011 A Modified Model for Flexible Workflow Access Control IEEE Computational Intelligence and Design 2011L2 279-81.

[6] Elmarzouqi.N and Garcia.E. ACCM 2007 a New Architecture Model for CSCW In: Proceedings of the 2007 11th International Conference on Computer Supported Cooperative Work in Design 2007 84-91.

[7] Convertino G, Billman G and Pirolli P 2008 The CACHE study: group effects in computer-supported collaborative analysis Computer Supported Cooperative Work (CSCW) 17L4 353-93.

[8] Khan SM,Sulaiman M and Tahir A M 2011 Domain-based classification of CSCW systems Research Journal of Applied Sciences 3L11 438-41.

Pollen Image Recognition Based on DGDB-LBP Descriptor

L P Han[1, *] and Y H Xie[1]

[1]College of computer and software, Nanjing University of Information Science and Technology, Nanjing 210044, China

*E-mail: hanliping93@163.com

Abstract. In this paper, we propose DGDB-LBP, a local binary pattern descriptor based on the pixel blocks in the dominant gradient direction. Differing from traditional LBP and its variants, DGDB-LBP encodes by comparing the main gradient magnitude of each block rather than the single pixel value or the average of pixel blocks, in doing so, it reduces the influence of noise on pollen images and eliminates redundant and non-informative features. In order to fully describe the texture features of pollen images and analyze it under multi-scales, we propose a new sampling strategy, which uses three types of operators to extract the radial, angular and multiple texture features under different scales. Considering that the pollen images have some degree of rotation under the microscope, we propose the adaptive encoding direction, which is determined by the texture distribution of local region. Experimental results on the Pollenmonitor dataset show that the average correct recognition rate of our method is superior to other pollen recognition methods in recent years.

1. Introduction

The classification and identification of pollen particle has a wide range of applications in allergic pollen index forecast, drug development, paleoclimatic reconstruction and other fields. At present, pollen identification is mainly done by artificial inspection under microscopy, which requires the operator to have a rich knowledge of sporopollen morphology. The identification process is time consuming and laborious, and the identification result is susceptible to subjective opinions, resulting in a relatively low accuracy. Since pollen images under microscope have similar contours, textures and structural features as common images, it is an effective way to extract pollen features and classify them automatically by computer.

Since Flenley suggested the use of algorithms to classify and recognize pollen in palynology in 1968, plenty of feature extraction and recognition algorithms for pollen images have been proposed. Texture, as an effective pollen image feature, has been extensively studied and focused. For example, E Cernadas et al.[1] proposed a texture and shape based descriptor (TSD), which obtained 89% recognition rate, however, the recognition performance partly depended on the type of pollen and the universality is not high. DS Guru et al.[2] proposed a surface texture based pollen image classification model, using Gabor wavelet, local binary patterns (LBP), gray-level co-occurrence matrix (GLCM) and gray-level difference matrix (GLDM) to extract the texture features of polllen. The combination of these features is used for identifying pollen, and the recognition rate is 91.66%, whereas, the computational cost of this method is high and the scale of the extracted feature is single. JV Marcos et al.[3] extracted the texture features of pollen images by combining gray-level co-occurrence matrix (GLCM), Log-Gabor filter (LGF), local binary patterns (LBP) and discrete orthogonal moments (DTM), and obtained 94.83% recognition rate. However, the method is time-consuming and the extracted feature has a large amount of redundant information. A Daood et al.[4] decomposed the pollen image into multiple layers (Multi-Layer Feature Decomposition), extracting the texture features and geometric features of each layer, respectively, which obtained 86.94% recognition rate, but this method is not robust to the rotation of pollen images.

In this paper, we improve the sampling strategy and encoding scheme of tradional LBP, and propose a new local binary pattern descriptor based on the pixel blocks in the dominant gradient direction. Differing from the traditional LBP and its variants, we calculate the gradient amplitude of pollen images in the dominant gradient direction, which avoids the influence of noise on the extracted feature [5] and eliminates some redundant and non-informative features. In order to fully describe the texture features of pollen images and analyze it under the multi-scales, we propose multi-directional sampling strategy, which uses three types of operators to extract the radial, angular and multiple texture features under different scales. Considering that the pollen images have some degree of rotation under the microscope, we propose the adaptive encoding direction, which is determined by the texture distribution of each local region. We calculate the adaptive encoding direction and set it as initial coded point for binary encoding, which enhances the feature descriptor's robustness to the rotation of pollen images.

2. DGDB-LBP descriptor

In view of the shortcomings of traditional LBP and its variants, we improve the sampling strategy and coding scheme, as shown in Figure 1, the specific steps are as follows:

2.1. *Multi-directional sampling*

2.1.1. Calculate the dominant gradient direction. We first calculate the gradient of all pixels in the image, then the gradient angle of each pixel is rounded down, so that the gradient direction of the whole image is discretized into eight directions. The gradient magnitude of all pixels is weighted to eight discrete directions, taking the longest for the dominant gradient direction of the image.

2.1.2. Calculate the gradient magnitude of the pixel block. We accumulate the gradient magnitude of all the pixels in the dominant gradient direction as the gradient magnitude of the pixel block, which is denoted as $G\left(B_{r,P,m,i}\right)$. $B_{r,P,m,i}$ is the pixel block on the circle with radius r and sampling points P, i is the serial number of pixel blocks with size $m \times m$, and $G\left(B\right)$ is the accumulation of the dominant gradient magnitude on pixel block B.

2.1.3. Extract radial, angular and multiple texture features. In order to obtain more texture information, we also extract the radial texture feature, angular texture feature and multiple texture feature of pollen image. The three operators for extracting different texture features are named DGDBLBP_RD, DGDBLBP_AD and DGDBLBP_MD, respectively.

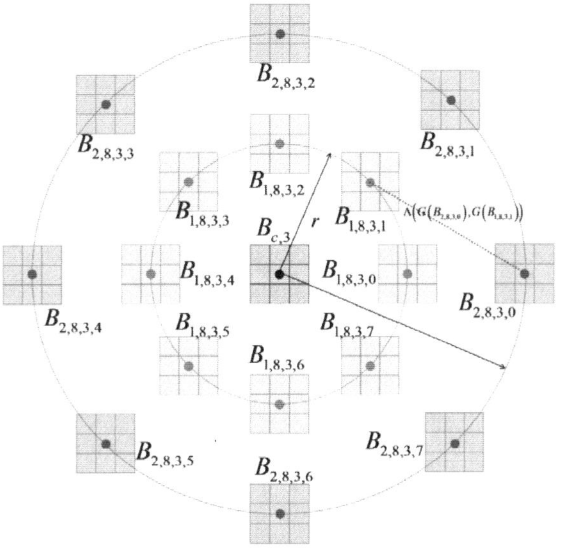

Figure 1. The sampling figure of DGDB-LBP.

2.2. *Adaptive encoding*

2.2.1. DGDBLBP_RD. The adaptive coding direction of DGDBLBP_RD is defined as the symbol of circular adjacent pixel block with the largest gradient difference from the center pixel block, which is denoted as D_R. The specific formula of radial texture feature is as follows:

$$DGDBLBP\_RD(x_c) = \sum_{i=0}^{P-1} S\left(G\left(B_{r,P,m,i}\right) - G\left(B_{c,m}\right)\right) 2^{\mathrm{mod}(i-D_R+P,P)} \tag{1}$$

$$D_R = \underset{i\in(0,1,\cdots,P-1)}{\arg\max} \left|G\left(B_{r,P,m,i}\right) - G\left(B_{c,m}\right)\right| \tag{2}$$

Where $S(x)$ is a symbolic function, which is defined as follows:

$$S(x) = \begin{cases} 0 & x<0 \\ 1 & x\geq 0 \end{cases} \tag{3}$$

2.2.2. DGDBLBP_AD. The adaptive coding direction of DGDBLBP_AD is defined as the symbol of pixel block with the largest gradient difference from the adjacent pixel blocks on the circular neighborhood, which is denoted as D_A. The specific formula of angular texture feature is as follows:

$$DGDBLBP\_AD_{r,P}(x_c) = \sum_{i=0}^{P-1} S\left(2G\left(B_{r,P,m,i}\right) - \left(G\left(B_{r,P,m,\mathrm{mod}(i+P-1,P)}\right) + G\left(B_{r,P,m,\mathrm{mod}(i+P+1,P)}\right)\right)\right) 2^{\mathrm{mod}(i-D_A+P,P)} \tag{4}$$

$$D_A = \underset{i\in(0,1,\cdots,P-1)}{\arg\max} \left|2G\left(B_{r,P,m,i}\right) - \left(G\left(B_{r,P,m,\mathrm{mod}(i+P-1,P)}\right) + G\left(B_{r,P,m,\mathrm{mod}(i+P+1,P)}\right)\right)\right| \tag{5}$$

2.2.3. DGDBLBP_MD. The adaptive coding direction of DGDBLBP_MD, we record as D_M, is defined as the symbol of pixel block on the outer ring that maximizes the gradient difference between the pixel block pairs on the outer ring and inner ring. The specific formula of multiple texture feature is as follows:

$$DGDBLBP\_MD_{r,r+1,P}\left(x_c\right)=\sum_{i=0}^{P-1}S\left(\Delta\left(G\left(B_{r+1,P,m,i}\right),G\left(B_{r,P,m,\mathrm{mod}(i+1,P)}\right)\right)-\Delta\left(G\left(B_{r+1,P,m,\mathrm{mod}(i+3,P)}\right),G\left(B_{r,P,m,\mathrm{mod}(i+4,P)}\right)\right)\right)2^{\mathrm{mod}(i-D_M+P,P)} \tag{6}$$

$$D_M=\underset{i\in(0,1,\cdots,P-1)}{\arg\max}\left|\Delta\left(G\left(B_{r+1,P,m,i}\right),G\left(B_{r,P,m,\mathrm{mod}(i+1,P)}\right)\right)\right| \tag{7}$$

3. Experimental results and analysis

We performed the experiments on the Pollenmonitor dataset with 22750 pollen grains from 33 taxa, in which the images were automatically collected, prepared and recorded by the first prototype of the pollenmonitor in Europe in 2006. The Correct Recognition Rate (CRR), False Recognition Rate (FRR) and Recognition Time (RT) were used to measure the recognition performance of the algorithm. Figure 2 shows the results of our method for classifying six representative pollen images on the Pollenmonitor dataset. It can be seen from the experimental results that most of the unpolluted and undeformed pollen images with rich texture information can be correctly classified and identified. In order to verify the effectiveness of the proposed method, we compare the DGDB-LBP with four recent pollen recognition methods. Table 1 shows the experimental results of our method and the other four methods on the Pollenmonitor dataset. It can be seen that the correct recognition rate of our method is higher than that of other four methods on the Pollenmonitor dataset, and the computational efficiency is higher than that of most methods.

Table 1. The results of our method and other methods.

Dataset	Method	CRR/%	RT/s
Pollenmonitor	TSD	84.00	17.3
	Combined Gabor-LBP-GLCM-GLDM	84.27	16.2
	Combined GLCM-LGF-LBP-DTM	89.75	18.1
	Multi-Layer Feature Decomposition	86.20	8.5
	DGDB-LBP	91.82	12.8

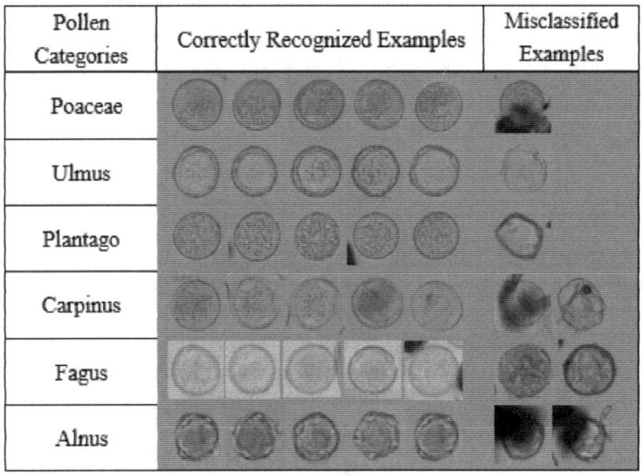

Figure 2. Experimental results of the Pollenmonitor dataset.

4. Conclusions

In this paper, we propose a local binary pattern descriptor based on the pixel blocks in the dominant gradient direction, which has a number of advantages. Firstly, the center pixel is quantified by comparing the gradient magnitude of each pixel block in the dominant gradient direction, which avoids saltations in the binary pattern caused by noise. Secondly, the analysis of pollen images under the multi-scales improves the discrimination of the descriptor. Thirdly, the adaptive encoding with reference to the texture features of adjacent regions enhances the descriptor's robustness to the rotation of pollen images.

References

[1] Rodriguez-Damian M, Cernadas E, Formella A, Fernandez-Delgado M and Sa-Otero PD. Automatic detection and classification of grains of pollen based on shape and texture 2006 *J. IEEE Transactions on Systems Man & Cybernetics*. 531-42

[2] Guru DS, Siddesha S and Manjunath S.Texture in classification of pollen grain images 2013 *J.Lecture Notes in Electrical Engineering*. 77–89

[3] Marcos JV, Nava R, Cristobal G, Redondo R, Escalante-Ramirez B, Bueno G, Deniz O, Gonzalez-Porto A, Pardo C, Chung F and Rodriguez T. Automated pollen identification using microscopic imaging and texture analysis 2015 *J.Micron*. 36–46

[4] Daood A, Ribeiro E and Bush M 2016 Pollen Grain Recognition Using Deep Learning *International Symposium on Visual Computing (Advances in Visual Computing vol 10072)* pp 321-30

[5] Ren J, Jiang X and Yuan J.Noise-resistant local binary pattern with an embedded error-correction mechanism 2013 *J.IEEE Transactions on Image Processing*. 4049-60

A No-Reference Image Sharpness Metric Based on Large-Scale Structure

Bingzhi Pei[1], **Xiaolin Liu**[1,*] **and Zicheng Feng**[1]

[1]College of Mechatronic Engineering and Automation, National University of Defense Technology, 109Deya Road, Changsha, China

*lxlchangsha@sina.com

Abstract. Most of no-reference image sharpness metrics suffer interference of tiny textures and the contrast around edges. Based on large-scale structure, this paper proposes a novel metric to avoid this problem. Firstly prominent edges are extracted by a weighted least-squares filter. Then, we calculate the statistics proportion of edges with different width and the mean value of the maximum gradients from the 51th to 250th as the features of the image. Finally, the support vector regression is used to get the relationship model between the features and the subjective assessment result. The metric can reduce the influence of textures, contrast and the content of images effectively, accordingly enhancing the assessment accuracy.

1. Introduction

In contrast to machines, assessing the sharpness of an image is easy for humans. However, subjective image quality assessment is easily interfered by evaluators and observation conditions. Besides, it is inefficient and can not be embedded in digital image processing systems. All of these shortcomings narrow their applicability. To tackle this challenge, full-reference (FR) quality assessment is considered. In practice, however, FR algorithms are only applicable when both sharp image and blurred image can be captured available at the same physical location and imaging conditions. Thus, no-reference objective image quality assessment is widely used.

In spatial domain, no-reference image sharpness metrics mainly focus on pixel intensity and edge width. Image sharpness can be estimated through the ratio of the grey difference between two edge ends and the edge width, as in [1]. But because the method is dependent on image contrast, it can not accurately evaluate the sharpness of different images. In [2], the notion of just noticeable blur (JNB) based on the characteristics of human visual system is introduced. On this basis, the authors in [3] propose the method of cumulative probability of blur detection (CPBD). The probability of detecting blur is calculated through the edge width and the JNB width, and the cumulative probability histogram is used to evaluate the sharpness of the image.

In frequency domain, metrics mainly associate sharpness with transform coefficients. In [4], the concept of local phase coherence (LPC) is proposed. For a sharp image, the phases of steerable pyramid coefficients exhibit a consistent relationship across scales, and the phases of the finest scale coefficients can be well predicted from the coarser scale coefficients. Therefore, image sharpness can be evaluated by the reduction of LPC. In [5], a generalized Gaussian distribution (GGD) is used to fit the distribution of the normalized steerable pyramid coefficients, and the model parameters can describe the image quality. The distribution of DCT coefficients can also be fitted by GGD as in [6]. In [7], the distribution of Fourier coefficients is similar to a negative exponential distribution. The combination of the exponential parameter and the total variation of the block represents image sharpness. However, due to the methods in [6] and [7] using overlapped blocks, they are time-consuming.

In most cases, edge width can represent image sharpness. The wider the edges, the more blurred the image. However, sometimes there is no obvious increase in edge width after an image is blurred, while gradients of the edge decrease. Besides, it's considered that textures and image content can interfere with evaluation accuracy.

To avoid these problems, a new metric based on large-scale structure is proposed. After extracting prominent edges by a weighted least-squares filter (WLS), we combine the statistics proportion of edges with different width and the mean value of the maximum gradients from the 51th to 250th to evaluate image sharpness. The flow chart of the proposed method is shown as figure 1. Denote the proportion of edges by $prop_i$, and the mean value by $GRAD_{in}$.

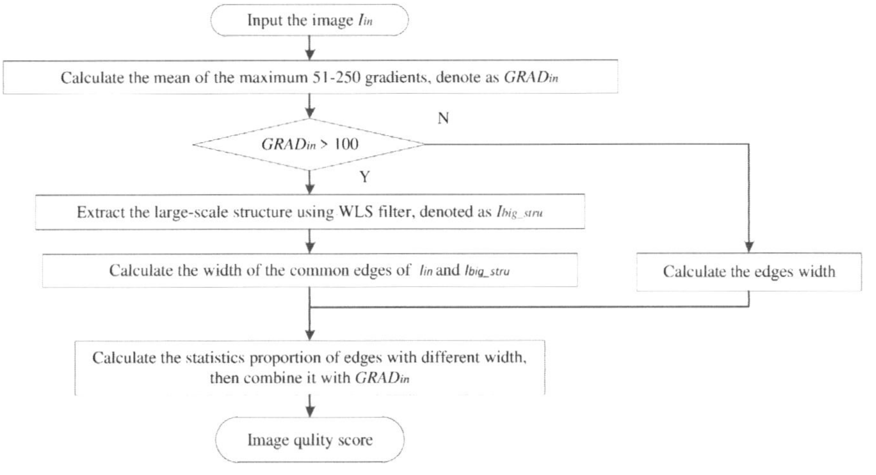

Figure 1. The flow chart of the proposed metric.

Extracting the large-scale structure reduces the interference of textures; calculating $prop_i$ reduces the interference of content; and adding $GRAD_{in}$ avoids the problem that sometimes there is no obvious increase in edge width after an image is blurred.

This paper is organized as follows. Section 2 mainly introduces the details of the proposed method. Performance results are presented in Section 3. A conclusion is given in Section 4.

2. No-reference sharpness metric

2.1. Edge width

In this paper, an edge point is defined as the pixel with the largest gradient value along the normal direction of the edge. On both sides of the edge point, when the pixel's gradient value is no longer decreasing or is the last one that is bigger than 10% of the edge point's gradient, the pixel is the end of the edge. Edge width is considered to be the distance between two ends. Figure 2 illustrates the width measurement by showing the variance of gradient values along the normal direction of two consecutive edges. The black diamonds represent the edge points, the red circles represent the edge ends, and the distances between the adjacent red circles represent the edge width. Sobel operator is chosen to calculate gradients.

When calculating the edge width, in order to avoid the interpolation operation in a certain direction, the normal direction is divided into 0° direction, 45° direction, 90° direction and 135° direction according to the phase of the gradient at the edge point.

In most cases, edge width can represent the sharpness of an image. The more blurred the image, the wider the edges, as shown in figure 3. In order to reduce the effect of the number of edges, this paper selects the statistics proportion of edges with different width as the sharpness feature of an image.

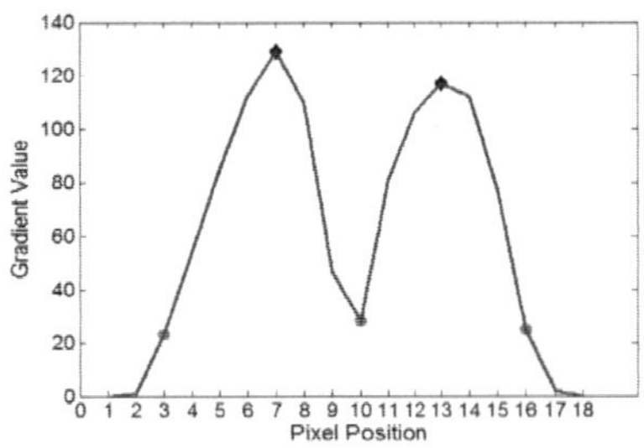

Figure 2. An illustration of the width measurement.

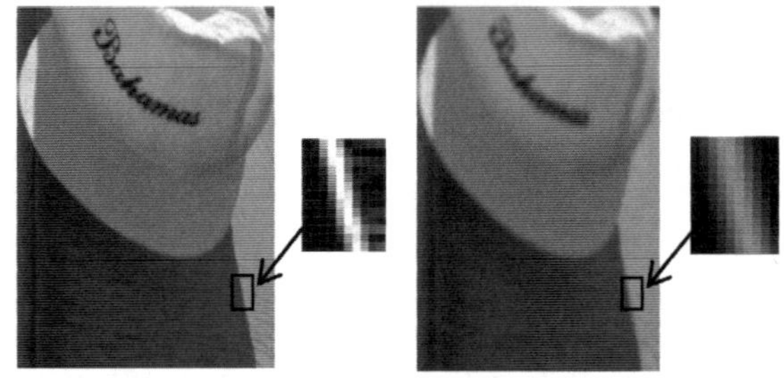

Figure 3. Edges become wider after the image is blurred.

There are so many tiny edges in texture area that will interfere with evaluation accuracy. As shown in figure 4(a) and figure 4(b), ripples on the surface of the lake bring about lots of edges, whose width are small and won't increase obviously after the image is blurred. And these edges get close to each other. In addition, humans are more concerned with the change of prominent edges while ignoring tiny textures. Therefore, WLS is used to extract the large-scale structure of an image and filter textures, as equation (1):

$$I_{big\_stru} = WLS * I_{in} \tag{1}$$

I_{in} represents the input image, and I_{big_stru} represents the large-scale structure. The result is shown as figure 4(c). Canny operator is chosen to extract edge points of I_{in} and I_{big_stru}. It can be found that edges of ripples are erased after WLS is used, as shown in figure 4(b) and figure 4(d).

(a) (b) (c) (d)

Figure 4. Large-scale structure and edge maps. (a) Blur image; (b) Edge map of the blur image; (c) Large-scale structure of the blur image; (d) Edge map of the large-scale structure.

For different images, in order to ensure that the extracted edge points are enough and their gradient values are relatively large, this paper chooses different Canny thresholds adaptively. $GRAD_{in}/3$ and $GRAD_{big\_edge}/3$ are selected as threshold. For large blurriness images whose $GRAD_{in}$ is less than 100, textures have been obscured. Thus edge width can be calculated directly in I_{in} without extracting the large-scale structure. And $3*GRAD_{in}/4$ is selected as threshold in order to prevent false edges around the true edges. The threshold is computed as equation (2):

$$T_{in} = \begin{cases} \dfrac{GRAD_{in}}{3} & GRAD_{in} > 100 \\ \dfrac{3*GRAD_{in}}{4} & GRAD_{in} \le 100 \end{cases} \qquad GRAD_{in} = \frac{\sum_{i=51}^{250} (grad_{in})_i}{200}$$

$$T_{big\_stru} = \frac{GRAD_{big\_stru}}{3} \qquad GRAD_{big\_stru} = \frac{\sum_{i=51}^{250} (grad_{big\_stru})_i}{200} \tag{2}$$

Since edge width of two images are different, we still calculate it in I_{in}, and only the common edge points are taken into consideration. Thus, the interference of textures can be reduced. It's found that the WLS filter makes the positions of edge points drift. Therefore, when we extract the common edge points, for an edge point A with the position (x_0, y_0) in I_{big_stru}, if there is an edge point B at the same position in I_{in}, B is considered as the common edge point. If B doesn't exist, one edge point in I_{in} is searched within the eight neighborhoods of (x_0, y_0) as the common edge point. Otherwise, A is discarded.

Based on width, the obtained edges are divided into nine parts: less than or equal to 3, equal to 4 to 9, and equal to or greater than 11, and the statistics proportion of each part is calculated as sharpness features of the image, as equation (3):

$$prop_i = \frac{count_i}{\sum count_i} \qquad i = 3,4,5\cdots 11 \tag{3}$$

2.2. Mean value of Maximum Gradients

It is difficult to evaluate image sharpness only via edge width, because in some cases, the edge width does not change after the image is blurred. However the gradients at edge points become smaller, as shown in figure 5. Thus, gradients can also describe image sharpness.

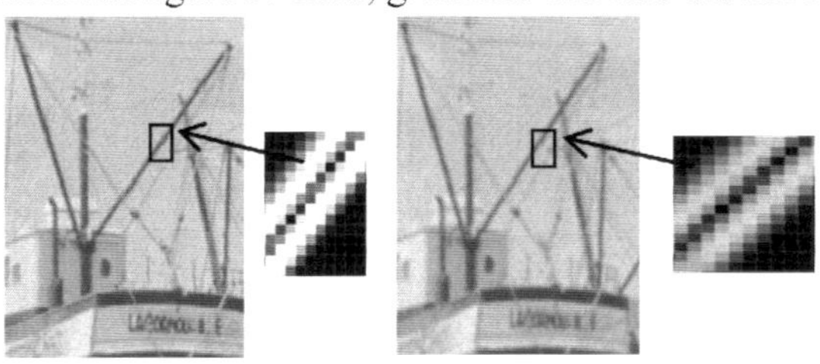

Figure 5. Edge width remain unchanged but gradients become smaller after the image is blurred

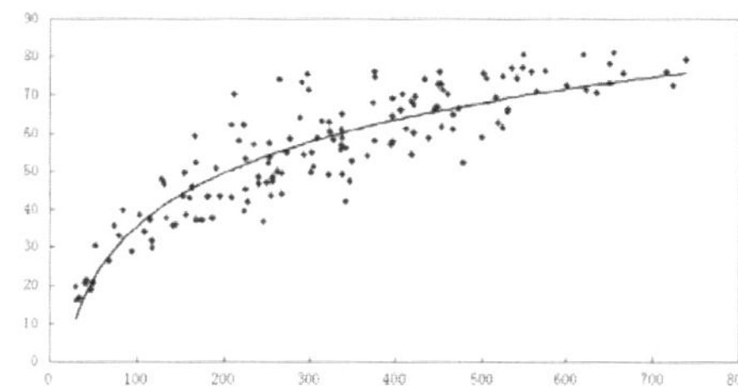

Figure 6. The relationship between $GRAD_{in}$ and image scores.

The relationship between $GRAD_{in}$ and image scores are shown as figure 6. So $GRAD_{in}$ is dealt as another sharpness feature of the image, as equation (4).

$$cont = 20\ln(GRAD_{in}) - 56.5 \tag{4}$$

2.3. Support vector regression

The relationship model style between image features $prop_i$, $cont$ and subjective assessment scores is unknown. Linear regression model and RBF-SVR model are chosen to fit the relationship severally.

SVR maps the input samples to a high-dimensional space through an inner product kernel function, and then build a linear model in the high-dimensional space to estimate the regression model. The computation complexity is independent of the dimension of the samples, which avoids the dimension disaster. SVR can avoid over-fitting caused by too many adjustable parameters, so that the evaluation model has better adaptability.

RBF-SVR is proved to have better results via experiments.

3. Performance result

3.1. Experiment setup

This paper uses the LIVE database, the IVC database and the CSIQ database to verify the validity of the proposed method. There are 29 sets of blurred images in the LIVE database, 4 sets in the IVC database, and 30 sets in the CSIQ database. Each set contains 5 different sharpness images with the same content. Images are compressed after blurred in CSIQ database.

Because the LIVE database contains a large number of images which are uncompressed, it is used as the training set and IVC and CSIQ are used as the test set. For the LIVE database itself, 28 sets of images are used for training, and the remained set is used for testing. The process is carried out circularly, in order to avoid that different sharpness images with the same content appear in the training set and the test set at the same time, which leads to a non-objective evaluation result.

IOP Conf. Series: Journal of Physics: Conf. Series **960** (2018) 012018 doi:10.1088/1742-6596/960/1/012018

3.2. Results and analysis

The proposed method is compared with three classical methods: S3, CPBD and LPC-SI. Scatter plots and four criteria between the objective scores and the subjective scores are employed to evaluate the metric effectiveness. These criteria are SRCC (Spearman rank-order correlation coefficient), KRCC (Kendall's rank-order correlation coefficient), PLCC (Pearson linear correlation coefficient) and RMSE (root mean squared prediction error), as in [4]. SRCC and KRCC are used to evaluate prediction monotonicity, the greater the value, the better the monotonicity. PLCC and RMSE are used to evaluate prediction accuracy, the larger the PLCC and the smaller the RMSE, the higher the accuracy.

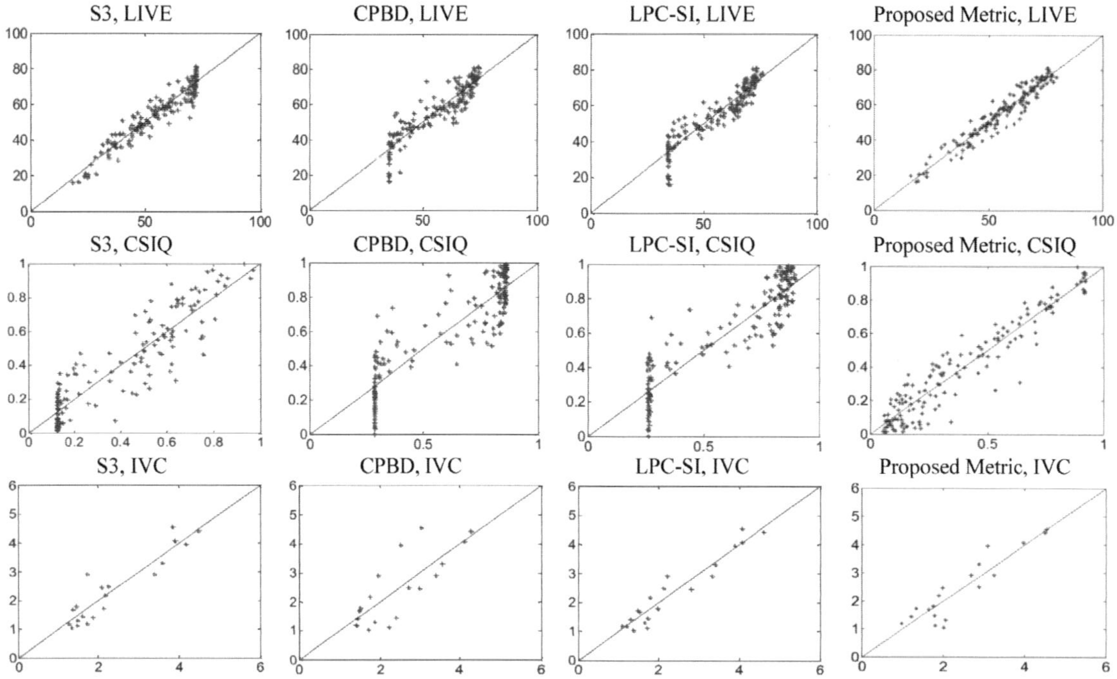

Figure 7. The scatter plots between the objective scores and the subjective scores for the four metrics on three databases.

The scatter plots for the four metrics on three databases are presented in figure 7. The left of the subtitle is the metric, and the right represents the database. Each point represents a test image, whose ordinate represents the subjective score and abscissa represents the objective score. The closer the point to the diagonal, the better the consistency between the objective score and the subjective score. It can be found from figure 7 that the proposed metric has good monotonicity and accuracy. When the image quality is poor, CPBD and LPC-SI can not distinguish the image quality, which is obvious in LIVE database and CSIQ database. This is because these two metrics only consider the edge width. However the proposed metric overcome this problem by extracting large-scale structure and using $GRAD_{in}$. The scatter plots of S3, CPBD and LPC-SI are obtained from [4].

Table 1 summarizes the four criteria of the four metrics on three databases. Table 2 shows the runtime of the four metrics when processing a 1024*1024 image. The computer configuration is i7 CPU, 8 GB RAM, Windows7 64-bit and Matlab 7.10. It can be found from table 1 that the proposed metric has good monotonicity and accuracy on LIVE and CSIQ, and the criteria are better than others. For IVC, the metric performs averagely. The proposed method is short-time running and takes most of the time in extracting large-scale structure. S3 performs excellently, but it's time-consuming because of too much fitting in overlapped blocks. The data of S3, CPBD and LPC-SI are obtained from [4].

Table 1. Four Criteria of the four metrics on three databases

LIVE database(145 images)				
Measure	SRCC	KRCC	PLCC	RMSE
S3	**0.9517**	**0.8157**	**0.9494**	**4.9503**
CPBD	0.9271	0.7714	0.9024	6.7943
LPC-SI	0.9501	0.7994	0.9219	6.1092
Proposed Metric	**0.9553**	**0.8219**	**0.9611**	**4.3606**
CSIQ database(150 images)				
Measure	SRCC	KRCC	PLCC	RMSE
S3	**0.9058**	**0.7290**	**0.9106**	**0.1184**
CPBD	0.8790	0.6905	0.8822	0.1349
LPC-SI	0.8931	0.7022	0.9061	0.1212
Proposed Metric	**0.9212**	**0.7663**	**0.9508**	**0.0888**
IVC database(20 images)				
Measure	SRCC	KRCC	PLCC	RMSE
S3	**0.8691**	**0.7090**	0.9274	0.4269
CPBD	0.7744	0.6105	0.8012	0.6832
LPC-SI	**0.9202**	**0.7831**	**0.9574**	**0.3295**
Proposed Metric	0.8322	0.6773	**0.9312**	**0.4165**

Table 2. Runtime of the four metrics for images of 1024*1024 resolution

Metric	S3	CPBD	LPC-SI	Proposed Metric
Runtime (s)	142.5	11.3	4.37	4.35

4. Conclusion

This paper proposes a novel no-reference image sharpness metric based on large-scale structure, which can reduce the interference of textures, contrast and content of images. After prominent edges are extracted by a weighted least-squares filter, the statistics proportion of edges with different width and the mean value of the maximum gradients from the 51th to 250th are calculated and combined to construct the evaluation model using SVR. The metric has shown relatively good evaluating performance for most images. However, for images that are compressed after blurred, the metric is not so effective and needs for further improvements.

Reference

[1] Wang X, Baofeng T, Chao L and Dongcheng S 2008 Blind image quality assessment for measuring image blur *CISP 08* **1** 467-470.

[2] Rony F and Lina J K 2009 A no-reference objective image sharpness metric based on the notion of just noticeable blur (JNB) *IEEE Trans. Image Processing* **18** 717 - 728.

[3] Niranjan D N and Lina J K 2011 A no-reference image blur metric based on the cumulative probability of blur detection (CPBD) *IEEE Trans. Image Processing* **20** 2678 - 83.

[4] Hassen R, Wang Z and Magdy S 2013 Image sharpness assessment based on local phase coherence *IEEE Trans. Image Processing* **22** 2798-2810.

[5] Anush K M and Alan C B 2011 Blind image quality assessment: From natural scene statistics to perceptual quality *IEEE Trans. Image Processing* **20** 3350-64.

[6] Michele A, Alan C B and Christophe C 2012 Blind image quality assessment: A natural scene statistics approach in the DCT domain *IEEE Trans. Image Processing* **21** 3339-52.

[7] Cuong T V,Thien D P and Damon M C 2012 S3: A spectral and spatial measure of local perceived sharpness in natural images *IEEE Trans. Image Processing* **21** 934-945.

Topology Data Analysis of 9×9 Range Image Patches

Wen Wang[1,*]

[1]School of Science, Shandong Jianzhu University, Jinan 250101 P.R.China.

*E-mail: wen@sdjzu.edu.cn

Abstract. We use the NEB method to deal with data with the property of high dimension from a range image database. From the Brown database we select a large random sample of log-valued, high contrast, normalized, 9×9 range image patches. We create a Morse function via a density estimator. we make 1-dimensional cell complexes from the image data. We detect some topological properties of 9×9 range image patches, find that there exist two types of subsets of 9×9 range image patches modelled as a ring.

1. Introduction

In data analysis Computational topology turns into an efficient method to analyse large sets of multidimensional data for all branches of science and engineering [1], [2], [3], [4]. To investigate multidimensional data, we create a series of complexes from the range image patch data to produce a simple bulletin. The constructed complexes usually compose of lots of simplexes, often they are too large to deal with. Adams [5] used the NEB to make cell complexes by density functions of the given data, they made models for nonlinear data sets with cell complexes, effectually detected the homology. It reveals that cell complex models are the ways for analyzing multidimensional data with high efficiency.

We find a ring model for 9 × 9 range image patches through the method of [5], and give some topological characteristics of 9 × 9 range image patches. With different parameters of σ and subsets of X^n, we compute and get the 0-cells and 1-cells with certain densities. All the 1-cells make various circles. This method can be used to deal with bigger size range image patches than 9 × 9.

2. Background

We briefly introduce some topics: CW complexes, Morse theory.

2.1. CW complexes.

A CW complex is a type of cell complex [5]. For k a nonnegative integer, a k-cell is the closed ball $\{y \in R^k \,|\, \|y\| \leq 1\}$ of dimension k. So a 0-cell is a point, a 1-cell is a line segment, a 2-cell is a disk, etc. We form the k-skeleton $W^{(k)}$ by gluing the boundaries of k-cells to the (k-1)-skeleton $W^{(k-1)}$. If W is a finite-dimensional CW complex, then this process terminates and we have $W = W^{(k)}$ for some k.

2.2. Morse theory.

The following introduction to Morse theory is informal; see [6] for thorough treatment. Suppose M is a compact manifold of dimension d and Morse function $f: M \to R$ is smooth with non-degenerate critical points $m_1, \ldots, m_k \in M$ satisfying

$$a_0 < f(m_a) < a_1 < f(m_2) < \cdots < a_{k-1} < f(m_k) < a_k \tag{1}$$

The index λ_i of critical point mi is the number of linearly independent directions around m_i in which f decreases. So a minimum has index 0, a maximum has index d, and a saddle point has index between 1 and (d−1). Let $M_a = f^{-1}((-\infty, a])$ be the sublevel set corresponding to a∈R. Morse theory tells us that each M_{ai} is homotopy equivalent to a CW complex with one λ_i-cell for each critical point mi. In particular, M_{ai} is homotopy equivalent to Mai-1 with a single λ_i-cell attached. For instance, M_{a1} is homotopy equivalent to a point and is obtained from M_{a0}, the empty set, by attaching a single 0-cell. Figure 1 contains an example.

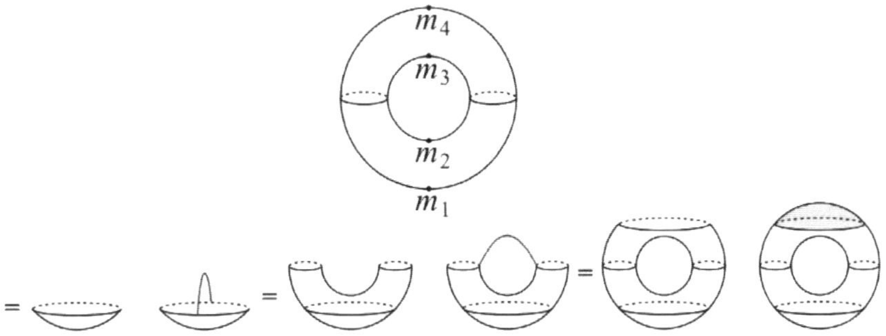

Figure 1.Morse theory example. Manifold M is a torus. There are four critical points m1, . . ., m4with indices 0, 1, 1, and 2 respectively. Let $a_0 < f(m_1) < a_1 \cdots < f(m_4) < a_4$. Moving from left to right, we attach cells of dimension 0, 1, 1, and 2 in order to obtain M_{ai} from $M_{a(i-1)}$.

3. Data set of range image patches

The Brown range image database by Lee and Huang is a set of 197 range images from indoor and outdoor scenes, mostly 444 × 1440 pixels. The operational range for the Brown scanner is typically 2–200 meters, and the distance values for the pixels are stored in units of 0.008 meters. The database can be found at the following webpage: http://www.dam.brown.edu/ptg/brid/index.html. From the Brown database we obtain a space of range image patches through the following steps [5]:

(1) We randomly select about $5 \cdot 10^5$ size 9×9 patches from the images in the database. Each patch is represented by a vector x $\in R^{81}$with logarithm values.

(2) We compute the contrast norm $\|x\|_D = \sqrt{\sum_{i\sim j}(x_i - x_j)^2}$ of each patch and select the patches with contrast norm in the top 20% of the entire sample.

(3) We subtract from each patch the average of its coordinates and divide by the contrast norm.

(4) We change to the DCT basis $\{e_1, \ldots, e_{80}\}$ for 9×9 patches, normalized to have contrast norm one. This maps the patches to an 80-dimensional sphere.

Let R be the resulting set of high-contrast, normalized, 9×9 range patches[7]. Our data set is a random subset X⊂R of size 5,0000. See Figure 2 for an example of range image patches from the Brown database.

Figure 2 Forest scene --Sample range image from the Brown database.

4. Computing method

In this part, the steps of calculation procedure are given. refer to the paper [5] for more details of the method. For $X \subset R^n$ from probability density function $f: R^n \rightarrow [0, \infty)$. We get superlevel sets $X^\alpha = f^{-1}([\alpha, \infty)) = \{x \in R^n | f(x) \geq \alpha\}$, important topological information is given by the high dense regions. We may make CW complex models Z^α to estimate X^α.

Only the 1-dimensional skeleton of the cell complex is given. We give a differentiable function to estimate the unknown density function. then we can get the partial maximum of the density estimate in order to make 0-cells, we make original bands randomly, then discover the convergent bands through NEB [8], so we can get the 1-cells.

5. Experimental results

Now we analyze subsets of X_9 with two types subsets of X_n: (1) $\overline{X}_n(50000)$of X_n with size 50000 randomly; (2) core subsets $\overline{X}_n(k,p)$,with k=200,300 and p=30.Table 1 gives the detailed parameters used in the experiments [9], [10].

Table 1. Data set information

	$\overline{X}_9(50000)$	$\overline{X}_9(200,30)$	$\overline{X}_9(300,30)$
Size of data set	50000	15000	15000
Dimension n	9×9	9×9	9×9
Standard deviation σ	0.37	0.31	0.34

We deal with the set $\overline{X}_9(50000)$ withσ=0.37, finding four 0-cells, whose densities are 960.6,1359,912.7,882.6, and four 1-cells with densities of 918.7,796.6,888.0,773.7, these cells make a ring. (Figure 3).

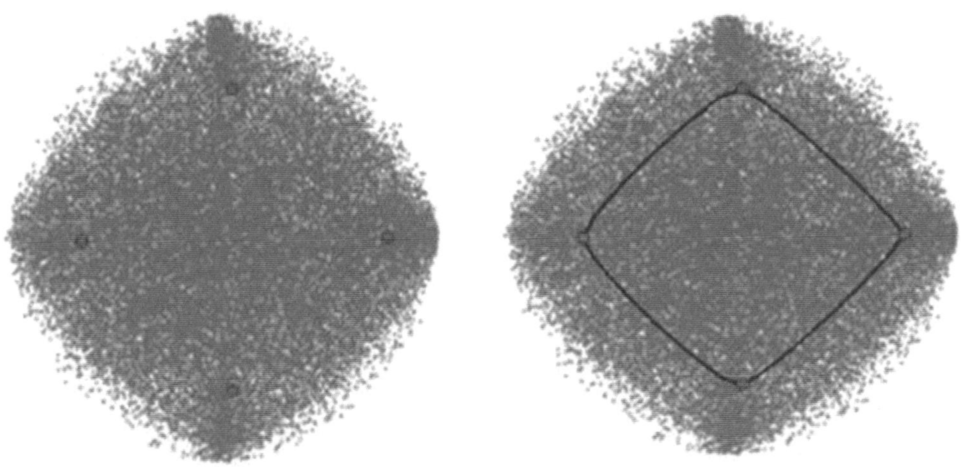

Figure 3 $\overline{X}_9(50000)$ and the ring $Z^{773.7}$, projected to a plane.

For $\overline{X}_9(200,30)$, with σ=0.31, there are four 0-cells with densities in 3713,1761,2379,2040 and four 1-cells with densities in 1903,1698,1277,1156, and they form a ring (Figure 4).

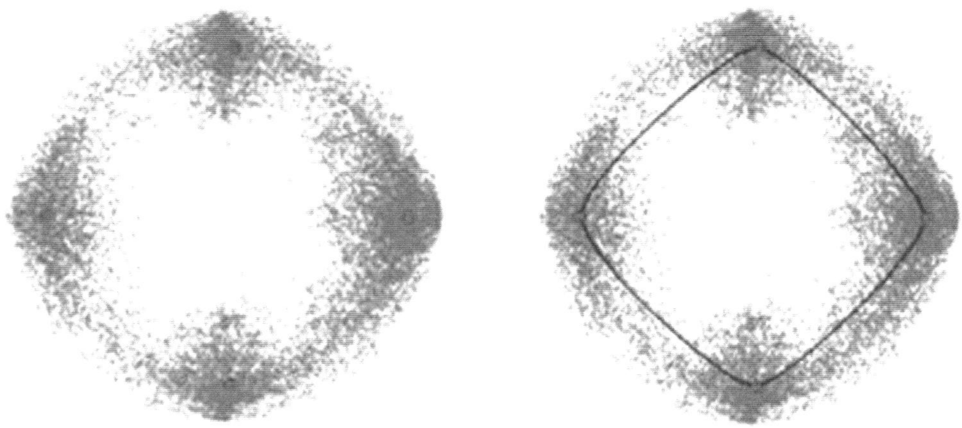

Figure 4. $\overline{X}_9(200,30)$ and the ring Z^{I156}, projected to a plane

For $\overline{X}_9(300,30)$, by σ = 0.34, there are four 0-cells with densities in 1770,2055,2402,3763 and four 1-cells with densities in 1157,1284,1717,1929, and they form a circle too (Figure 5).

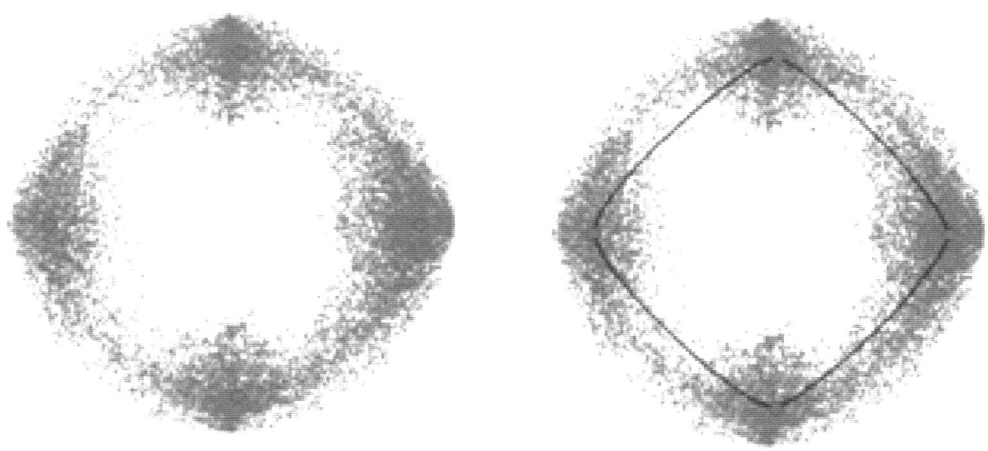

Figure 5. $\overline{X}_9(300,30)$ and the ring Z^{I157}, projected to a plane.

6. Conclusions

We utilize the NEB method to analyze the data of 9×9 Range image patches, the experimental results indicate that the spaces of 9×9 range image patches having different subsets modeled as a ring. There are four quarter-circular 1-cells forming a loop. We test the way on range image patch data sets and find compact complexes showing important nonlinear topology structures. The results got with this method are native properties of Range image patches, and they do not change with different methods and data sets. But there are still lots of problems to make higher dimensional cells with this method [5].

Acknowledgements

This project is supported by the National Natural Science Foundation of China (Grant No.61471409).

References

[1] H. Adams and G. Carlsson, *On the nonlinear statistics of range image patches*, SIAM J. Imag. Sci., 2 (2009), 110–117.

IOP Conf. Series: Journal of Physics: Conf. Series **960** (2018) 012019 doi:10.1088/1742-6596/960/1/012019

[2] G. Carlsson, T. Ishkhanov, V. de Silva, A. Zomorodian, *On the local behavior of spaces of natural images*, Internat. J. Computer Vision, 76 (2008), pp. 1–12.
[3] G. Carlsson, *Topology and data*, Bulletin (New Series) of the American Mathematical Society, 46 (2009), pp. 255–308.
[4] A. Zomorodian, *Topological data analysis*, Advances in Applied and Computational Topology (Proceedings of Symposia in Applied Mathematics), A. Zomorodian, eds., American Mathematical Society, 70 (2012), pp. 1–39.
[5] H. Adams, A. Atanasov, and G. Carlsson, *Nudged elastic band in toplogical data analysis*, Topological Methods in Nonlinear Analysis, 45 (2015), pp. 247–272.
[6] J. Milnor. *Morse Theory*. Princeton University Press, Princeton, 1965.
[7] A. B. Lee, K. S. Pedersen, and D. Mumford, *The non-linear statistics of high-contrast patches in natural images*, Internat. J. Computer Vision, 54 (2003), pp. 83–103.
[8] G. Henkelman, B. Uberuaga, H. Jonsson, *A climbing image nudged elastic band method for finding saddle points and minimum energy paths*, Journal Chemical Physics, 112 (2000), pp. 9901–9904.
[9] S. Xia, *A topological analysis of high-contrast patches in natural images*, J. Nonlinear Sci. Appl., 9 (2016), pp. 126–138.
[10] Y. Cheng, *Mean shift, mode seeking, and clustering*, IEEE Trans. Pattern Anal., 17 (1995), pp. 790–799.

Image Registration Algorithm Based on Parallax Constraint and Clustering Analysis

Zhe Wang[1], Min Dong[1, a], Xiaomin Mu[1, b] and Song Wang[2]

[1]School of Information Engineering, Zhengzhou University, China
[2]Industrial Technology Research Institute, Zhengzhou University, China

[a]iemdong@zzu.edu.cn, [b]iexmmu@zzu.edu.cn

Abstract. To resolve the problem of slow computation speed and low matching accuracy in image registration, a new image registration algorithm based on parallax constraint and clustering analysis is proposed. Firstly, Harris corner detection algorithm is used to extract the feature points of two images. Secondly, use Normalized Cross Correlation (NCC) function to perform the approximate matching of feature points, and the initial feature pair is obtained. Then, according to the parallax constraint condition, the initial feature pair is preprocessed by K-means clustering algorithm, which is used to remove the feature point pairs with obvious errors in the approximate matching process. Finally, adopt Random Sample Consensus (RANSAC) algorithm to optimize the feature points to obtain the final feature point matching result, and the fast and accurate image registration is realized. The experimental results show that the image registration algorithm proposed in this paper can improve the accuracy of the image matching while ensuring the real-time performance of the algorithm.

1. Introduction

Image processing is to obtain valid information as much as possible by processing a limited number of images. Image registration is a basic problem in image processing. It refers to the process of matching two or more pictures containing the same information through spatial domain transform. To get more information about a scene, the priority is to register two or more images that contain the same scene. However, when images with large amount of information have a high real-time requirement, the results of image registration are often unsatisfactory. Therefore, how to ensure the registration speed and improve the registration accuracy is the focus of this paper.

The image registration method has been appreciated by researchers. Professor Lowe first proposed the Scale-Invariant Feature Transform (SIFT) algorithm and improved it in 2004 [1]. The feature descriptor extracted by this algorithm has good stability for scale transform, illumination, and rotation, so the registration accuracy is high and the robustness is good. However, there is a problem that the registration efficiency is not high due to the large amount of calculation in the feature point description. Qizhi Xu et al. proposed a blob-like structure to improve the registration accuracy of the SIFT algorithm on satellite imagery [2]. Bin Li et al. improved the computational efficiency of image matching using the Euclidean distance strategy and the improved RANSAC based on the existing energy function [3]. Jianwei Fan et al. combined with sparse representation when registering SAR images [4].

According to the literature [5], feature matching is the most important and indispensable step in the overall image registration process. The quality of the matching is related to the success of the final registration results. Therefore, most scholars focus on finding a better way to complete the work of feature matching. The method of feature matching can be divided into two categories: region-based method and feature-based method. Normally, the image contains a large amount of available information, including feature points (high curvature points, corner points, etc.), straight lines, edges, contours, closed areas and statistical features (center of gravity, geometric moments), and so on. Among them, the feature points in the image registration process have their own unique advantages, so most feature-based image registration algorithms are based on point features [6].

Corner point is a very important local feature in the image, that is in the image of the larger curvature of the two places or the intersection of multiple lines. Corner points contain a wealth of two-dimensional structure information, easy to be detected, and can adapt to changes in ambient light, it is preferred feature of many of the features matching algorithm. Therefore, this paper uses the corner point feature to research the image matching algorithm .

The main chapters of the paper are arranged as follows:

(1) Combining the research background and current status of image registration, the research direction is determined as the feature matching algorithm based on corner point by analyzing the current research hotspot. (2) Introduce the main idea in each feature matching stage of the improved algorithm and give the algorithm flow. (3) Compare the proposed method with the unimproved algorithm, it is proved that the proposed algorithm has obvious improvement in matching speed and accuracy, and then the conclusion is drawn.

2. Methodology

2.1. Harris Corner Detection

Harris feature point is the point where the brightness of the image changes abruptly in multiple directions. By detecting the change of the brightness of the image in a window, the feature point can be separated from other pixels, usually using an autocorrelation matrix which is similar to the image brightness differential point detection operator:

$$A(x) = \sum_{x,y} \omega(x,y) \begin{bmatrix} I_x^2(x) & I_x(x)I_y(x) \\ I_x(x)I_y(x) & I_y^2(x) \end{bmatrix} \tag{2.1}$$

In (2.1), I_x and I_y are the partial derivatives of x and y directions, $\omega(x,y)$ is the normalized kernel function, and the higher weight is given to the pixels near the center point.

In the experiment, the Gaussian kernel is used, that is

$$\omega(x,y) = g(x,y,\sigma) = \frac{1}{2\pi\sigma}\exp(-\frac{x^2+y^2}{2\sigma}) \tag{2.2}$$

where σ is the variance of the Gaussian kernel.

Let λ_1 and λ_2 detect the eigenvalues of the operator. When the matrix A is at the feature point position, it will has two nonnegative eigenvalues at the same time, and the Harris discriminate is

$$R = \det(A) - \alpha tr^2(A) = \lambda_1\lambda_2 - \alpha(\lambda_1+\lambda_2)^2 \tag{2.3}$$

The constant α usually evaluate to 0.04. If a pixel is a local maximum and R is greater than the set threshold, it is a Harris feature point.

2.2. Approximate Matching by NCC Function

Let I_1, I_2 be the gray scale of two images, $n \times n$ is the size of the matching window, the first image of the feature points is d_i, $i = 1,2\cdots,m_1$, the second image of the feature point is d_j, $j = 1,2\cdots,m_2$, (u_i^1, v_i^1) and (u_i^2, v_i^2) are the i-th and j-th feature points to be matched in the two graphs respectively, The NCC function is used to describe the degree of correlation between the feature points[7]. The NCC function is defined as follows:

$$\begin{aligned} C_{ij} &= \Big(\sum_{m=-\omega}^{\omega}\sum_{n=-h}^{h}[I_1(m+u_i^1, n+v_i^1) - \bar{I}_1(u_i^1, v_i^1)] \times \\ &[I_2(m+u_j^2, n+v_j^2) - \bar{I}_2(u_j^2, v_j^2)] / \\ &(\{\sum_{m=-\omega}^{\omega}\sum_{n=-h}^{h}[I_1(m+u_i^1, n+v_i^1) - \bar{I}_1(u_i^1, v_i^1)]^2\} \times \\ &\{[I_2(m+u_j^2, n+v_j^2) - \bar{I}_2(u_j^2, v_j^2)]^2\})^{1/2} \\ &\omega = h = (n-1)/2\,. \end{aligned} \tag{2.4}$$

Obviously, the value range of C_{ij} is (-1,1), the higher the value of C_{ij}, the higher the degree of correlation between d_i and d_j. For a given feature point, the matching set of points with the highest degree of correlation in all matching points are given.

2.3. Improved Approximate Matching and Preprocessing Algorithm

After the initial feature point matching, there will be a lot of false match points, to a large extent affected the fine match. The matching point pairs are mapped to the corresponding feature quantity space, such as the distance feature vector between the pair of points and the direction feature vector. In the eigenvector space, there is no obvious rule of the mismatched point pair distribution, and the correct matching point distribution is more concentrated. Therefore, the K-means clustering method is used to eliminate the misplaced point pairs.

The K-means clustering algorithm is to find a partition scheme of k clusters by iteration, and make the global error is minimal when the mean of the k clusters is used to represent the corresponding samples[8]. The K-means clustering algorithm is based on the criterion of least sum of square error. The cost function is:

$$J(c,\mu) = \sum_{i=1}^{k}\left\| x^{(i)} - \mu_{c^{(i)}} \right\|^2 \tag{2.5}$$

Where $\mu_{c^{(i)}}$ represents the mean of the i-th cluster.

The basic idea of K-means clustering algorithm based on parallax constraint is described below: It can be observed that the distance feature vector of the correct matching point pair is very similar when the feature point matching is carried out. If the distance of the matching point pair is expressed by Euclidean distance, no more than 10 pixels; and the distribution of correct matching points is concentrated, and it is suitable to use K-means clustering to eliminate false matching points.

Assuming that the set of matches obtained in Section 2.1 is M :

$$\begin{aligned} &M = \{h_k = (d_i, d_j)\}, \\ &i = 1,2\cdots,m_1; j = 1,2,\cdots,m_2 \end{aligned} \tag{2.6}$$

Where (d_i, d_j) is a set of matching pairs of two images to be registered.

Distance feature vector D :

$$D = \{d_k = deu(d_i, d_j) \mid h_k^{\cdot} = (d_i, d_j) \subset M\} \tag{2.7}$$

Where d_k is the Euclidean distance of the k-th pair of matching points:

$$deu(d_i, d_j) = \sqrt{(d_{ix} - d_{jx})^2 + (d_{iy} - d_{jy})^2} \tag{2.8}$$

The K-means clustering algorithm is greatly influenced by the selection of the initial clustering center and the number of categories. In the case of value clustering, for example, the difference between the correct matching points is no more than 10 pixels. Therefore, 10 pixels is used as the classification limit, so that the difference between *deu* is less than 10 pixels, until the aggregation can not be continued. The K-means clustering is performed using the mean of each category as the center of the initial cluster.

Algorithm K-means clustering algorithm based on distance feature

Step 1 Calculate the distance feature vector D according to the matching point set M, and the classification limit is d_{clu} = 10.

Step 2 Perform a monotonically increasing sort of distance feature vector D.

Step 3 For each component d_k in D, if $d_k - d_{k-1} < d_{clu}$,

d_k, d_{k-1} is marked as a class; otherwise, d_k is marked as a new class. After determining all the components, we get the number of categories c, and the mean of each category is taken

as the initial mean value $\mu_1, \mu_2, \cdots, \mu_c$, that is, $\mu_i = \frac{1}{|C_i|}\sum_{k \in C_i} d_k$,

where $C_i (i = 1, 2, \cdots, c)$ represents the rest of position vectors in the i-th class after reclassification.

Step 4 For each component d_k in D, calculate its distance from μ_i and merge it into the μ_i class with the smallest distance, $\min_{j \in (1,2 \cdots, c)} \|d_k - \mu_j\|^2$, Recalculate $\mu_1, \mu_2, \cdots, \mu_c$.

Step 5 If the center points $\mu_1, \mu_2, \cdots, \mu_c$ no longer change and select the category μ_i with the largest number of components in the category, then the matching point set corresponding to the class μ_i is the correct match point set; otherwise, Step4 is recalculated until meeting the conditions.

2.4. Fine Matching by RANSAC

There are often some false match pairs in the initial match, so the mismatch is further eliminated by fine matching. Common optimization algorithms are Least Median of Squares (LMedS), M-estimator, RANSAC. In this paper, RANSAC algorithm is used for fine matching.

The basic idea of the RANSAC algorithm is as follows: first construct an objective function, then extract the initial value of the parameter in the function by extracting the minimum point set multiple times, divides all the data into "inliers" and "outliers" with the initial parameters, "inliers" is to meet the estimated parameters of the point, "outside point" on the contrary, finally use all the "inliers" to re-calculate the objective function parameters.

In the RANSAC algorithm, F is the basic transformation matrix; m is the current number of samples; r is the current number of four points that are randomly selected to meet the condition (three points are non-collinear); M is the maximum number of samples; R is the maximum number of four points that are randomly selected to meet the condition (three points are non-collinear); N is the number of samples required to get p, p is the probability that at least one random choice contains s "inliers" after N times sampling, and $p > 0.95$ usually. N and p satisfy the following relation:

$$p = (1-\varepsilon'^{s})^{N} \tag{2.9}$$

Where s is the number of feature points per sample, and $s = 8$; ε' represents the probability that a selected feature point is an "inliers". In the algorithm; t is the distance threshold; Gui is the criterion of "inliers" and "outliers"; bin is a set of most "inliers"; and $bnin$ is the number of bin.

3. Experimental Results And Analysis

In order to verify the effectiveness of the proposed algorithm, we selected two images that were rotated and translated to carry out simulation experiments.

3.1. Experiments on Harris corner detection and approximate matching

As shown in Figure 1 and Figure 2, the number of feature points of the improved algorithm is greatly reduced. In the preprocessing of the improved algorithm, some of the correct matching points are also removed along with the wrong match point, but the number of remaining matching points is enough, it does not affect subsequent matching process.

(a) (b)

Figure 1. Feature point of unimproved algorithm

a.Image 1 b.Image 2

Figure 2. Feature point of improved algorithm

Figure 3. Approximate matching of unimproved algorithm

Figure 4. Approximate matching of improved algorithm

Figure 3 and Figure 4 show the effect of rough matching between two algorithms. In the rough matching result of the improved algorithm, the distance feature vectors of most matching points are similar, which will be beneficial to the process of fine matching.

3.2. Experiments on fine matching

The fine matching results are given in Figure 5 and Figure 6, it can be seen that the improved algorithm promote the precision of image registration and reduce the impact of the wrong matching point pairs effectively.

Figure 5. Fine matching of unimproved algorithm

Figure 6. Fine matching of improved algorithm

3.3. Analysis of experimental results

Table 1-3 show the time, number of points in each feature matching stage and the accuracy rate of the two algorithms.

Table 1. Time in each feature matching stage of two algorithms

	Approximate matching	K-means clustering algorithm	RANSAC algorithm	Total
Unimproved algorithm	0.9146s	0s	0.3215s	1.2361s
Improved algorithm	0.9146	0.0185	0.0310s	0.9641s

It can be observed from Table 1 that the time of K-means clustering matching algorithm based on Euclidean distance of parallax constraint is reduced by 22% compared with the unimproved algorithm, and after the features matching of K-means clustering, time of RANSAC matching is greatly shortened, only 9.64% of the unimproved algorithm. The proposed algorithm improves the speed of image registration.

Table 2. The number of points in each feature matching stage of two algorithms

	Approximate matching	K-means clustering algorithm	RANSAC algorithm
Unimproved algorithm	163	0	127
Improved algorithm	163	133	120

It can be seen from Table 2 that the number of points in approximate matching is 163 pairs. After running the K-means clustering algorithm, 30 pairs of wrong matching points were removed, 133 pairs

IOP Conf. Series: Journal of Physics: Conf. Series **960** (2018) 012020 doi:10.1088/1742-6596/960/1/012020

were left. Finally, the number of matching points by the two algorithms is 127 pairs and 120 pairs respectively.

Table 3. The accuracy rate of two algorithms

	Total number of feature points	The number of correct matching points	The number of wrong matches points	Accuracy rate
Unimproved algorithm	127	123	4	96.85%
Improved algorithm	120	118	2	98.33%

The accuracy rate of two algorithms are presented in Table 3. The wrong matching points of the improved algorithm are reduced by half compared with the unimproved algorithm, and the accuracy rate is increased to 98.33%.

4. Conclusion

In order to overcome the shortcomings of traditional feature point matching method in real time and poor precision, firstly Harris corner instead of SIFT DoG corner is uesd to improve registration speed, and feature point matching optimization scheme based on parallax constraint and clustering analysis is proposed. The K-means clustering algorithm is introduced into the feature point rough matching, which effectively simplifies the initial matching set and eliminates most of the wrong matching. The experimental results show that the matching algorithm is correct and efficient, and the precision and speed of feature matching process are improved in image registration.

References

[1] Gonzalez R C, Woods R E, Eddins S L. Digital Image Processing Using MATLAB[M]. Publishing House of Elec, 2013.

[2] Xu Q, Zhang Y, Li B. Improved SIFT match for optical satellite images registration by size classification of blob-like structures[J]. Remote Sensing Letters, 2014, 5(5):451-460.

[3] Li B, Ming D, Yan W, et al. Image Matching Based on Two-Column Histogram Hashing and Improved RANSAC[J]. IEEE Geoscience & Remote Sensing Letters, 2014, 11(8):1433-1437.

[4] Fan J, Wu Y, Li M, et al. SAR Image Registration Using Multiscale Image Patch Features With Sparse Representation[J]. IEEE Journal of Selected Topics in Applied Earth Observations & Remote Sensing, 2016, (99):1-11.

[5] Izadi M, Saeedi P. Robust weighted graph transformation matching for rigid and nonrigid image registration[J]. IEEE Transactions on Image Processing, 2012, 21(10):4369-82.

[6] Lee C Y, Wang H J, Chen C M, et al. A Modified Harris Corner Detection for Breast IR Image[J]. Mathematical Problems in Engineering,2014,(2014-7-17), 2014, 2014(8):1-12.

[7] Satoh S. Simple low-dimensional features approximating NCC-based image matching[J]. Pattern Recognition Letters, 2011, 32(14):1902-1911.

[8] Lee E, Schmidt M, Wright J. Improved and simplified inapproximability for k -means[J]. Information Processing Letters, 2015, 120:40-43.

Fast Time-domain Imaging Method for Bistatic Circular SAR

Hongtu Xie[1,2,3,4,*], Shaoying Shi[3], Junfa Mao[1,*], Daoxiang An[4], Fuhai Li[2], Zhimin Zhou[4] and Guoqian Wang[5,*]

[1]Key Laboratory of Ministry of Education of Design and Electromagnetic Compatibility of High Speed Electronic Systems, Shanghai Jiao Tong University, Shanghai, China
[2]College of Electrical and Information Engineering, Hunan University, Changsha
[3]Air Force Early Warning Academy, Wuhan, China
[4]College of Electronic Science, National University of Defense Technology, Changsha
[5]Hunan Institute of Traditional Chinese Medicine, Changsha, China
[*]Hongtu Xie and Guoqian Wang are the co-first authors, and Junfa Mao is the corresponding author.

*e-mail: xht20041623@163.com

Abstract. Fast time-domain algorithm (FTDA) for the one-stationary bistatic circular synthetic aperture radar (OS-BCSAR) imaging is presented. In this paper, the subimage is represented in the slant-range plane instead of ground plane. FTDA includes the fast backprojection algorithm (FBPA) with the subaperture and elliptical polar grid processing and the FBPA with the subaperture and polar grid processing. Experimental results are given to prove its validity.

1. Introduction

Based on the synthetic aperture radar (SAR), bistatic SAR (BSAR) system was developed by several countries, and some experiment results were obtained [1]. One-stationary bistatic SAR (OS-BSAR) [2] [3] is a special BSAR, which has a moving radar and a stationary radar. OS-BSAR has the advantage of the reducing vulnerability, improving the detectability, getting additional information, and lower cost. Circular SAR (CSAR) [4] is a SAR with a circular aperture, which is able to provide the higher resolution image and increased information in comparison to the SAR with a linear aperture (LSAR).

Bistatic CSAR (BCSAR) has advantages of the BSAR and CSAR, so it has gained wide attention [5] [6]. Australian Defence Science and Technology Organisation conducted a one-stationary BCSAR (OS-BCSAR) experiment [5], and the monostatic and bistatic data were collected by operating the *Ingara* radar in a circular spotlight mode in conjunction with a stationary ground-based receiver. A BCSAR experiment has been carried out by Chinese Academy of Sciences [6], and the transmitter and receiver fixed in different platforms move in parallel with equal radiuses but the different heights.

BSAR imaging algorithm includes frequency-domain algorithm (FDA) and time-domain algorithm (TDA). FDA usually aims to minimize processing time, but it leads to some limitations. Conversely, TDA doesn't face to the limitations, but it has the high computation load. Time-domain backprojection algorithm (BPA) is considered as a linear transformation from the echo into SAR images, so it can be used directly to the mono and bistatic CSAR. To reduce the computation load, the fast implementation of the BPA was used for the monostatic CSAR imaging, i.e., fast backprojection algorithm (FBPA) [4]. However, the FBPA has not been investigated for the BCSAR imaging in the earlier publications.

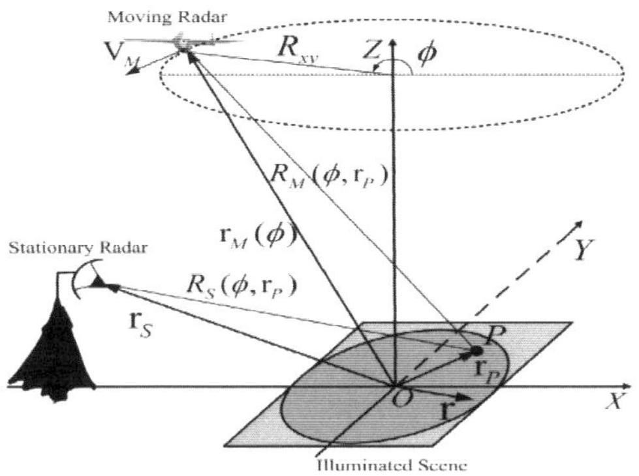

Figure 1. Imaging geometry of the OS-BCSAR system.

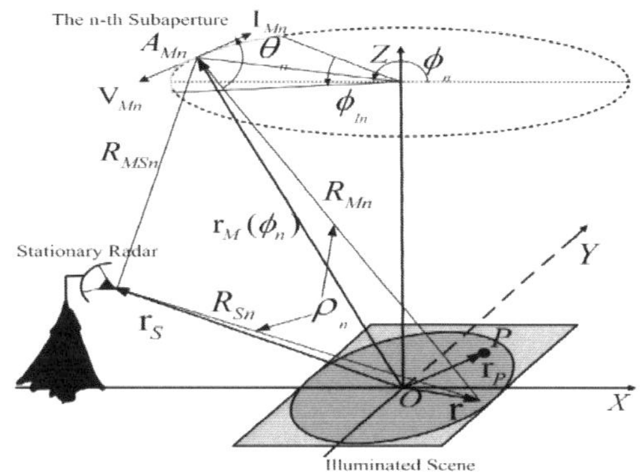

Figure 2. Imaging geometry for FBPA with subaperture and elliptical polar grid processing.

This paper explores two FBPA for OS-BCSAR imaging, including the FBPA based on subaperture and elliptical polar grid processing and the FBPA based on subaperture and polar grid processing.

2. BPA for OS-BCSAR imaging

OS-BCSAR imaging geometry is shown in Figure 1, which has a moving radar fixed on an airplane and a stationary radar placed on a mountain. Solid circle is the moving radar's track, with the position $\mathbf{r}_M(\phi) = (x_M(\phi), y_M(\phi), z_M) = (R_{xy}\cos(\phi), R_{xy}\sin(\phi), z_M)$. $\phi \in [0, 2\pi]$ is the angular variation of moving radar, i.e. $\phi(\eta) = \mathbf{V}_M \eta / R_{xy}$ at slow time η. $\mathbf{V}_M$ is the moving radar's speed at ϕ. R_{xy} and Z_M are radius and altitude of the moving radar's circular track, respectively. $\mathbf{r}_S = (x_S, 0, z_S)$ is the stationary radar position. P is an arbitrary scattering target in the scene, with the position $\mathbf{r}_P = (x_P, y_P, 0)$. The ranges from the moving and stationary radars to the scattering target P at ϕ are $R_M(\phi, \mathbf{r}_P)$ and $R_S(\phi, \mathbf{r}_P)$, respectively.

The traveling distance from the moving and stationary radars to the scattering target P at ϕ is $R(\phi, \mathbf{r}_P) = R_M(\phi, \mathbf{r}_P) + R_S(\phi, \mathbf{r}_P)$. The transmitted baseband signal is $p(\tau)$, so the range compressed signal of the scattering target P is $s_{rc}(\tau, \phi) = \sigma_p p_r\left[B\left(\tau - R(\phi, \mathbf{r}_P)/c_0\right)\right]$. τ is the fast time, σ_P is the scattering coefficient of the scattering target P, c_0 is the speed of light, B is the transmitted signal bandwidth, $p_r[\cdot]$ is the range compressed pulse envelope. The backprojection (BP) of the BPA for the OS-BCSAR imaging is performed over the elliptical mapping. $\mathbf{r} = (x, y, 0)$ is the position of an arbitrary sample in the scene, f is the radar frequency, then the value of the OS-BCSAR image at the sample $\mathbf{r}$ is

$$\begin{aligned} I(\mathbf{r}) &= \int_0^{2\pi} s_{rc}\left(R(\phi,\mathbf{r})/c_0, \phi\right)\exp\left[j2\pi f R(\phi,\mathbf{r})/c_0\right]d\phi \\ &= \int_0^{2\pi} \sigma_P p_r\left[B\left(\left(R(\phi,\mathbf{r}) - R(\phi,\mathbf{r}_P)\right)/c_0\right)\right]\cdot\exp\left[j2\pi f R(\phi,\mathbf{r})/c_0\right]d\phi \end{aligned} \tag{1}$$

3. FBPA with subaperture and elliptical polar grid processing

Imaging geometry for FBPA with subaperture and elliptical polar grid processing is given in Figure 2. For the n-th subaperture, ϕ_n is the angle corresponding to the n-th subaperture center A_{Mn}, and ϕ_{In} is its integration angle. $\mathbf{r}_M(\phi_n) = (x_M(\phi_n), y_M(\phi_n), z_M)$ is the moving radar's position at ϕ_n. The ranges from the moving and stationary radars to the sample $\mathbf{r}$ at ϕ_n are $\mathbf{R}_{Mn}$ and $\mathbf{R}_{Sn}$, with the magnitudes R_{Mn} and R_{Sn}. ρ_n is the bistatic range from the moving and stationary radars to sample $\mathbf{r}$, and θ_n is the angle between the vectors $\mathbf{R}_{Mn}$ and $\mathbf{I}_{Mn}$ at ϕ_n. The elliptical polar coordinates (ρ_n, θ_n) of the sample $\mathbf{r}$ are

$$\rho_n = R_{Mn} + R_{Sn} = R_M(\phi_n, \mathbf{r}) + R_S(\phi_n, \mathbf{r}); \qquad \theta_n = \arccos\left(\left(\mathbf{R}_{Mn} \cdot \mathbf{I}_{Mn}\right)/\left|\mathbf{R}_{Mn}\right|\right) \tag{2}$$

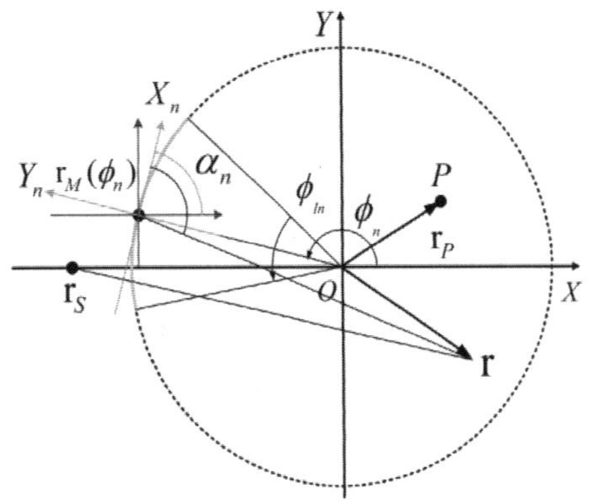

Figure 3. Top view in Figure 2.

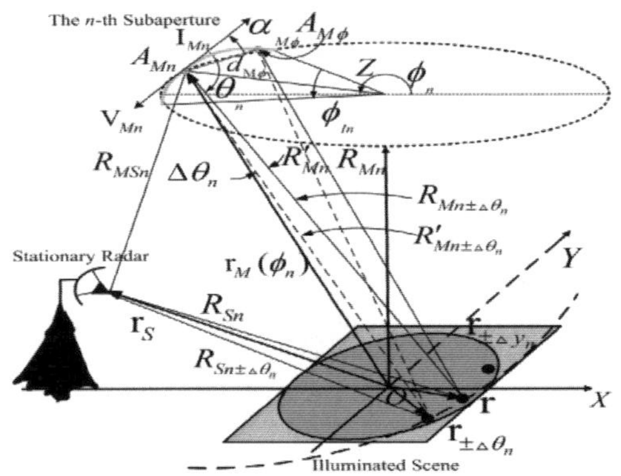

Figure 4. Bistatic range error.

$\mathbf{I}_{Mn}$ is $(\sin(\phi_n), -\cos(\phi_n), 0)$, with the origin at A_{Mn}, which is anti-parallel to $\mathbf{V}_{Mn}$. Figure 3 shows the top view of the imaging geometry in Figure 2. In Figure 3, we define a local coordinate system (X_n, Y_n, Z_n) with the origin at A_{Mn}. It is the tangent to the circumference, which is translated to A_{Mn} and rotated an angle α_n with respect to the global Cartesian coordinate system (X, Y, Z). α_n is the angle between the axes X and X_n, and it is found that $\alpha_n = \phi_n - \pi/2$. The sample $\mathbf{r}$ in the Cartesian coordinate system is

$$\mathbf{r} = \mathbf{B}\mathbf{r}_0 + \mathbf{r}_M(\phi_n) \tag{3}$$

Where, $\mathbf{r}_0 = (x_0, y_0, z_0) = \left[R_{Mn}\cos\theta_n, -\sqrt{(R_{Mn}\sin\theta_n)^2 - z_M^2}, -z_M \right]$ is the position of the sample $\mathbf{r}$ in the local coordinate system, and $\mathbf{B} = [\cos\alpha_n \ \ -\sin\alpha_n \ \ 0;\ \sin\alpha_n \ \ \cos\alpha_n \ \ 0;\ 0 \ \ 0 \ \ 1]$ is the rotation matrix.

Similarly, the elliptical polar coordinates (ρ_{np}, θ_{np}) of the scattering target P at ϕ_n are defined. Define that $R(\phi, \rho_{np}, \theta_{np}) = R(\phi, \mathbf{r}_P)$ and $R(\phi, \rho_n, \theta_n) = R(\phi, \mathbf{r})$, then the BPA for the n-th subaperture is

$$I_n(\rho_n, \theta_n) = \int_{\phi_n - \phi_{In}/2}^{\phi_n + \phi_{In}/2} \sigma_P p_r \left[B\left(R(\phi, \rho_n, \theta_n) - R(\phi, \rho_{np}, \theta_{np})\right)/c_0 \right] \cdot \exp\left[j2\pi f R(\phi, \rho_n, \theta_n)/c_0 \right] d\phi \tag{4}$$

First, the bistatic range sampling requirement for elliptical polar grids of subimages is $|\Delta\rho_n| \le c_0/B$.

Second, the angular sampling requirement for elliptical polar grids of subimages needs to calculate the bistatic range error between two consecutive angular samples, which is depicted in Figure 4. $A_{M\phi}$ is the moving radar's position at ϕ, with $\phi \in (\phi_n - \phi_{In}/2, \phi_n + \phi_{In}/2)$. $d_{M\phi}$ is the range from A_{Mn} to $A_{M\phi}$. $\alpha_{M\phi}$ is the angle from $d_{M\phi}$ to $\mathbf{I}_{Mn}$. $\mathbf{r}_{\pm\Delta\theta_n}$ and $\mathbf{r}$ are two consecutive angular samples, with the elliptical polar coordinates $(\rho_n, \theta_n \pm \Delta\theta_n)$ and (ρ_n, θ_n). The ranges from the moving and stationary radars to the sample $\mathbf{r}_{\pm\Delta\theta_n}$ at ϕ_n are $\mathbf{R}_{Mn\pm\Delta\theta_n}$ and $\mathbf{R}_{Sn\pm\Delta\theta_n}$, with the magnitudes $R_{Mn\pm\Delta\theta_n}$ and $R_{Sn\pm\Delta\theta_n}$. The ranges from the position $A_{M\phi}$ to the samples $\mathbf{r}$ and $\mathbf{r}_{\pm\Delta\theta_n}$ can be calculated and approximated as

$$\begin{cases} R'_{Mn} = \left[R_{Mn}^2 + d_{M\phi}^2 - 2R_{Mn}d_{M\phi}\cos(\theta_n - \alpha_{M\phi}) \right]^{1/2} \approx R_{Mn} - d_{M\phi}\cos(\theta_n - \alpha_{M\phi}) \\ R'_{Mn\pm\Delta\theta_n} = \left[R_{Mn\pm\Delta\theta_n}^2 + d_{M\phi}^2 - 2R_{Mn\pm\Delta\theta_n}d_{M\phi}\cos(\theta_n \pm \Delta\theta_n - \alpha_{M\phi}) \right]^{1/2} \approx R_{Mn\pm\Delta\theta_n} - d_{M\phi}\cos(\theta_n \pm \Delta\theta_n - \alpha_{M\phi}) \end{cases} \tag{5}$$

According to $R_{Mn\pm\Delta\theta_n} + R_{Rn\pm\Delta\theta_n} = R_{Mn} + R_{Sn}$, the bistatic range error between the samples $\mathbf{r}$ and $\mathbf{r}_{\pm\Delta\theta_n}$ is

$$\Delta R_n = R'_{Mn\pm\Delta\theta_n} + R_{Rn\pm\Delta\theta_n} - (R'_{Mn} + R_{Sn}) \approx d_{M\phi}\left(\cos(\theta_n - \alpha_{M\phi}) - \cos(\theta_n - \alpha_{M\phi} \pm \Delta\theta_n)\right) \approx \pm d_{M\phi}\sin(\theta_n - \alpha_{M\phi})\Delta\theta_n \tag{6}$$

d_{Mn} is the length of the n-th subaperture, $-d_{Mn}/2 \le d_{M\phi} \le d_{Mn}/2$, thus the upper bound of ΔR_n is

$$|\Delta R_n| \le d_{M\phi}\left|\sin(\theta_n - \alpha_{M\phi})\right||\Delta\theta_n| \le d_{Mn}|\Delta\theta_n|/2 \tag{7}$$

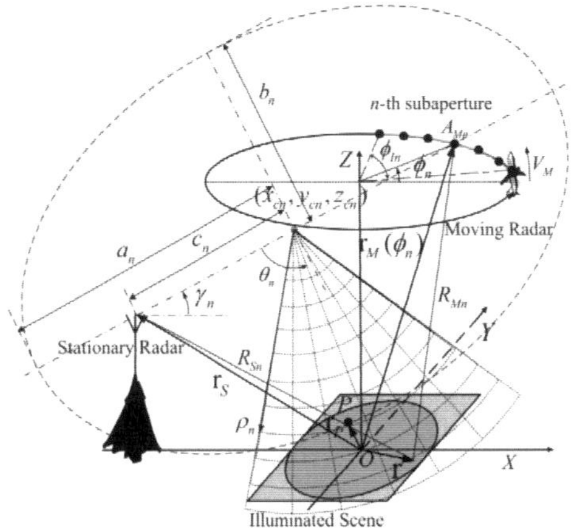

Figure 5. Imaging geometry for FBPA with the subaperture and polar grid processing.

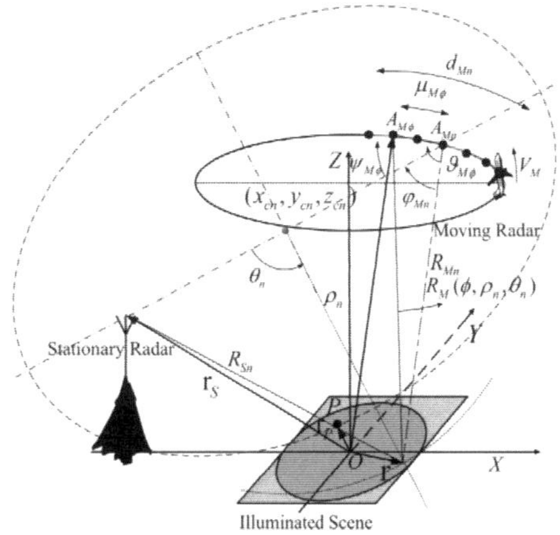

Figure 6. Calculation of the bistatic range.

The maximum phase error is $\Delta\Phi_{n,\max} = 2\pi|\Delta R_n|_{\max}/\lambda_{\min}$, $\lambda_{\min}$ is the minimum of the wavelength. If $\Delta\Phi_{n,\max} \le \pi/8$, the phase error effect can be neglected. Thus, the angular sampling requirement is

$$|\Delta\theta_n| \le \lambda_{\min}/8d_{Mn} \tag{8}$$

4. FBPA with subaperture and polar grid processing

Imaging geometry for FBPA with subaperture and polar grid processing is shown in Figure 5. For the n-th subaperture, a_n and b_n are major and minor axes of the dashed ellipse, and the linear eccentricity is defined as $c_n = \sqrt{a_n^2 - b_n^2}$. The polar coordinates (ρ_n, θ_n) of the sample $\mathbf{r}$ are defined. First, the polar grid's origin is the center point of the stationary radar's position and the considered moving radar's subaperture center, i.e., (x_{cn}, y_{cn}, z_{cn}). Second, the polar range ρ_n is the distance from the polar grid's origin to sample $\mathbf{r}$, and the polar angle θ_n is the angle from a_n to ρ_n. The polar coordinates (ρ_n, θ_n)

$$\begin{cases} \rho_n = \sqrt{[(x_S + x_M(\phi_n))/2 - x]^2 + [y_M(\phi_n)/2 - y]^2 + [(z_S + z_M(\phi_n))/2]^2} \\ \theta_n = \arccos\left(c_n^2 + \rho_n^2 - \left((x_S - x)^2 + y^2 + z_S^2\right)/2\rho_n c_n\right) \end{cases} \tag{9}$$

Similarly, the polar coordinates (ρ_{np}, θ_{np}) of the target P are defined. Since $R(\phi, \mathbf{r}_P) = R(\phi, \rho_{np}, \theta_{np})$ and $R(\phi, \mathbf{r}) = R(\phi, \rho_n, \theta_n)$, the value of the polar subimage at (ρ_n, θ_n) for the n-th subaperture imaging is

$$I_n(\rho_n, \theta_n) = \int_{\phi_n - \phi_{In}/2}^{\phi_n + \phi_{In}/2} \sigma_P p_r\left[B\left(R(\phi, \rho_n, \theta_n) - R(\phi, \rho_{np}, \theta_{np})\right)/c_0\right] \bullet \exp\left[j2\pi f R(\phi, \rho_n, \theta_n)/c_0\right] d\phi \tag{10}$$

To investigate the sampling requirements for the polar grids, we need to calculate the bistatic range from the moving and stationary radars to the sample (ρ_n, θ_n), which is shown in Figure 6. $A_{M\phi}$ is the moving radar position at ϕ, with $\phi \in (\phi_n - \phi_{In}/2, \phi_n + \phi_{In}/2)$. $R_M(\phi, \rho_n, \theta_n)$ is the distance from $A_{M\phi}$ to (ρ_n, θ_n). $\mu_{M\phi}$ is the range between A_{Mn} and $A_{M\phi}$. $\vartheta_{M\phi}$ is the angle between the lines $\mu_{M\phi}$ and R_{Mn}, and $\psi_{M\phi}$ is the angle between straight lines $\mu_{M\phi}$ and a_n. φ_{Mn} is the angle between the lines R_{Mn} and a_n. d_{Mn} is the length of the n-th subaperture, $-d_{Mn}/2 \le \mu_{M\phi} \le d_{Mn}/2$. Thus, the bistatic range at ϕ is

$$R(\phi,\rho_n,\theta_n)=R_M(\phi,\rho_n,\theta_n)+R_S(\rho_n,\theta_n)=\sqrt{R_{Mn}^2+\mu_{M\phi}^2-2R_{Mn}\mu_{M\phi}\cos(\vartheta_{M\phi})}+R_{Sn}\approx R_{Mn}+R_{Sn}-\mu_{M\phi}\cos(\vartheta_{M\phi})$$
$$\approx\sqrt{\rho_n^2+c_n^2+2\rho_n c_n\cos(\theta_n)}+\sqrt{\rho_n^2+c_n^2-2\rho_n c_n\cos(\theta_n)}-\mu_{M\phi}\cos(\vartheta_{M\phi}) \tag{11}$$

From the imaging geometry in Figure 6, $\cos(\vartheta_{M\phi})$ can be calculated by as follows

$$\cos(\vartheta_{M\phi})=\cos(\varphi_{Mn}+\psi_{M\phi})=c_n\cos(\psi_{M\phi})+\rho_n\cos(\theta_n+\psi_{M\phi})\Big/\sqrt{\rho_n^2+c_n^2+2\rho_n c_n\cos(\theta_n)} \tag{12}$$

Two-dimensional Fourier transforms of $I_n(\rho_n,\theta_n)$ with respect to ρ_n and θ_n is

$$I_{FTn}(k_{\rho_n},k_{\theta_n})=\iint I_n(\rho_n,\theta_n)\exp\left[-j2\pi(k_{\rho_n}\rho_n+k_{\theta_n}\theta_n)\right]d\rho_n d\theta_n \tag{13}$$

Where k_{ρ_n} and k_{θ_n} are wavenumbers corresponding to polar range ρ_n and polar angle θ_n, respectively. Substituting (17) into (20), then the stationary phase condition is given by

$$\begin{cases}\partial\left(2\pi fR(\phi,\rho_n,\theta_n)/c_0-2\pi k_{\rho_n}\rho_n\right)\big/\partial\rho_n=0\\ \partial\left(2\pi fR(\phi,\rho_n,\theta_n)/c_0-2\pi k_{\theta_n}\theta_n\right)\big/\partial\theta_n=0\end{cases} \tag{14}$$

Solving the above equations shows that $I_{FTn}(k_{\rho_n},k_{\theta_n})$ is nonzero, when

$$\begin{cases}2f_{\min}\Big/\left(c_0\sqrt{1+\delta_n^2}\right)\le k_{\rho_n}\le 2f_{\max}\Big/\left(c_0\sqrt{1+\delta_n^2}\right)\\ -f_{\max}d_{Mn}\Big/\left(2c_0\sqrt{1+\delta_n^2}\right)\le k_{\theta_n}\le f_{\max}d_{Mn}\Big/\left(2c_0\sqrt{1+\delta_n^2}\right)\end{cases} \tag{15}$$

Where δ_n is the ratio of c_n to ρ_n (i.e., $\delta_n=c_n/\rho_n$). The bounds of k_{ρ_n} and k_{θ_n} can be translated into the sampling requirements of ρ_n and θ_n for the *n*-th polar subimage, which are given by

$$\begin{cases}|\Delta\rho_n|\le c_0\sqrt{1+\delta_n^2}\big/2(f_{\max}-f_{\min})=c_0\sqrt{1+\delta_n^2}\big/2B\\ |\Delta\theta_n|\le c_0\sqrt{1+\delta_n^2}\big/f_{\max}d_{Mn}=\lambda_{\min}\sqrt{1+\delta_n^2}\big/d_{Mn}\end{cases} \tag{16}$$

5. Experimental Results

To give a comparison of two proposed FBPA, the experimental results are shown in this section. The simulation parameters of the OS-BCSAR system are shown in Table 1. Five point targets labeled as A~E are located in the scene (both 50m in the X and Y axes), which are shown in Figure 7.

Figure 7 gives the imaging results by two proposed FBPA, and it is shown that the simulated scene is well reconstructed. From the top row, the CSAR feature for the targets A, C and D is seen by the FBPA with the subaperture and elliptical polar grid processing. From the bottom row, it is seen that the imaging results of targets A, C and D by the FBPA with the subaperture and polar grid processing is similar to those at the top row. However, the focusing quality of targets A, C and D at the bottom row is slightly degraded compared with those at the top row, due to the different grids of subimages.

Measured parameters (Resolution (RES) and PSLR) of the selected targets are computed in Table 2, which shows that they are close for two proposed FBPA. Processing time of two proposed FBPA is 49.9s and 46.6s, respectively, which shows that the imaging efficiency of two proposed FBPA is close.

Table 1. Simulation parameters of the OS-BCSAR system.

Parameter	Value	Parameter	Value
Carrier frequency	400MHz	Signal bandwidth	200MHz
Sampling frequency	220Hz	Pulse duration	1us
Pulse repetition frequency	100Hz	Stationary radar position	(-1500, 0, 100)m
Moving radar angular speed	6°/s	Radius and altitude of the circular track	1400m (500m)

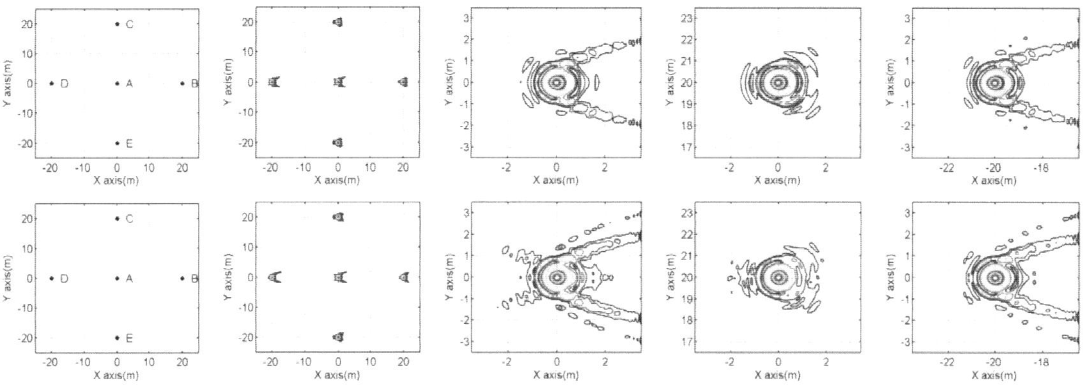

Figure 7. Imaging results obtained by two algorithms. Top row: FBPA with the subaperture and elliptical polar grid processing; Bottom row: FBPA with the subaperture and polar grid processing.

Table 2. Measured parameters of the selected targets.

Algorithm	Measured parameters		Target A	Target C	Target D
FBPA with the subaperture and elliptical polar grid processing	RES (m)	X axis	0.279	0.270	0.280
		Y axis	0.271	0.278	0.281
	PSLR (dB)	X axis	-8.417	-7.958	-8.373
		Y axis	-8.228	-8.731	-8.241
FBPA with the subaperture and polar grid processing	RES (m)	X axis	0.284	0.275	0.281
		Y axis	0.271	0.278	0.283
	PSLR (dB)	X axis	-7.197	-7.882	-8.326
		Y axis	-7.167	-8.833	-8.321

6. Conclusion

In this paper, we present the FTDA for the OS-BCSAR imaging. It shows that the FBPA with the subaperture and elliptical polar grid processing and the FBPA with the subaperture and polar grid processing are generally available for the OS-BCSAR imaging.

Acknowledgments

This work was supported in part by the National Natural Science Foundation of China (No. 61571447), in full by Scientific Research Project of Education Department of Hunan (No. 16C1233), and in part by the Domain Foundation of Equipment Advance Research (No. 6140410301 and No. 6140410304).

References

[1] Rodriguez-Cassola M et al 2012 First bistatic spaceborne SAR experiments with TanDEM-X *IEEE Geosci. Remote Sens. Lett.* 9 33-37.

[2] Xie H, An D, Huang X and Zhou Z 2016 Efficient raw signal generation based on equivalent scatterer and subaperture processing for one-stationary bistatic SAR including motion errors *IEEE Trans. Geosci. Remote Sens.* 54 3360-3377.

[3] Xie H, An D, Huang X and Zhou Z 2015 Research on spatial resolution of one-stationary bistatic ultrahigh frequency ultrawidebeam-ultrawideband SAR based on scattering target wavenumber domain support *IEEE J. Sel. Topics Appl. Earth Observ.* 8 1782-1798.

[4] Ponce O et al 2014 Fully polarimetric high-resolution 3-D imaging with circular SAR at L-band *IEEE Trans. Geosci. Remote Sens.* 52 3074-3090.

[5] Goh A, Preiss M, Stacy N and Gray D 2010 Bistatic SAR experiment with the Ingara imaging radar *IET Radar Sonar Navig.* 4 426-437.

[6] Wang Y, Tan W, Hong W, Lin Y and Wu Y 2009 Focusing bistatic circular SAR data using polar format algorithm *Proc. Asian Pacific Synth. Aperture Radar* (Xian, China) pp 989-992.

An Accurate Framework for Arbitrary View Pedestrian Detection in Images

Y Fan[1,*] , G Wen[1] and S Qiu[1]

[1]Science and Technology on Automatic Target Recognition Laboratory (ATR), National University of Defense Technology, Changsha, China

*E-mail: fanyaxiang@126.com

Abstract. We consider the problem of detect pedestrian under from images collected under various viewpoints. This paper utilizes a novel framework called locality-constrained affine subspace coding (LASC). Firstly, the positive training samples are clustered into similar entities which represent similar viewpoint. Then Principal Component Analysis (PCA) is used to obtain the shared feature of each viewpoint. Finally, the samples that can be reconstructed by linear approximation using their top- k nearest shared feature with a small error are regarded as a correct detection. No negative samples are required for our method. Histograms of orientated gradient (HOG) features are used as the feature descriptors, and the sliding window scheme is adopted to detect humans in images. The proposed method exploits the sparse property of intrinsic information and the correlations among the multiple-views samples. Experimental results on the INRIA and SDL human datasets show that the proposed method achieves a higher performance than the state-of-the-art methods in form of effect and efficiency.

1. Introduction

Inspired by robot vision, intelligent transportation system, security monitoring and other applications, detecting object from the image and video is a hot topic in computer vision research. As one of the typical sub-problem, because of its different applications, pedestrian detection has attracted a surge of interest from both academia and industry. Despite the fact that the research of pedestrian detection has made significant progress, where the camera has a fixed viewpoint and captures a static background, it is still far away to solve this problem under more complex challenge such as viewpoint changes.

Most of the current methods regard pedestrian detection as a pedestrian/non-pedestrian two categories of classification problem. Current methods contain two steps [1]: feature extraction and classification. In [2], the histograms of oriented gradients (HOG) descriptors are proposed as the feature extraction method and linear SVM is used as the classification. However, the continuous change viewpoints of pedestrians constitute a Manifold structure, which resulting in poor detection performance for the traditional linear classifier. To deal with this problem, cascade classifiers [3][4] are proposed. In [3], an effective and efficient feature selection method based on Gentle Adaboost (GAB) cascade and the Four Direction Feature (FDF) are proposed. As a result, the irrelevant and redundant features are removed to improve the performance. Ye et al. [4] combined cascade Adaboost classifier and linear SVM for multi-view pedestrian detection. The first four layers of the classifier used the Adaboost and the last layer used the linear SVM. These methods have achieved good results, but the relevant information among the various viewpoints is ignored. In addition, the viewpoint of the test sample is not limited to the training set of several discrete viewpoints in practical applications, and one of the most difficult problems is to detect the unseen view samples. Joint multi-view model [5] is proposed to address this issue by establishing a certain relationship among the various viewpoints. Another way to deal with this problem is to use the shared feature (or called joint feature) [6][7] by finding common features that can be shared across the viewpoints. In this way, the latent information among different viewpoints is utilized and the compute cost is reduced.

Recently, by combining the advantages of both sparse representation based classification (SRC) and joint sparse representation based classification (JSRC), Zhang et al. [8] proposed a mixed-norm sparse representation model for multi-view face recognition. Their work assumed that the sample of unseen view can be regarded as the sparse representation of samples from all viewpoints. However, this method tried to use the samples of all the viewpoints to identify the arbitrary view sample, which ignored the shared feature among the training sample. In light of this, we propose a new method for detecting pedestrian under arbitrary view from images collected under various views based on the locality-constrained affine subspace coding (LASC) [9]. In the training stage, we first cluster the positive training samples into similar entities according to the viewpoints. Then the Principal Component Analysis (PCA) is utilized to learn the shared features of each viewpoint. In the testing stage, sliding window scheme is adopted to find the correct detection window that can be reconstructed by linear approximation using their top- k nearest shared features with a small error. No negative samples are required for our method. We conduct experiments on two widely available public datasets. The experiment results demonstrate the superior performance of the proposed approach compared to several state of-the-art algorithms

The remainder of this paper is organized as follows. Section II is a detailed presentation of the proposed algorithm. Experimental results and discussions are given in Section III. In the end, Section IV concludes this paper.

2. Method

An overview of the proposed method is illustrated in Figure 1. First, the positive training samples are clustered into several clusters according to the similarity of viewpoints and PCA is employed to obtain the subspace of each cluster which referred to the shared feature. Then, given the query window, the reconstruction cost is computed by locality-constrained affine subspace coding (LASC) with the shared feature, and then, the detection is determined with the small reconstruction cost.

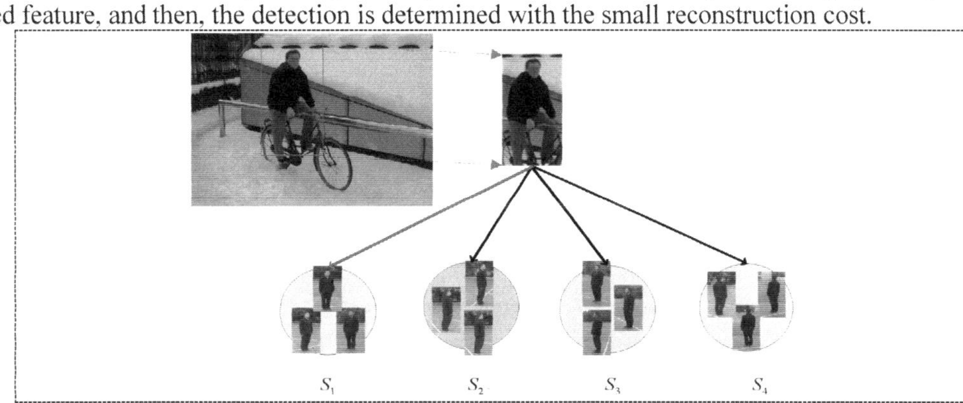

Figure 1. Overview of the proposed method for anomalous event detection.

2.1. Feature Extracting

In our work, HOG features [2] are extracted to represent pedestrian. Due to the ability to characterize the edge feature and insensitivity to light and small offset, HOG features have achieved excellent results in pedestrian detection task. A 64×128 training image is divided into 16×16 overlapped image blocks, which contain 2×2 cells of size 8×8 pixels. In each small cell, gradient is calculated according to calculating the nine-bin histogram of gradient orientations in the range of $0°$ to $360°$. As a result, each sample is represented by 105 blocks and a 3780-dimensional feature vector is extracted and normalized.

2.2. Clustering

Suppose the positive training samples can be represented as $\mathbf{D}=[\mathbf{d}_1,\mathbf{d}_2,\ldots,\mathbf{d}_l]\in\mathbf{R}^{m\times l}$ where m=3780 is the dimension of feature vector and l is the number of training samples. The training sample contains various viewpoints and it can be clustered into n similar entities $C=\{C_1,C_2,\ldots,C_n\}$ based on the similarity of viewpoint, which can be performed by employing K-means clustering algorithm as follow

$$\arg\min\sum_{i=1}^{n}\sum_{\mathbf{d}\in C_i}\|\mathbf{d}-\boldsymbol{\mu}_i\|_2^2 \tag{1}$$

where $\boldsymbol{\mu}_i$ is the mean vector of cluster C_i.

2.3. Shared Feature Obtaining

After clustering the positive training samples into several similar viewpoint entities, we try to find the shared feature of each viewpoint. It is assumed that each cluster has the geometry of a low-dimensional linear subspace, which is referred to the shared feature. Principal Component Analysis (PCA) is utilized to preserve the directions that have larger variances for each cluster. Suppose that $\mathbf{u}_{i,t}$ is the t-th orthogonal of eigenvalues of covariance matrix $\boldsymbol{\Sigma}_i$, then the shared feature of cluster C_i can be represented as:

$$\mathbf{A}_i=\mathbf{U}_i=[\mathbf{u}_{i,1},\cdots,\mathbf{u}_{i,p}] \tag{2}$$

where $i=1,2,\ldots,n$ and p is the dimension of the shared feature.

2.4. Pedestrian Detection via Locality-constrained Affine Subspace Coding

With the shared feature mentioned above, we present the detail of pedestrian detection utilizing locality-constrained affine subspace coding (LASC). LASC [9] is an improved model from the locality-constrained linear coding (LLC) method [10], which leverages locality constraints to achieve both high efficiency and local smooth sparsity. The LASC scheme focuses on the local geometric structure of various viewpoints samples and uses top-k nearest shared feature to encode the test sample. It is obvious that a normal pedestrian of arbitrary view can be linearly represented by its neighbouring subspaces of various viewpoints samples while the reconstruction error is small. In contrast, the negative sample is different from the normal pedestrian of arbitrary view. As a result, it is often difficult to find the linear representation of the neighbouring subspaces around the negative sample.

In detail, suppose $\mathbf{A}\in\mathbf{R}^{m\times p}$ is the linear subspace projection matrix, the geometry of the feature space $\mathbf{D}=[\mathbf{d}_1,\mathbf{d}_2,\ldots,\mathbf{d}_l]\in\mathbf{R}^{m\times l}$ can be represented as a new local coordinate scheme: $S_i=\{\mathbf{d}_i+\mathbf{A}_i\boldsymbol{\beta}_i,\boldsymbol{\beta}_i\in\mathbf{R}^p\},i=1,\ldots,n$. With a test sample $\mathbf{y}\in\mathbf{R}^m$, LASC attempts to represent it by making a linear combination of the top-k nearest affine subspaces which is referred to the shared feature. Specifically, LASC can be formulated based on the following optimization:

$$\min_{\forall x_i}\sum_{S_i\in N_k^S(\mathbf{y})}\left\{\|(\mathbf{y}-\mathbf{d}_i)-\mathbf{A}_i\boldsymbol{\beta}_i\|_2^2+\lambda\mathrm{dis}(\mathbf{y},S_i)\|\boldsymbol{\beta}_i\|_2^2\right\} \tag{3}$$

where $N_k^S(\mathbf{y})$ is the neighbourhood of $\mathbf{y}$, which is defined by the k closest shared features of $\mathbf{y}$. Here, $dis(\mathbf{y},S_i)$ is the approximate distance of $\mathbf{y}$ to the shared feature S_i. Use the Gaussian-distribution as a kernel to measure the similarity between the point $\mathbf{y}$ and cluster centroid $\boldsymbol{\mu}_i$, $\mathrm{dis}(\mathbf{y}, S_i)$ can be represented as:

$$\mathrm{dis}(\mathbf{y}, S_i) = \frac{\exp(-\frac{1}{2\delta^2}\|\mathbf{y}-\boldsymbol{\mu}_i\|_2^2)}{\sum_{i'=1}^{n}\exp(-\frac{1}{2\sigma^2}\|\mathbf{y}-\boldsymbol{\mu}_{i'}\|_2^2)} \tag{4}$$

where $\boldsymbol{\mu}_i$ is the mean vector of the cluster C_i.

According to [9], the solution of (3) is as follows:

$$\underset{S_i \in N_k^S(\mathbf{y})}{\boldsymbol{\beta}_i} = (\mathbf{A}_i^T\mathbf{A}_i + \lambda \mathrm{dis}(\mathbf{y}, S_i)\mathbf{I})^{-1}\mathbf{A}_i^T(\mathbf{y}-\mathbf{d}_i) \tag{5}$$

where $\mathbf{I}$ is the unit matrix of order p. If $S_i \notin N_k^S(\mathbf{y})$, then $\boldsymbol{\beta}_i$ is equal to zero. The code result of $\mathbf{y}$ can be represented as $\boldsymbol{\beta} = [\boldsymbol{\beta}_1^T, \ldots, \boldsymbol{\beta}_i^T, \ldots, \boldsymbol{\beta}_n^T]^T$.

Then we detect the pedestrian that is based on the reconstruction error by

$$LRC(\mathbf{y}) = \sum_{S_i \in N_k^S} \|(\mathbf{y}-\mathbf{d}_i) - \mathbf{A}_i\boldsymbol{\beta}_i\|_2^2 \tag{6}$$

where $\tilde{\boldsymbol{\beta}}_i$ is the solution of (3) according to (4) and (5), $\mathbf{A}_i$ is obtained from (2). Here, $\mathbf{y}$ is detected as a correct detection if the following criterion is satisfied:

$$LRC(\mathbf{y}) > \hat{\varepsilon} \tag{7}$$

where $\hat{\varepsilon}$ is the threshold that decides the sensitivity of the detection method.

3. Experiments

In this section, we describe the datasets used in our experiments, and then compare the detection results with the state-of-the-arts on these datasets. Details are given as follows.

3.1. Datasets and Evaluation

We collected 8000 positive samples of different viewpoints for training from INRIA dataset [1], SDL dataset [11] and Multi-view Pedestrian Recognition (MVPR) dataset [12]. All the samples are trimmed to size of 64×128. Some of training samples of different viewpoints are shown in Figure 2.

Figure 2. Training samples of different viewpoints

The INRIA test set and the SDL test set are chose for testing. For the INRIA test set, there are 288 images with 589 multi-view pedestrian samples. And for the SDL test set, there are 258 images with 1688 pedestrian samples for testing, in which pedestrians are almost situated in multi-view appearance. Both of the two datasets contain a variety of viewpoints, to demonstrate that our method could detect the pedestrian of arbitrary view.

An overlap threshold [1] for determining a correct detection for can be represented as $\frac{area(BB_{gt} \cap BB_{dt})}{area(BB_{gt} \cup BB_{dt})} \geq \vartheta$, where BB_{gt} and BB_{dt} stand for the detected bounding box and annotated bounding box, respectively. ϑ is often set to 0.5.

Miss rate against False Positives Per Window (FPPW) is chose for evaluation, which are defined as $MissRate = \frac{\#Missed\ positive\ detection}{\#Total\ positives}$ and $FPPW = \frac{\#False\ positive\ detections}{\#Total\ image\ windows}$.

3.2. Experiments result

The number k of shared feature and the shared feature dimension p were set empirically to 3 and 64, respectively. The parameter σ *in* (4) and the regularization parameter λ in (5) are set to 0.1 and 1, respectively. We first experimented with the INRIA dataset in order to evaluate the effect of the number n of cluster centroids. In detail, we set $n = 4,6,8,10,12$ and check the experiment result test on the INRIA test set. The results are shown in Figure 3 (a), given different parameter settings in terms of Miss rate against False Positives Per Window (FPPW). The performance increases with the number of cluster centroids when the latter is less than 8. This is due to the fact that if the number of cluster centroids is too few, the samples of different viewpoints are confused. Equally, performance drops quickly with increasing number of cluster centroids when the latter is beyond 8. Here the reason is that more cluster centroids are more likely to produce fewer characteristic shared features. Based on the above, we chose $n = 8$ as the parameters for the succeeding experiments.

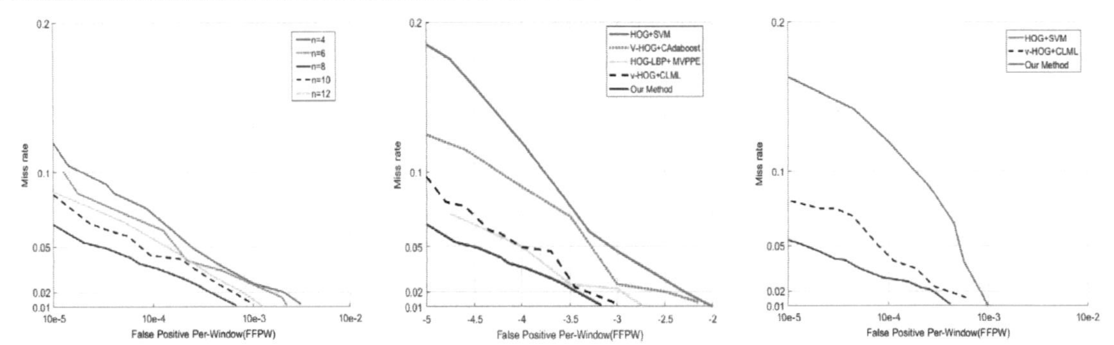

(a) Performance under different number of cluster centroids

(b) Detection result for INRIA Dataset

(c) Detection result for SDL Dataset

Figure 3. The parameters analysis and comparison of ours method with recent state-of-the-art methods .

When evaluating the detection performance on the INRIA Dataset, four algorithms specifically designed for Multi-View pedestrian-detection are chose for comparable, including HOG+SVM [2], v-HOG+CAdaboost [13], HOG-LBP+MVPPE method [14] and v-HOG+CLML [15]. The results of the contrast methods are obtained in their respective papers. It can be observed in Figure 3 (b) that the Miss rate decreases with the increase if the False Positive Per-Window (FFPW), and our method achieves superior performance compared with other four algorithms in all situations. In detail, our method achieves 3.5% miss rate at FPPW of 10^{-4}, which is about 9% lower than the HOG+SVM method, about 5% lower than the v-HOG+CAdaboost method, about 3.5% lower than HOG-LBP+MVPPE method and 1.5% lower than v-HOG+CLML method.

For the SDL dataset, which is a new dataset, we compare the results of our method with the HOG+SVM [2] and v-HOG+CLML [14]. The results of the contrast methods are obtained in [14]. As shown in Figure 3.(C), our method is superior to the other state-of-the-arts methods, with 5.5% miss rate at FPPW of 10^{-5} and 2.5% miss rate at FPPW of 10^{-4}. Therefore, these findings demonstrate that our algorithm achieves better results than state-of-the-art methods for arbitrary view pedestrian detection task.

4. Conclusions

We have presented a new method for arbitrary view pedestrian detection in images by employing locality-constrained affine subspace coding (LASC). Notwithstanding that the method only uses the positive samples for training, the learning model is capable of detecting pedestrian under arbitrary view. Experimental results on two datasets prove that the proposed method achieves a higher performance than the state-of-the-art methods in form of effect and efficiency. In the future, we will extend the approach to other objects detection task.

References

[1] Nguyen D T, Li W, Ogunbona P O. Human detection from images and videos: A survey. Pattern Recognition, 2015; 51:148-175.
[2] Dalal N, Triggs B. Histograms of oriented gradients for human detection. Proceedings of the IEEE Conference on Computer Vision and Pattern Recognition, 2005, 886–893.
[3] Kuang H, Chong Y, Li Q, Zheng C. Mutual cascade method for pedestrian detection. Neurocomputing, 2014, 137: 127–135.
[4] Ye Q, Jiao J, Jiang S, Fast and robust pedestrian detection algorithm with multi-Scale orientation features. Journal of Software, 2011, 22(12): 3004–3014.
[5] Xue Z, Li G, Huang Q, Joint multi-View representation learning and image tagging. Proceedings of the Thirtieth AAAI Conference on Artificial Intelligence, 2016.
[6] Torralba A, Murphy K P, Freeman W T, Sharing visual features for multiclass and multiview object detection. IEEE Transactions on Pattern Analysis and Machine Intelligence. 2007, 29(5): 854-869.
[7] Zhu Y, Wang S, Yue L, Ji Q, Multiple-Facial action unit recognition by shared feature learning and semantic relation modeling, International Conference on Pattern Recognition, 2014.
[8] Zhang X, Pham D, Venkatesh S, Liu W and Phung D. Mixed-norm sparse representation for multi view face recognition. Pattern Recognition. 2015, 48(9): 2935-2946.
[9] Li P, Lu X, Wang Q. From dictionary of visual words to subspaces: locality-constrained affine subspace coding. Proceedings of IEEE Conference on Computer Vision and Pattern Recognition, 2015.
[10] Wang J, Yang J, Yu K, Lv F, Huang T S, and Gong Y, Locality-constrained linear coding for image classification. Proceedings of IEEE Conference on Computer Vision and Pattern Recognition (CVPR), 2010.
[11] Ye Q, Liang J, Jiao J, Pedestrian detection in video images via error correcting output code classification of manifold subclasses, *IEEE Transaction on Intelligent Transportation Systems*, 2012, 13(1): 193-202.
[12] Wu L, Sun H, Ji K, Fan Y, Zhang Y, Multi-view pedestrian recognition via non-negative least-squares. Proceedings of the Chinese Intelligent Automation Conference, 2015, 199-208.
[13] Zhu Q, Avidan S, Yeh M C, Cheng K T. Fast human detection using a cascade of histograms of oriented gradients. Proceedings of IEEE Conference on Computer Vision and Pattern Recognition, 2012.
[14] Liu W, Yu B, Duan C, Cha L, Yuan H, Zhao H, A pedestrian-detection method based on heterogeneous features and ensemble of multi-view–pose parts, *IEEE Transaction on Intelligent Transportation Systems*, 2015, 16(2): 813-823.
[15] Xu R, Jiao J, Zhang B, Ye Q. Pedestrian detection in images via cascaded L1-norm minimization learning method. Pattern Recognition. vol. 2012, 45(7). 2573–2583.

Enhanced Imaging of Building Interior for Portable MIMO Through-the-wall Radar

Yongping Song[1,*], Jiahua Zhu[1], Jun Hu[1], Tian Jin, Zhimin Zhou[1]

[1]College of Electronic Science, National University of Defense Technology, Changsha, China

* sypopqjkl@163.com

Abstract. Portable multi-input multi-output (MIMO) radar system is able to imaging the building interior through aperture synthesis. However, significant grating lobes are invoked in the directly imaging results, which may deteriorate the imaging quality of other targets and influence the detail information extraction of imaging scene. In this paper, a two-stage coherence factor (CF) weighting method is proposed to enhance the imaging quality. After obtaining the sub-imaging results of each spatial sampling position using conventional CF approach, a window function is employed to calculate the proposed "enhanced CF" adaptive to the spatial variety effect behind the wall for the combination of these sub-images. The real data experiment illustrates the better performance of proposed method on grating lobes suppression and imaging quality enhancement compare to the traditional radar imaging approach.

1. Introduction

Through-the-wall radar system is widely employed in radar detection by using low-frequency signal for the target detection behind the wall. Besides the detection of human target, through-the-wall radar system is also able to evaluate the structure of building and obtain the interior information. Synthesis aperture radar (SAR) are usually applied for through-the-wall system to illuminate the building from several directions [1,2], but this operation mode is more suitable for outdoor imaging due to the large and complex system size. For indoor illumination (like the room imaging in a narrow stairway), a portable multi-input multi-output (MIMO) through-the-wall radar system is needed.

The portable MIMO through-the-wall radar system is also able to achieve the illumination of detection scene through MIMO-SAR imaging, while the synthesis aperture is limited in narrow space detection. Behzad *et al.* propose a separated transmitting-receiving system for all directions of illumination to the building in [3], which makes the direction of synthetic aperture no longer parallel to the wall, but a circle along the radar system. This method can obtain an imaging result with identical resolution in all directions. However, it is hardly to apply in practical because of the extreme requirement to the system precision.

Since multi-directions illumination is not always available in practical detection, the aperture synthetic pattern of portable MIMO through-the-wall radar system is usually along a single direction parallel to the illumination wall. As to the indoor radar imaging, the distribution of antennas may not satisfy the spatial sampling theorem, which induces significant grating lobes as interference to real targets and decreases the imaging quality. Coherence factor (CF) is one of the most common used corrections to suppress grating lobes, including traditional CF approach, phase CF approach [4], sign

CF approach [5], etc. While these methods are effective, they may also blank the real targets in the MIMO-SAR system due to the spatial fluctuation of targets in the imaging results of different channels.

In this paper, we proposed a two-stage CF method for the enhancement of imaging quality. The imaging results through traditional CF method is further processed by a window function to compute the "enhanced CF" adaptive to the spatial variety effect behind the wall for the combination of these results. The method is validated by real data experiments for a better performance.

2. MIMO-SAR Imaging Method

As mentioned before, portable MIMO through-the-wall radar employs the synthesis aperture technique to combine the images obtain from each spatial sampling position and output the final imaging results. The schematic figure of the system is shown in Figure 1. Different from traditional imaging, the existence of refraction in the wall makes it difficult to calculate the accurate delay of target in MIMO-SAR imaging.

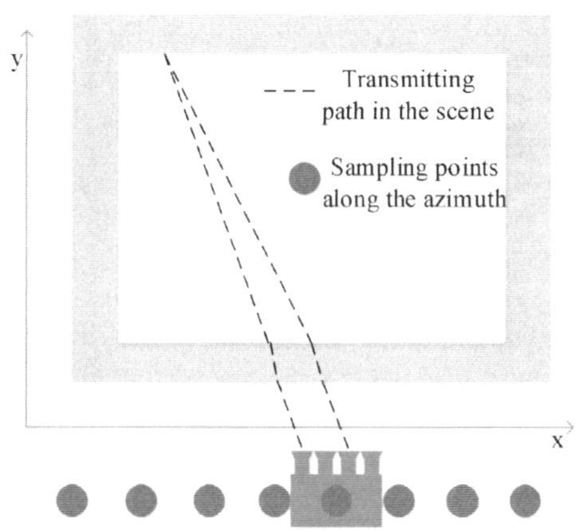

Figure 1. The schematic figure of MIMO-SAR through-the-wall imaging. Due to the refraction in the front wall, the delay of a target behind the wall will be larger than the situation without the wall. Actually, multi-path effect [6] of transmitted signal exists in a closed indoor illumination, but we do not consider this problem here for simplicity. Besides, this problem can be addressed by the system design of MIMO radar.

Conventionally, the back projection (BP) method is used for through-the-wall imaging [7]. Without loss of generality, when transmitting a step frequency signal, the imaging magnitude of point q in the detection scene at the k th spatial sampling point can be expressed as

$$I(q,k)=\frac{1}{M}\sum_{n=0}^{N-1}\sum_{m=0}^{M-1}R(m,n,k)\exp(\mathrm{i}2\pi f(m)t(q,n,k)) \tag{1}$$

where M is the frequency sampling number, N is the number of channel, $R(m,n,k)$ is the radar return of channel n corresponding to the sampling point m at the k th spatial sample, i represents the imaginary sign, $f(m)$ denotes the operation frequency of sampling point m, $t(q,n,k)$ is the corresponding delay of point q. Note that the accuracy of $t(q,n,k)$ depends on the thickness and permittivity of the wall, and it can be precisely calculated when these two parameters are accurately known. However, the parameters are almost impossible to measure in a random and unfamiliar detection scene. Besides, although several auto-focusing algorithms [8] are proposed for the calculation of delay considering the refraction, they are all employed for the wall with homogeneous dielectric and structure, while this is hard to guarantee during the MIMO-SAR illumination at different directions. Therefore, we still calculate the $t(q,n,k)$ without considering the refraction of wall for simplicity. And ignoring the refraction in the wall, $t(q,n,k)$ can be calculated as

$$t(q,n,k) = \left(\left|p(q) - p_{tx}(n,k)\right| + \left|p(q) - p_{rx}(n,k)\right|\right)/c \tag{2}$$

where c is the speed of light, $p(q)$ is the location of point q in the imaging scene, $p_{tx}(n,k)$ and $p_{rx}(n,k)$ are the locations of the transmitting and receiving antenna of channel n at the k th spatial sample, respectively. Note that it is reasonable to ignore the defocus caused by the refraction in the wall [9] since the aperture of the portable MIMO through-the-wall radar system is usually small. Next, the final imaging result of point q, denoted as $I_s(q)$, is the coherence sum of the sub-images obtained at all the K spatial samples

$$I_s(q) = \sum_{k=0}^{K-1} I(q,k) \tag{3}$$

3. Two-stage CF Enhancement Method

The antenna number of portable MIMO through-the-wall radar system is limited by its size and cost, while large antenna synthetic aperture is desired for a better azimuth resolution. Besides, the illumination scene of through-the-wall radar usually does not have the character of "far-field" illumination, and the beams along the azimuth is usually not a uniform distribution in the MIMO radar that the transmitting and receiving antennas are separated, which finally invoke obvious grating lobes in the image. The grating lobes may be falsely detected or interfere the imaging of real targets. CF approach [4] is one of the simplest and most widely used methods to suppress the grating lobes. The CF weight of point q in the multi-channel imaging at the k th spatial sample is defined as

$$CF(q,k) = \frac{\left|\sum_{n=0}^{N-1}\sum_{m=0}^{M-1} R(m,n,k)\exp(\mathrm{i}2\pi f(m)t(q,n,k))\right|^2}{N\sum_{n=0}^{N-1}\left|\sum_{m=0}^{M-1} R(m,n,k)\exp(\mathrm{i}2\pi f(m)t(q,n,k))\right|^2} \tag{4}$$

And the sub-image after CF processing is

$$I_{CF}(q,k) = I(q,k)CF(q,k) \tag{5}$$

Note that the traditional CF approach only suppresses the grating lobes at a certain spatial sample. However, the non-uniform distribution of spatial sampling to the MIMO array may also generates new grating lobes during the synthesis aperture imaging, even if the azimuth samples are uniform distributed. Therefore, it is necessary to add a further CF processing for the suppression of grating lobes when integrating the final imaging result.

It is well studied that traditional CF weighting method is suitable for ideal point targets, while it may also suppress the main lobe of extended targets like walls, household appliances and other strong reflectors, since these targets may fluctuate in spatial in the target profiles obtained from different spatial samples, which results in the decreasing of target coherence. The Parseval theorem illustrates that $CF(q,k)$ represents the ratio between $I(q,k)$ and the energy of radar returns at the corresponding locations in channels (ignore the constant N). However, $I(q,k)$ only represents the energy with zero-frequency, but the frequency spectrum of target profile may spread during the integration of imaging results in spatial, thus it is better to extend the frequency interval of the numerator and includes the extended energy of target for the second stage of CF calculation.

It is shown in (4) that the weights of traditional CF method for any channels are all 1, which means that by changing these weights, it is able to change the frequencies included in the numerator. As a consequence, a window function with larger pass band like Hanning or Hamming window may use to replace the rectangle window, for the extension of frequency interval included in the CF. Denote $CF_s(q)$ as the weight of point q during the integration of the sub-images, then

$$CF_s(q) = \frac{\left|\sum_{k=0}^{K-1} w(k)CF(q,k)I(q,k)\right|^2}{K\sum_{k=0}^{K-1}\left|CF(q,k)I(q,k)\right|^2} \tag{6}$$

where $w(k)$ is the value of the k th element in the chosen window function with length equals to K , then the final output image with enhanced imaging quality can be obtained as

$$I_E(q) = CF_s(q)\sum_{k=0}^{K-1} I_{CF}(q,k) \tag{7}$$

4. Practical Experiment and Results Discussion

The effectiveness of the proposed method is validated by a portable MIMO through-the-wall radar system with 4 transmitting antennas and 5 receiving antennas, shown in Figure 2. The length of antennas array is 0.4m, a step-frequency signal from 1.5GHz to 2.1GHz is transmitted and the step frequency is 4MHz. Figure 2 also demonstrates the structure of the illumination scene, a meeting room containing some common furniture. Set the left-bottom corner of the scene as the origin of coordinates, and there are 46 azimuth sampling points with the interval 0.1m. The distance between the radar system and the front surface of wall is 2.1m.

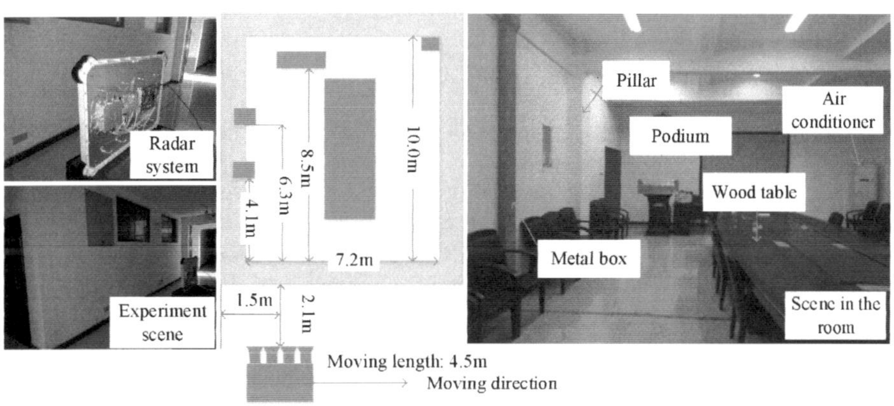

Figure 2. The illumination scene of the experiment. During the experiment, the radar moves along the wall with a constant speed for 4.5m, and makes a spatial sample every 0.1m. The strong reflectors are marked in this figure.

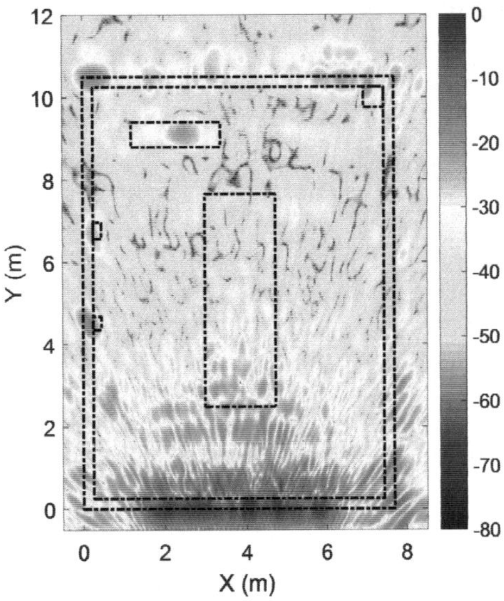

Figure 3. Imaging result of BP approach.

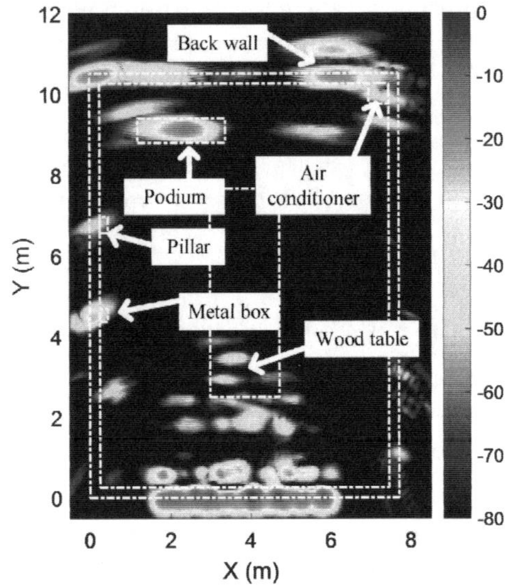

Figure 4. Imaging result of proposed method.

The imaging results of traditional BP approach and our proposed method are shown in Figure 3 and Figure 4, respectively, output in normalized dB. The imaginary lines illustrate the actual layout of the scene. Note that the imaging results have an overall range delay to the actual layout since we do not consider the refraction in the front wall. Significant grating lobes can be observed in the BP approach, which are tens of dB higher than the returns of reflectors like wood table, air conditioner, etc. due to the serious signal attenuation when penetrating the front wall. These reflectors are hardly recognized in such strong grating lobes.

On the other hand, the imaging quality is apparently enhanced employing the proposed two-stage CF method (the Hamming window is chosen in (5) for this experiment). The dihedral angles of metal box and pillar are demonstrated by strong magnitude of radar returns shown in Figure 4, and the wood table is well separated from the grating lobes of front wall. The air conditioner is also able to be detected, which is totally submerged in Figure 3. Though the portable MIMO through-the-wall radar certainly performs not as well as those large vehicle-based through-the-wall radar systems, due to the limitation of synthetic aperture size, directions of illumination and the hardware complexity, this experiment still validates its effectiveness of building interior imaging using our proposed imaging quality enhancement method.

5. Conclusions

The paper proposed a quality enhanced method for the imaging of building interior using portable MIMO through-the-wall radar. The pattern of MIMO-SAR is employed for the portable MIMO radar antennas to achieve a larger aperture as well as viewing angle. Besides, a two-stage CF correction is applied for the suppression of grating lobes. The practical experiment results show a better imaging quality than traditional imaging process in terms of lower grating lobes, which is better for the reconstruction of building environment. While different from the optical photo, further detection and classification is needed for the recognition of detail target information, and this will be researched in the future.

Acknowledgments

This work was supported in part by the National Natural Science Foundation of China under Grant 61271441 and 61372161

References

[1] Moeness A, 2014, *Compressive Sensing for Urban Radar* (Crc Press). 71-2.

[2] Chen B, Jin T, Lu B, *et al.*, 2014, "Building interior layout reconstruction from through-the-wall radar image using MST-based method," *EURASIP J. Adv. Signal Process.*, **31**. doi:10.1186/1687-6180-2014-31.

[3] Yektakhah B, and Sarabandi K, 2016, "All-directions Through-the-wall radar imaging using a small number of moving transceivers," *IEEE Trans. Geosci. Remote Sens.*, **54(11)**, 6415-28.

[4] Anwar N S N and Abdullah M Z, 2016, "Sidelobe suppression featuring the phase coherence factor in 3-D through-the-wall radar imaging," *Appl. Wireless Commun.*, **25**, 730-40.

[5] Liu J, Jia Y, Kong L, *et al.*, 2016, "Sign-coherence-factor-based suppression for grating lobes in through-wall radar imaging," *IEEE Geosci. Remote Sens. Lett.*, **13**, 1681-5.

[6] Chen X, Chen W, 2016, "Multipath ghost elimination for through-wall radar imaging," *IET Radar Sonar Navi.*, **10(2)**, 299-310.

[7] Sun X, 2015, *Research on Method and Technique of Ultra-wideband Through-the-wall Radar Imaging* (National University of Defense Technology).

[8] Hoorfar A, Thajudeen C and Zhang W, 2014, "Recent advances in wall characterization and fast target imaging in through-the-wall radar," *2014 IEEE Benjamin Franklin Symp. Microwave and Antenna Sub-systems for Radar, Telecommun., and Biomedical Appl. (BenMAS)*, Philadelphia, PA, 1-3. doi: 10.1109/BenMAS.2014.7529452.

[9] Liu X, Leung H, Lampropoulos G A, 2012, "Effect of wall parameters on ultra-wideband synthetic aperture through-the-wall radar imaging," *IEEE Trans. Aeros. Electron. Syst.*, **48(4)**, 3435-49.

Fast Detection of Airports on Remote Sensing Images with Single Shot MultiBox Detector

Fei Xia[1] and HuiZhou Li[1,*]

[1]Institute of Electronic Information Warfare, Naval University of Engineering, Wuhan, China

[*]xcyphoenix@nudt.edu.cn

Abstract. This paper introduces a method for fast airport detection on remote sensing images (RSIs) using Single Shot MultiBox Detector (SSD). To our knowledge, this could be the first study which introduces an end-to-end detection model into airport detection on RSIs. Based on the common low-level features between natural images and RSIs, a convolution neural network trained on large amounts of natural images was transferred to tackle the airport detection problem with limited annotated data. To deal with the specific characteristics of RSIs, some related parameters in the SSD, such as the scales and layers, were modified for more accurate and rapider detection. The experiments show that the proposed method could achieve 83.5% Average Recall at 8 FPS on RSIs with the size of 1024*1024. In contrast to Faster R-CNN, an improvement on AP and speed could be obtained.

1. Introduction

Object detection on remote sensing images plays a more and more important role in our lives, e.g. urban planning, disaster prevention, etc. Bridges, ships and aircrafts are becoming significant objects in these applications. With great developments of remote sensing technologies in recent years, airport detection has caused widespread concern due to its civil and military value. Simultaneously, the various structures of airports and complex backgrounds make airport detection challenging [1].

The usual airport detection methods are composed of two steps: region proposal generation and feature recognition. In both steps, airport features influence the quality of region proposals and the performance of airport recognition directly. Besides, these two steps are processed independently. As a result of different features used in two steps, the airport could not be recognized exactly without accurate region proposals and computing different features used in these two steps cost too much. Only using one kind of robust feature and making two steps cooperate are helpful to improve the airport detection performance.

As for the features used in previous works on airport detection, they can be divided into two classes: hand-designed features and machine-learning features. hand-designed features are some designed feature descriptors based on the nature of images and the expert knowledge, such as lines, textures, scale-invariant feature transform (SIFT [2]) and so on. For example, many methods extracted the long straight lines as the runways to generate the region proposals for further detection [3, 4]. However, rivers and roads produced wrong proposals due to the same linear features. It is the same with other hand-designed features because the target object is hard to exactly describe from limited knowledge and finite angles. As a result of many differences between the human perception and the potential feature, the hand-designed features may decrease the detection performance when the application scene changed.

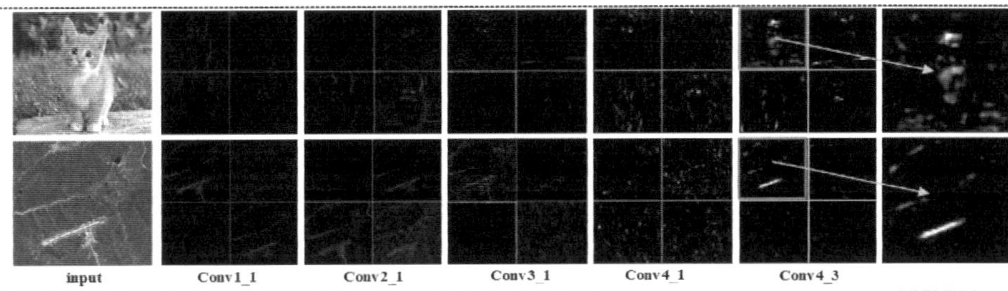

Figure 1. Feature maps of natural images and RSIs from the low layers of pre-trained VGG-16

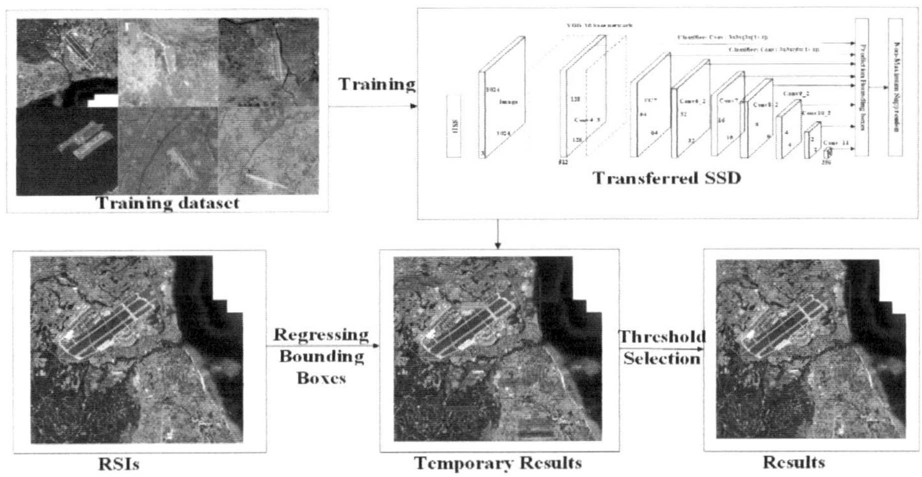

Figure 2. The framework of the proposed airport detection method

On the other hand, machine learning features make good use of some prior knowledge, automatically learning low-level and high-level image features that humans do not know how to represent. Since 2012, due to the success of deep CNN Alexnet [5], deep learning has shown up the excellent potential for superior image classification and object detection. However, large amounts of labeled data are required to train an ideal CNN model while the annotated RSIs are in scarce. Based on the common features between RSIs and natural images, transferring the feature representation of a CNN model trained on natural images to airport detection in RSIs is feasible. The method [6] used the Alexnet to extract and recognize the airports achieved a good performance. However, the existing airport detection methods using CNN still separate the region proposal generation and feature recognition.

As for object detection algorithms using CNN on natural images are consisted of two aforementioned steps. The methods of generating region proposal are BING [7], EdgeBox [8] or Selective Search [9], that all of them spent high computation to process large-sized RSIs without the guarantee of achieving high-quality proposals. Appeared with SPP-NET [10] and the subsequent Faster R-CNN [11], region proposal generation with hierarchical convolutional layers and ROI pooling for decreasing the computation of each region proposal become the mainstream. However, the two steps in these methods are still independent. SSD [12], the state-of-the-art detection algorithm, used a single deep neural network and generated object positions and categories from the network directly. SSD combined two separate steps into one and was faster than YOLO [13] with high performance. In the paper, SSD was employed in airport detection on RSIs.

SSD used a group of default boxes, which are in different scales and aspect ratios, to discretize the output space of bounding boxes for per feature map location by a fully convolutional network. During the airport detection task, the structures of airports are flexible and the sizes are various. By combining the feature maps attained from different layers at different scales, we can achieve the fixed-length airport features with different scales without considering the airport sizes. SSD was firstly designed to detect objects in natural images and some modifications were necessary to deal with the differences between RSIs and natural images. At first, the size of remote sensing images is much larger than the natural images, while RSIs usually covers a wide area and the natural images only covers a small place. Moreover, the proportion the target took up in natural images is much bigger than RSIs. Secondly, the remote sensing images, which were taken from a very high altitude point along with various uncertainties, contain complex backgrounds with the illumination intensity changing. Nevertheless, the low-level visual features in RSIs were in common with natural images and the pre-trained VGG-16 was used in transferred SSD.

As far as we know, it may be the first study which introduced SSD into airport detection on RSIs. The paper proceeds as follows. We detailed the airport detection method in Section II. Section III shows some experiments and analyses. In section IV, we make the conclusion.

2. Methodology

The proposed airport detection method used the convolutional network to produce the bounding boxes of airports and scores directly. Based on SSD, some auxiliary layers were added to process RSIs with large sizes, without cutting slices. Benefiting from the pre-trained neural model VGG-16 [14] and the end-to-end way of detection, the airport features extracted were robust enough and the detection is fast.

2.1. Framework

Based on the pre-trained model VGG-16, the model could extract stronger features from airports than training a new one. Training a deep CNN model effectively is to compute the suitable values for millions of parameters, which depended on a big amount of annotated data for training. Though the data of RSIs is abundant, the annotated images are in scarce. Because of that RSIs are usually very large and the targets (airports) are much smaller than the image itself, manual labeling costs very much work. It is a fact that the rich low and middle level feature representation learned by a trained convolutional neural networks could be transferred to other visual recognition tasks [15]. Figure 1 contained several visual feature maps generated by the convolution layers from *conv1_1* to *conv4_3* from a RSIs and a natural image with the pre-trained VGG-16 network. Based on the hypothesis that RSIs and natural images share plenty of common low and middle level visual features, we transferred the object detection framework SSD and VGG-16 from natural images to RSIs domain. Figure 2 shows the transferred airport detection framework. The transferred SSD architecture built on a normal feed-forward CNN. And the feature representations of low and middle level are generated by the pre-trained VGG-16 network, which was called as *Base network*. Then a small amount of annotated RSIs was used to finetune the initial CNN and estimate the extra added parameters, that would guide the model learn better features of airports.

2.2. Modification for large input

In the structure of VGG-16, the convolution layers were kept and the full-connected layers were re-implemented with convolution. Therefore, both *fc6* and *fc7* would produce feature maps with specific scales. In the original SSD, extra convolutional layers were added to generate the coordinates and confidences of bounding boxes. The sizes of each extra feature maps, which decreased with the increasing number of convolution layers, corresponded to different scales of the original RSIs. The fixed structure of network decided the sizes of input images. As for SSD, there are two different network structures processing the input size of 300 and 512 respectively. SSD-512 contained an extra *Conv_10* branch, which guaranteed that the last layer of feature maps is in the size of 1x1.

In our method, the *Conv_11* branch, consisting of convolution layers and 'PriorBox' layer, was added to process larger size of input images (1024). Without changing the other layers, the *Conv_11* branch took the output of *Conv_10* as input and the results were combined with others for regression. The parameters of *Conv_11*, such as *kernel, pad*, was depended upon the size of the input image. The *min_size*, *max_size*, *step* were decided by the target size and we needed to add the extra *aspect ratio* in the network manually. It is different for the reception filed of feature maps in each layer. Combining these feature maps from different layers can help detect the multi-scale targets. As for each layer after

Conv4_3 (included), SSD used several convolutional filters with the size of 3x3 to generate a group of detection predictions with fixed number, including the offsets of the bounding boxes and the scores for each class. Airports in RSIs usually varied in size and structure. Figure 4 shows the statistic information of the airport lengths in RSIs, which indicates the distribution feature of the airport scales. The default aspect ratio need not be changed for airport detection. As seen from Figure 2, there are seven convolution layers for predicting the bounding boxes, called *Conv4_3*, *Conv6* (*FC6*), *Conv7* (FC7), *Conv*8_*2*, *Conv9_2*, *Conv10_2* and *Conv11_2*, which represent seven features in different scales. In the end, the step called non-maximum suppression (NMS) removed the redundant boxes and achieved the final detections.

Figure 3. Four detection samples by transferred SSD where airports were in mountains and cities.

Table.1 Evaluation results of transferred SSD and Faster R-CNN on training dataset.

Method	AP	Time (s)
Transferred SSD (8m GSD, Input 1024)	0.835	0.11
Transferred SSD (16m GSD, Input 512)	0.796	0.06
Transferred SSD (24m GSD, Input 512)	0.771	0.07
Faster R-CNN (8m GSD, Input 1024)	0.806	0.52s
Faster R-CNN (16m GSD, Input 512)	0.780	0.24s

3. Experiment

The proposed method was tested on RSIs downloaded from GoogleEarth. The dataset contained 217 RSIs, covering most Chinese airports and several international airports. All the RSIs were with 8m ground sample distance (GSD) and the sizes are about 3000 x 3000. The training set was made up of 800 images cropping from 100 RSIs with some augmentation. 17 RSIs were used as the validation set and the other 100 RSIs were the test set. We added the training data into VOC2007 for fine-tuning the SSD. The proposed method was conducted on a computer containing a CPU of Intel E5-2640 and NVIDIA TITAN X Pascal.

3.1. Airport Detection Using Transferred SSD

In the proposed transferred SSD, the sizes of input images were 1024x1024. All the original RSIs would be resized to the fixed size. The large input size could keep more airport features. Or the regions of airports in images were too small to extract features when resizing the RSIs to 512x512 or 300x300. To estimate the proposed method, the public criterion like Precision Rate, Recall Rate are defined as follows.

$$\begin{cases} PR = \dfrac{number\ of\ detected\ airports}{number\ of\ detected\ objects} \\ RR = \dfrac{number\ of\ detected\ airports}{number\ of\ airports} \end{cases} \tag{1}$$

The intersection-over-union (IoU) is calculated by the detected bounding box and the ground-truth of objects. If IOU is bigger than 0.5, we considered that the airport is detected. Figure 3 shows some airport detection results on the test dataset. The IoU scores of prediction results are high, which is above 0.6. In Figure 4, the airports are beside the mountains and rivers or alongside the beach. In Table 1, the evaluation results of airport detection using transferred SSD and Faster-RCNN are shown with the given recall rate. Figure 4 depicts the Recall-Precision-Curve of two methods on test dataset. AP and Time indicate the detection performance and speed respectively. The VGG-16 network was used as Base network for both methods. The experimental results indicate that the transferred SSD has an improvement than Faster R-CNN in terms of accuracy and time-consumption. At the speed of 8 FPS, the transferred SSD could get an average performance (AP) of 83.5%. Contrast to Faster R-CNN, the speed is much faster with 3% improvement achieved on AP.

For the airport detection task on RSIs, the sizes of airports are varied and some airports are so small only to take up a little area. And Faster R-CNN only employed the last layer's feature maps, which is different from SSD. The features of small airports may disappear and weaken in the process of information transmitting between so many layers, result in some misrecognitions. However, the transferred SSD utilized the feature maps from different layers and handle the multi-scale target properly. Different from Faster R-CNN which is consisted of region proposal extraction and feature recognition, SSD uses the single end-to-end CNN added with convolutional filters with small kernels to predict offsets of the default bounding boxes and classes directly, which reduced the computational cost at large.

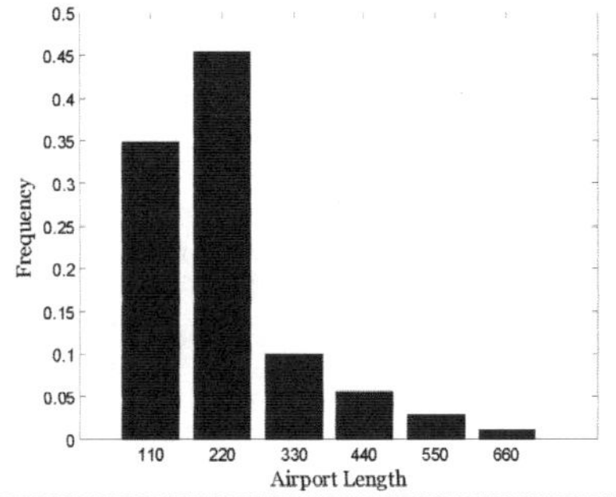

Figure 4. Statistic of the airport lengths in RSIs of 1024x1024.

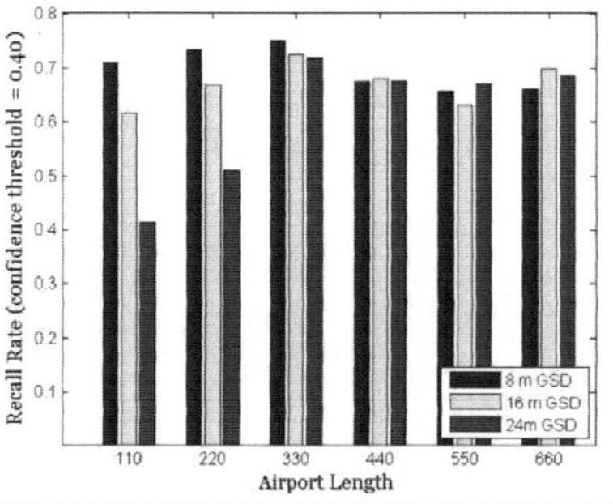

Figure 5. The Recall Rate of each set with different resolution.

3.2. *Effect of Different Resolution on Airport Detection*

To demonstrate the impact of different resolution on airport detection, we built other datasets by cropping the original RSIs into the size of 1024x1024 and 1500x1500 individually. And resizing the cropped images to the sizes of 512x512 and 1024x1024 respectively as the input of SSD. The original collected RSIs are 8 m GSD. After these resize operations, we gained three databases. In *Database1*, the images with the size of 1500 are resized to 512. In *Database2*, the images with size of 1024 are resized to 512. In *Database3*, the images with size of 1024 do not change. The resolution of the images in different databases is 24 m, 16m and 8m GSD individually. Table 1 shows that, with the spatial resolution increasing, the AP of airport detection declined at the same time. To make a further study on this phenomenon, the airports were grouped into 6 sets based on the length of airports and the recall rate of each set was calculated, which is shown in Figure 5. According to the chart, the recall rate of the groups in which the length is less than 330, decreased obviously. The recall rate declined gradually when the airport length is over 330. The regions with short airports were not large enough to generate abundant features for transmitting in the deep network. It is further proved that the visual feature of small objects may lose when it passed through hierarchical layers.

4. Conclusion

This paper proposes a transferred SSD which transferred visual feature representations from natural images to RSIs. The method locates and recognizes the targets in one step. The transferred SSD predicts the offset of default boxes with feature maps from different layers, which help detect the multi-scale airports more accurately. In Comparison with other methods, this method utilizes a fully convolutional network and integrates the region proposal generation and feature extraction into one stage, which makes the computational cost decrease a lot. The experiments indicated that our method achieves better airport detection performance and satisfies time requirements of most applications.

Acknowledgement

We acknowledge support by the National Natural Science Foundation of China under Grant 61572515 and the China Postdoctoral Science Foundation of Grant 2016M593023.

References

[1] O. Aytekin, U. Zongur, and U. Halici, "Texture-based airport runway detection," IEEE Geosci. Remote Sens. Lett., vol. 10, no. 3, pp. 471–475, May 2013
[2] Lowe D G. Distinctive Image Features from Scale-Invariant Keypoints[J]. International Journal of Computer Vision, 2004, 60(2):91-110.
[3] Z. Li, Z. Liu, and W. Shi, "Semiautomatic airport runway extraction using a line-finder-aided level set evolution," IEEE J. Sel. Topics Appl. Earth Observ. Remote Sens., vol. 7, no. 12, pp. 4738–4749, Dec. 2014. 375
[4] W. Wu et al., "Recognition of airport runways in FLIR images based on knowledge," IEEE Geosci. Remote Sens. Lett., vol. 11, no. 9, pp. 1534–1538, Sep. 2014.
[5] Krizhevsky, Alex, I. Sutskever, and G. E. Hinton. "ImageNet classification with deep convolutional neural networks," International Conference on Neural Information Processing Systems Curran Associates Inc. 2012:1097-1105.
[6] Zhang P, Niu X, Dou Y, et al. Airport Detection on Optical Satellite Images Using Deep Convolutional Neural Networks[J]. IEEE Geoscience & Remote Sensing Letters, 2017, PP(99):1-5.
[7] M. M. Cheng, Z. Zhang, W. Y. Lin, et al. Cheng, Ming Ming, et al. "BING: Binarized Normed Gradients for Objectness Estimation at 300fps," (2014):3286-3293.
[8] C. L. Zitnick and P. Dollár. "Edge Boxes: Locating Object Proposals from Edges," 8693(2014):391-405.
[9] J. Uijlings, K. van de Sande, T. Gevers, and A. Smeulders. "Selective Search for Object Recognition[J]," International Journal of Computer Vision, 2013, 104(2):154-171.
[10] K. He, X. Zhang, S. Ren, et al. "Spatial Pyramid Pooling in Deep Convolutional Networks for Visual Recognition," He, Kaiming, et al. "Spatial Pyramid Pooling in Deep Convolutional Networks for Visual Recognition." IEEE Transactions on Pattern Analysis & Machine Intelligence 37.9(2015):1904-1916.
[11] S. Ren, K. He, R . Girshick, et al. "Faster R-CNN: Towards Real-Time Object Detection with Region Proposal Networks[J]," Computer Science, 2015:1-1.
[12] W. Liu, D. Anguelov, D. Erhan, et al. "SSD: Single Shot MultiBox Detector[J]. 2016.
[13] J. Redmon, S. Divvala, R. Girshick, et al. "You Only Look Once: Unified, Real-Time Object Detection[J]," Computer Science, 2015.
[14] Simonyan K, Zisserman A. Very Deep Convolutional Networks for Large-Scale Image Recognition[J]. Computer Science, 2014.
[15] M. Oquab, L. Bottou, I. Laptev, and J. Sivic, "Learning and transferring mid-level image representations using convolutional neural networks," in Proc. IEEE Comput. Vis. Pattern Recog., 2014, pp. 1717–1724.

A Plane Target Detection Algorithm in Remote Sensing Images based on Deep Learning Network Technology

Li Shuxin[1,*], Zhang Zhilong[1] and Li Biao[1]

[1]ATR National Key Laboratory, National University of defence technology, Changsha, province of Hunan, China

*570938695@qq.com

Abstract. Plane is an important target category in remote sensing targets and it is of great value to detect the plane targets automatically. As remote imaging technology developing continuously, the resolution of the remote sensing image has been very high and we can get more detailed information for detecting the remote sensing targets automatically. Deep learning network technology is the most advanced technology in image target detection and recognition, which provided great performance improvement in the field of target detection and recognition in the everyday scenes. We combined the technology with the application in the remote sensing target detection and proposed an algorithm with end to end deep network, which can learn from the remote sensing images to detect the targets in the new images automatically and robustly. Our experiments shows that the algorithm can capture the feature information of the plane target and has better performance in target detection with the old methods.

1. Introduction

As remote sensing technology develops continuously, the resolution of the remote sensing images is very high which can provide wider observations of the detailed information and changes on the surface of the earth. Using these remote sensing images, we can develop applications on geological monitoring, city planning, environmental monitoring, precision agriculture, geographic information system and so on[1], it can also be used in spying in the military field. As the amount of remote sensing data is becoming larger and larger, it provide more and more stress on the work of manual interpretation, which pushed the technology of detecting targets automatically developed continuously. Plane target is an mobile target on the surface of the earth, it is of great value to detect the plane targets automatically. And the high resolution remote sensing images provide the possibilities to detect the plane targets automatically.

To interpretation the remote sensing targets, the old method is based on the feature of the target and develop heuristic detectors, such as the cross feature of the plane target[2][3]. But this matching method has poor stability, which can not adapt to the complex changes in the real scene. As machine learning technology develops continuously, more and more methods based on machine learning come out[4][5][6], which extract features from sliding window and train classifier to detect targets on sliding window. These methods learn from the labeled samples from the training set, which can focus on the key high level features through learning to detect targets more robustly. The problem of these method is that they need process on large amount of sliding windows, which leads to great amount of computation that they can not adapt to the applications in large number of remote sensing data processing.

Deep learning technology comes from the study of artificial neural network in the last century. In the 21th century, as the emergence of large data sets and the continuous improvement in software and hardware, there comes an new research fever on the neural network, namely deep learning network technology. The method brings more depth to the neural network with new network structure to realize the leaning and processing of complex problems. The deep learning network method brings great improvement on many computer vision fields, such as object detection[7][8][9], recognition[10] segmentation[11]. They improved the performance on the complex big data sets continuously and make the methods more and more robust. Combining the new hardware, such as GPU, the methods can process images faster and faster to work in real time.

Combining the deep learning technology and remote sensing plane targets detection, we proposed a new algorithm to detect the plane targets. The method used the deep learning network to extract features from the images and get the key features through learning , which can detect the plane targets more robust and faster. It can bring great help to the automatically detection of the plane targets in the remote sensing images.

2. Related work

To detect the plane targets in remote sensing images automatically, there are 3 main methods. (1) heuristic method. This method used the key cross feature of the plane targets to match cross feature in the image in order to detect plane[2]. The shortage of the method is that it can not adapt to complex changes of the plane targets in the images, such as the change of target size, direction and brightness. (2) Visual saliency method[12][13]. This method use the difference of the low level features between the targets and background to capture the most salient region in the image, which always contain the plane targets. The method has some shortage that there always exists a lot of clutter in the ground scene, which lead to amount of wrong targets in the salient region. This bring some problems to detect the true targets. (3) machine learning method[4][5][6]. These methods extract features(such as HoG feature description) from the sliding windows in the training set to train a plane target classifier(such as SVM classifer), which can be used to the sliding windows in a new image to find if there exist a plane in the window. These methods can learn from large amount of training data to capture the robust feature of the target, which can lead to better performance. But they only use single type of feature, which may loss some information. And the large amount of sliding windows made the method slow.

We proposed a new plane targets detection method based on deep learning, which can dig deep features of the image to get better performance. The method use a network structure with many layers, such as convolution layers, pooling layers and full connection layers. There are many convolution temples to get different features, which flow through many convolution layers and pooling layers to get many feature combinations , so that we can classify the target in higher dimensional space to get better performance. This method can also realize the end-to-end training and testing. Combining the hardware of GPU to accelerate the convolution, the method can process images faster and avoid processing on large amount of sliding windows.

3. Plane target detection algorithm based on deep learning network

Using the deep learning network technology to detect plane target, the algorithm extract the features and detect the targets in a single end-to-end network. So we should design a network with proper network structure based on the goal of detecting the plane targets, which can capture the key features of the plane targets and detect them accurately. During training, we design proper structure of output layer and loss layer to make sure the training can converge and the network can be used to detect the targets accurately.

3.1. Network structure

The deep learning network technology can extract the feature of the target and detect the target in a single end-to-end network, which make the algorithm faster and more effective. The network structure diagram is shown in figure 1.

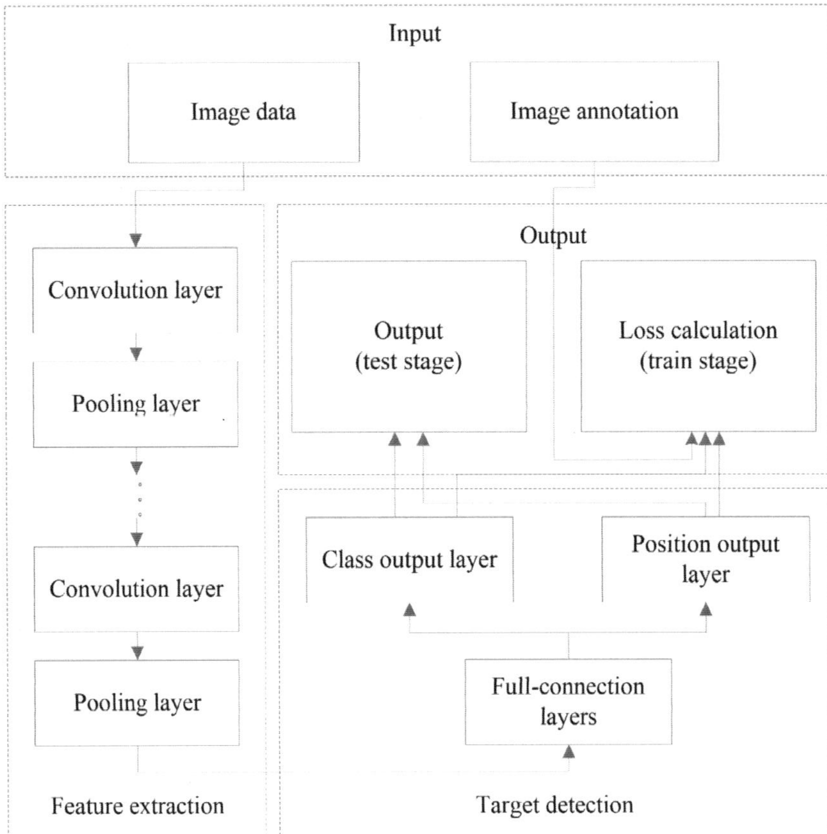

Figure 1. Network structure diagram.

The first part of the network is the input part which put the image data and annotations to the network. The second part is the feature extraction part , which contains convolution layers and pooling layers. This part can extract many kinds of features using many kinds of convolution temples and concentrate the features through pooling layers. After the processing, the dimension of the feature space can be very high so that it can express the features of the target more accurate and comprehensive. In our work, we use the feature extraction part of the VGG16 network[14] as our feature extraction part, which shows great feature extraction capability in our experiments.

After the feature extraction part, there will usually be full connection layers to do more processing and special output layers for special tasks. For the target detection task, we use full connection layers to combine the high dimensional features and distinguish the features to help detection. Then we use the class output layer and the position output layer to output the target class(foreground or background) and the position of the target in the image. To ensure the algorithm detecting the target more accurately, we should divide the image into grids and use each grid to predict the final target position and class. The grids are divided follow the rules below: (1) the center of each grid represent the different position of the image and divide the image into grids of 8x8 and 4x4; (2) for each position, add new boxes of different sizes and aspect ratio so that a box can match the groundtruth target properly. During the training, we only focus the loss generated from the grid matched to a groundtruth target, and propagate back the loss to update the weights until the network training reached convergence. Figure 2 shows the groundtruth target location and the corresponding divided grids.

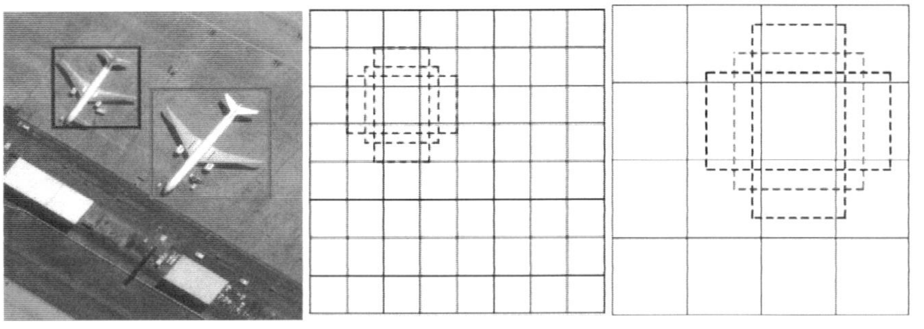

Figure 2. The groundtruth target location and the corresponding divided grids

In our work, to detect the targets in more sizes, there are more convolution layers and pooling layers to produce feature maps of different sizes, which will each predict target class and position. The output layer contains class output layer(softmax layer) and position output layer.

3.2. Network training

After designing the network structure, we should also choose the method of loss computation and the strategy of network weight update, so that the network will reach convergence faster and more effectively.

To compute the loss, we should compute the loss of the class output layer and the position output layer. Using $x_{ij}^p = \{1,0\}$ to represent that the *i*th divided grid is matched to the *j*th groundtruth target in class *p*. So all the loss is the weighted sum of the class loss and the position loss:

$$L(x,c,l,g) = \frac{1}{N}(L_{conf}(x,c) + \alpha L_{loc}(x,l,g)) \quad \text{* MERGEFORMAT (1)}$$

In which, *N* is the number of the matched divided grids. If *N*=0, we set the loss to 0. The parameter α usually set to 1.

We use the Smooth L1 loss function to calculate the loss of position, which calculated between the parameters of the predict box(*l*) and the groundtruth box(*g*).

$$L_{loc}(x,l,g) = \sum_{i \in Pos}^{N} \sum_{m \in \{cx,cy,w,h\}} x_{ij}^k \text{smooth}_{L1}(l_i^m - \hat{g}_j^m) \quad \text{* MERGEFORMAT (2)}$$

$$\hat{g}_j^{cx} = (g_j^{cx} - d_i^{cx}) / d_i^w, \hat{g}_j^{cy} = (g_j^{cy} - d_i^{cy}) / d_i^h \quad \text{* MERGEFORMAT (3)}$$

$$\hat{g}_j^w = \log(\frac{g_j^w}{d_i^w}), \hat{g}_j^h = \log(\frac{g_j^h}{d_i^h}) \quad \text{* MERGEFORMAT (4)}$$

As the output of the predict *l* is the offset of (cx, cy, w, h) , so we should calculate the offset $\hat{g}$ between the groundtruth information g_i and the divided grid information d_i.

The loss of the class output is calculated using the softmax loss function of *c* classes:

$$L_{conf}(x,c) = -\sum_{i \in Pos}^{N} x_{ij}^p \log(\hat{c}_i^p) - \sum_{i \in Neg} \log(\hat{c}_i^0), \hat{c}_i^p = \frac{\exp(c_i^p)}{\sum_p \exp(c_i^p)} \quad \text{* MERGEFORMAT (5)}$$

To update the network weights, we use the gradient descent method. We set the initial learning rate to 0.001 and reduce it by 10 times every 5000 iterations.

In our experiments, we showed that the training can reach the convergence effectively and obtain good performance using the parameters above.

3.3. Network test

In our testing, we used the remote sensing images of big size. In order to control the performance loss of different size, we cut the images into small patches to process in the network. Finally, we combined the detection results and processed using the non-maximum suppression to get the final target detection results.

4. Experiments and analysis

4.1. Image sets description

The image sets used in the experiments contain colour remote sensing images from google earth and some panchromatic remote sensing images. The colour remote sensing images contain plane target from different airport background and different illumination. The panchromatic remote sensing image are gray image with big size. The resolution of the images are around 1m. We divide the images into 3 sets: colour image set 1 with 1190 images; colour image set 2 with 200 images; gray image set with 4 images. The sample images are showed in figure 3.

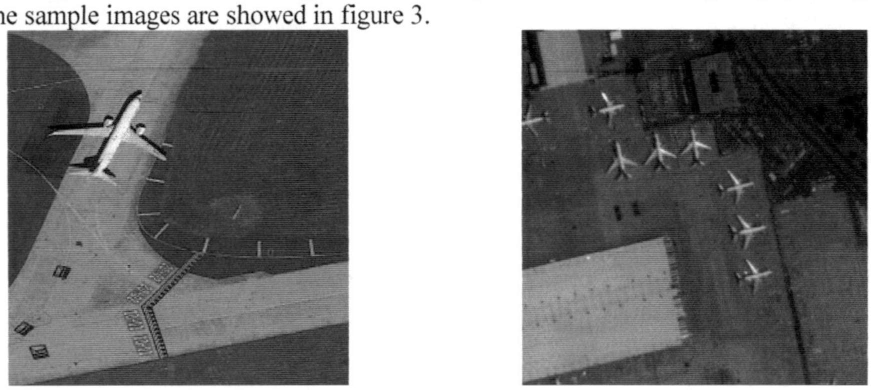

(a) (b)

Figure 3. Remote sensing images in the datasets

4.2. Experiment result and analysis

We used the 200 images randomly selected from the colour image set 1 to train the deep network and used another 200 randomly selected images as the verification data. We trained 20000 iterations to get the final model and used it to test on the 3 image sets, colour image set 1 with 1190 images; colour image set 2 with 200 images; gray image set with 4 images. The test results are shown in figure 4.

(a) (b) (c)

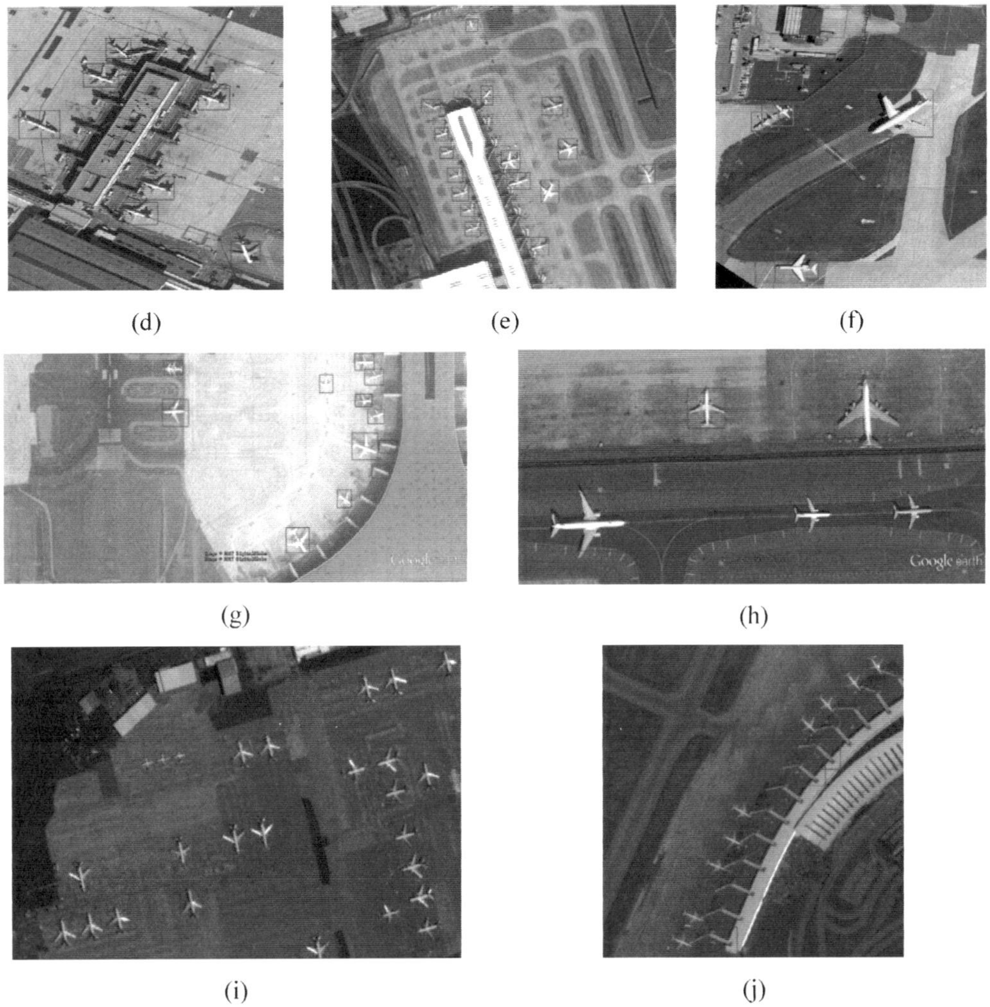

Figure 4. Plane target detection result

In figure 4, figure (a)-(f) shows the result of images from colour image set 1, figure (g)-(h) shows the result of images from colour image set 2, figure (i)-(j) shows the screenshot result of images from gray image set. The results show that the algorithm can detect the plane targets in complex background effectively without the influence of changes in target size, shape, orientation and luminance, and the false alarm rate is low.

After the subjective evaluation, we analysed the objective evaluation of the results quantitatively. The main evaluation indicators for target detection is Recall, Precision, false alarm rate and AP(Average Precision). Recall is the ratio of the number of the targets detected correctly and the number of the groundtruth targets; Precision is the ratio of the number of the targets detected correctly and the number of the targets detected; false alarm rate is the ratio of the number of the targets detected incorrectly and the number of the targets detected, which can be calculated by 1-Precision. If the Recall and Precision is higher and the false alarm rate is lower , the performance of the algorithm is better. To calculate the above 3 indicators, we need to set the threshold of the output score. AP calculates the area below the Precision-Recall curve, which is a better indicator that can reflect the comprehensive performance under different threshold of the output score. When the AP is higher, the performance is better.

In the table 1, we shows the evaluation indicators of the test results on 3 different image sets. The threshold used to calculate the Recall, Precision and False alarm rate is set to 0.25.

Table 1. Objective evaluation of the test results.

Performance indicator	Recall	Precision	False alarm rate	AP
Colour image set 1	0.907	0.876	0.124	0.806
Colour image set 2	0.852	0.601	0.399	0.760
Gray image set	0.577	0.608	0.392	0.634
mean	0.779	0.695	0.305	0.733

From the objective evaluation showed in the table 1, we can see that the best performance is the results tested on the colour image set 1, of which the feature distribution is most close to the training images. The performance of the colour image set 2 is still high but a little drop, as there are some plane with special shape which has big difference from the plane in the training set. The performance of the gray image set is also a little lower, as there are big amount of background in the big size image which leads to some false alarms. The mean AP is 0.733 in all the test data, which shows the overall performance is good.

4.3. Results compared with other methods

In this part, we compare the performance of our method with the conventional method, such as the method based on saliency and the method based on machine learning. In the saliency method, we calculate the difference between the low level features of different regions to get the saliency map. Then we combine the meanshift segmentation results to get the saliency region. From the saliency region, we analyse the region feature to get the final target region. In the machine learning based method, we first extract HoG features on the sliding windows of the images in training set and train a SVM(support vector machine) classifer, then we test a new image through predicting every sliding window on the image to get the final detection result. The evaluation results of the 3 method on the colour image set 1 are shown in table 2.

Table 2. Objective evaluation of the test results of 3 different methods.

	Recall	Precision	False alarm rate	AP
Saliency method	0.661	0.352	0.648	
HoG+SVM	0.671	0.483	0.517	0.464
Our method	0.907	0.876	0.124	0.806

The results show that our method has better performance on the 4 objective evaluation indicators, which means our method has a better detection performance. In the saliency method, it only extract the low level features of the image to detect the target without combining the high level knowledge, which will consequentially detect more false alarm targets and miss more targets. In the machine learning method using HoG features and the SVM classifer, it can learning from the large numbers of image samples in the training set to extract the high level features of the target, but the performance is limited as it only uses the single HoG feature. In our method based on deep learning network, it uses the deep network to extract more features and feature combinations, which can lead to focusing on the key character of the target through learning and reach a more robust performance.

5. Conclusion

This paper has proposed a plane targets detection method based on deep learning network. The method can detect the plane target faster and more robust as it effectively used the end-to-end deep network. The experiments show that the method can extract more comprehensive and rich features from the image to get the key character though leaning, which make the method detecting the plane targets faster and more robust. The shortcoming of the method is the need of a large amount of training target images to train the deep network, which can't be used to the noncooperative targets with little image samples. So it's a good research direction to detect targets with fewer training image samples.

References

[1]. Cheng G and Han J 2016 *Isprs Journal of Photogrammetry & Remote Sensing.* **117** 11-28
[2]. Alatan A and Arslan D 2012 *SPIE Defense, Security, and Sensing.* **Vol.8398** 839808-839808-7.
[3]. Cai H and Su Y 2005 *Proceedings of Spie the International Society for Optical Engineering.* **5985** 529-34.
[4]. Cheng G, Han J, Guo L, Qian X, Zhou P, Yao X and Hu X 2013 *Isprs Journal Of Photogrammetry And Remote Sensing.* **85** 32-43.
[5]. Zhang W, Sun X, Wang H and Fu K 2015 *Isprs Journal of Photogrammetry & Remote Sensing.* **99** 30-44.
[6]. Cheng G, Han J, Zhou P and Guo L 2014 *Isprs Journal of Photogrammetry & Remote Sensing.* **98** 119-32.
[7]. Girshick R, Donahue J, Darrell T and Malik J 2014 *IEEE Conference on Computer Vision and Pattern Recognition, CVPR.* 580-7.
[8]. Ren S, He K, Girshick R and Sun J 2015 *Annual Conference on Neural Information Processing Systems, NIPS.* 91-9.
[9]. Liu W, Anguelov D, Erhan D, Szegedy C, Reed S, Fu C-Y and Berg A 2016 *14th European Conference on Computer Vision, ECCV.* 21-37.
[10]. Krizhevsky A, Sutskever I and Hinton GE 2012 *26th Annual Conference on Neural Information Processing Systems, NIPS* .1097-105.
[11]. Long J, Shelhamer E and Darrell T 2015 *Computer Vision and Pattern Recognition.* 3431-40.
[12]. Han J, Zhou P, Zhang D, Cheng G, Guo L, Liu Z, Bu S and Wu J 2014 *Isprs Journal Of Photogrammetry And Remote Sensing.* **89**:37-48.
[13]. Li W and Pan C 2011 *Communications in Computer & Information Science.* 327-33.
[14]. Simonyan K and Zisserman A 2014 *Computer Science.* arXiv preprint arXiv:1409.1556.

Color Image Enhancement Using Multiscale Retinex Based on Particle Swarm Optimization Method

F Matin[1,*], Y Jeong[1], K Kim[2] and K Park[1]

[1] School of Electronics and Engineering, Kyungpook National University, Daegu, South Korea
[2] Etri, Electronics and Telecommunications Research Institute, South Korea

*E-mail: matin.academic@gmail.com

Abstract. This paper introduces, a novel method for the image enhancement using multiscale retinex and practical swarm optimization. Multiscale retinex is widely used image enhancement technique which intemperately pertains on parameters such as Gaussian scales, gain and offset, etc. To achieve the privileged effect, the parameters need to be tuned manually according to the image. In order to handle this matter, a developed retinex algorithm based on PSO has been used. The PSO method adjusted the parameters for multiscale retinex with chromaticity preservation (MSRCP) attains better outcome to compare with other existing methods. The experimental result indicates that the proposed algorithm is an efficient one and not only provides true color loyalty in low light conditions but also avoid color distortion at the same time.

1. Introduction

Image and video captured in different lights conditions like low brightness and various weather climates always result in unfavorable illumination. Moreover, it can proceed to lose significant details and information from the captured image. The propose of color image enhancement which is an unavoidable part of digital image processing [1] is to modify interpretability and perception of details in an image to prepare improved input for computer machines or human vision system. Image and video enhancement have turned into important research area owing to the prevalent use of the color image in many applications. Many image enhancement techniques in particular histogram equalization [2], gamma correction [3], homomorphic filtering [4], filtering intensity transformation [5] were designed to get distinguished details of an image which is not visible resulting from different lighting conditions. The most challenging part of image enhancement is to tune the parameters and lack of unified algorithms. In other words, when it comes to human vision system there is no quantitative standard theory to determining what an acceptable enhanced image is to human perception. Vast majority of the image enhancement algorithms focus on the input image rather than accommodate local aspects and also the image parameters need manual tuning in order to achieve an acceptable result [6]. Retinex is a nonlinear image enhancement algorithm that simulates the human visual system. The retinex algorithm has color constancy, high dynamic range and also it can sharpen the detail [7]. In this study Particle Swarm Optimization (PSO) is used to optimize the parameters of retinex image enhancement. The proposed algorithm performance has been measured with standard images, and the result compares with other retinex based algorithms. In the following section, introduce the basic theory of retinex and PSO. In section 3 the proposed method (PSO-MSRCP) has been described, in detail. The experiment result and conclusion are represented in section 4 and section 5 respectively.

2. Theory of Retinex and Particle Swarm Optimization

Edwin Land and McCann introduced the retinex theory to model the human visual system [8]. The retinex theory has seen various development and alteration by many researchers. More specifically, multiscale retinex with chromaticity preservation [9] prepare satisfactory result which can be used in image processing. Although, the primary challenge with the retinex algorithm is selecting the values of the parameter such as Gaussian scales, gain and offset, etc. These parameters require being adjusted in order to procure reasonable effect. The principal purpose of the proposed method is to obtain the optimal values of the parameters which used in MSRCP algorithm via utilizing the optimization algorithm PSO. The proposed method (PSO-MSRCP) improves greater administration to compare with other retinex algorithms.

2.1. Multiscale Retinex Algorithm

Jobson et al. [10] developed Single-scale Retinex (SSR), Multiscale Retinex (MSR) and Multiscale Retinex with color (MSRCR) based on Land's retinex. SSR is a center-surrounded algorithm. In the center-surrounded algorithm, the illumination is considered by the difference between the input value (center) and an average of its neighborhood. The formula of SSR can be represented as equation (1).

$$R_i(x,y) = \log I_i(x,y) - \log[C] \tag{1}$$

where $R_i(x,y)$ is the retinex output image for the channel i^{th}, x and y specified the coordinates of a pixel in image, $I \in$ {R, G, B}, log $I_i(x,y)$ is the image pixel value on the ith color channel, $C = F(x,y) * I_i(x,y)$, the symbol * indicates convolution and $F(x,y)$ is Gaussian surrounded function. The convolution $[F(x,y) * I_i(x,y)]$ operation depicts the ith of each color channel. Dynamic range compression and tonal range can provide by SSR simultaneously. But the result seems unnatural. To address this problem SSR extend to MSR. MSR can provide acceptable exchange between color rendition and dynamic range compression. The MSR output is specified as a weighted sum of various different SSR outputs.

The formula of MSR can be express as equation (2).

$$R_{MSR_i} = \sum_{n=1}^{N} \omega_n R_{n_i} = \sum_{n=1}^{N} \omega_n \left[\log I_i(x,y) - \log[F_n(x,y) * I_i(x,y)]\right] \tag{2}$$

where n is the scales numbers, ω_n is each scale weight, R_{n_i} is the SSR output of the ith component of the nth and $F_n(x,y) = C_n \exp[-(x^2+y^2)/2\sigma_n^2]$. The MSR output is multiplying by a color restoration function. Mathematically the MSRCR restoration is described by equation (3).

$$R_{MSRCR_i}(x,y) = C_i(x,y)\, R_{MSR_i}(x,y) \tag{3}$$

where $C_i(x,y) = f(I_i^{'}(x,y))$ is the ith band color of restoration function (CRF). Jobson et al. [9] provided the finest general color restoration is defined by equation (4).

$$C_i(x,y) = \beta \log[\alpha I_i^{'}(x,y)] \tag{4}$$

where β is a gain constant and the nonlinearity strength can be controlled by α.

The final formula of MSRCR is described by equation (5).

$$R_{MSRCR_i}(x,y) = G_i[R_{MSRCR_i}(x,y) - b] \tag{5}$$

where the G is the gain value and b is offset value.

2.2. Particle Swarm Optimization

Particle swarm optimization is an evolutionary computational algorithm [11] was developed by Kennedy and Eberhart in 1995. The PSO is an initiative inspired algorithm which imitates bird's social behavior and fish schooling that optimize a problem by iteratively trying to improve a candidate solution concerning given measure of quality. In this algorithm every potential solution known as a particle and every particle through the random search space according to the set of rules for a given random velocity. Each particle tracks the prior best position *P*best and its equivalent fitness. In the swarm, there will be many *P*best particles, and the particle with the greatest fitness know as a *G*best which is the best global in the swarm and guide the movement. The basic concept of PSO algorithm is to contribute to Pbest and Gbest for each particle. This process is repeated until a satisfactory solution discovered [6]. The particle updates its velocity and position after detecting beast values of *P*best and *G*best by following equation (6) and (7).

$$\mathrm{v}_i(\tau+1) = \omega\mathrm{v}_i(\tau) + c_1 r_1(p_i(\tau) - x_i(\tau)) + c_2 r_2(g(\tau) - x_i(\tau)) \tag{6}$$

$$x_i(\tau+1) = x_i(\tau) + \mathrm{v}_i(\tau+1) \tag{7}$$

where $x_i(\tau)$ illustrates position and $\mathrm{v}_i(\tau)$ demonstrate velocity of the particle i at time representative, ω is inter used to gained balance between global and local search, c_1 and c_2 are positive acceleration constants, and r_1 and r_2 are random values is the range of [0,1]. $p_i(\tau)$ is the best solution of the ith particle and $g(\tau)$ is the global best solution. In equation (6) in the first stage demonstrates the inertia velocity of the particle, in the second stage, represents the decision of the particle which made by its own experience and in the third stage shows the swarm's experience.

3. Proposed Method

In this section multiscale retinex with chromaticity preservation [9] based on particle swarm optimization is presented. The color restoration block in the retinex algorithm (MSRCR) is at risk of inverting color. Color inversion can be observed in image pixels with about 0 values, which may jump to 255 or image pixels with about 255 values may drop to 0. This issue followed by color saturation in the image. To address this problem one of the solutions is to delete the color restoration part. However, eliminating the color restoration part turn the image in to grayish. In Particular, each image pixel is compared to the average of neighbourhood pixels and the difference is kept. In the result, the output can be grayish. A possible settlement to this problem is to use multiscale retinex on the intensity channel which results in the chromatically become identical to the input image. In the intensity channel $I = \frac{\sum_{j=1}^{s} I_i}{s}$, where s shows the number of channels, apply to the MSR formula and for stretching the result to [0,255] that is appropriate to concerning good image a linear transformation enforce to intensity output. MSRCP can define by equation (8) and (9).

$$MSR = [1/3(\log(I_R + I_B + I_G)/3) - \log(1/3(I_R + I_B + I_G) * G_{\sigma_i})] \tag{8}$$

$$F'_{MSRCP_i} = [\min(\frac{255}{\max(I_{R(i)}, I_{B(i)}, I_{G(i)})}, \frac{int_1(i)}{int(i)})] \tag{9}$$

where I is the input image, σ_i is the scales, G is the Gaussian, int_1 is simplest color balance (MSR, s_1, s_2) and s_i is the percentage of clipping pixel on each side. In the final formula gain and offset are added to obtain an enhanced image following equation (10).

$$F_{MSRCP_i} = G \times [F'_{MSRCP_i} + b] \tag{10}$$

3.1. *Objective Function*

To measure the quality of an enhancement image automatically and without human intermediation, it is preferable to use an objective function [12]. The objective function is shown by equation (11).

$$F(I_e) = \log(\log(E(I_s))) \times \frac{n\_edgels(I_s)}{M \times N} \times H(I_e) \tag{11}$$

where $F(I_e)$ is the fitness function, I_e is the enhanced image equivalent to $g(x, y)$, $E(I_s)$ illustrate the sum of M × N pixel intensities of Sobel edge image, $n\_edgels(I_s)$ represents the number of pixels, whose the intensity value is above a threshold in the Sobel edge image. $H(I_e)$ is the entropy value which is based on the histogram, computed on the enhanced image as equation (12).

$$H(I_e) = \sum_{i=0}^{255} e_i \text{ where } e_i = \begin{cases} h_i log_2, & if\ h_i \neq 0 \\ 0, & otherwise. \end{cases} \tag{12}$$

And h_i is the probability of occurrence of the xth intensity value of enhanced image (I_e) between 0 to 255, shadows grey level in input image.

The multiscale retinex parameters in particular number of Gaussian scales σ_1,σ_1,σ_1, gain G, offset b, percentage of clipping pixels on the left and right side (simplest color balance) optimized with PSO. For each particle, the number of particles P valued 30, the number of dimension d set to 7 and number of iteration installed 30 to realize pleasant result. The range of parameters suggested for proposed method are as follows: Gaussian scale σ_1 [0, 50], σ_2 [51, 100], σ_3 [101, 255], gain [0, 30], offset [-15, 15] and percentage of clipping pixel are set with [1,5] for both side respectively. The parameters choose for NASA'S MSRCR are α [100, 150] G [150, 200], β [0, 25] respectively. The steps required in the optimization of above parameters are as follows:

Step1: Initialize P number of particles with dimension d.
Step2: The particle position and corresponding velocity for σ_1, σ_1, σ_1, gain, offset and color balance saturation is randomly initialized within their range.
Step3: The number of iterations is initialized.
Step4: The enhanced image is obtained for every particle using equation (10).
Step5: The fitness value of every particle is computed using equation (11).
Step6: The fitness values of every particle is compared with its *P*best value; if fitness value of particle is better than *P*best, then it becomes the *P*best.
Step7: Next, the fitness values of every particle are compared with the '*G*best' value; if fitness value of particle is better than '*G*best', then it becomes the '*G*best'.
Step8: The particle's velocity and position are updated using equation (6) and (7).
Step9: If the number of iterations is less than the chosen, steps from 4 to 8 are repeated. Otherwise, the algorithm is terminated.

4. Experimental Result and Analysis

This section presents the experimental result to appraise the performance of the proposed method with a number of standard image enhancement techniques in particular, MSRCR [7] and NASA's MSRCR [10]. The input images resolution resize to 512×336 in order to reduce the operation time. The proposed method was coded and verified using C++. Several experiments were presented by considering many test images with poor quality. To illustrate the result of proposed method two different images are shown in Figure 1 and 2. Figure 1(a) and Figure 2(a) show the input images. In Figure 1(a) the background of the image is pretty dark, and the details of the picture are not visible. In Figure 2(a) the face of the girl and the reflection in the car window have bad quality and hardly noticeable. The image enhancement MSRCR used in Figure 1(b) and Figure 2(b) as a result the contrast of the images are improved, and the backgrounds parts of both images are more visible, but images suffer from the color distortion. In Figure 1(c) and Figure 2(c) NASA's MSRCR applied to the images and the result is far better compared to MSRCR. In Figure 1(c) the details about the clothes of the first person and her face are not clear. In Figure 2(c) the reflection in the windows of the car and the eyes of the girl are over enhanced and barely visible. Figure 1(d) and Figure 2(d) show the results of enhanced images by the proposed method MSRCP based PSO. In Figure 1(d) most of the details, in addition, the clothes of the first person, the water color and the background of the image all improved much better compared with NASA'S MSRCR. In Figure 2(d) the enhanced image preserves the contract and the sharpness of the image. Moreover, the girl eyes, the reflection in the windows of the car and the trees and sky in the background are suitably enhanced compared to other methods.

The Table 1 shows the proposed method parameters acquired the test images optimized with PSO.

Table 1. Optimized parameters of MSRCP by using PSO

Image	Low scale [0,50]	Medium scale [51,100]	High scale [101,255]	Gain [0,30]	Offset [-15,15]	color balance saturation left [1-5]	color balance saturation right [1-5]
Girl	46	57	134	19	1	1.6	1.8
Boat	14	97	194	21	8	1.4	1.5

Figure 1. Boat image and enhancement result: (a) input image. (b) MSRCR. (d) NASA's MSRCR. (e) result of the proposed method.

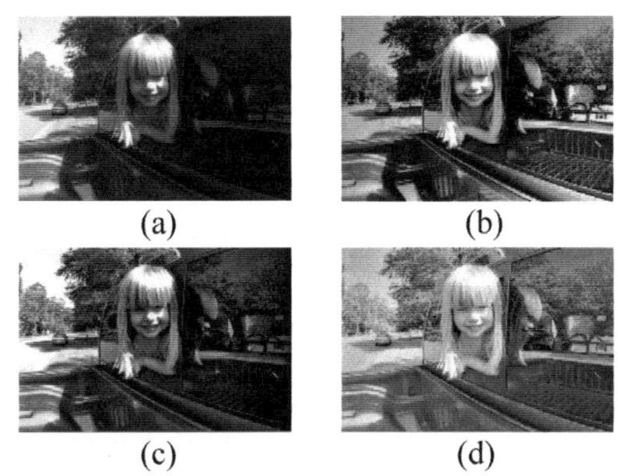

Figure 2. Girl image and enhancement result: (a) input image. (b) MSRCR. (c) NASA's MSRCR. (d) result of the proposed method.

In order to evaluate the efficiency of the proposed method, it compared with other methods based on histogram plot, luminance enhancement performance, and contrast enhancement performance. Table 1 and 2 shows the contrast and luminance values of Figure 1 and 2 respectively. It is clear that the luminance value of the enhanced image with the proposed method determined 55% or greater details. Besides proposed method performance offers a balance between luminance and contrast compare to MSRCR and NASA's MSRCR.

Histogram plot obtained by the input image, MSRCR, NASA'S MSRCR and proposed method are shown in corresponding images from Figure 3(a)-3(d) to Figure 4(a)-4(d) respectively. Based on histogram plot proposed method offers better pixel distribution, dynamic range of color and balance between contrast and luminance compare to MSRCR and NASA's MSRCR.

Table 2. Performance of Figure 1

Boat Image	Input Image	MSRCR	NASA's MSRCR	Proposed method
Luminance	0.546	1.130	1.066	1.331
Contrast	0.352	0.640	0.682	0.664

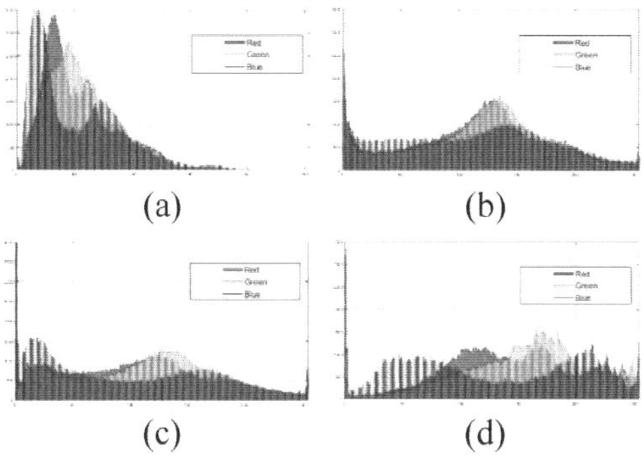

Figure 3. Histogram plot of boat image: (a) input image. (b) MSRCR. (d) NASA's MSRCR. (e) result of the proposed method.

Table 3. Performance of Figure 2

Girl Image	Input Image	MSRCR	NASA's MSRCR	Proposed method
Luminance	0.649	0.884	0.746	1.183
Contrast	0.714	0.731	0.786	0.710

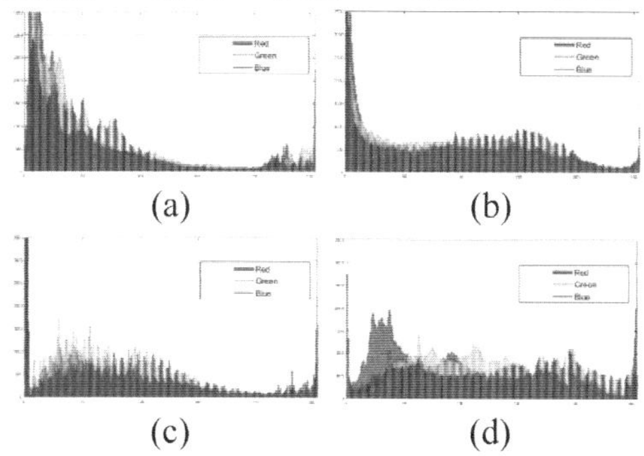

Figure 4. Histogram plot of girl image: (a) input image. (b) MSRCR. (c) NASA's MSRCR. (d) result of the proposed method.

5. Conclusion

In this paper a novel method for the image enhancement using multiscale retinex and practical swarm optimization has been introduced. An efficient, objective function rooted at edge and entropy information of an image has been used in this work to measure the quality of a provided image. The retinex parameters such as Gaussian scales are optimized by PSO to achieve the best exist parameters for image enhancement. The proposed method results compare to standard and existing methods which used the retinex algorithm, improve quality of the color image and achieve high performance in both color quality and clarification. The image captured in different lights conditions and various weather climates can be enhanced by presented method.

Acknowledgments

This study was supported by the BK21 Plus project funded by the Ministry of Education, Korea (21A20131600011).

References

[1] Gonzalez R and Woods R 2017 *Digital Image Processing* (New Jersey: Prentice–Hall)
[2] Singh R and Dixit M 2015 *J. Signal Proccesing.* **8** 345
[3] Cao G, Zhao Y, Ni R and Li X 2014 *J. IEEE. Transaction on Information.* **9** 515
[4] Xiao L, Li C, Wu Z, Wang T, 2016 *J. Neurocomputing.* **195** 56
[5] Zhao Y, Wang X, Wang X and Shih F 2014 *J. Pattern Recognition.* **47** 3537
[6] Hanumantharaju M, Ravishankar M, Rameshbabu D and Aradhya V 2014 *J. Swarm Intel.* **1** 133
[7] Hanumantharaju M, Ravishankar M, Rameshbabu D and Ramachandran V 2011 *Int. Conf on Emerging Applications of Information Technology* (Kolkata:India) p 93
[8] Land E and McCann 1971 *J.Optical Society of America* **61** 1
[9] Petro A, Sbert C and Morel J 2014 Multiscale retinex *J. Imagr Proc. On line.* IPOL(2014)107
[10] Jobson D and Rahman Z 1997 *J. IEEE. Transaction on Image Proc.* **6** 965
[11] Mohan S and Mahesh T 2013 *Int. Conf on Intel. Systems and Control* (Coinbatore:India) p 384

An Improved Text Localization Method for Natural Scene Images

JIANG Mengdi[1,*], CHENG Jianghua[1], CHEN Minghui[2] and KU Xishu[1]

1 College of Electronic Science and Engineering, National University of Defense Technology, Changsha 410073, China
2 Military Representative Office of the PLA Rocket Force in Changsha Area, Changsha 410073, China

* jiangmengdi11@163.com

Abstract. In order to extract text information effectively from natural scene image with complex background, multi-orientation perspective and multilingual languages, we present a new method based on the improved Stroke Feature Transform (SWT). Firstly, The Maximally Stable Extremal Region (MSER) method is used to detect text candidate regions. Secondly, the SWT algorithm is used in the candidate regions, which can improve the edge detection compared with tradition SWT method. Finally, the Frequency-tuned (FT) visual saliency is introduced to remove non-text candidate regions. The experiment results show that, the method can achieve good robustness for complex background with multi-orientation perspective, various characters and font sizes.

1. Introduction

Text detection and localization in natural images serves is considered to be an important aspect of overall image understanding. It has gained considerable attention in both academia and industry in the last decade. Despite the tremendous effort devoted to solving this problem, text localization remains to be challenging due to unconstrained color, sizes, alignment of characters, lighting, various shapes of fonts and the complexity of scenes in natural images. Furthermore, text-like background objects, such as bricks, windows and leaves, often lead to many false alarms in text detection.

Existing text localization can be broadly classified into four main groups: edge based[1,2], texture based[3,4], connected component (CC)-based[5,6] and stoke based methods[7,8,9,10]. Among them, stroke based method has been shown to be effective and is heavily used in many recent approaches due to its insensitivity to the sizes, orientations, color, fonts and languages of the text. The stroke-based method utilizes a feature that distinguishes between the text and other elements of the scene - the nearly constant width of the stroke, which can be used to detect the area that contains text areas. The text can be regarded as a model of the combination of stroke elements in each direction, and the characteristics of the text can be extracted by the combination and distribution of stroke elements. Jung C et al. [7] proposed the concept of Stroke filter (SF), which defines the characteristics of strokes. The article thinks that the features like edge (gradient), connected component, texture, and so on are the external features of the text, and the stroke is the intrinsic characteristic of the text, so the stroke-based method is adaptive not only to all languages but also to the handwritten text. Motivated by it, Epshtein et al.[8] proposed Stroke Width Transform (SWT). The algorithm first performs edge detection by canny operator, and searches for the opposite pixel in the edge gradient direction. Then define the distance between the pair pixels as the stroke width, and use the feature of text strokes to detect the text. This algorithm is dependent on the effect of edge detection. However, it's hard to obtain effective edge as the background is usually complex in natural scene images. Pan et al.[9] presented a hybrid method for detecting and localizing texts in natural scene images by stroke segmentation, verification and grouping. The author proposed a scale adaptive segmentation method to extract stroke candidate and a CRF model with pair-wise weight by local fitting to verify stroke. Since the stroke width transform relies on a precise edge detection scheme, Mosleh et al. [10] introduced a novel bandlet-based edge detector which is quite effective at obtaining text edges in images as well as dismissing noisy and foliage edges.

The SWT algorithm can detect most of the text area, but when the image background is complex, it will produce serious false alarm, such as leaves, stripes, bushes, signs, houses which will often be mistaken for candidate text area. The performance of the SWT algorithm highly depends on the edge detector, which is often not feasible in many challenge cases. Compared to the SWT based methods, MSER operator is capable of detecting more true text components. Therefore, we perform SWT detection in the candidate regions detected by MSER in this paper. The maximally stable extremal Regions contain most of the text and filter a lot of noise in the same time. So, perform SWT in the regions detected by MSER can effectively improve the edge detection which is suitable for low resolution text detection and localization. Moreover, the Frequency-tuned (FT) visual saliency is introduced to improve the robustness of the algorithm to complex background. An important feature of visual saliency is to achieve the suppression of background and the highlighting the significant areas. With this property, we can remove the non-text regions effectively.

2. Related Technology.

2.1. Maximally Stable Extremal Region

The concept of Maximally Stable Extremal Regions (MSERs) was proposed by Matas et al. [11]. MSERs define an extremal region as a connected component of an image whose pixels have intensity contrast against its boundary pixels. For a grayscale image, consider all possible thresholds, set the pixels below the threshold to black, and set the pixels above the threshold to white. Set the threshold small to large: $t = \{0,1,...,255\}$, and obtain the corresponding images. First, we can see a full white image. Then the image will slowly appear with some small grayscale black spots, and gradually merged into small areas. In the last, the image will become full black. In the process of the increasing of the threshold, the number of pixels within some areas is almost constant during the change of several thresholds, this area is called the maximum stable extremal region (MSER).

Since the individual text in the scene image generally has equal color intensity and obvious intensity contrast with the background. The MSER algorithm can effectively detect the text area in the scene image. The test results are shown in figure 1:

(a) (b) (c)

Figure 1.MSER method. (a) original image. (b) Maximally Stable Extremal regions. (c) candidate regions detected by MSER method

2.2. Stroke Width Transform

The stroke width transformation is the method proposed by Boris Epshtein[7] in the 2010 CVPR conference. The text elements in the natural scene have an almost constant stroke width, and the stroke width of the text in its adjacent area is approximately equal, so we can use the stroke width of each pixel to obtain the text areas. The input of SWT is the original image, and the output is a stroke width image with a same size of original image, where the value of each pixel is its stroke width value. The process of SWT method is as follows: Firstly, initialize the value of SWT image to ∞. Then, use canny operator to detect the edge of the image, and each edge pixel has a gradient value in the gradient direction. Take an edge pixel p (the gradient direction is dp.), and look for another edge pixel q of the opposite gradient direction (dq) of dp. When dp and dq satisfied the condition: $|dp + dq| < \pi/6$, calculate the Euclidean distance $\|p - q\|$ between p and q, and set all pixels on the path of p and q to this distance, that is, stroke width value. Figure 2 is the process of SWT method.

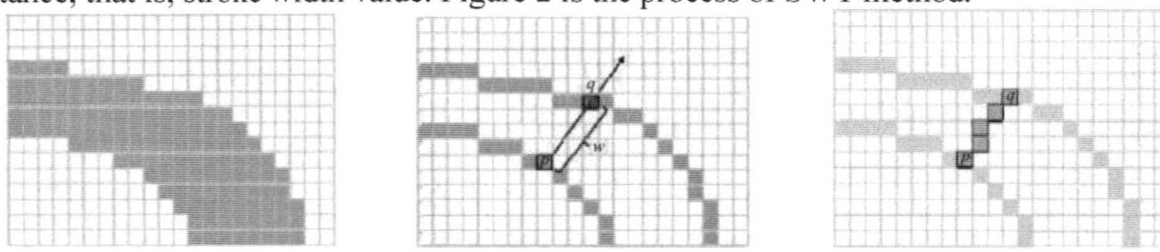

(a) (b) (c)

Figure 2. SWT: (a) A typical stroke. (b) Find the edge pixels p and q. (c) Determine the stroke width.

If the width ratio of two adjacent pixels is within $[1/3, 3]$, the adjacent pixels are considered to be connected. Then, we filter the candidate regions follow the heuristic rules, and grouping letters into text lines.

2.3. Frequency-tuned (FT) visual saliency

Visual saliency analysis is a kind of innate ability of human beings, which is the basic function of visual information processing. Texts in the natural scenes images are mostly slogans, advertising and signs, and their visual perception is generally prominent and prominent, which is more attractive to attract people's attention. The Frequency-Tuned Saliency model is a kind of bottom-up saliency detection method proposed by Achanta et al. [12], and the saliency value of the pixel is obtained by a multi-scale method of the contrast of local color and luminance feature. The algorithm has three advantages over the existing method: 1) the salient regions have a clear boundary; 2) the saliency graph has full resolution; 3) the computational efficiency is high. The model is calculated in the Lab color space, which is defined as:

$$S(x, y) = \|I_\mu - I_{\omega hc}(x, y)\| \tag{1}$$

where I_μ is the mean image feature vector, $I_{\omega hc}$ is the corresponding image pixel vector value after the Gaussian blur in Lab space (using a 5×5 separable binomial kernel) of the original image, and $\| \ \|$ is the L_2 norm. The FT visual saliency results are shown in figure 3.

(a) (b) (c)

Figure 3.FT visual saliency example: (a) original image (b) FT saliency image (c) saliency region segmented by FT saliency thresholding

3. The proposed method

In this section, we describe the text detection algorithm. The key point of SWT is to precisely calculate the stroke width value of each pixel in the image, which depends on the good edge detection. In this paper, we first use MSER to detect candidate text areas, and perform SWT in the candidate areas, which will improve the effect of edge detection. Then, we use some heuristic rules to delete non-text areas. Finally, the FT visual saliency is introduced to further delete the false alarm areas. The algorithm has achieved good results, and its structure is shown in figure 4.

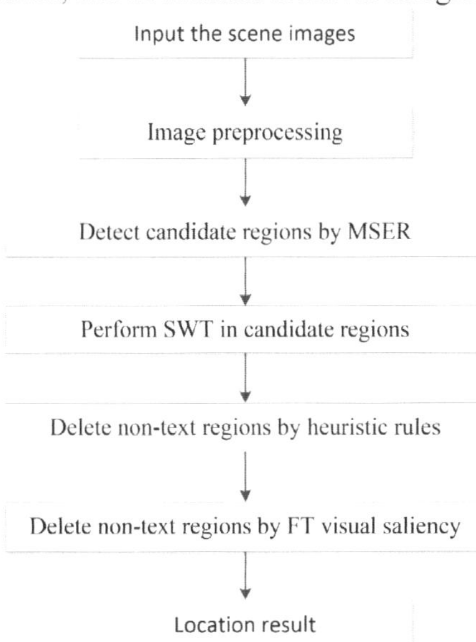

Figure 4. Structure of the proposed method

3.1. Preprocessing and MSER detection

The principle of the MSER algorithm has been introduced in 2.1. As the individual character in the scene image generally has uniform color intensity and obvious intensity contrast against background, the MSER method can effectively detect text regions. However, there still exist some problems: (1) For the low contrast image, the image is usually very vague, which makes difficult to extract the text region as the maximum stable extremal region. (2) For the image with non-uniform illumination, the effect of MSER extraction is poor. In this paper, histogram equalization is used to improve image contrast and contrast compensation is used to correct the effect of non-uniform illumination.

(1) Improve the low contrast image with histogram equalization. Histogram equalization is computationally fast, and the improved is shown in figure 5. The experimental results show that the MSER algorithm can detect the text area well in low contrast image after enhanced by histogram equalization.

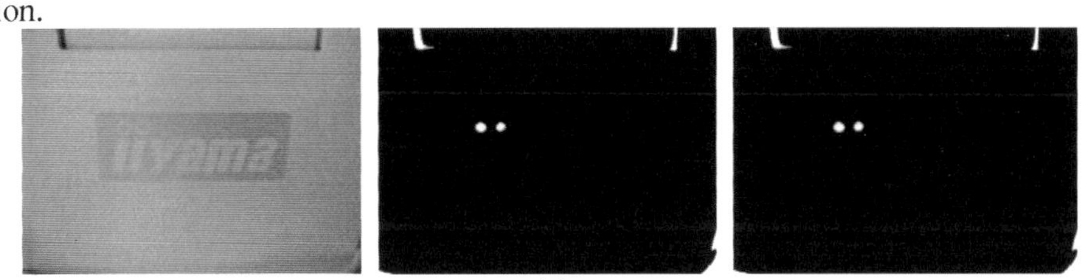

(a)

(b)

Figure 5. From left to right: (a) original image and its MSER image and the text region detected by MSER. (b) the image enhanced by histogram equalization and its MSER image and the text region detected by MSER.

(2) Correct the influence of non-uniform illumination by contrast compensation. The algorithm first estimates the background image of the original image. we can simply understand that the uneven illumination background can be improved by the original figure minus the background map. However, it is conceivable that the contrast between the background and text is smaller when the background color is darker. Therefore, we will compensate the contrast according to the depth of the background color. As shown in figure 6, we can see that the text regions in original image can't be detected by MSER algorithm because of the non-uniform illumination. However, after the contrast compensation, the text region has been detected, and the false alarm areas will be deleted in the last step.

(a)

(b)

Figure 6. From left to right: (a) original image and its MSER image and the text region detected by MSER. (b) the image improved by contrast compensation and its MSER image and the text region detected by MSER.

3.2. Perform SWT in the candidate regions

The proposed method takes SWT in the candidate text regions detected by MSER algorithm. Since the MSER algorithm can locate most of the text areas and filter out a large number of background, the edge detection in MSER is significantly improved compared to the edge detection in the whole image This step improves the accuracy of the calculation of the stroke width. Then we use SWT method to locate text regions and a series of heuristic rules to delete false alarm areas. The heuristic rules are as follows:

(1) The area of candidate regions. If the area of the candidate region is too small, the region is judged to be an alarm region. We set the region whose area is less than 20 pixels to be deleted.

(2) The aspect ratio of candidate regions. Many natural processes may generate long and narrow components that may be mistaken for possible letters. Therefore, we limit the aspect ratio to a value between 0.1 and 10 and the regions where the aspect ratio is too large or too small will be removed.

(3) The stroke width information of candidate regions. If the stroke width within a region changes too large, the regions should be removed. The text regions should satisfy the following conditions:

$$Var_{SWT} < Mean_{SWT}/2 \tag{2}$$

where Var_{SWT} is the variance of stroke width in the candidate region; $Mean_{SWT}$ is the average stroke width in the candidate region.

The experimental results are shown in figure 7. We can see that the edge detection in the regions detected by MSER has been improved obviously. Compared with directly conducting SWT on the original images, the improved SWT location result is better with removing a non-text region. However, there are still some false alarm areas, and we will introduce FT visual saliency for the further judgments.

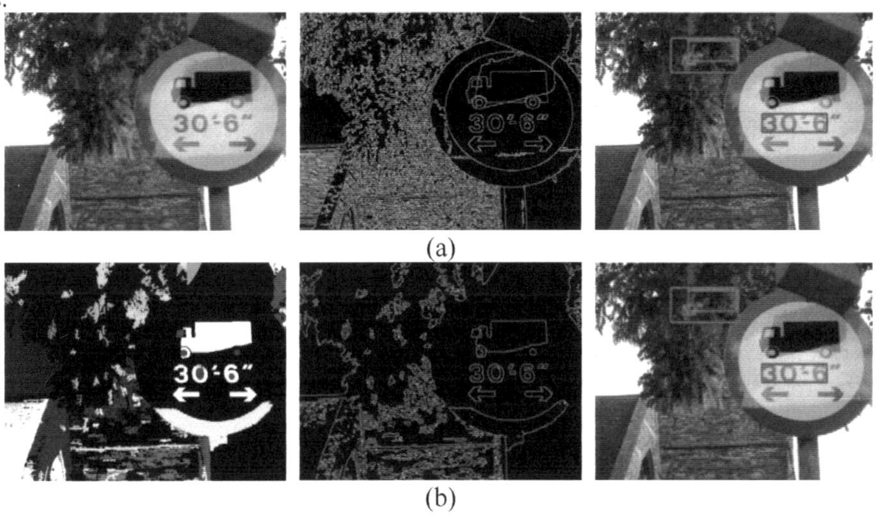

Figure 7. Comparison of SWT in whole image and candidate regions detect by MSER. From left to right: (a) original image and its edge image and the SWT location result. (b) MSER image and its edge image and the SWT location result.

3.3. Text location based on SWT and FT visual saliency

In this section, we proposed a method combined with SWT and FT visual saliency to further remove the false alarm regions. The texts in natural scene image mostly are slogans, advertising and signs which usually capture our attention. Thus the visual saliency value of text region is relative higher than the background. Therefore, we use this property to judge the candidate regions.

The value of each pixel in the visual saliency image indicates the saliency of the corresponding pixel of the original image, and the value is between 0 and 1. So we can set threshold to achieve the salient area detection. In short, if we get the visual saliency image, the pixels of salient object can be calculated by:

$$O(x) = \begin{cases} 1 & if\ S(x) > T_s \\ 0 & otherwise \end{cases} \tag{3}$$

where $O(x)$ is the binarized image of the visual salient object regions; The threshold T_s is based on empirical statistics, which is generally set as:

$$T_s = n * E(S(x)) \tag{4}$$

Where n is the coefficient of T_s and the average saliency value. The principle based on SWT and FT visual saliency is described as follows: First, calculate the average saliency value of each candidate region SWT_i detected by SWT algorithm.

$$aveSaliency_i = \frac{\sum_x S(x)}{num(x)}, \qquad \forall x \in SWT_i(\bullet) \tag{5}$$

where x is the pixel in the candidate region SWT_i ; $S(x)$ is the FT visual saliency image. If the average saliency value of the candidate region is greater than the threshold, it is retained. Otherwise, the candidate region whose average saliency value is less than the threshold will be removed.

$$SWT_i = \begin{cases} text\ regions & if\ aveSaliency_i \geq T_s \\ false\ alarm\ regions & otherwise \end{cases} \tag{6}$$

4. Experiments and results

The performance of the proposed method was evaluated on the ICDAR'11/13 dataset, which remains the most widely used benchmark for text detection in natural scenes. The performance of our method is quantitatively measured by precision (P), recall (R) and F-measure (F), which are defined as follows:

$$R = C/T \tag{7}$$

$$P = C/E \tag{8}$$

$$F = \frac{1}{\frac{\alpha}{P} + \frac{1-\alpha}{R}} \tag{9}$$

where C is the number of the correctly detected text regions, T is the number of the actual text regions, and E is the number of the detected text regions (including the false alarm regions). F-measure is the weighted average of precision and recall, where is the weighting factor, and in this paper. The algorithm is compared with the classic SWT and MSER. The results are shown in Table 1. The evaluation of the first two algorithms is measured on the ICDAR'03/05 dataset. The algorithm is measured in the ICDAR'11/13 dataset. The ICDAR'11/13 dataset is extended on the ICDAR'03/05 dataset, containing multi-directional and distorted scene text images, which is more difficult to detect the text. From Table 1 we can see that this method has high extraction accuracy and good overall performance. Figure 8 shows some comparison of the results of classic SWT algorithm the proposed method.

Table 1. The comparison of the results of different algorithms

author	P	R	*F*	property
Epshtein[9]	0.73	0.60	0.66	SWT
Matas[13]	0.59	0.55	0.57	MSER
Our method	0.78	0.66	0.71	MSER,SWT,FT visual saliency

(a) (b) (c) (d)

Figure 8. The comparison of the result of the classic SWT and the improved method

5. Conclusion

In this work we have presented a novel system for detecting and localizing text-lines in natural scene images. First, we preprocess the image to improve the influence of low contrast and non-uniform illumination, which makes the subsequent MSER locate more accurate. Second, we conduct SWT in the candidate text regions detected by MSER. Finally, we introduce FT visual saliency to remove non-text candidate regions. Experimental results show that our approach has achieved the state-of-the-art results on ICDAR'11/13 dataset.

References

[1] Q. Ye, J. Jiao, J. Huang, H. Yu. Text detection and restoration in natural scene images [J]. Journal of Visual Communication and Image Representation, 2007, 18(6) 504-513.

[2] X.Liu, J.Samarabandu. Multisccale edge-based text extraction from complex images [C]//International Conference on Multimedia and Expo, IEEE, 2006:1721–1724.

[3] Y.F. Pan, C.L. Liu, X. Hou. Fast scene text localization by learning-based filtering and verification [C]// 17th IEEE International Conference on Image Processing ,ICIP, IEEE, 2010:2269-2272.

[4] G. Zhou, Y. Liu, Q. Meng, Y. Zhang. Detecting multilingual text in natural scene [C]// International Symposium on Access Spaces, ISAS, IEEE, 2011:116–120.

[5] H. Zhang, C. Liu, C. Yang, X. Ding, et al . An improved scene text extraction method using conditional random field and optical character recognition [C]//International Conference on Document Analysis and Recognition ,ICDAR, IEEE, 2011:708-712.

[6] Y. Pan, X. Hou, C. Liu. A hybrid approach to detect and localize texts in natural scene images [J]. IEEE Transactions on Image Processing , 2011,20:800-813.

[7] Jung C, Liu Q, Kim J.A stroke filter and its application for text localization [J].Pattern Recognition Letters, 2009,30(2):114-122.

[8] Epshtein B, Ofek E, Wexler Y. Detecting text in nature scenes with Stroke Width Transform [C]//Proceedings of Computer Vision and Pattern Recognition, CVPR, IEEE,2010:2963-2970.

[9] Pan Y F, Zhu Y, Sun J, et al. Improving Scene Text Detection by Scale-Adaptive Segmentation and Weighted CRF Verification[C]// International Conference on Document Analysis and Recognition. IEEE, 2011:759-763.

[10] Mosleh A, Bouguila N, Hamza A B. Image text detection using a bandlet-based edge detector and stroke width transform [C]// British Machine Vision Conference. 2012.

[11] L. Neumann, J. Matas. A method for text localization and recognition in real world images [J]. Computer Vision,ACCV, 2011: 770-783.

[12] Achanta R, Hemami S, Estrada F, et al. Frequency-tuned salient region detection[J]. 2009, 22(9-10):1597-1604.

Blur Length Estimation Based on Image Energy Spectrum Statistics

ZHAO Xia[1,a] and HUANG Kuihua[2,b]

[1]College of Arts and Science, National University of Defense Technology, Changsha, China
[2]College of Information Systems and Management, National University of Defense Technology, Changsha, China

Email: [a]zxmdi@163.com; [b]kuihh@163.com

Abstract. In order to resolve the non-smooth blur problem for coded exposure imaging, this paper proposes an automatically motion blur estimation method based on image energy spectrum statistics, and expands successfully this method to estimate motion blur length for traditional motion blurred images. This paper finds that the residual sums of squares of polynomial fitness result of the deblurred image energy spectrum statistics can reach the minimum, only using correct blur length in the deconvolution process. Given an initial value, using simulated annealing algorithm to implement the iteration, the correct blur length can be obtained automatically in a very short time. Both the real and simulated experiment results demonstrate the effectiveness and robustness of the proposed method.

1. Introduction

As the relative motion between the imaging device and the scene, the resulted image will appear motion blur. The motion blur can seriously reduce the quality of the image, which brings great difficulties to the follow-up researches and application of the image. The restoration of blurred image has always been a hot and difficult problem in the field of image processing [1].

From the imaging process, Raskar proposed a novel and effective coded exposure technology to solve the problem of large length motion blur [2]. Although the technique has many advantages, the deconvolution process of the coded exposure image requires firstly the accurate estimation of image blur length. Due to the exposure coded image blur information is discrete, the current motion blur estimation methods are less effective, such as the cepstrum method[3] and Radon transform method [4], etc, which are based on the analysis of the spectrum of the blurred image.

The researchers have done a lot of effective work on the motion blur estimation of the coded exposure images. Raskar et al. used artificial calibration methods to estimate manually the blur length of coded exposure images [1]. Subsequently, Agrawal and Xu extracted the motion information of the encoded exposure image from the blurred edges of the motion objects on the idea of Alpha cutgraph [5]. The Alpha cutgraph method requires the obvious contrast between the image foreground and background, which restricts its application. Based on motion blur projection model, Tai proposed a method for motion information estimation of the coded exposure image [6]. However, this method is a semi-automatic method, which requires the artificial participation of certain professional image processing knowledge. Xu Shukui constructed a mixture visual acquisition system [7] , and used the motion measurement technique to obtain the motion information of the object. This technology obviously increases a lot of hardware costs, and the multi-camera calibration itself is a problem.

Without loss of generality, this paper studies the blurred length estimation of image motion in the condition of uniform linear motion between imaging equipment and shooting scene. Based on the nature image energy spectrum statistics, an automatically motion blur estimation method is proposed for the coded exposure image. And simulation and real image experiment results verify the accuracy and robustness of this method. In addition, through a large number of experiments and the statistics analysis of the energy spectrum of traditional motion blurred image, a little adjustment to the statistical parameters in the proposed method, this method will be expand effectively to the traditional camera motion blur image blur length estimate.

2. Natural image energy spectrum statistics

The motion blur process of image can be expressed as the convolution of clear image and bluring kernel:

$$g(x,y) = f(x,y) \otimes h(x) + n(x,y) \tag{1}$$

Here, $g(x,y)$, $f(x,y)$ and $h(x)$ represents respectively the blurred image, clear image and bluring kernel function. $\otimes$ means the convolution operator, and $n(x,y)$ is the additive noise. If the noise is ignored, according to the convolution theorem, the Fourier transform of the formula (1) can be obtained as follows:

$$|G| = |FH| = |F||H| \tag{2}$$

Where, $|G|$, $|F|$ and $|H|$ epresent respectively the powerspectrum of the Fourier transform of the blurred image, clear image and bluring kernel function

The study of Van der Schaaf and Van Hateren found that the spectra of powerspectrum of clear natural images without motion blurring obey $1/\omega$ exponential distribution [8].

The natural image powerspectrum is converted to polar coordinates $|F(\omega,\phi)|$, where, $u = \omega\cos\phi$, $v = \omega\sin\phi$, ω is the radius of polar coordinates, ϕ is polar Angle. The powerspectrum statistics of natural images is as follows:

$$S_\omega(|F(\omega,\phi)|) = \frac{1}{360}\sum_{\phi=1}^{360}|F(\omega,\phi)| \approx \frac{A}{\omega^\gamma} \tag{3}$$

Where, $\omega = 1,2,\cdots,M$, A and γ are constants.

In the actual operation, "$M = 127$" is usually set, because the high frequency part of the image powerspectrum is affected badly by the noise.

After taking the logarithm of the formula (3), the following linear expression can be obtained:

$$\log\left(\tilde{S}_\omega(\mathrm{F})\right) = \log(A) - \gamma\log(\omega) = \log(A) - \gamma x \tag{4}$$

where, $x = \log(\omega)$.

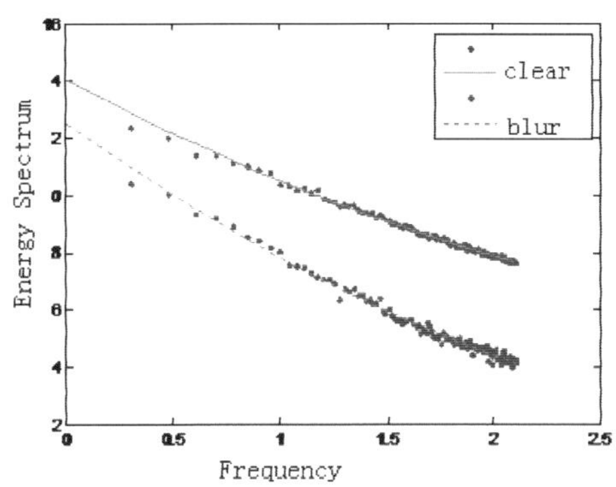

Figure 1. Energy spectrum statistics and polynomial fitmess of latent sharp image and blurred image

The formula (4) shows that the powerspectrum of the clear image obeys the one-dimensional linear distribution (also known as the power law distribution) under the double logarithm axis. However, when the image appears motion blur, its energy spectrum statistics will deviate from the power law distribution, and show more divergence, as shown in Figure 1. It is found in this paper that the larger the image blurs length, the greater the divergence of the statistical data points of his energy spectrum.

3. Blur length automatic estimation

3.1. Automatic estimation of blur length of coded exposure image

Based on the analysis to the relation between the powerspectrum of natural image statistics and image blur length, the polynomial fitting is used in this paper to fit the energy spectrum statistics data, and the residual sum of squares is used to interpret the degree of divergence of statistics data points. Suppose the data points of the image powerspectrum statistics in the double logarithmic coordinates are respectively (ω_1, S_1), (ω_2, S_2), $\cdots$, (ω_n, S_n), and the corresponding polynomial fitting data points are $(\omega_1, \hat{S}_1)$, $(\omega_2, \hat{S}_2)$, $\cdots$ $(\omega_n, \hat{S}_n)$, then the residual sum of squares can be shown by the formula (5):

$$RSS = \sum_{i=1}^{n} \left\| S_i - \hat{S}_i \right\|^2 \quad (5)$$

A large number of experimental results showed that when the image was reconstructed using the correct blur length, the minimum RSS of the powerspectrum statistical data points based on the reconstructed image was obtained.

In this paper, a coded exposure image with a blur length of 50 pixels is simulated, and the anti-convolution parameters, the blur length of 30 to 70 pixels, are used in turn to reconstruct the image. Then to do the polynomial fitting of the energy spectrum of the 41 reconstructed images statistics, and calculate the corresponding RSS of the corresponding image energy spectrum statistics. The results (Figure 2) show the reconstructed image with the right blur length has the minimum RSS.

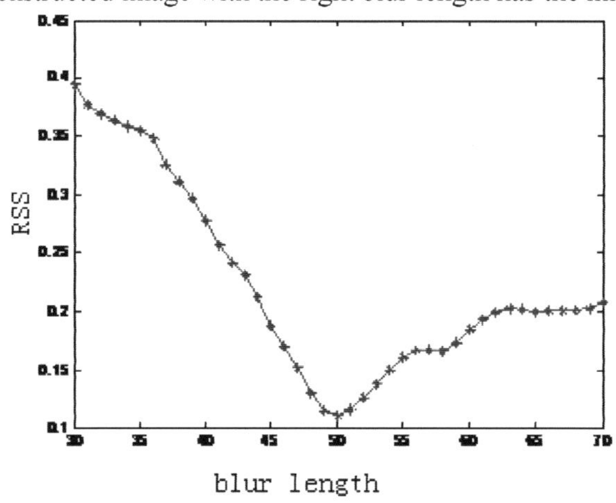

Figure 2. Relations between the RSS and blur length

3.2. Blur length automatic estimation for the traditional motion blurred image

In this paper, an iterative optimization method is used to estimate the blur length of the coded exposure image. Because of the reversibility of the blur kernel function of the coded exposure image, this paper directly uses the generalized matrix inversion approach to recover the image, which greatly improves the efficiency of the algorithm, and makes the proposed method approach practical. Another important reason for using matrix inversion approach is that the recovery process doesn't join any nonlinear optimization operation, which makes the restored image can accurately reflect the blur length is correct or not.

However, the direct matrix inverse approach has a severe ringing effect on the restoration of traditional motion blur images, as shown in Figure 3 (a), which affects its energy spectrum statistics. Because the blur kernel function of the traditional motion blurred image contains high frequency zeros. The bright stripes like "waves" will appear in the corresponding locations of these high frequency zeros of the energy spectrum of the restored images, as shown in Figure 3 (b), when the matrix inverse approach is used directly in the image restoration. These bright stripes can seriously interfere with the energy spectrum statistics of the reconstructed image.

Due to the bright stripes mainly deriving from the high-frequency information, in order to eliminate the ringing effect on the influence of the energy spectrum statistics of the restored image, this paper proposes, based on a large number of experimental analysis, that only using the data points with a frequency of less than 50 (above the dotted line as shown in Figure 3(b)) can get better results.

(a)Deblurred image

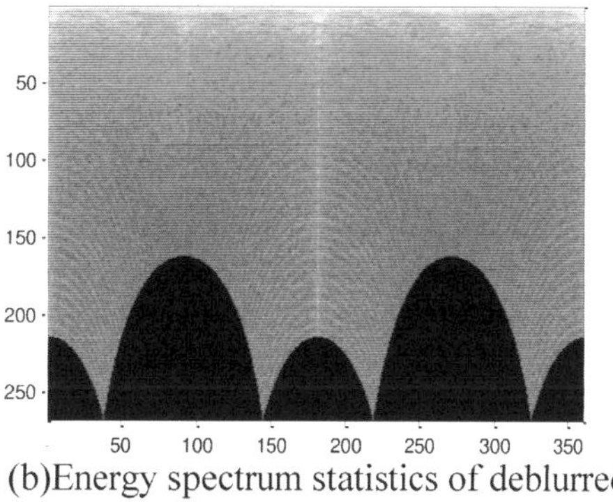

(b)Energy spectrum statistics of deblurred image

Figure 3. Deblurred image using pseudo matrix inversion algorithm and the corresponding energy spectrum statistics in polar coordinates

A traditional motion blurred image with a blur length of 50 pixels is simulated to be used to test. In the same way, in order to show intuitively the RSS of the energy spectrum statistics in the different blur length, the anti-convolution parameters, the blur lengths of 30 to 70 pixels, are used in turn to reconstruct the image. The results (shown in Figure 4) also show that the reconstructed image with the right blur length have the minimum RSS.

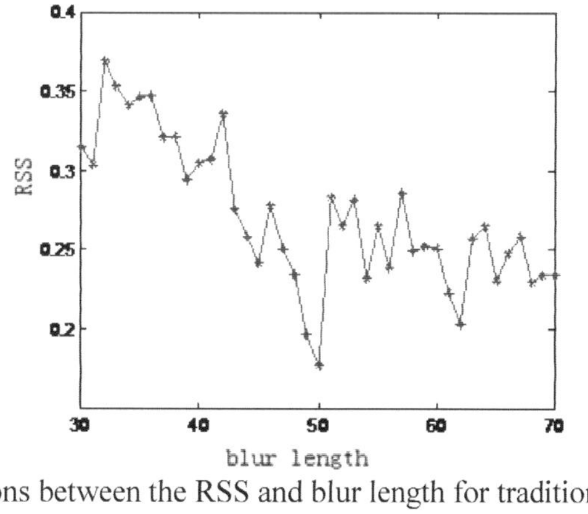

Figure 4. Relations between the RSS and blur length for traditional blurred image

3.3. Image cropping

Image restoration is a process of deconvolution. As a result, there will be a serious ringing effect on the edge parts of the restored image, which affects the energy spectrum statistics in the proposed algorithm. In order to eliminate the influence of the ringing effect, in this paper, a simple image cropping method, as shown in Figure 5, is used to cut the left and right sides of the restored image by 10% of the image length or make the cropped length equal to the blur length of the input image.

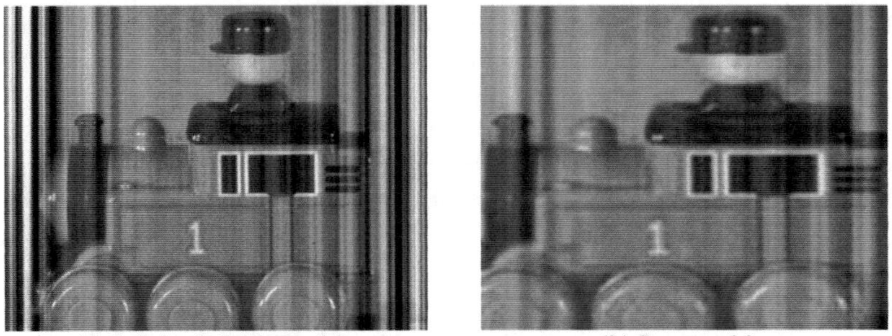

(a)Original deblurred image (b)Cropped image

Figure 5. Cropping operator

3.4. The process of algorithm

In this paper, the simulated annealing algorithm is used to automatically estimate the blur length of motion blurred images. The detailed Process of Algorithm is shown in Figure 6.

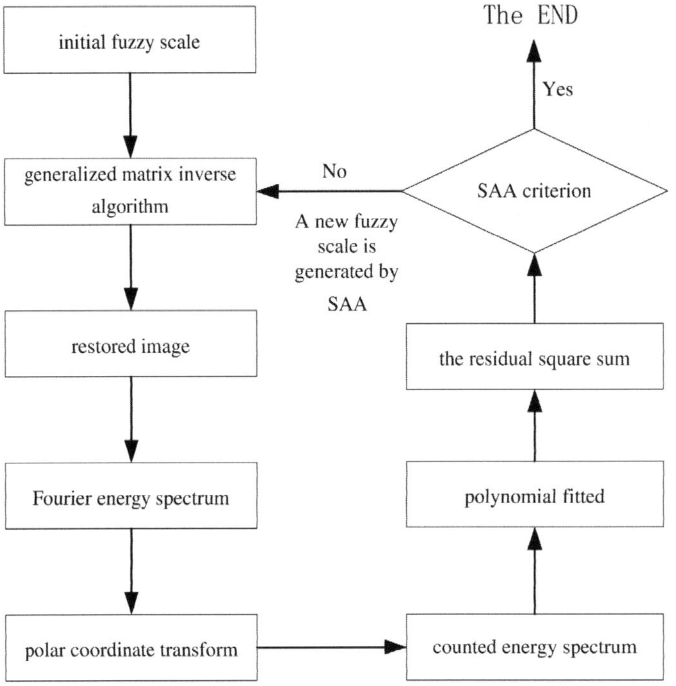

Figure 6. Flow of the proposed method

First, an initial blur length is inputted, and then the generalized matrix inverse algorithm is used to obtain the restored image, and the Fourier energy spectrum of the restored image is obtained. Then, the energy spectrum data are counted and polynomial fitted under the double logarithmic polar coordinate, and the residual square sum is obtained. And after that, the simulated annealing algorithm (SAA) is used to determine whether the optimal residual square sum is obtained or the algorithm has reached the endding condition of the simulated annealing algorithm. If yes, end the algorithm. If no, the new blur lengths are given by the simulated annealing algorithm, and automatically loop iteration.

4. Results and analysis

The experimens are runned in the laptop with intel Core(TM) i5-3230M 2.6GHz, dual-core processor, and 4G DDR3 RAM. The experiment platform is Matlab 2011a. The annealing algorithm used in this paper is provided by Matlab global optimization toolbox. The end condition of the algorithm: the maximum iterations is less than 100 or the RSS is less than 5×10^{-5}.

4.1. Blur length estimation for real coded exposure image

In order to verify the validity of the proposed algorithm in this paper, two real coded exposure blurred images "Tomas" (830*613) and "CabCar" (992*366) provided by Amit Agrawal automatic, as shown respectively in Figure 7(a) and figure 8 (a), are used for the blur length estimation experiments [9]. These two images were all manually calibrated and validated by Agrawal. The blur length in figure 7(a) is 118 pixels and the blur length in figure 8(a) is 66 pixels.

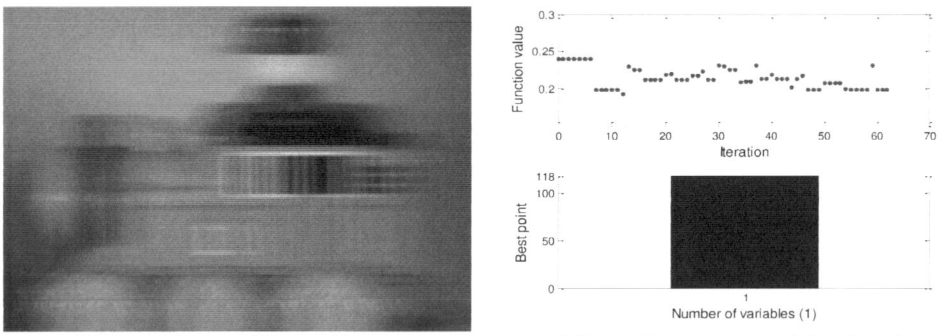

(a)Blurred image "Tomas" (b)Iteration process of simulated annealing algorithm and the corresponding result

Figure 7. Blur length estimation of "Tomas"

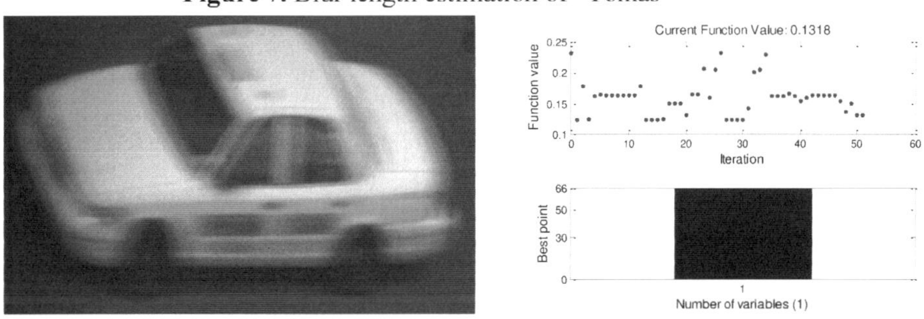

(a) Blurred image "CabCar" (b) Iteration process of simulated annealing algorithm and the corresponding result

Figure 8. Blur length estimation of "CabCar"

Using the algorithm of this paper, the automatic estimation process and result of the blur length for Fig7(a) is shown in Fig7(b), which shows the estimated blur length is 118 pixels. The validity of the proposed algorithm is verified. The time of the proposed method for "Tomas" for running the algorithm 1~10 times is respectively shown in Table 1, which shows the high time efficiency of the proposed method.

Table 1. The time (unit:second) of the proposed algorithm for "Tomas" in 1~10 times

No.	time	No.	time
1	30.8617	6	27.1035
2	33.3275	7	28.4581
3	31.6435	8	28.1503
4	31.6796	9	33.0002
5	27.8680	10	37.0472

Using the algorithm of this paper, the automatic estimation process and result of the blur length for Figure 8(a) is shown in Figure 8(b), which shows the estimated blur length is 66 pixels. The validity of the proposed algorithm is also verified.

4.2. Blur length estimation for the traditional motion blurred image

Only a lillte adjustment to the parameters of the energy spectrum statistics, the proposed blur length estimation algorithm can be applied effectively to the traditional motion blur image. Figure 9(a) is a simulation image blurred by motion. Figure 9(c) is the automatic estimation process and the result of the proposed algorithm. The estimated blur length is 29.2 pixels, which has the error 2.67%. The deblurred result after 20 iterations is shown in Figure 9(b). Figure 9(c) shows that the proposed algorithm for the traditional motion blur image restoration need more time than for the coded exposure blurred image restoration.

(a)Traditional motion blurred image (b) Deblurred result (c)Iteration process and result of SAA

Figure 9. Blur length estimation for the traditional motion-blurred image

4.3. Robust analysis of the algorithm

In order to verify the robustness of the algorithm in noise disturbance, A coded exposure blurred images with blur length 70 pixel is simulated, and the gaussian white noise with the mean value of 0 and the variance of 0.001 and 0.005 is added respectively, shown in Figure 10(a) and (b). Figure 10(d) is the original image. Figure 10 (c) is the reconstructed image of Figure 10(a).

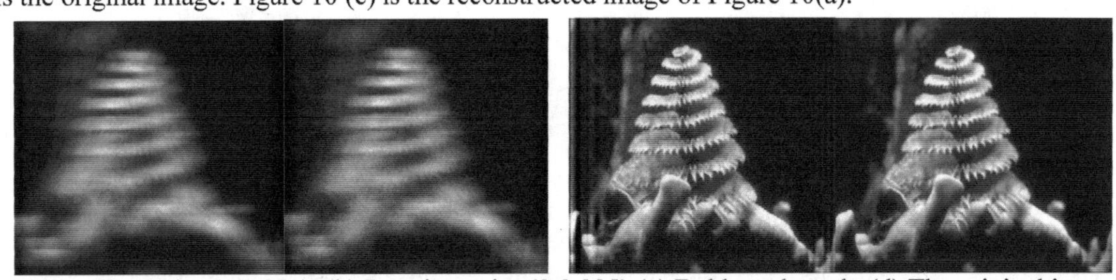

(a)Gaussian noise (0,0.001) (b)Gaussian noise (0,0.005) (c) Deblurred result (d) The original image

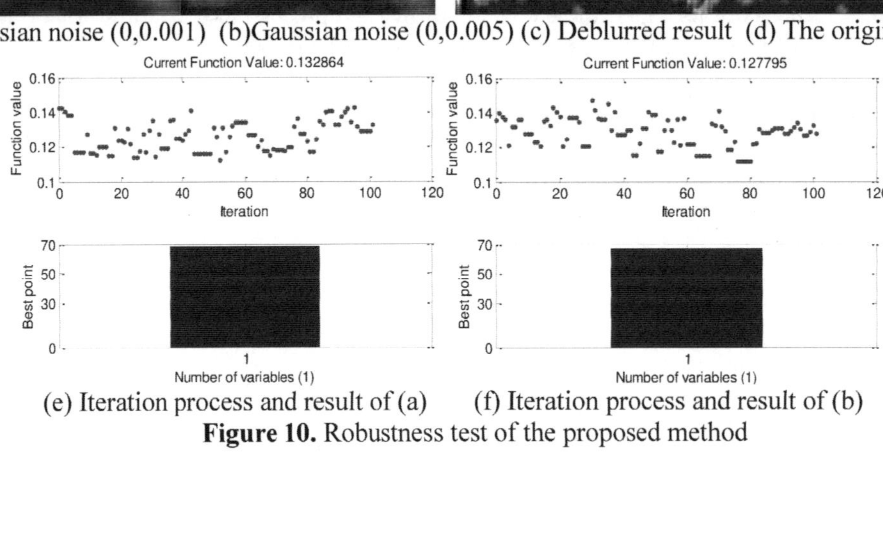

(e) Iteration process and result of (a) (f) Iteration process and result of (b)

Figure 10. Robustness test of the proposed method

Using the proposed algorithm, the estimated two blur length is respectively 68.69 pixels and 67.11 pixels. The errors are respectively 1.87% and 4.13%. Figure 10(e) and (f) are respectively the estimation processes and the corresponding results of Figure 10(a) and (b).

The above results show that the proposed algorithm has strong robustness. Although the estimation accuracy of is a little affected by noise interference, even under the strong noise interference with the variance of 0.005, the error of the algorithm is still in the reasonable range of 5% (proposed by McCloskey, etc. [10]).

5. Conclusions

The discontinuity of the blurred information of the coded exposure image makes the current popular motion blur length estimation method not applicable. In this paper, an automatic method of the blur length estimation is proposed based on the statistics of image energy spectrum statistics. The method can also be applied to the blur length estimation of the traditional motion blurred image. The accuracy of this method is reduced to a certain extent when the strong Gaussian noise is disturbed. The better robustness of the method will be further studied.

Acknowledgments

This work was supported by the National Natural Science Foundation of China under grants 61271437.

References

[1] ZOU Mouyan. *Deconvolution and Signal Recovery* [M]. Beijing: National Defence Industry Press, 2001. (in Chinese)

[2] Raskar R, Agrawal A, Tumblin J. Coded exposure photography: motion deblurring using fluttered shutter [J]. ACM Trans Graph, 2006, 25(3): 795-804.

[3] Ji H, Liu C. Motion blur identification from image gradients[C]. In: 2008 IEEE Conference on Computer Vision and Pattern Recognition (CVPR), 2008: 1-8.

[4] Moghaddam M E, Jamzad M. Finding point spread function of motion blur using Radon transform and modeling the motion length[C]. In: Proceedings of the Fourth IEEE International Symposium on Signal Processing and Information Technology, 2004: 314-317.

[5] Agrawal A, Xu Y. Coded exposure deblurring: Optimized codes for PSF estimation and invertibility[C]. In: IEEE Conference on Computer Vision and Pattern Recognition (CVPR), 2009: 2066-2073.

[6] Tai Y-W, Kong N, Lin S, et al. Coded exposure imaging for projective motion deblurring[C]. In: 2010 IEEE Conference on Computer Vision and Pattern Recognition (CVPR), 2010: 2408-2415.

[7] XU Shukui, ZHANG Jun, TU Dan, et al. A constant acceleration motion blur image deblurring based on hybrid coded exposure [J]. Journal of National University of Defense Technology, 2011, 33(6): 78-83. (in Chinese)

[8] Schaaf A v d, Hateren J H v. Modelling the power spectra of natural images: Statistics and information [J]. Vision Research, 1996, 36: 2759-2770.

[9] Agrawal A, Xu Y, Raskar R, et al. Motion Blur Datasets and Matlab Codes [EB/OL]. http://www.amitkagrawal.com/MotionBlur/index.html. 2009.

[10] McCloskey S, Yuanyuan D, Jingyi Y. Design and Estimation of Coded Exposure Point Spread Functions [J]. IEEE Transactions on Pattern Analysis and Machine Intelligence (TPAMI), 2012, 34(10): 2071-2077.

Optimization of Stereo Matching in 3D Reconstruction Based on Binocular Vision

Qiyang Gai[1]

[1]Electronic Information Engineering, Glasgow College, University of Electronic Science and Technology of China, Chengdu, Sichuan 611731, China

Email: *15511392991@163.com

Abstract. Stereo matching is one of the key steps of 3D reconstruction based on binocular vision. In order to improve the convergence speed and accuracy in 3D reconstruction based on binocular vision, this paper adopts the combination method of polar constraint and ant colony algorithm. By using the line constraint to reduce the search range, an ant colony algorithm is used to optimize the stereo matching feature search function in the proposed search range. Through the establishment of the stereo matching optimization process analysis model of ant colony algorithm, the global optimization solution of stereo matching in 3D reconstruction based on binocular vision system is realized. The simulation results show that by the combining the advantage of polar constraint and ant colony algorithm, the stereo matching range of 3D reconstruction based on binocular vision is simplified, and the convergence speed and accuracy of this stereo matching process are improved.

1. Introduction

Binocular vision is an important branch in the computer vision field [1, 2]. It refers to the shooting of the same scene or object by movement or rotation of two or more cameras at different positions. By calculating the parallax between the two images, the three-dimensional geometric description of the objects is restored and the spatial properties of objects in the scene are quantitatively determined.

1.1. The basic principle of binocular vision

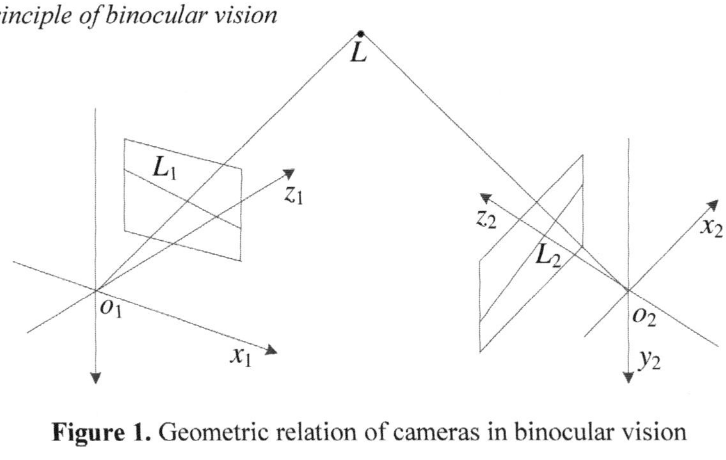

Figure 1. Geometric relation of cameras in binocular vision

According to the principle of camera imaging, the spatial point is L(x, y, z). L0(x0, y0) is the intersection point between the connection line of L and camera's optical center and camera screen. If the image points of L in the left and right camera are Ll (x1, yl) and L2 (x2, y2), the original space point L is located at the intersection of the extension line of the spatial ray O1L1 and of O2L2 (O1, O2 are respectively the optical center of the left and right camera.). The intersection of the spatial ray O1L1 and of O2L2 can be inverted to find the 3D coordinate of L, which is shown in Figure 1. Therefore, binocular vision can be viewed as a mapping process from the image coordinates to the 3D world coordinates [3].

1.2. Binocular vision system

A complete binocular vision system can usually be divided into image acquisition, camera calibration, feature extraction, stereo matching, and depth recovery and depth interpolation.

1.2.1 Image acquisition. There are many ways to obtain images, which mainly depends on the application of the occasion and purpose, but also considers the point of view differences, lighting conditions, camera performance and scene characteristics and other factors to facilitate the three-dimensional calculation [4].

1.2.2 Camera calibration. Camera calibration is used to determine the camera location, attribute parameters, and to establish an imaging model to determine the corresponding relationship between the object points in space coordinate system and the image points in the image plane [5].

1.2.3 Feature extraction. Feature extraction is to obtain the required image feature matching. Because there is not a universally applicable theory to be applied to image feature extraction, it leads to the diversity of feature matching in stereo vision research [6].

1.2.4 Stereo matching. Stereo matching is the one-to-one correspondence between the pixels that seek the same spatial scene at different points of view [7].

1.2.5 Depth recovery. It is easy to recover 3D distance after stereo imaging model and matching parallax [8].

1.2.6 Depth interpolation. The ultimate goal of stereo vision is to restore the complete information of the visual surface of the scene. For a complete stereo vision system, the final surface interpolation reconstruction must be carried out [9].

2. Stereo matching

Stereo matching is one of the key steps of 3D reconstruction. The so-called stereo matching is to seek the corresponding relations between image pixels from the same space scene under different viewpoints [10].

In order to remove the ambiguity, and reduce the computational complexity, the polar constraint technique is used to reduce the search range of stereo matching; at the same time, due to the reduced stereo matching search range matching results bring non optimal solution, the ant colony algorithm for stereo matching calculation process optimization and optimal solution for stereo matching is used to overcome the negative effects, and to improve the computational efficiency of stereo matching [11].

2.1. Reducing search range by epipolar constraint

As shown in Figure 2, if L1 point in the left image corresponds to L2 point in the right image, L1, L2, O1, O2 and L are located in the D plane, d1 and d2 are respectively the intersection line of the D plane and the two image planes, L1 is a point on D1, L2 is D2 the point, d1 is called polar of the point L2, d2 is called polar of the point L1, p1 and p2 are the pole. It can be seen that the projection of the corresponding point L of the L1 on the right image is always on the d2. So, L2, which is the corresponding point of the any point L1, can be only found in its corresponding d2. This is the calculation logic of the polar constraint.

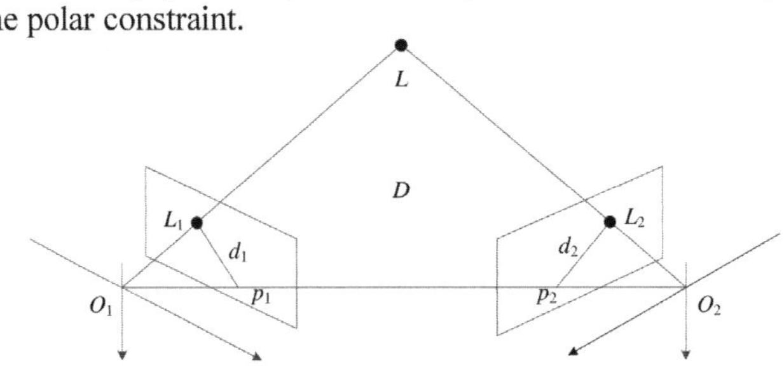

Figure 2. Pole and polar in stereo images

For the two images obtained from different angles on the same target, the previous matching search method is to obtain a feature point on the first image and then find the corresponding feature point of the feature point on the other image. And the principle of polar constraint can be known: if there is no occlusion, the feature point on the first image must be located in the corresponding polar line on the second image. So, the two-dimensional search on the two-dimensional plane can be transformed into a one-dimensional search on the one-dimensional. If the distance between the image acquisition device and the target is obtained, the search range can be limited on a specific small interval of the polar line, as shown in Figure 3.

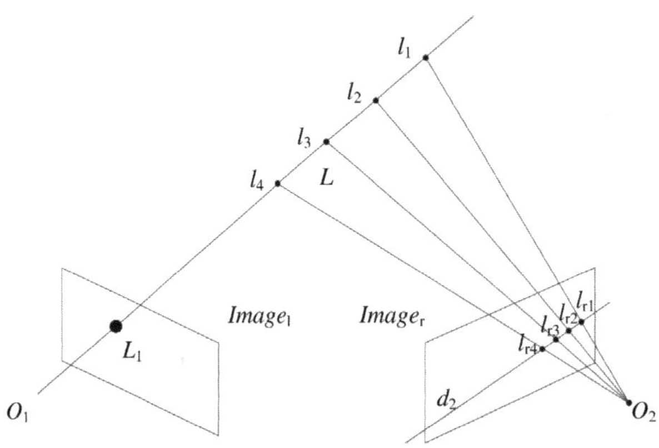

Figure 3. Space someone distance a rectilinear writing recorder segment pairs in the zone should the most on-line limited zone

2.2 application of ant colony algorithm optimization in polar constraint

Because the ant colony algorithm has the characteristics of self-heuristic search, the ant colony algorithm is used to realize the optimal solution of the stereo matching search function under the constraint of the polar line [12].

The stereo matching search is performed in the corresponding polar lines of the left and right images. Then, the pole set of ith row in the left image is PLi={pl1,pl2,…,pln} and in the right image, the pole set of ith row is PRi={ pr1,pr2,…,prn }.

The objective of ant colony optimization is to find the pole points matching relation between the set PLi and PRi. The ant colony algorithm is used to solve the pole matching problem on the corresponding polar line of the left and right images [13]. Its design is as follows.

- Initialize the stereo matching ant colony algorithm optimization parameters, such as setting the number of ants, the number of iterations, the total amount of pheromone and so on.
- Between the ith line pole sets PLi of the left image and the ith line pole sets PRi of the right image, the pole set ,which has less pole points, is used as the, and the other pole set is the search point set.

- All ants are placed at the first feature point of the model point set. Each ant randomly selects a feature point of the search point set, and adds the feature point is added into each ant's marked table.
- The ants are placed to the next feature point sets. According to the change rules, the feature point is selected by each ant in search point set and not in marked table. It is assigned to the feature points of the model where the ants are, that is, the selected feature points constitute a mapping with the model points of the ants. The change rules that ants follow when selecting feature points are:
The ant a current model feature is the point i, which selects the feature point j in probability p_{ij}^{a} to form the mapping from point j to point i. The formula of p_{ij}^{a} is described in (1).

$$p_{ij}^{a} = \begin{cases} \dfrac{(\tau_{ij})^{\alpha} * (\eta_{ij})^{\beta}}{\sum\limits_{i,j\in R} (\tau_{iR})^{\alpha} * (\eta_{iR})^{\beta}} & i, j \in R \\ 0 & otherwise \end{cases} \tag{1}$$

The ant a selects the search j as the path to the model i in this iteration. The pheromone τij is accumulated on the search. ηij is the heuristic information that maps the search feature point j to the feature point i of the model set, ηij=1/Sij. Sij is the search function for selecting the search j to reach the model i, which is equivalent to the search length from j to i, and the Sij calculation is determined by the specific line constrained geometric model. α and β represent the different effects of the accumulated pheromone τij and the heuristic information ηij j in the process of ants select the search set point. R denotes a set of feature points that the ant a can currently select from a set of search points, which does not include the points in the marked table. The above process is cycled until all the ants walk through all the feature points in the model points, that is, all the ants find the mapping from the feature points in the search point to the feature points in the set of model point.
- The pheromone updates when an iteration is completed. Only the pheromone on the optimal search is adjusted. The sum of the search lengths W (W = ΣSij) is compared in the search for each ant in this iteration. The pheromone on the search for an ant with minimal W is updated as follows:

$$\tau_{ij}(t+1) = \rho\tau_{ij}(t) + \Delta\tau_{ij}^{best}(t), \rho \in (0,1) \tag{2}$$

Where τij (t+1) and τij (t) are respectively the pheromones on the search j on the model i at the (t+1) th iteration and the tth iteration. $\Delta\tau_{ij}^{best}$ Represents the pheromone increment of the current an iteration of the ants in this search. $\Delta\tau_{ij}^{best}(t) = Q / S^{best}$ is the total length of each search for the current iteration's optimal ant. Q is the total amount of pheromones of ants, and ρ is the pheromone residual coefficient.

- If the iteration termination condition is satisfied, the iteration ends; otherwise, empty the marked table and turn to the step c. At the end of the iteration, the ant with the minimum length of S is chosen to achieve the global optimal matching solution of the stereo matching under the constraint condition of the polar constraint.

3. Design of stereo matching process based on ant colony algorithm under polar constraint

The optimization workflow design of stereo matching in the 3D reconstruction based on the ant colony algorithm under polar constraint is as follows, shown in Figure 4.

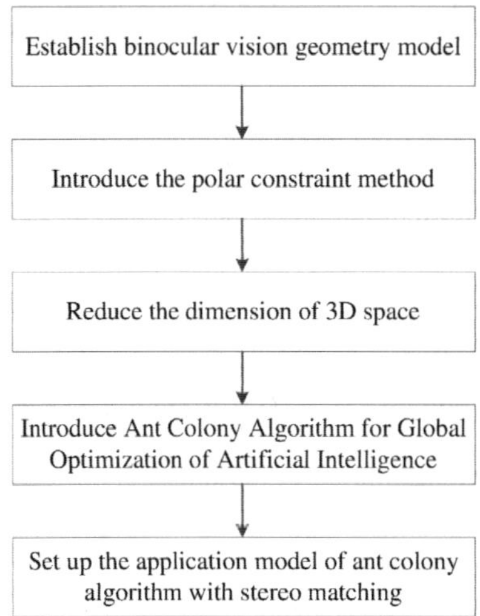

Figure 4. The optimization workflow design of stereo matching in the 3D reconstruction based on the ant colony algorithm under polar constraint

3.1 Establish binocular vision geometry model

The spatial target points are locked by the left and right visual sources, and the geometric relation is used to gradually establish the geometric plane of the spatial target points in the left and right views.

3.2 Introduce the polar constraint method

The concept of polar line is defined according to the feature points projected from the space target point to the left and right planes, and the geometric calculation method is constructed to realize the polar constraint range of the space target point.

3.3 Reduce the dimension of 3D space

According to the searching range of polar lines, the work of the 3D to 2D and 2D to 1D is implemented. The complexity of the algorithm is reduced by the simplified work, in order to define the search range and simplify the analysis process of the stereo matching data.

3.4 Introduce Ant Colony Algorithm for Global Optimization of Artificial Intelligence

In order to avoid the incompleteness of data information caused by polar constraint, ant colony algorithm is introduced to realize the optimal solution of global information of stereo matching.

3.5 Set up the application model of ant colony algorithm with stereo matching

According to the geometric information of the spatial target points in the binocular vision system, the number of ants, the number of iterations and the total amount of pheromone are set up to realize the automatic optimization of stereo matching.

4. Simulation and analysis

4.1 Simulation environment

In order to verify the stereo matching optimization process in the binocular vision system using the "polar constraint + ant colony algorithm" combination mode, the simulation environment is set as follows.

Space range V = length * width * height = 2 * 1 * 1.5 (m3);

Number of space target points: 50;

Maximum length of polar line: 1 m;

Training sample (ants) Number: 200;
Number of iterations: 100;
Pheromone value: (0, 1] ;

It is assumed that the spatial information of 50 target points is known. There are four cases of three-dimensional matching effect comparison, as shown in Figure.5.

Case 1: there is no polar line constraints and no ant colony algorithm to participate in the three-dimensional match.

Case 2: Stereo matching is only realized under polar constraint.

Case 3: Stereo matching is only realized under ant colony algorithm.

Case 4: both polar constraint and ant colony algorithm are involved in the stereo matching.

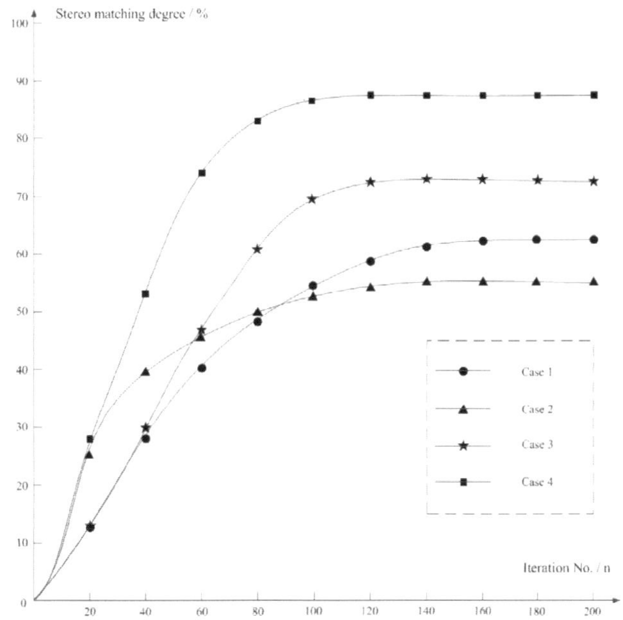

Figure 5. The Comparison of stereo matching effects

By using the method of polar constraint in stereo matching, the initial iterative stage effect is obvious. Because the polar constraint is simplified to a one-dimensional search range, reduce the amount of data to be processed, the initial growth rate of the stereo matching curve is higher. Through constant iterative process of stereo matching with ant colony algorithm, on the one hand, it's faster to achieve the stereo matching final state; on the other hand, the degree of stereo matching is improved. Therefore, the stereo matching using ant colony algorithm can be faster and higher to achieve the matching state in the middle and late stages of the stereo matching curve.

5. Conclusion

The algorithm model and workflow design of stereo matching in 3D reconstruction based on binocular vision provide a feasible theoretical basis for the experimental simulation.

Through the experimental simulation, it proves the technical advantages of the combination of the polar constraint and the ant colony algorithm. On the one hand, the stereo matching range of 3D reconstruction under binocular vision is clarified; on the other hand, the convergence speed and accuracy of the stereo matching process are improved.

References

[1] Sedlackova A, Szczecinski N S, Quinn R D. Binocular Vision Using Synthetic Nervous Systems [M]// Biomimetic and Biohybrid Systems. 2017.

[2] Ciuffreda K J. Binocular vision in the twenty-first century. [J]. Journal of Optometry, 2017, 10(3):139-140.
[3] Lin Y H, Shou K P, Huang L J. The initial study of LLS-based binocular stereo-vision system on underwater 3D image reconstruction in the laboratory [J]. Journal of Marine Science & Technology, 2017(3):1-20.
[4] Koprowski R. Image Acquisition [M]// Processing of Hyperspectral Medical Images. Springer International Publishing, 2017.
[5] Godding R, Hornberg A. Camera Calibration [M]// Handbook of Machine and Computer Vision: The Guide for Developers and Users. Wiley - VCH Verlag GmbH & Co. KGaA, 2017.
[6] Zhu J, Narabu Y, Tanaka T, et al. Manufacturing feature extraction system of 3D-model for process planning [J]. Proteomics, 2017, 11(19):3779.
[7] Liu J, Yu L, Li X, et al. Research on Joint Segment Optimization and Stereo Matching [J]. Iet Communications, 2017.
[8] Chen C, Hai X P, Pavlovic V, et al. Using 3D Face Priors for Depth Recovery[J]. Journal of Visual Communication & Image Representation, 2017.
[9] Wang Y, Yan P. Depth image interpolation algorithm based on confidence map [J]. Proceedings of the Spie, 2017, 255:102550D.
[10] Oh J, Lee C. Extraction of digital elevation model using stereo matching with slope-adaptive patch transformation [J]. Ksce Journal of Civil Engineering, 2017, 20(7):1-8.
[11] Ohyama J, Ohta N, Oagwa N. Correct correspondence selection between points on two asteroid images using epipolar constraint[C]// Jpgu-Agu Joint Meeting. 2017.
[12] Alaya I, Solnon C, Ghdira K. Ant Colony Optimization for Multi-Objective Optimization Problems[C]// IEEE International Conference on TOOLS with Artificial Intelligence. IEEE, 2017:450-457.
[13] Sammoud O, Solnon C, Ghédira K. Ant Algorithm for the Graph Matching Problem [J]. Lecture Notes in Computer Science, 2017, 3448:213-223.

A Survey on Sentiment Classification in Face Recognition

Jingyu Qian[1]

[1]Engineering School, University of Michigan, Michigan MI 48105, US

*egbert_qian@163.com

Abstract. Face recognition has been an important topic for both industry and academia for a long time. K-means clustering, autoencoder, and convolutional neural network, each representing a design idea for face recognition method, are three popular algorithms to deal with face recognition problems. It is worthwhile to summarize and compare these three different algorithms. This paper will focus on one specific face recognition problem–sentiment classification from images. Three different algorithms for sentiment classification problems will be summarized, including k-means clustering, autoencoder, and convolutional neural network. An experiment with the application of these algorithms on a specific dataset of human faces will be conducted to illustrate how these algorithms are applied and their accuracy. Finally, the three algorithms are compared based on the accuracy result.

1. Introduction

Face recognition has become an important problem for a long time. Commercial and law enforcement applications makes it critical for both business industry and public security. In addition, researchers from different fields including image processing and computer visions are continuously applying different algorithms for face recognition problems [1].

Till now, typical methods for face recognition include holistic methods, feature-based methods, and hybrid methods [1]. Holistic methods include algorithms such as principal component analysis (PCA) and its derivation two-dimensional principal component analysis (2DPCA) [1], [2]. Feature-based methods include algorithms such as hidden Markov model (HMM) [1], [3]. Hybrid methods include algorithms such as modular eigenfaces [1], [4]. In addition, k-means clustering and its derivation, due to their computational efficiency, are also used in face recognition. Autoencoders, including deep autoencoder [5], are widely used in complicated face recognition problems, too. Despite its long training time, convolutional neural network has also become a widely used algorithm to do image processing work including face recognition problems due to its consideration of spatial characteristics.

In this paper, we will focus on three different algorithms to deal with a specific problem of face recognition - sentiment classification from still images of facial expressions. The three algorithms include k-means clustering, autoencoder, and convolutional neural network. We will first summarize these three algorithms in Section 2 by showing details and mathematical backgrounds of them. Then, we will generally compare these three algorithms in Section 3. Next, we will practically apply these three algorithms on a specific labelled dataset of still images of facial expressions to train models and compare their accuracy in Section 4. Section 5 will conclude the paper.

2. Three ways of sentiment classification from images of facial expressions

2.1 K-means clustering

K-means clustering [6] is an unsupervised learning algorithm in machine learning, meaning that it can be used to conduct classification without labels. The goal of this algorithm is to classify all the samples into different clusters so that in the same cluster, the sum of squares is minimized [7]. Suppose there are n samples defined by feature vectors $(\boldsymbol{x}_1, \boldsymbol{x}_2, \ldots, \boldsymbol{x}_n)$ to be classified into k clusters $(C_1, C_2, \ldots, C_k)$. In this case, we need to find k centroids $(\boldsymbol{c}_1, \boldsymbol{c}_2, \ldots, \boldsymbol{c}_k)$ with respect to k clusters, and the assignment of each sample to one cluster such that it minimizes:

$$\sum_{j=1}^{k}\sum_{x_i \in C_j}\left\|\boldsymbol{x}_i - \boldsymbol{c}_j\right\|_2^2. \quad (1)$$

Since the centroids and assignment are unknown initially, traditional k-means clustering initializes randomly k centroids [8]. With the k centroids, find the assignment for each feature vector $\mathrm{x_i}$ to one centroid $\mathrm{c_j}$ that minimizes $\|\mathrm{x_i} - \mathrm{c_j}\|_2$. After that, recalculate each centroid with the means of the samples assigned to it. Then find the assignment again and iterate until the centroid does not change after recalculation. The iteration is guaranteed to end.

There are several extensions of k-means. K-means++ is an algorithm of carefully choosing initial centroids [9]. It first chooses one sample randomly from n samples as a centroid. Then, with previously chosen centroids, assign a probability to each sample that has not been chosen such that the higher the shortest distance of that sample to the previously chosen centroids is, the higher the probability is. Then, choose next centroid according to the probability. The initialization ends when all the k centroids are chosen. Other extensions include penalized and weighted k-means, which uses the weighted distances and penalty term in the loss function [10].

2.2 Autoencoder

Autoencoder is an unsupervised learning algorithm using neural networks [11]. It is used to learn the structure of the input feature vector instead of predicting the output label. An autoencoder consists of two parts: encoder and decoder. Suppose F is the input space, and $F \in \mathbb{R}^k$ is a subspace of the k-dimensional real space. The encoder and decoder is defined as follows.

$$\text{encoder: } F \to \mathbb{R}^d \quad (2)$$

$$\text{decoder: } \mathbb{R}^d \to F \quad (3)$$

As we do not know the labels for unsupervised learning, we learn an function defined by decoder ∘ encoder, whose domain and image are both in F. This function first encodes the input and then reconstructs the input feature vector by decoding the compressed-dimension feature vector. Typically the encoder decreases the dimensionality of the original input feature vector, i.e. $\mathrm{k < d}$ [11]. One way of decreasing the dimensionality is to use pooling layers in the encoder. It is intended to throw away irrelevant dimensions of data and help reduce the likelihood of overfitting. For each input feature vector $\mathrm{f \in F}$, the loss function is given by the minimum square error:

$$\|\boldsymbol{f} - \text{decoder} \circ \text{encoder}(\boldsymbol{f})\|_2. \quad (4)$$

Both the encoder and decoder can contain single or multiple layers of neural network. To learn the function decoder ∘ encoder, the weights related to the neural network for the encoder and decoder are learned.

Extensions of autoencoder includes denoising autoencoder, sparse autoencoder, and so on. Denoising autoencoder is intended to recover useful information from corrupted input [12]. Sparse autoencoder constrains the number of active hidden layers so that useful information can be extracted when the number of hidden layers is large [13].

2.3 Convolutional neural network (CNN)

Convolutional neural network (CNN) is a supervised learning algorithm [14]. It consists of four parts: convolutional, pooling, activation, and fully connected layers. The convolutional layer uses different filters to produce several feature maps for the input. The feature maps have smaller dimension than the original input. The pooling layer is intended to downsample previous feature maps and avoids overfitting [14]. For example, the max pooling will apply the max filter to the sub-region of the previous feature maps, while the average pooling will apply the average filter. In this way, the dimensionality is reduced further. Between each successive two layers, the activation layer can apply activation functions to do more complicated calculation than linear combination. Activation functions include rectifier, sigmoid, and so on. The fully connected layer is typically applied at last to produce a combination of features of the previous layer for prediction. Linear combination and activation function can be used in this layer. Figure 1 shows a simple example of a convolutional neural network. In this example, two filters are applied to the original input, creating two feature maps. After the pooling layer, the dimensionality of the two feature maps further decreases. Finally, the fully connected layer will combine the two feature maps.

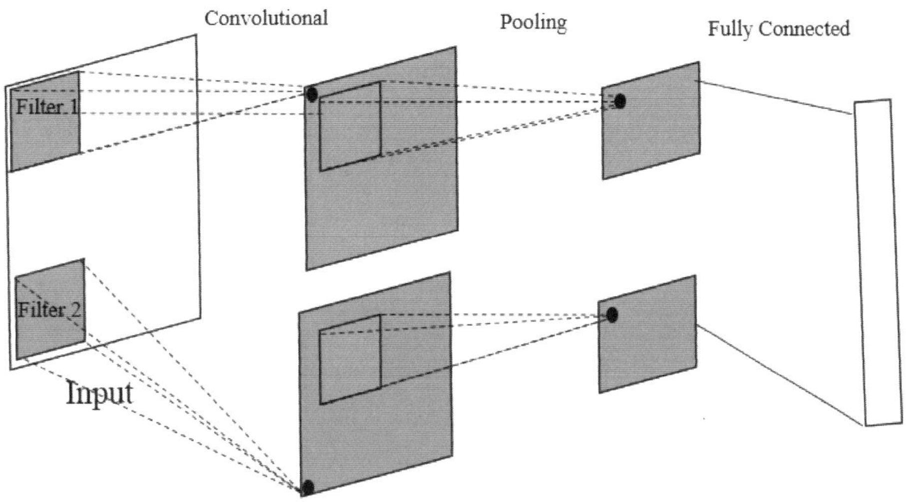

Figure 1. An Example of Convolutional Neural Network

During the training process, the weights in the convolutional and fully-connected layers are learned. Different design of hidden layers can be applied to deal with different machine learning problems.

3. Advantages and disadvantages of K-means clustering, autoencoder, and convolutional neural network

As two unsupervised leaning algorithms, k-means clustering and autoencoder does not require the label for the input. Thus, no human efforts are required to manually generate the labels for a large number of input. Besides that, k-means clustering is computationally fast [15]. However, k-means clustering often produces clusters of relatively uniform size. Thus, if the input has different true cluster size, k-means clustering can fail to learn that. Moreover, the result of k-means clustering depends on initialization a lot [15]. Poor initialization can produce poor clustering result. Autoencoder can be used to compute good representation of the input data. The combination of encoder and decoder can reduce irrelevant part of the input data. However, if the input has only limited essential information, the autoencoder will fail to throw away most of the irrelevant aspects of the data. Moreover, it is computationally expensive to train an autoencoder. Convolutional neural network takes advantage of locality of input data, and works well to learning problem that exhibits locality including sentiment detection from image. However, it requires large training dataset, and the training time is long. Table 1 gives a general comparison between the three algorithms.

Table 1. Comparison of K-Means, Autoencoder, and CNN

Method	Label Free	Computational Efficiency	Accurate	Model-Based
K-means	√	√		
Autoencoder	√			√
CNN			√	

4. Experiment

4.1 Experiment settings

Our dataset contains images of facial expressions with labels. Using parts of them as training data, we need to find a classification of the images of facial expressions into seven sentiments: Anger, Disgust, Fear, Happiness, Sadness, Surprise, and Neutral. We denote these 7 sentiments sequentially as image class 0 to 6.Three algorithms of classifications, k-means clustering, autoencoder, and convolutional neural network, are evaluated.

We first preprocess the data by splitting data samples into training, validation, and test sets. Next, we need to remove the blank image. Our definition of a blank image with feature matrix X is given by Equation (5).

$$\|X - \mathrm{mean}(X)\|_2 < 1 \quad (5)$$

Here mean(·) takes a matrix X and returns a matrix of the same shape whose entries are the same and equal to the mean of the sum of entries of X. Finally, we normalize the feature matrix X by making its mean of sum of entries zero and variance one by shifting and scaling values of each entry of X.

We use three different algorithms (k-means clustering, autoencoder, and convolutional neural network) to train the models and test the accuracy of each model by using cluster purity. Figure 2 shows a general procedure of the experiment.

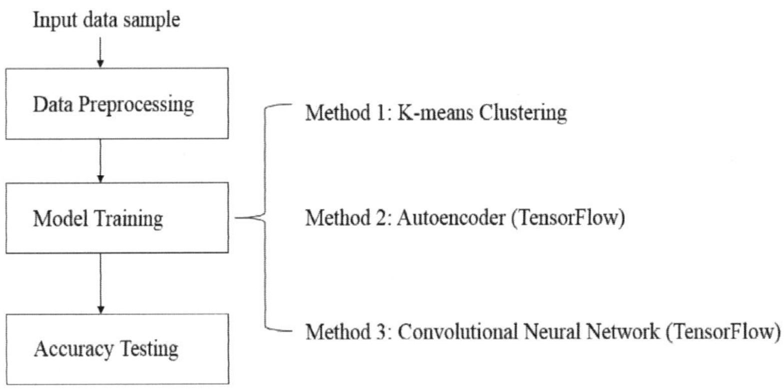

Figure 2. Experiment Setup

4.2 Experiment description

4.2.1 K-means clustering. Instead of randomly initializing the cluster centroids, we use k-means++ initialization. The details of k-means++ is given below [9]:

1. Randomly select one point as the first centroid.
2. For each point x, calculate the distance to the nearest centroid that has been selected. The distance is denoted as D_x.
3. Select the next centroid with the probability of selecting point x proportional to D_x^2.
4. Repeat step 2 and step 3 until k centroids are chosen.

We test all the possible k from [1, 10] for random initialization and k-means++ initialization. The clustering performance is evaluated by using cluster purity. After the performance comparison, we will choose the best k as the cluster number, and compare the performance of random initialization and k-means++ initialization.

4.2.2 Autoencoder. The autoencoder neural network consists of 6 layers. Layer 1 is an average pooling layer to reduce dimensionality. Layer 2 is a fully-connected layer to further reduce the image dimensionality to one dimension. Rectifier function $f(x) = \max(x, 0)$ is used as the activation function. Layer 1 and 2 works as the encoder. Layer 3 is a fuller-connected layer with rectifier function as activation function. Layer 4 is a convolutional layer intended to grow the dimensionality of the image. Layer 3 and 4 works as the decoder. Layer 5 is intended to crop the image to have the same dimensionality as the original input image, and layer 6 carries out color normalization, making the mean of the pixel values zero and the variance of the pixel values one for the image. Layer 5 and 6 are used to post-process. The weights are initialized by independently randomly sampling from Gaussian distribution given mean and standard deviation.

When training the data, we use root mean square error as the loss function. Then, we use logistic regression with L2-regularization and regularization strength equal to 1 to classify the sentiment of the faces by the representational vectors achieved by the autoencoder.

4.2.3 Convolutional neural network (CNN). The architecture of the convolutional neural network consists of 5 layers. The first three layers are convolutional layers with the rectifier function (ReLU) as the activation function. The fourth layer is a fully connected layer, and the fifth layer is the output layer, which outputs the label for the sentiment. We then train the model on the training data and test it on the validation data. We can then achieve the accuracy (cluster purity) of the classification.

4.3 Experiment results

For k-means clustering algorithm, Figure 3 shows a general increase trend of the cluster purity for both k-means and k-means++ clustering. The purity reaches maximum when the number of clusters k = 10 for both cases. When k = 10, the random initialization of k-means produces cluster purity = 0.2306. On the other hand, k-means++ initialization produces cluster purity = 0.2358. Both cluster purities are close to each other with k-means++ initialization a bit better than random initialization. The highest accuracy achieved is 0.2358.

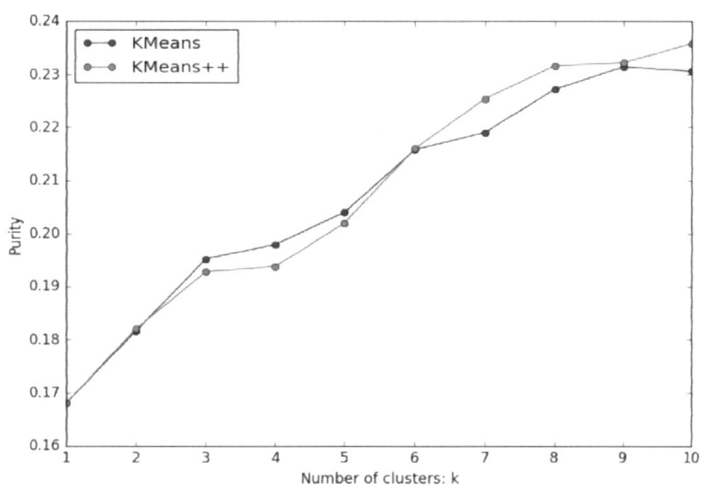

Figure 3. Cluster Purity vs. Number of Clusters

After classification using autoencoder, the cluster purity is achieved and equal to 0.268. Figure 4 gives an example of the original face and the reconstructed face after autoencoder. The autoencoder keeps track of the outline of the original face, and also some critical facial characteristics such as eyes and nose position.

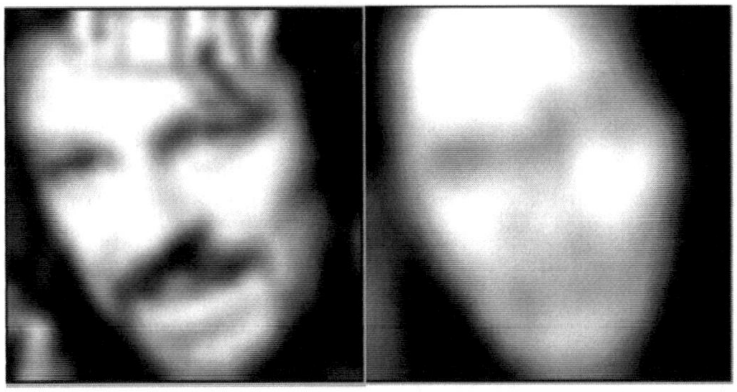

Figure 4. The original face and the reconstructed face after autoencoder

Classification using CNN produces the cluster purity as 0.444. Figure 5 shows that classification using autoencoder have a slightly higher accuracy than classification using k-means++. Classification using convolutional neural network will produce the highest accuracy.

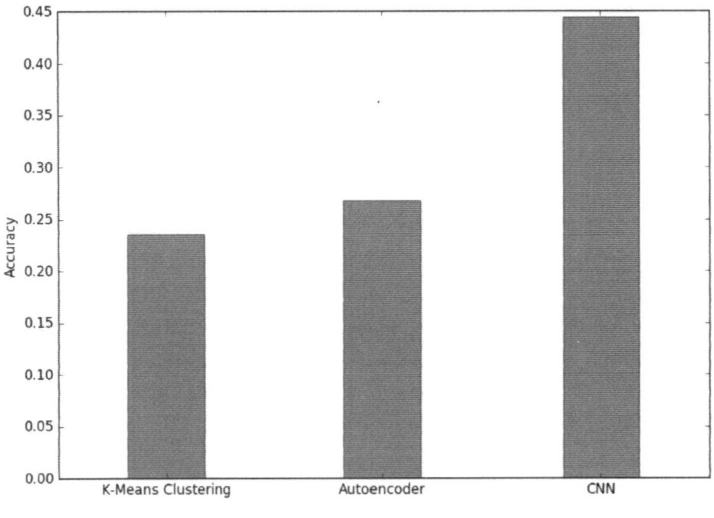

Figure 5. Accuracy of Three Models

5 Conclusions

By comparing the classification using three different algorithms, we find out that classification using convolutional neural network can produce the highest accuracy. In case of requirement of fast training speed with limited dataset, we might choose to use k-means clustering, which will sacrifice accuracy; otherwise, convolutional neural network is a nice algorithm to use. In addition, autoencoder can be combined with convolutional neural network to improve deep neural network [16]. Thus, to pursue a high accuracy when classifying sentiment from human face images, the combination of autoencoder and convolutional neural network is better than k-means clustering.

References

[1] Zhao W, Chellappa R, Phillips P. J and Rosenfeld A 2003 Face recognition: a literature survey *ACM. Compu. Surveys (CSUR)* **35** 399-458

[2] Yang J, Zhang D, Frangi A. F and Yang J. Y 2004 Two-dimensional PCA: a new approach to appearance-based face representation and recognition IEEE Transactions on Pattern Analysis and Machine Intelligence 26.1 131-137
[3] Nefian A. V and Hayes M. H 1999 An embedded HMM-based approach for face detection and recognition Proc. IEEE International Conference on Acoustics, Speech, and Signal Processing 6 3553-56
[4] Pentland A, Moghaddam B and Starner T 1994 View-based and modular eigenspaces for face recognition Computer Vision and Pattern Recognition 94 84-91
[5] Le QV 2013 Building high-level features using large scale unsupervised learning IEEE International Conference on Acoustics, Speech and Signal Processing 8595-98
[6] MacQueen J 1967 Some methods for classification and analysis of multivariate observations Proc. of the fifth Berkeley symposium on mathematical statistics and probability 1 281-297
[7] Hartigan J. A and Wong M. A 1979 Algorithm AS 136: A K-means clustering algorithm J. of the Royal Statistical Society. Series C (Applied Statistics) 28 100–108
[8] Khan SS and Amir A 2004 Cluster center initialization algorithm for K-means clustering Pattern recognition letters 25.11 1293-1302
[9] Arthur D and Vassilvitskii S 2007 K-means++: The advantages of careful seeding Proc. of the eighteenth annual ACM-SIAM symposium on discrete algorithms 1027-35
[10] Tseng G. C 2007 Penalized and weighted K-means for clustering with scattered objects and prior information in high-throughput biological data Bioinformatics 23.17 2247- 55
[11] Hinton G. E and Salakhutdinov R. R 2006 Reducing the dimensionality of data with neural networks Science 313.5786 504-507
[12] Vincent P, Larochelle H, Bengio Y and Manzagol P. A 2008 Extracting and composing robust features with denoising autoencoders Proc. of the 25th Int. Conf. on Machine learning 1096-1103
[13] Ng A 2011 Sparse autoencoder CS294A Lecture notes 72.2011 1-19
[14] Kryzhevsky A, Sutskever I and Hinton G. E 2012 Imagenet classification with deep convolutional neural networks Advances in neural information processing systems 1097-1105
[15] Singh K, Malik D and Sharma N 2011 Evolving limitations in K-means algorithm in data mining and their removal Int. J. of Computational Engineering & Management 12 105-109.
[16] Le QV 2015 A Tutorial on Deep Learning Part 2: Autoencoders, Convolutional Neural Networks and Recurrent Neural Networks Google Brain

An Integrated Tone Mapping for High Dynamic Range Image Visualization

Lei Liang[1, 3, *], Jeng-Shyang Pan[1, 2] and Yongjun Zhuang[3]

[1] Department of Computer Science and Technology, Shenzhen Graduate School, Harbin Institute of Technology, Shenzhen, China
[2] Fujian Provincial Key Lab of Big Data Mining and Applications, Fujian University of Techology, Fujian, China
[3] Qihan Research, Qihan Technology Co., LTD, Shenzhen, China

Email:[*]lianglei8568@163.com

Abstract. There are two type tone mapping operators for high dynamic range (HDR) image visualization. HDR image mapped by perceptual operators have strong sense of reality, but will lose local details. Empirical operators can maximize local detail information of HDR image, but realism is not strong. A common tone mapping operator suitable for all applications is not available. This paper proposes a novel integrated tone mapping framework which can achieve conversion between empirical operators and perceptual operators. In this framework, the empirical operator is rendered based on improved saliency map, which simulates the visual attention mechanism of the human eye to the natural scene. The results of objective evaluation prove the effectiveness of the proposed solution.

1. Introduction

In order to solve the problems of the mismatch between high dynamic range (HDR) image and low dynamic range (LDR) display device, many scholars have proposed a variety of tone mapping operators in recent years. Existing tone mapping operators are mainly divided into two categories: one is perceptual operators and the other is empirical operators. Perceptual operators are based on human visual system (HVS) model, such as photoreceptor adaptation model [1], threshold-versus-intensity (TVI) model [2], and adaptive logarithmic model [3] and so on. Physiological mechanisms of the human eye are generally added to the operator. The advantage is the mapped images are realistic and the disadvantage is the lack of visibility of local details (As shown in Figure 1). Empirical operators are not based on HVS model, but it tries to create aesthetically-pleasing or visual impact images inspired by other fields, such as histogram adjustment [4], bilateral filtering [5] and gradient domain [6] and so on. These operators can maintain the visibility of local details, but the image looks unrealistic (As shown in Figure 2). Nowadays, there is no common tone mapping operators suitable for all applications. Therefore, the suitable tone mapping operators should be selected based on the specific application. To solve this problem, this paper proposes a novel tone mapping framework which can achieve conversion between empirical operators and perceptual operators, called integrated tone mapping. The proposed framework is based on improved saliency map and has a user parameter, by adjusting the parameter, it can be conversion between detail visibility and realism of the result image.

Figure 1. Result of the HDR images (bottle and church) processed by the tone mapping algorithm based on perceptual models.

Figure 2. Result of the HDR images (bottle and church) processed by the tone mapping algorithm based on empirical models.

2. Proposed framework

Overall procedure of the proposed framework, depicted in Figure 3. The steps of our framework are summarized as follows:

Step 1: The input HDR image was tone mapped by empirical operators. This paper takes gradient operator as example. Fattal's gradient operator is the most well-known empirical tone mapping operators. Fattal *et al.* [6] conduct multi scale attenuation for luminance image in the image gradient domain, and then use compressed gradient distribution to restore the luminance of the image by solving the Poisson equation. This operator is simple, efficient and can almost display all the details. However, the effectiveness of the operator does not match the human visual perception and the realism of the result image is poor.

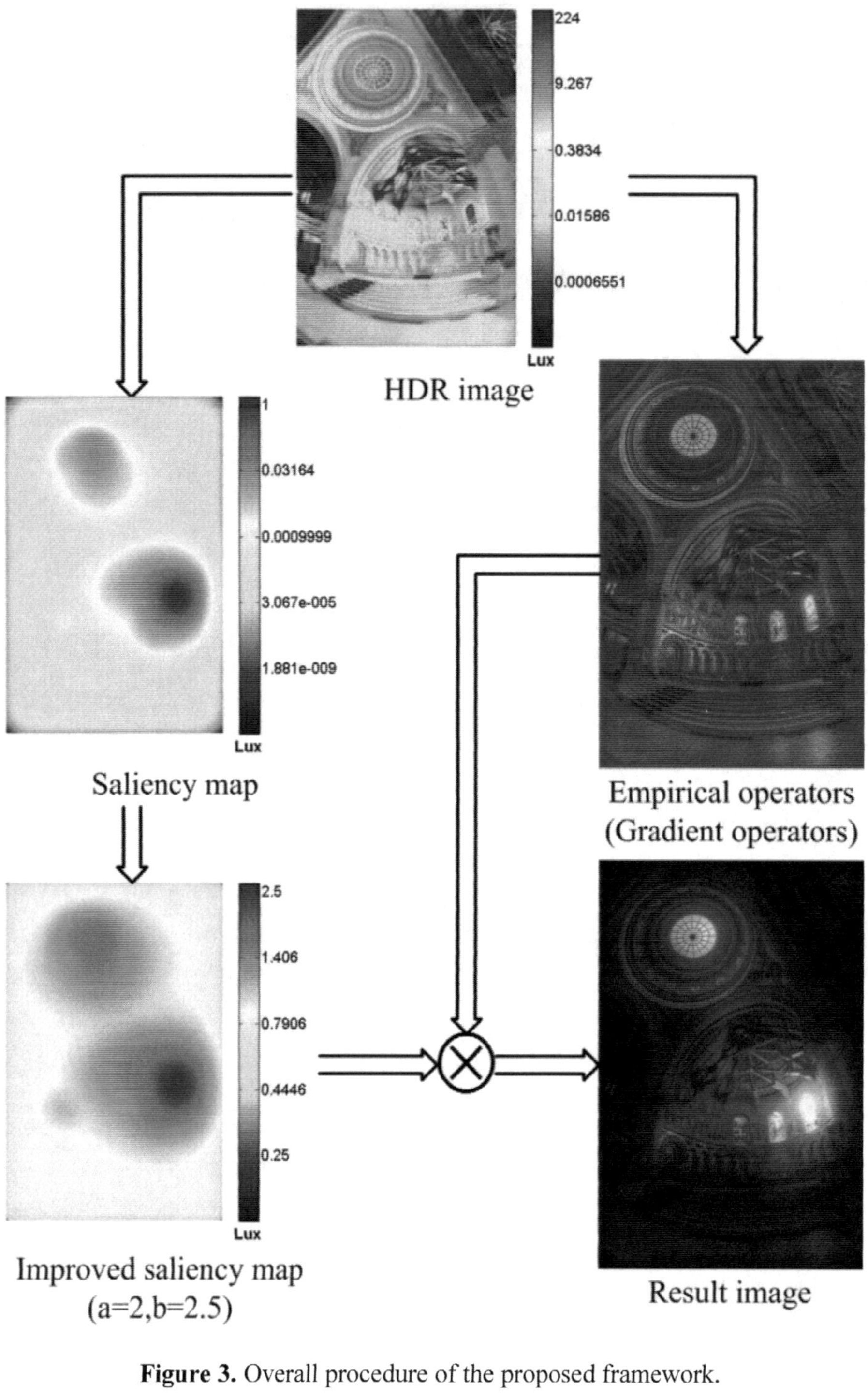

Figure 3. Overall procedure of the proposed framework.

Step 2: Calculating improved saliency map. Itti and Koch [7] proposed "Itti model" based on the neurobiological framework. Itti model provides a bottom-up mechanism with each feature map parallel computed. The model combines a variety of features in a multi-scale, and synthesizes into a comprehensive visual saliency map. But itti model is not suitable for calculating HDR image and we propose an improved saliency map calculation model for HDR image. Firstly the saliency map generation by calculating the itti model, and then it transform according to the Eq. (1-3),

$$s_1(i) = log_{10}[s(i)] \tag{1}$$

$$s_2(i) = a \cdot \frac{s_1(i) - max[s_1(i)]}{max[s_1(i)] - min[s_1(i)]} \tag{2}$$

$$s_3(i) = b \cdot 10^{s_2(i)} \tag{3}$$

where i is the position of the pixels in the map, $s(i)$ is saliency map values, s_3 is improved saliency map values, s_1,s_2 are intermediate variables. $max(s_1)$, $min(s_1)$ is the maximum and minimum pixel values in s_1. a and b are user parameters which can adjust the shape of the mapping curve. As shown in Figure 4. Above the black line represents the enhanced and below represents attenuation. Studies have shown that human eye perceive luminance in the form of a logarithmic curve [3]. The above transformation equation is equivalent to a logarithmic compression. It can simulate the human eye's visual sensitivity change in HDR scene.

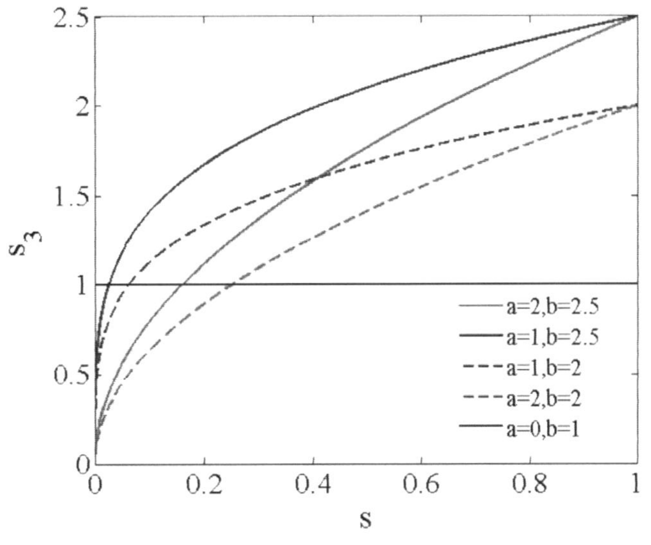

Figure 4. Improved saliency map curve.

Step 3: The result of step 1 and step 2 synthesis. As shown in Eq. (4).

$$R(i) = E(i) \cdot s_3(i) \tag{4}$$

Where $E(i)$ is the tone mapped image by empirical operators, $R(i)$ is the result image.

3. Experimental results

In this paper, Experimental results are evaluated by Dynamic Range Independent (DRI) image quality assessment model [8]. DRI model produces a distortion probability map, and the green pixels represent the pixel value is too small, the blue pixels represent the pixel value is too large, and the red pixels represent gradient reversal. As shown in Figure 5. When user parameters $a=0$ and $b=1$, the HDR image mapped by empirical operators. There are a large number of green pixels and blue pixels existing in bright and dark region, and the image is flat. When user parameters $a=2$ and $b=2.5$, the HDR image mapped by Perceptual operators, both green and blue pixels are reduced and the image looks more layered.

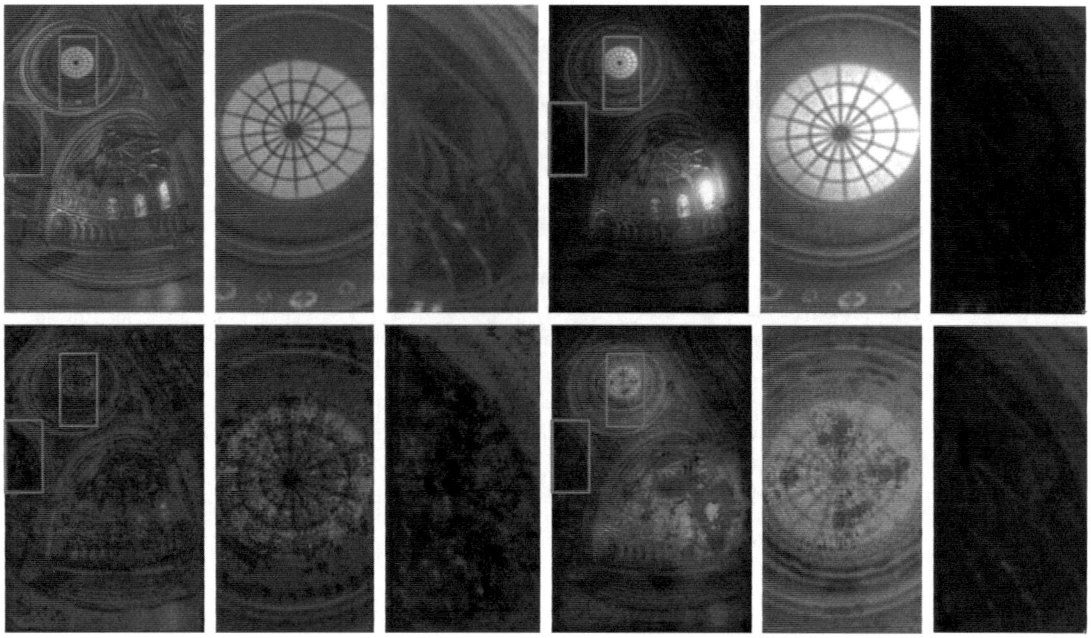

Figure 5. Top row: Experimental results. Bottom row: Evaluate results.
From left to right: Empirical operators and its bright, dark region. ($a=0$, $b=1$). Perceptual operators and its bright, dark region. ($a=2$, $b=2.5$).

4. Conclusion

Based on improved saliency map, a novel framework for integrated tone mapping is introduced. By adjusting the user parameter, it can be conversion between detail visibility and realism of the result image. Evaluate results of DRI model prove the effectiveness of the proposed solution.

References

[1] E. Reinhard and K. Devlin, Dynamic range reduction inspired by photoreceptor physiology, IEEE Trans. Visualization and Computer Graphics, vol. 11, pp. 13-24, 2005.
[2] J. A. Ferwerda, S. N. Pattanaik, P. Shirley, and D. P. Greenberg, A model of visual adaptation for realistic image synthesis, Proc. ACM SIGGRAPH '96, pp. 249-258, 1996.
[3] F. Drago, K. Myszkowski, T. Annen, and N. Chiba, Adaptive logarithmic mapping for displaying high contrast scenes, Computer Graphics Forum, pp. 419-426, 2003.
[4] J. Duan, M. Bressan, C. Dance, G. Qiu. Tone-mapping high dynamic range images by novel histogram adjustment, Pattern Recognition, vol. 43, pp. 1847-1862, 2010.
[5] F. Durand and J. Dorsey, Fast bilateral filtering for the display of high-dynamic-range images," ACM Trans. Graphics, vol. 21, pp. 257-266, 2002.
[6] R. Fattal, D. Lischinski, and M. Werman, Gradient domain high dynamic range compression," ACM Trans. Graphics, vol. 21, pp. 249-256, 2002.
[7] L. Itti, C. Koch, and E. Niebur, A model of saliency-based visual attention for rapid scene analysis, IEEE Trans. Pattern Analysis and Machine Intelligence, vol. 20, pp. 1254-1259, 1998.
[8] T. O. Aydin, R. Mantiuk, K. Myszkowski, S. Hans-Peter, Dynamic range independent image quality assessment, ACM Trans. Graphics, vol. 27, pp. 69-78, 2008.

Image Interpolation Scheme based on SVM and Improved PSO

X F Jia[1], B T Zhao[1,*], X X Liu[1] and H P Song[1]

[1]School of Electrical and Information Engineering, Anhui University of Science & Technology, Huainan 232001, China

[*]zhaobaiting@gmail.com

Abstract: In order to obtain visually pleasing images, a support vector machines (SVM) based interpolation scheme is proposed, in which the improved particle swarm optimization is applied to support vector machine parameters optimization. Training samples are constructed by the pixels around the pixel to be interpolated. Then the support vector machine with optimal parameters is trained using training samples. After the training, we can get the interpolation model, which can be employed to estimate the unknown pixel. Experimental result show that the interpolated images get improvement PNSR compared with traditional interpolation methods, which is agrees with the subjective quality.

1. Introduction

Image interpolation is designed to reproduce a high-resolution image with a low-resolution image, which is a key technology in digital image and video processing. Image interpolation techniques are divided into traditional interpolation, edge-based interpolation and region-based interpolation. Traditional interpolation is simple to implement, but the image edge is blurred. Edge-based interpolation has a certain enhancement on the edge of the image, but requires an accurate extraction of the edges of the image. Region-based interpolation requires accurate segmentation of the region, and different regions need to design different interpolation formula. In recent years, various learning-based image interpolation methods have been proposed, such as neural network image interpolation [1], SVM image interpolation [2-3].

The choice of parameters is an important problem in the field of SVM research, and its essence is an optimization search process. Phan AV [4] et al designed a novel hybrid model to enhance SVM's classification efficiency by using genetic algorithms. A SVM classifier [5] based on particle swarm optimization was proposed to increase the accuracy of classify electrical faults in radial distribution systems. Ji [6] et al introduced an iterative optimization algorithm applying the EnKF method to optimize SVM's parameters.

SVM is a hot research topic in the field of machine learning and data mining. The research of parameter selection is not uncommon, but few studies on the background of image interpolation. In this paper, we improve the particle swarm optimization algorithm to optimize the parameters of the SVM, and then use the SVM to optimize the image interpolation. A high precision image interpolation method is proposed. The design flow is shown in Figure 1.

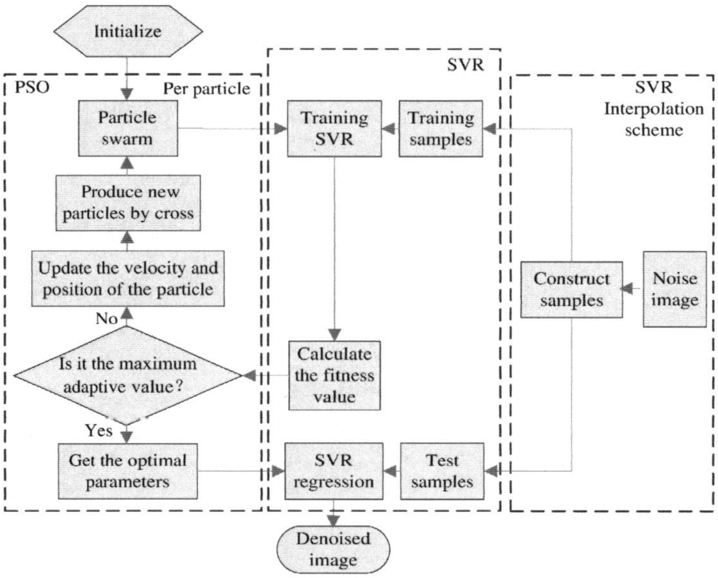

Figure 1. The flow chart of image interpolation

2. Theory

2.1. γ -Support vector regression machine

The basic idea of the vector regression algorithm is to map the data x to the high-dimensional feature space through a nonlinear mapping ϕ and linear regression in the space. The function regression problem can be described as: given a set of training samples $T = \{(x_i, y_i)\}_{i=1}^{n}$, construct a linear regression function in the feature space:

$$f(x) = \omega^T \phi(x) + b \tag{1}$$

Which, ω and b represents the regression factor, is obtained by solving the minimization of the regularization risk function, i.e. [7]:

$$\min \frac{1}{2} \| \omega \|^2 + C[\gamma\varepsilon + \frac{1}{n}\sum_{i=1}^{n} (\xi_i + \xi_i^*)] \tag{2}$$

$$s.t.\begin{cases} y_i - (\omega \cdot \phi(x_i)) - b \le \varepsilon + \xi_i \\ (\omega \cdot \phi(x_i)) + b - y_i \le \varepsilon + \xi_i^* \\ \varepsilon, \xi_i, \xi_i^* \ge 0, \cdots i = 1, \cdots n \end{cases}$$

Which, $\| \cdot \|$ on behalf of the regularization operator, C as a penalty factor and γ for the insensitive parameter.Solving the formula (1), the final regression function can be obtained:

$$f(x) = \sum_{i=1}^{m} (\alpha_i - \alpha_i^*) < \phi(x_i), \phi(x) > +b = \sum_{i=1}^{m} (\alpha_i - \alpha_i^*) k(x_i, x) + b \tag{3}$$

Among them, α_i and α_i^* is the Lagrangian multiplier. $k(x_i, x) =< \phi(x_i), \phi(x) >$ as a kernel function, used to achieve non-linear mapping.

2.2. Particle swarm optimization

Particle swarm optimization (PSO) [8] algorithm is a new branch of group intelligence, which is an efficient stochastic optimization technique. In solving the PSO optimization problem, the solution lies in the search for the location of a bird in space, saying that these birds are "particles". Each particle has its own position and velocity and individual fitness corresponding to the objective function. According to their own and group flight experience to adjust their flight path, to move closer to the best. Each adjustment of the flight trajectory is achieved by tracking two "extreme values", one is the optimal solution of the particle itself, called the individual extreme value $p_i = (p_{i1}, p_{i2}, \cdots, p_{in})$; the other is the optimal solution of the whole group, called the global extreme value $p_g = (p_{g1}, p_{g2}, \cdots, p_{gn})$. After the particles find the two optimal values, update their speed and position according to the formula (4-5):

$$v_{id}(t+1) = w v_{id}(t) + c_1 rand()(p_{id}(t) - x_{id}(t)) + c_2 Rand()(p_{gd}(t) - x_{id}(t)) \quad (4)$$

$$x_{id}(t+1) = x_{id}(t) + v_{id}(t+1) \quad (5)$$

Where w is the inertia weight, $d = 1, 2, \cdots, n$, t is the number of current evolutionary iterations, $i = 1, 2, \cdots, m$, m is the population size, $rand()$ and $Rand()$ are the random number distributed between [0,1]; c_1 and c_2 is the two positive numbers, called the acceleration factor; $v_{id}(t)$ is the first d dimension of the flight velocity in the first t iteration of the first i particle; $x_{id}(t)$ is the first d dimension component of the position of the first t iteration of the first d particle; $P_{id}(t)$ is the first d dimension component of the best position of the particle i; $P_{gd}(t)$ the first d dimension component of the best position of the population.

3. SVM optimization parameters based on PSO

C is used to balance the training error and the model complexity. The model complexity of SVM increases as C increases, but, the training error decreases. When *C* increases to a certain extent, the performance of SVM will not change any more. When C small, the punishment of empirical error is little, low complexity training, poor accuracy of fitting. The γ reflects the sensitivity of the regression model to the noise contained in the input variable, The greater the insensitive parameter, the lower the accuracy of the model fitting, the lower the complexity, prone to over-fitting. The kernel function σ represents the mean square error of the Gaussian function. When σ is small, the fitting function of the RBF kernel function is better, but the generalization ability is deteriorated.

3.1. Parameter optimization process

In this paper, we discuss the optimization of parameters C , γ and σ in the case of use γ -SVR and the RBF kernel function.

The PSO optimization algorithm is as follows:

① Initialization. Initialize the acceleration factor c_1 , c_2 and the inertia weight w and the maximum number of evolutionary iterations T to determine the population size. The parameters to be optimized are mapped to particle swarm, and the initial position and velocity of each particle are obtained in a random manner;

② Calculate the fitness value. The fitness value of each particle is calculated according to the fitness function;

③ Determine the individual extremes and global extremes;

④ Update the speed and location of each particle. Using the formula (4), (5) to generate a new generation of population;

⑤ Update individual extremes and global extremes. Calculating the fitness value of each particle in the new particle group according to the fitness function, and then updating the individual extreme and the global extreme value according to the fitness value;

⑥ If $f(p_i) > P_{id}$, update the individual extreme , if $f(p_i) > P_{gd}$, update the global extreme ;

⑦ Cross operation. The particles $m \times P_c$ are randomly selected according to the preset probability P_c , For each pair of particles were bit by bit "take large" operation and "take small" operation to produce two new particles;

⑧ Calculate the fitness values for new particles. If the fitness value is better than the fitness value of the corresponding particle before the cross operation, accept them and update the best position of the individual and the group; otherwise return to the state before crossing;

⑨ Check the termination condition. When the required evolutionary algebra or the maximum adaptation of the 15th generation did not change, terminate the operation. The particles corresponding to the global optimal position are mapped into parameters c , γ and σ otherwise go to step 4.

3.2. Fitness function

The fitness function is the only deterministic indicator of the merits of the particle position. Select the function that reflects the SVR regression performance as a fitness function, defined as:

$$F(x_i) = 1/(\frac{1}{n}\sum_{i=1}^{m} | y_i - \hat{y}_i |) \tag{6}$$

$F(x_i)$ is the fitness value of the first i particle in the population, y_i and $\hat{y}_i$ are expected output value and training output value corresponding to the SVR corresponding to the parameter in the first i particle respectively.

4. SVR interpolation algorithm

The advantages and disadvantages of image interpolation are closely related to the correlation of images. Correlation refers to the same local area of the adjacent pixels of an image have similar gray values. The gray value of the image to be interpolated is closely related to the average gray value of the pixel in the neighborhood and the change in the gray value of each direction pixel [3]. Therefore, SVR training samples should include them.

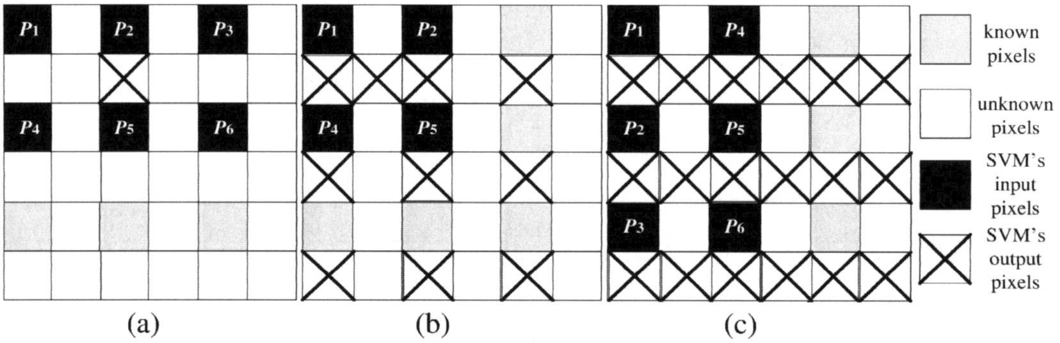

Figure 2. Interpolation mode

The interpolation process is illustrated by taking the image enlarged in 2 times in both the row direction and the column direction. The input of the training sample and the test sample are constructed as 22 dimensions, the former output is the known pixel value, and the latter output is the pixel value to be interpolated. The 22-dimensional inputs are the known pixel values $P_i, i = 1, 2 \cdots, 6$ contained in Figure 2, the mean avg of the six known pixels, and the difference between the two pixels in the horizontal, vertical, and diagonal directions $d_m, m = 1, 2 \cdots, 15$:

$$avg = \frac{1}{6}\sum_{i=1}^{6} P_i,\ d_1 = P_1 - P_2,\ d_2 = P_1 - P_3,\ d_3 = P_1 - P_4,\ d_4 = P_1 - P_5,\ d_5 = P_1 - P_6$$

$$d_6 = P_2 - P_3,\ d_7 = P_2 - P_4,\ d_8 = P_2 - P_5,\ d_9 = P_2 - P_6,\ d_{10} = P_3 - P_4,\ d_{11} = P_3 - P_5$$

$$d_{12} = P_3 - P_6,\ d_{13} = P_4 - P_5,\ d_{14} = P_4 - P_6,\ d_{15} = P_5 - P_6 \tag{7}$$

Using all the training samples to train the SVR to get the interpolation model, and then use the test sample as the input of the interpolation model to estimate the gray value to be interpolated. The interpolation is done in the following three steps:

① Interpolate even rows odd columns: according to Figure 2(a) interpolation model construction training samples;

② Interpolate even rows and columns: according to Figure 2(b) interpolation model construction training samples , when constructed, p_3 and p_6 are zero;

③ Interpolate odd rows even columns: according to Figure 2(c) interpolation model construction training samples.

5. Experimental results

The experiments were performed by LIBSVM [9] and MatlabR2009b. First, 1/2 under-sampling the standard image, then, interpolating the obtained image with different methods, such as linear interpolation, cubic cubic interpolation, nn interpolation, SVM interpolation and PSO optimization interpolation. In the simulation process, real-valued codes are used to randomly generate 3-dimensional particles as the initial positions of the particle swarm. Randomly generating random numbers in [0, 1] as the initial velocity of the particles. In order to solve the problem of position and velocity update of different parameters, the three parameters are normalized before optimization, and then reverse the results of the optimization. The range of the search parameters is $C \in (0.001, 150)$, $\gamma \in (0, 0.99)$, $\sigma \in (0.001, 2)$. PSO population size M=50, maximum evolutionary algebra T=100, crossover probability P_c=0.4, inertia weight w=1, acceleration factor $c_1 = c_2 = 2$. The global optimal particles obtained after optimization were (31.6208, 0.9204, 0.0714).

From table 1 we can find that the proposed interpolation effect is better than other methods, the PSNR values are higher than those of linear interpolation, cubic interpolation, nn interpolation and SVM interpolation. Figures 3 show the interpolation results of Lena in different methods. It can be seen that the result of linear interpolation, cubic interpolation and nn interpolation in Lena's hair, hat edge appear obvious saw-tooth. nn interpolation has an edge blur on the right side of Lena and the right side of the hair. SVM interpolation effect than the previous three have significantly improved, but in the shoulder of Lena there are obvious jagged effect, as shown in figure 4, it is the effect of enlarging the residual image of the shoulder in the white frame in Lena standard image in figure 3 by 350 times.

Table 1. Comparison of interpolation results for different interpolation methods

	Linear interpolation		Cubic interpolation		nn interpolation		SVM interpolation		PSO optimized interpolation	
	PSNR	NMSE	**PSNR**	NMSE	**PSNR**	NMSE	**PSNR**	NMSE	**PSNR**	NMSE
Lena	**23.26**	0.0175	**22.86**	0.0192	**24.05**	0.0152	**29.96**	0.0037	**30.31**	0.0035
Cameraman	**21.42**	0.0261	**20.72**	0.0307	**21.51**	0.0263	**28.03**	0.0057	**28.42**	0.0056
Pepers	**23.63**	0.0158	**23.28**	0.0172	**25.20**	0.0110	**30.34**	0.0034	**30.69**	0.0033
Boattest	**21.49**	0.0242	**20.73**	0.0289	**21.90**	0.0220	**27.69**	0.0058	**27.93**	0.0056
Clock	**22.21**	0.0103	**21.72**	0.0116	**23.20**	0.0082	**31.14**	0.0013	**31.55**	0.0012
House	**25.30**	0.0090	**25.07**	0.0095	**24.12**	0.0119	**31.94**	0.0020	**32.47**	0.0018
Average	**22.89**	0.0172	**22.40**	0.0195	**23.33**	0.0158	**29.85**	0.0037	**30.23**	0.0035

Standard image Linear interpolation Cubic cubic interpolation nn interpolation SVM interpolation PSO optimized interpolation

Figure 3. Lena interpolation results of each method

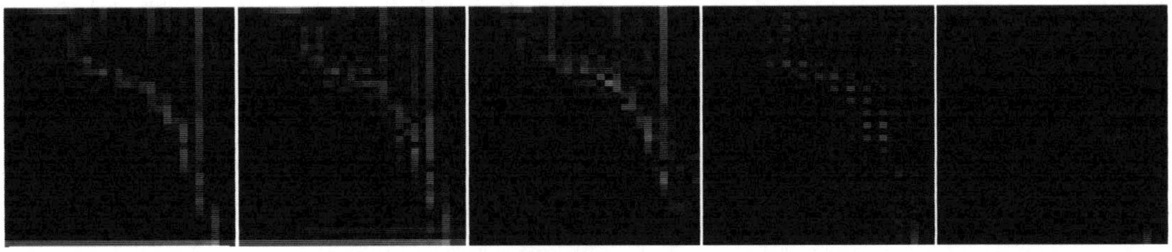

Linear interpolation residuals | Cubic interpolation residuals | nn interpolation residual | SVM interpolation residuals | PSO optimization interpolation residuals

Figure 4. The interpolation residual image of Lena's shoulder

6. Conclusion

In the PSO algorithm, the crossover operator based on the principle of "taking large" and "taking small" can solve the problem of updated the new speed and the new position beyond the boundary. The improved SVM parameters can be obtained by optimizing the PSO parameters. The SVM with optimal parameters can be used to obtain interpolation of images with significant improvement of subjective and objective indexes in the case of full consideration of image correlation. It can effectively prevent the presence of zipper effects in the image, and reduces edge blurring.

Acknowledgement

This research was financially supported by the following funds: The National Natural Science Foundation of China (61501006), the Postdoctoral Fund National Projects (2016M592035) and Anhui Provincial Natural Science Research Projects in Central Universities (KJ2017A076).

References

[1] Plaziac N 1999 *IEEE Transactions on Image Processing* vol 8 p 1647
[2] X F Jia, B T Zhao, M R Zhou and W Gao 2015 Application Research of Computers vol 32 p 623
[3] L Y Ma, Y Shen and J C Ma 2008 Journal of Systems Engineering and Electronics vol 19 p 618
[4] Phan AV, Nguyen ML and Bui 2017 Applied Intelligence vol 46 p 455
[5] Ming-Yuan C and Thi T H 2017 Computational Intelligence and Neuroscience p 1
[6] Y S Ji, Y S Chen, H H Fu and G W Yang 2017 Pattern Recognition vol 62 p 202
[7] Vapnik V 1995 Springer
[8] J Kennedy R Eberhart 1995 IEEE International Conference on Nerual Networks p 1942
[9] C C Chang and C J Lin 2001 Software available at http://www.csie.ntu.edu.tw/~ cjlin /libsvm

Deep Marginalized Sparse Denoising Auto-Encoder for Image Denoising

Hongqiang MA[1,*], Shiping MA[1], Yuelei XU[1] and Mingming ZHU[1]

[1]Aeronautics Engineering College, Air Force Engineering University, Xian, China

[*]18049025189@163.com

Abstract. Stacked Sparse Denoising Auto-Encoder (SSDA) has been successfully applied to image denoising. As a deep network, the SSDA network with powerful data feature learning ability is superior to the traditional image denoising algorithms. However, the algorithm has high computational complexity and slow convergence rate in the training. To address this limitation, we present a method of image denoising based on Deep Marginalized Sparse Denoising Auto-Encoder (DMSDA). The loss function of Sparse Denoising Auto-Encoder is marginalized so that it satisfies both sparseness and marginality. The experimental results show that the proposed algorithm can not only outperform SSDA in the convergence speed and training time, but also has better denoising performance than the current excellent denoising algorithms, including both the subjective and objective evaluation of image denoising.

1. Introduction

During the process of formation and transmission, digital images are often disturbed by different levels of noises, which degrade image quality. Image denoising is a very fundamental but crucial procedure in image processing, and its processing results directly influence the final effect of various tasks such as edge detection, target recognition and so on. Therefore, it is of great theoretical and practical value to study the image denoising algorithms and optimize the performance of the algorithms.

In recent years, a variety of image denoising methods has been proposed, typically including the following 3 categories. One approach is linear or non-linear filtering methods which are a relatively simple approach based on smoothing, such as median filtering and Wiener filtering. Another one is methods based on wavelet or dictionary decomposition of the image. Wavelet decomposition is to transform the image signal into wavelet domain for multi-layers decomposition, where they can be more easily separated from the noise, such as BLS-GSM [1]. The dictionary-based methods are to denoise by approximating the noisy patch using a sparse linear combination of atoms, including KSVD, which is an iterative algorithm that learns a dictionary on the noisy image at hand. [2]. The last one is methods based on global image statistics or other image properties, such as self-similarity. BM3D is generally considered the state-of-the-art algorithm for image noise removal. It uses a collaborative form of Wiener filter for high dimensional block of patches by grouping similar 2D blocks into a 3D data array [3].

While these methods have been successfully applied in practice, they share a shallow linear structure. When the noise intensity is small, the denoising effect is remarkable; and when the noise intensity is large, the denoised images often lose the edge, texture and other details information. Therefore, it is necessary to find a method to denoise effectively in high noise environment. Recent research suggests, however, that non-linear deep models can achieve superior performance in various realistic problems. A few of deep models have also been applied to image denoising [4,5,6]. Deep learning, as a deep network, through the establishment of a hierarchical model similar to the human brain structure and the feature extraction of input data from bottom to top step by step, which can well establish the mapping relation from bottom signal to high-level semantics [7]. The auto-encoder is a classic model of deep learning. We can use it to learn the compression and distributed feature representation of a given dataset to reconstruct the input data [8]. Later, the researchers added some constraints on it, and derived a variety of deformation auto-encoders, including Sparse Auto-Encoder(SAE), Denoising Auto-Encoder(DAE), Marginalized Denoising Auto-Encoder(MDA)[9] and Stacked Sparse Denoising Auto-Encoder(SSDA). Xie et al. [10] had success at removing noise from corrupted images with SSDA, especially when the noise intensity is large, the performance of noise reduction is better than that of KSVD and BM3D. However, as a deep structure neural network model, it needs to be trained layer by layer, resulting in a large amount of calculation, a slow training speed and a long time to optimize the parameters.

In this paper, we marginalize the loss function of SSDA network and form Deep Marginalized Sparse Denoising Auto-Encoder (DMSDA), which is applied to image denoising. The experiment results show that the proposed algorithm not only improves the training speed, but also has a higher PSNR and the more detail information.

2. Related model description

2.1 Denoising auto-encoder (DAE)

The traditional AE can easily copy the input directly to the output without any constraints, the model usually has very poor performance [11]. Therefore, Vincent et al. proposed the DAE algorithm to improve the robustness of the network. DAE is a 3 layers neural network, composed of input layer, hidden layer and output layer. The algorithm flow of DAE is shown in Figure 1. The original input data x first adds noise to get corrupted version $\tilde{x}$, then through the coding function g to get the feature representation h of the input, and last through the decoding function f maps h to the output layer to obtain the reconstructed input data y .The reconstruction error is defined by a loss function $L(x, y)$.

The encoding and decoding process of the DAE network is respectively $h = g(\tilde{x}) = s_g(w\tilde{x} + b)$, $y = f(h) = s_f(w'h + b')$ .Where, s_g and s_f is a nonlinear activation function. w and b are respectively encoding weights and biases. w' and b' are respectively the decoding weights and biases. Given training data $D = \{x^{(i)}\}_{i=1}^{N}$, the DAE is trained by backpropagation to minimize the reconstruction loss given by

$$L_{\mathrm{DAE}}(\theta) = \frac{1}{N}\sum_{x \in D} J(x, y) + \frac{\lambda}{2}(\|w\|_F^2 + \|w'\|_F^2) \quad (1)$$

Where, $\theta = \{w, w', b, b'\}$ are the parameters of the model, $J(x, y) = \frac{1}{2}\|x - y\|_2^2$, λ is the weight of decay term, which can reduce the magnitude of the weight and prevent overfitting.

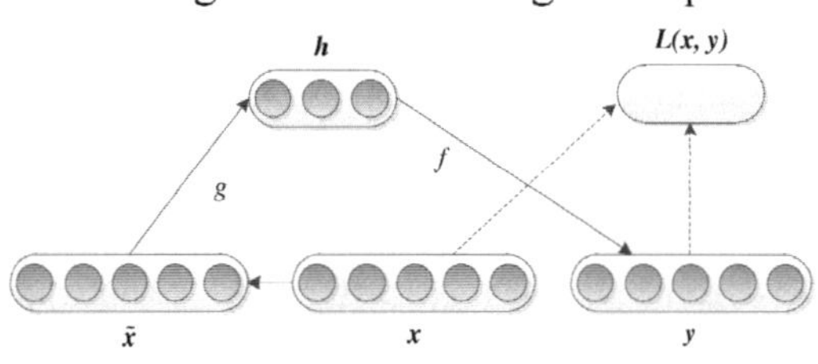

Figure 1. The algorithm flow of DAE

2.2 Sparse auto-encoder (SAE)
In addition to making the output equal to the input as much as possible, the auto-encoder requires that the hidden layers must satisfy a certain sparsity (i.e., the output value of most of the neurons on the hidden layers is close to 0). Thus, the input data can be compressed and reduce dimensionality.

Supposed ρ is a sparsity parameter, $\hat{\rho}_j$ is the average activation value of hidden unit j-th (averaged over the training set). In order to satisfy the sparsity of each neuron of the hidden layer, we approximately enforce the constraint $\hat{\rho}_j = \rho$, where, ρ is typically a small value close to zero (say $\rho = 0.05$). To achieve this, an extra penalty term KL divergence is added to loss function that penalizes deviating $\hat{\rho}_j$ significantly from ρ . $KL(\hat{\rho}_j \| \rho) = \rho \log \frac{\rho}{\hat{\rho}_j} + (1-\rho)\log \frac{1-\rho}{1-\hat{\rho}_j}$, This penalty function has the property that $KL(\hat{\rho}_j \| \rho) = 0$, if $\hat{\rho}_j = \rho$, and otherwise it increases monotonically as $\hat{\rho}_j$ diverges from ρ . So the penalty term is added to the loss function of the basic auto-encoder to form SAE.

2.3 Stacked Sparse Denoising Auto-Encoder(SSDA)
To make the auto-encoder both sparse and robust, a Sparse Denoising Auto-Encoder algorithm is proposed by the combination of SAE and DAE. The loss function of SDA is as follows:

$$L_{SDA}(\theta) = \frac{1}{N}\sum_{x \in D} J(x, y) + \beta \sum_{j=1}^{k} KL(\hat{\rho}_j \| \rho) + \frac{\lambda}{2}(\|w\|_F^2 + \|w'\|_F^2) \tag{2}$$

Where, β controls the weight of the sparsity penalty term.

However, SDA is still a shallow neural network with poor learning ability. It is difficult to learn the deep essential characteristics of the data. Therefore, a series of SDAs are connected to form a Stacked Sparse Denoising Auto-Encoder——a deep network formed by feeding the hidden layer's activations of one SDA into the input of the next SDA. New researches show that SSDA network with too many layers is easy to appear gradient dispersion phenomenon but also overfitting. In this paper, the network structure of 2 SDA is selected.

A good way to obtain good parameters for SSDA is to use Hinton's unsupervised greedy layer-wise training method proposed in [12]. The main idea of this method is: first train the first layer on raw input to obtain parameters, then use the representation of the input data to train the second layer to obtain parameters, repeat for subsequent layers, using the output of each layer as input for the subsequent layer. This method trains the parameters of each layer individually while freezing parameters for the remainder of the model. To produce better results, after this phase of training is complete, fine-tuning using backpropagation can be used to improve the results by tuning the parameters of all layers are changed at the same time. This entire network—the SSDA—is trained again by back-propagation in a fine-tuning stage, minimizing the loss given by

$$L_{SSDA}(\theta) = \frac{1}{N}\sum_{x \in D} J(x, y) + \frac{\lambda}{2}\sum_{l=1}^{2}(\|w^{(l)}\|_F^2 + \|w'^{(l)}\|_F^2) \tag{3}$$

The sparsity-inducing term is not needed for this step because the sparsity was already incorporated in the pre-trained SDAs [13].

3. Algorithm

3.1 DMSDA network model
To restore the original image to the maximum extent and improve the performance of image denoising, the SDA network loss function is marginalized to form Marginalized Sparse Denoising Auto-Encoder(MSDA).

In order to improve the robustness of the SDA network, normally, the training samples are repeatedly added with the same distributed noise to obtain as much training set as possible to train the optimal parameters. For a dataset $D = \{x_1, x_2, \cdots, x_N\}$, it can yield corrupted copy $\{x_i^1, x_i^2, \cdots, x_i^m\}$ by adding noise to each x_i m-times. In equation (3), let $m_\theta(\hat{x}) = y$, and minimize the averaged reconstruction loss

$$L_{\mathrm{SDA}}(\theta) = \frac{1}{N}\sum_{i=1}^{N}\frac{1}{m}\sum_{j=1}^{m} J(\mathrm{x}_i, m_\theta(\hat{x}_i^j)) + \beta\sum_{j=1}^{k} KL(\hat{\rho}_j \| \rho) + \frac{\lambda}{2}(\|w\|_F^2 + \|w'\|_F^2) \tag{4}$$

When m is large, this increase directly computational cost and training time. Therefore, the idea of limit is introduced, as $m \to \infty$, the average loss function for formulation (4) is:

$$L_{\mathrm{SDA}}(\theta) = \frac{1}{N}\mathrm{E}(J(\mathrm{x}_i, m_\theta(\hat{x}_i)) + \beta\sum_{j=1}^{k} KL(\hat{\rho}_j \| \rho) + \frac{\lambda}{2}(\|w\|_F^2 + \|w'\|_F^2) \tag{5}$$

The loss function of SDA by its Taylor expansion with respect to up to the second-order:

$$J(x, m_\theta(\tilde{x})) \approx J(x, m_\theta(\mu_x)) + (\tilde{x} - \mu_x)^{\mathrm{T}}\Delta_{\tilde{x}} J + \frac{1}{2}(\tilde{x} - \mu_x)^{\mathrm{T}}\Delta_{\tilde{x}}^2 J(\tilde{x} - \mu_x) \tag{6}$$

Where, $\mu_x = \mathrm{E}[\tilde{x}]$ is the expected value of $\tilde{x}$, $\Delta_{\tilde{x}} J$ and $\Delta_{\tilde{x}}^2 J$ are respectively the first-order derivative and second-order derivate about $\tilde{x}$ .we take the mean with regard to $J(x, m_\theta(\tilde{x}))$

$$\mathrm{E}(J(x, m_\theta(\tilde{x})) \approx J(x, m_\theta(\mu_x)) + \frac{1}{2}\mathrm{tr}(\mathrm{E}\left[(\tilde{x} - \mu_x)(\tilde{x} - \mu_x)^{\mathrm{T}}\right]\Delta_{\tilde{x}}^2 J) \tag{7}$$

Here, we supposed the matrix $\Sigma_x = \mathrm{E}\left[(\tilde{x} - \mu_x)(\tilde{x} - \mu_x)^{\mathrm{T}}\right]$, and obtain

$$E(J(x, m_\theta(\tilde{x})) \approx J(x, m_\theta(\mu_x)) + \frac{1}{2}\mathrm{tr}(\Sigma_x \Delta_{\tilde{x}}^2 J) \tag{8}$$

The matrix Σ_x can simplify to diagonal matrix, because the noise is added to each dimension of the input vector independently. The diagonal terms of the Hessian $\Delta_{\tilde{x}}^2 J$ is given by

$$\frac{\partial^2 J}{\partial \tilde{x}_d^2} = (\frac{\partial z}{\partial \tilde{x}_d})^{\mathrm{T}} \frac{\partial^2 J}{\partial z^2}\frac{\partial z}{\partial \tilde{x}_d} + (\frac{\partial J}{\partial z})^{\mathrm{T}}\frac{\partial^2 z}{\partial \tilde{x}_d^2} \tag{9}$$

Here, d represents the dimension of the input vector.

LeCun et al. [14] remove the second term in (9). The matrix $\partial^2 J / \partial z^2$ is the Hessian of J with respect to z, and is positive definite. Therefore, the non-negative diagonal term of the matrix can be further simplified by using the positive definite property, here, we give the final approximation

$$\frac{\partial^2 J}{\partial \tilde{x}_d^2} \approx \sum_{h=1}^{n}\frac{\partial^2 J}{\partial z_h^2}(\frac{\partial z_h}{\partial \tilde{x}_d})^2 \tag{10}$$

Where, n is the number of hidden neurons, z is the feature representation of hidden layers. After the above simplified calculation, the marginalized loss function

$$\mathrm{E}(J(x, m_\theta(\tilde{x})) \approx J(x, m_\theta(\mu_x)) + \frac{1}{2}\sum_{d=1}^{n}\sigma_{xd}^2\sum_{h=1}^{m}\frac{\partial^2 J}{\partial z_h^2}(\frac{\partial z_h}{\partial \tilde{x}_d})^2 \tag{11}$$

Where, σ_{xd}^2 is the corruption variance of the dth input dimension. The final loss function of MSDA is deduced:

$$J_{MSDA}(\theta) = \frac{1}{N}\sum_{x\in D}(J(x, m_\theta(\mu_x)) + \frac{1}{2}\sum_{d=1}^{n}\sigma_{xd}^2\sum_{h=1}^{m}\frac{\partial^2 J}{\partial z_h^2}(\frac{\partial z_h}{\partial \tilde{x}_d})^2) + \beta\sum_{j=1}^{k} KL(\hat{\rho}_j \| \rho) \tag{12}$$

Table 1 summarizes two loss function and their corresponding derivatives. In this paper, we choose square error as reconstruction loss function. In the case of additive Gaussian white noise, the second item in formulation (11) is transformed to: $R_0(\mu_x) \propto \sum_h z_h^2(1-z_h)^2 \sum_d y_d(1-y_d)w_{hd}^4$

In contrast to typical measuring model parameters with L2 norms, the regularizer is faster than w_{hd}^2 .So it accelerates the convergence speed of the network, makes the output closer to the input.

Table 1. Reconstruction loss function and the relevant derivatives.

Type	$J(x, y)$	$\partial^2 J/\partial z_h^2$	$\partial z_h/\partial \tilde{x}_d$
Squared loss	$\|x - y\|^2$	$2\sum_d w_{hd}^2$	$z_h(1-z_h)w_{hd}$
Cross-entropy loss	$-x^T\log(y) - (1-x)^T\log(1-y)$	$\sum_d y_d(1-y_d)w_{hd}^2$	$z_h(1-z_h)w_{hd}$

MSDA is still a shallow neural network, and it is difficult to effectively extract the potential hierarchical features of data. Compared with the SSDA algorithm, in this paper, the two MSDA are stacked to form DMSDA. The network diagram of DMSDA is shown in Figure 2.

The network DMSDA is still trained by greedy layer-wise method. First, each MSDA is trained individually, and then the whole deep network is fine-tuned. DMSDA after fine-tuning has the basic characteristics of the biological nervous system, which reflects some functions of the human brain to some extent and simulates the biological system better.

We choose the following loss function to fine-tune the whole parameters of the DMSDA.

$$J_{DMSDA}(\theta) = \frac{1}{N}\sum_{i=1}^{N}\frac{1}{2}\left\|x^{(i)} - f(g(\tilde{x}^{(i)}))\right\|_2^2 \tag{13}$$

There is no weight of decay term and the sparsity-inducing term in eq. (13). This is due to the fact that both have been taken into account when training a single MSDA. There is not a significant change in performance when they are included.

3.2 Specific algorithm steps

The specific implementation process of image denoising based on DMSDA is as follows. The algorithm flow is shown in Figure 3.

Step 1 70000 patches are randomly selected from natural images, and the corresponding noisy image patches are obtained as training samples.

Step 2 The optimal network parameters are obtained by using the training samples to train the first MSDA and minimizing its loss function. At the same time, we need to calculate the output of the first hidden layer.

Step 3 The output obtained by step (2) is used as the input of second MSDA, and the parameters of second hidden layer network is trained by the same method.

Step 4 These parameters of the two layers are used as the initialization weights of the DMSDA network, and then use the BP algorithm to fine-tune all the layers to obtain the optimal parameters of the whole network. At this point, DMSDA training is over.

Step 5 The noisy image is divided into patches that are consistent with the size of the samples to input the trained DMSDA network for testing. Finally, the denoised image patches are then combined into a denoised image.

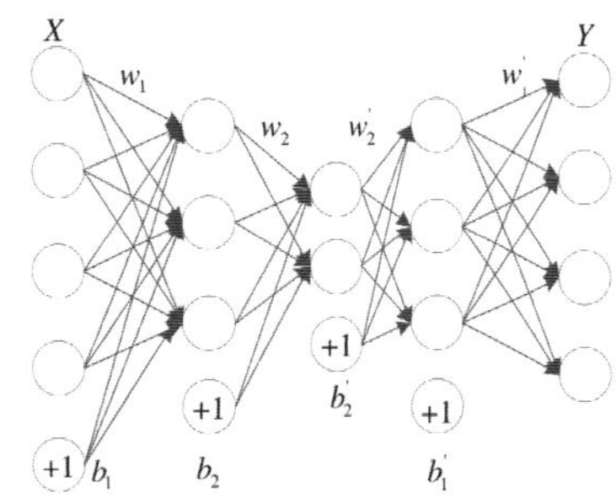

Figure 2. The network diagram of DMSD

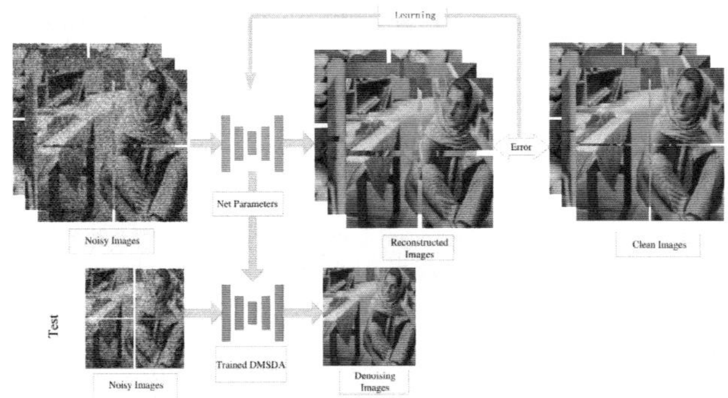

Figure 3. The algorithm flow based on DMSDA

4. Results and discussion

In this paper, we will demonstrate the effectiveness of the proposed algorithm from two aspects of convergence speed and denoising performance. 80 pieces of 512×512 natural images are selected as training sets. To denoise the whole image is extremely difficult. In view of this, this paper will select small image patches from the entire image as training samples. The number of hidden neurons in the experiment was 40, $\lambda = 10^{-5}$, $\beta = 10^{-2}$.

4.1 Convergence speed

SSDA and DMSDA are usually trained to get a better deep network by three stages, including the first layer, the second layer and the fine-tuning. In order to show that DMSDA has faster convergence rate than SSDA, this paper will carry out simulation experiments from the above three stages. The experiment results are shown in Figure 4.

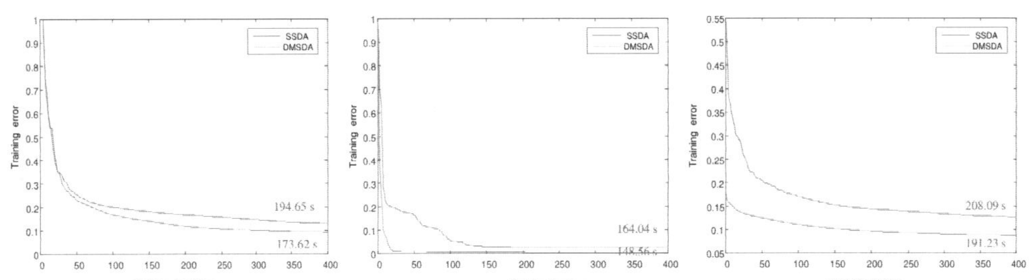

(a) (b) (c)

Figure 4. Training iterative error curve of SSDA and DMSDA. (a) the first layer; (b) the second layer; (c) the fine-tuning.

It can be seen from Figure 4 that when the experimental parameters are consistent, the DMSDA algorithm requires less iterations and training time than the SSDA in the same training phase. This is because the DMSDA algorithm is to marginalize the loss function of the SDA. The average loss function is approximately expressed by Taylor first order and two order expansion.

4.2 Denoising performance

In order to verify the better denoising effect of DMSDA, this paper will evaluate the quality of denoised images from both objective and subjective aspects. At present, an important objective criterion for the evaluation of denoising performance is PSNR, the higher the PSNR, the better the effect of noise reduction. Subjective evaluation is to see whether the denoised image is too smooth, to lose the details information of the original images.

4.2.1 Objective evaluation .5 standard natural images are selected in the experiment to effectively illustrate the image denoising algorithm DMSDA has higher PSNR. The standard deviation of Gaussian white noise is respectively set to 25, 35, 50, 75. Then use the above four methods for noise reduction, and take the average PSNR. Table 2 is the average PSNR of various algorithms under different noise standard deviations.

Table 2. The average PSNR of various algorithms under different noise standard deviations.

Algorithms	$\sigma = 25$	$\sigma = 35$	$\sigma = 50$	$\sigma = 75$
Noisy images	22.18	19.30	16.27	13.02
KSVD	31.38	29.71	27.48	26.37
BM3D	32.22	30.93	28.29	26.53
SSDA	30.61	29.69	28.55	27.02
DMSDA	30.66	29.73	28.86	27.33

It is shown from Table 2 that the PSNR of KSVD and BM3D is higher than that of SSDA and DMSDA, when the noise intensity is small (such as $\sigma=25$, $\sigma=35$). But with the noise intensity increasing ($\sigma \geq 50$), the PSNR of DMSDA is higher than the former three algorithms.

4.2.2 Subjective evaluation. In this paper, 4 natural images with rich texture and detail are selected to test, when $\sigma=25$ .The experiment results are compared with three algorithms of KSVD, BM3D and SSDA, which have better denoising effect. The experimental results are shown in Figure 5.

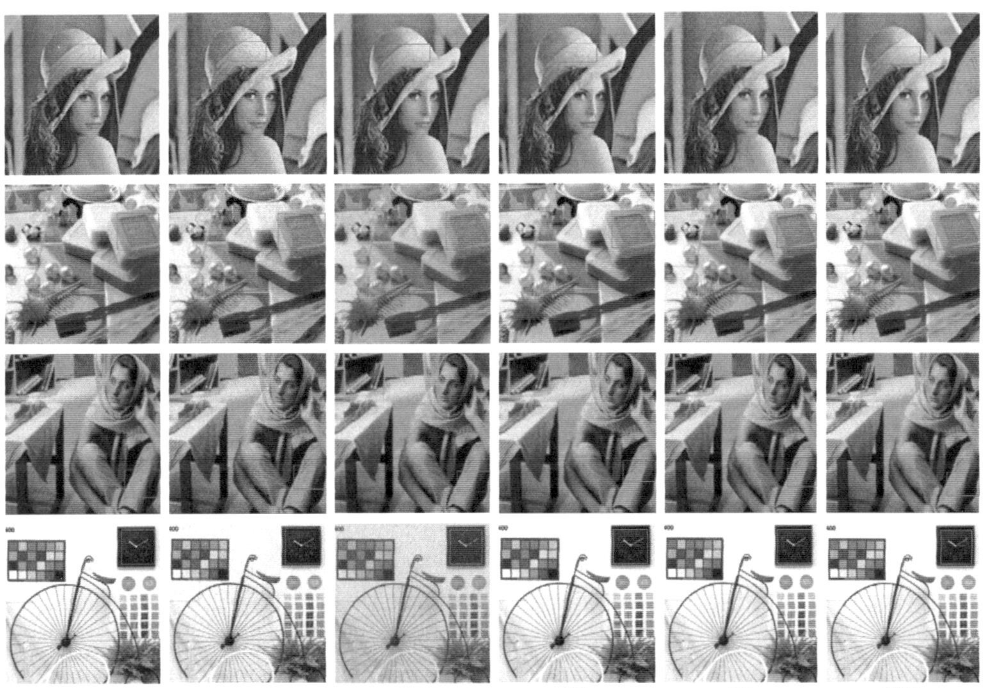

Figure 5. The details comparison of the different denoising algorithms

5. Conclusion and future work

Aiming at the problem that the SSDA algorithm is not easy to converge, the computational complexity is high and the training time is long, we present DMSDA deep network by marginalizing the loss function of the SDA. Results show that the proposed algorithm not only in the convergence rate is faster than that of SSDA, but also can effectively remove the noise and keep original image details, compared with the excellent denoising algorithm, such as KSVD, BM3D and the classical SSDA in deep learning.

References

[1] Portilla J, Strela V, Wainwright M J, et al.Image denoising using scale mixtures of Gaussians in the wavelet domain[J]. *IEEE Transactions on Image Processing*, 2003, **12(11)** pp 1338-51.
[2] Hua Zhisheng, Fu Lihua. K-SVD image denoising based on noisy image blocks classification and dictionary optimization. Computer Engineering and Applications,2017,53(16) pp 187-192.
[3] Dabov, A. Foi, V. Katkovnik and K. Egiazarian, "Image denoising by sparse 3-D transform-domain collaborative _ltering", IEEE Transactions on Image Processing (TIP), vol. 16, no. 8, (2007) pp 2080-95.
[4] Tang Y, Salakhutdinov R, Hinton G E, et al. Robust Boltzmann Machines for recognition and denoising[C]// Proceedings of the 19th IEEE International Conference on computer vision and pattern recognition(CVPR).
[5] Vincent P, Larochelle H, Bengio Y, et al. Extracting and composing robust features with denoising auto encod-ers[C]//Proceedings of the 25th international conference on Machine learning. New York, USA: ACM Press, pp 1096-1103.
[6] Keyvanrad M A, Pezeshki M, Homayounpour M A, et al. Deep Belief Networks for Image Denoising[C]// Proceed-ings of the International Conference on Learning Repre-sentation. Berlin, Germany: Springer, 2014.
[7] Yu Kai, Jia Lei, et al. Deep Learning: Yesterday, Today, and Tomorrow[J].Journal of Computer Research and Development, 50(9) pp 1799-1804.
[8] Rumelhart D E, Hinton G E, Williams R J. Learning Rep-resentations by Back-propagating Errors[J].Nature, 1986, 323(9) pp 533-6.
[9] Chen M, Weinbercer K, Sha F, et al. Marginalized de-noising auto-encoders for nonlinear representations [C]//Proceedings of the 2014 31th International Confer-ence on Machine Learning. New York: ACM Press, pp 1476-84.
[10] Xie Junyuan, Xu Linli, Chen Enhong, et al. Image de-noising and inpainting with deep neural networks[C]// Proceedings of the 25th International Conference on neural information processing systems. New York, USA: ACM Press, pp 341-9.
[11] Vincent P, Latochelleh, Lajoie I, et al. Stacked denoising autoencoders: learning useful representations in a deep network with a local denoising criterion[J]. The Journal of Machine Learning Research,2010(9999) pp 3371-3408.
[12] Hinton G E, Osindero S, The Y W. A fast learning algorithm for deep belief nets[J]. Neural Computation, 2006, 18(7) pp 1527 -54.
[13] Agostinelli F, Anderson M R, Lee H, et al. Adaptive mul-ti-column deep neural networks with application to robust image denoising[C]//neural information processing sys-tems, 2013 pp 1493-1501.
[14] Lecun Y, Bottou L, Orr G B, et al. Efficient BackProp[M]. Orr G B, M Ller K-R. Neural Networks: Tricks of the Trade, LNCS 1524. Berlin: Springer, 1998.

Near-duplicate Video Detection Algorithm Based on Global GSP Feature and Local ScSIFT Feature Fusion

Xidao Luan[1], Yuxiang Xie[2,*], Jingmeng He[2], Lili Zhang[2], Chen Li[2] and Xin Zhang[2]

[1]College of Computer Engineering and Applied Mathematics, Changsha University, Changsha 410003, China
[2]College of Information Systems and Management, National University of Defense Technology, Changsha 410073, China

*Email: yxxie@nudt.edu.cn

Abstract. The main problem with near-duplicate video detection is the high computational complexity and the low efficiency. Near-duplicate video detection methods based on global feature is running fast but with low accuracy, on the contrary, methods based on local feature is accurate, but the calculation is large and time-consuming. Therefore, a near-duplicate video detection algorithm combining global GSP feature and local ScSIFT feature is proposed. Firstly, the video clips and the query ones are discretized into a set of key frame sequences, and the temporal information is recorded at the same time. Secondly, by filtering the video clips on global Gaussian-Scale pyramid feature, similar video clips are selected as the candidates to assure the high recall. Then, combined with the temporal features of the keyframes, the candidate videos are further detected by the local ScSIFT feature to obtain a higher precision. Experimental results show that the proposed algorithm can improve the accuracy of near-duplicate video detection on the basis of guaranteeing the timeliness of the algorithm.

1. Introduction

Near-duplicate video is generally the same from the original video but is compressed to different quality, redefined for different sizes and frame rates, or in space and time to make different edits to modify the videos. In literature [1], by statistics of YouTube's near-duplicate video in the amateur production and professional production, it finds that the most popular video are mostly produced by the professional. Professional production of video is more easily to be copied and modified; obviously in these shared video can find a considerable number of near-duplicate video clips. Therefore, many research scholars have begun to pay attention to how to quickly and effectively remove near-duplicate videos in the video search results. The core technology is near-duplicate video detection, which has already had a lot of mature algorithm. Near-duplicate video detection mainly focuses on two factors like the detection accuracy and efficiency of the algorithm. These two factors usually have inversely proportional relationship; higher detection accuracy will usually bring a higher computational burden, but if the algorithm is fast, it will obviously reduce the accuracy of algorithm detection.

Judging the similarity of video content based on keyframes is a commonly used method. This type of methods is usually measured by the number of matching similar keyframes. At present, most of the near-duplicate keyframe detection research is based on visual features, or the fusion of other modal features. Early researches only consider the matching of video data attributes. These methods are simple, but the video similarity is calculated only by the proportion of the number of similar keyframes in the videos. Since the temporal information of the video keyframe is ignored, the similarity of near-duplicate video clips is not consistent with human cognitive. In [2], a large amount of redundant video is eliminated by clustering the similar video clips, thus the efficiency of near-duplicate video detection is improved. In literature [3], videos which are generated by different camera shooting angle are also considered in the detection, and a near-duplicate video detection system called UQLIPS is established according to the temporal information of the keyframe. The system can detect the similar video clips quickly and effectively. [4] proposes a method of matching video for internal changes (such as different shooting angles, different camera motions, etc.).

In the related work of near-duplicate shot detection, the researchers are more concerned with the performance of video sequence matching [5]. For example, literature [6] uses the editing distance to match the video. However, this method is only applicable to small video data sets. The algorithm proposed in [7] and [8] detects the feature points of keyframes by Harris corner detection algorithm, establishes a 20-dimension feature vector with temporal information for each feature point, and searches for similar feature points by Hilbert curves with sublinear search complexity. The algorithm can be applied to both near-duplicate image and video detection, but the algorithm is less real-time.

From the above analysis, it can be seen that the main disadvantage of near-duplicate video detection based on keyframes is that the amount of calculation is so large that the algorithm is inefficient. Therefore, it is possible to improve the efficiency of the algorithm by reducing the amount of test keyframe data, reducing the number of feature points and the dimension of the feature vector in each keyframe. Near-duplicate video detection based on the global feature of the keyframe is fast but inaccurate; however, near-duplicate video detection based on local feature of the keyframe is accurate but takes too long. This makes it possible to fuse two features in near-duplicate video detection. Combined global gray-scale pyramid feature with SIFT feature based on sparse coding acceleration, a near-duplicate video detection algorithm which is based on fusion of global GSP and local ScSIFT feature called GLC-NDVD is proposed.

2. GLC-NDVD Algorithm

In the GLC-NDVD algorithm, global feature selection of the keyframe is the color histogram feature, which is the most commonly used and fast to be calculated feature. In order to avoid the influence of illumination and clipping on the keyframe image, the gray-scale pyramid is established for the keyframe, and the local feature adopts the ScSIFT. The algorithm first discretizes the video clips and the query video into a set of keyframe sequences, and records the temporal information of the keyframe at the same time. Second, the global GSP feature is used to do coarse filter in the query library, with a higher recall to ensure that the keyframe similar to the video can be included in the candidate set. Then, combined with the temporal feature of the key sequence, the candidate set is further detected by the local ScSIFT feature to obtain a higher detection precision. The algorithm can quickly improve the efficiency of the algorithm by filtering the dissimilar video based on the GSP feature and reducing the amount of the candidates by local feature detection.

2.1. Process of the algorithm

Near-duplicate video detection algorithm based on the global GSP feature and the local ScSIFT feature is divided into two parts, one is the off-line index extraction and construction of query video library features, the other is the real-time video detection. The algorithm of off-line index extraction and construction of query video library features is a combination of gray-scale pyramid construction algorithm and SIFT feature extraction algorithm based on sparse coding. The basic process is as follows:

(1) Discretize the query video into a keyframe sequence and record its temporal information;
(2) Establish the gray-scale pyramid of the query video keyframes;
(3) Extract the color histogram of the gray-scale pyramid image in each keyframe;
(4) Extract the SIFT feature of each keyframe, sparse all SIFT features with the sparse dictionary

D, save the ScSIFT feature α, and build the index index;

(5) End.

The basic process of real-time video detection algorithm is as follows:

(1) The color histogram of the test keyframe is extracted, and the video candidate set with similar histogram of keyframe is selected by comparing with the gray-scale pyramid feature of the query library image;

(2) Read the ScSIFT features α_c of all keyframes in the candidate set, and the ScSIFT feature index $index_c$ of the candidate set;

(3) Extract SIFT feature of the test video keyframe, sparse represent the extracted SIFT feature by sparse dictionary D, and record it as β;

(4) Set the similarity distance threshold of the ScSIFT feature, keyframe similarity threshold, video similarity threshold, and read the first column of ScSIFT feature $x = \beta_0$ of the test video keyframe;

(5) According to the index $index_c$, query the nearest k column coefficients α_{ck} of the ScSIFT feature x;

(6) Calculate the distance d_{ck} between x and each column α_{ck}, if $d_{ck} \le \sigma$, the feature matching amount Σ_i of the column which belongs to keyframe i will add 1. The distance can be calculated as follows:

$$\|\alpha_X - \alpha_Y\|_2 = \sqrt{\sum_i \alpha_{Xi}^2 + \sum_j \alpha_{Yj}^2 + \sum_k (\alpha_{Xk} - \alpha_{Yk})^2} \quad (1)$$

Where i is the element ordinal for those value non-zero in vector α_X and value zero in vector α_Y. Similarly, j is the element ordinal for those value zero in vector α_X and value non-zero in vector α_Y. k is the element ordinal for those value non-zero in both vector α_X and α_Y.

(7) Repeat steps (4), (5) until all columns of ScSIFT feature β are cycled;

(8) For each feature matching amount $\Sigma_i \ne 0$ of image i, the total number of sparse feature points $Total_i$ is counted and the similarity degree $\Sigma_i / Total_i$ is calculated. If $\Sigma_i / Total_i \ge \theta$, keyframe i in the query library is a similar key frame for the test image;

(9) If the number of similar video frames of the two videos exceeds the proportion δ of the number of test keyframes, then the first n video with the most similar temporal information is selected as the near-duplicate video detection result;

(10) End.

2.2. GSP feature extraction

In order to solve the shortcomings of the color histogram on the change of illumination, clipping and poor distinction in images, the gray-scale pyramid (GSP) is built on the idea of establishing Gaussian scale pyramid in SIFT algorithm. Through the scale transform of each keyframe, the pyramid model is constructed by making the gray scale transformation of each key frame at different scales. Unlike Gaussian pyramid, the scale of the gray-scale pyramid achieves different scales by continuously cutting the image area of the fixed image in the current image to ensure that the color features of the image are invariant to the cropping. In this paper, the current image centre height is 70% of the original image area as the next scale image. The specific process of GSP feature extraction algorithm is described in [9].

2.3. ScSIFT feature extraction

SIFT feature acceleration detection algorithm based on sparse coding (ScSIFT) focuses on the sparsing of SIFT feature vectors. After SIFT feature extraction, the algorithm does sparse coding for the 128 dimension SIFT feature vector and establishes the sparse coding index to improve the matching speed. The main idea is using the SIFT feature extracted from the keyframe of the query library as a training sample to train the over-complete dictionary to obtain a set of over-complete bases. The SIFT feature vector of the keyframe in the query library is sparsely encoded with the over-complete dictionary, and the sparse feature vector is indexed. Furthermore, the SIFT feature of the test keyframe is matched with the feature index of the query frame by using the over-complete dictionary to obtain a set of similar candidate sets, compare the sparse coefficient between the test keyframe and the candidate, and finally get similar keyframe detection results. The sparse coefficient vector which is sparsely coded from SIFT feature is called ScSIFT feature of the keyframe.

3. Experimental results and analysis

In order to test the effectiveness of GLC-NDVD algorithm, a near-duplicate video detection algorithm based on global GSP feature and a near-duplicate video detection algorithm based on ScSIFT feature are selected as the control. The experimental query library contains 1000 clips of video, and a total of 10816 frames are extracted. The original video clips and 10 video clips which have been performed some editing styles like Gaussian blur, adding logo, gray scale transformation and scale cropping are selected as the test video. The near-duplicate detection is carried out with the query library video, and the experimental results are compared with each other. The results of three near-duplicate video detection algorithms based on different features are shown in Table 1.

Table 1. Comparison of the effectiveness of three algorithms.

	Recall			Precision			Run time/s		
	GLC	GSP	ScSIFT	GLC	GSP	ScSIFT	GLC	GSP	ScSIFT
Original video	98.9%	72.2%	98.9%	92.7%	96.5%	49.7%	29.41	4.29	192.37
Gaussian blur	98.9%	72.2%	99.1%	93.6%	96.5%	50.4%	30.34	4.29	193.98
Adding Logo	98.6%	55.5%	98.2%	92%	96.9%	49.3%	30.75	4.37	188.23
Gray+1	97.1%	82.1%	98.7%	78.3%	96.7%	48.5%	22.92	4.40	183.55
Gray+2	17.4%	43.7%	98.8%	2.6%	13.8%	46.7%	14.40	4.05	172.40
Gray-1	92.8%	28.6%	99.2%	20.2%	24.7%	49.6%	30.74	4.26	184.91
Gray-2	95.4%	40.0%	99.2%	25.9%	27.1%	48.4%	25.25	4.22	174.83
Scale 0.7	98.6%	51.3%	98.9%	76.2%	98.2%	49.2%	40.18	4.36	141.55
Scale 0.5	81.5%	44.3%	98.9%	73.5%	99.3%	44.5%	25.29	4.60	97.06
Scale 0.35	82.2%	47.1%	99.0%	69.8%	91.5%	28.9%	15.54	4.40	61.94
Average	86.1%	53.7%	98.9%	62.5%	74.1%	46.5%	26.48	4.32	159.08

Figure 1 shows the results corresponding to Table 1. In the figure, the abscissa indicates the precision or recall rate. "*", "o" and "□" refers to the test results of GLC algorithm, GSP algorithm and ScSIFT algorithm respectively.

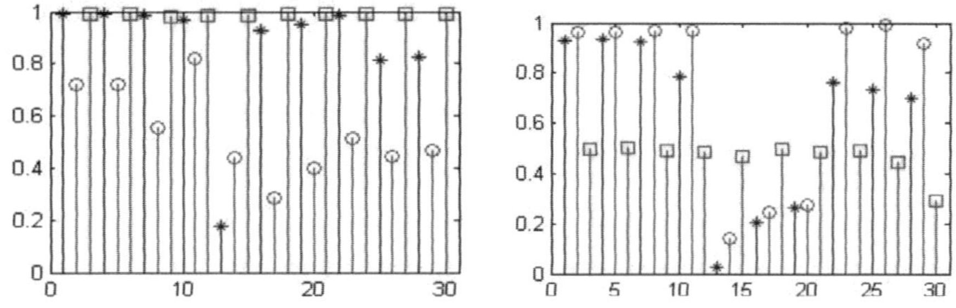

(a) Precision (b) Recall

Figure 1. Precision and recall of three algorithms with different transformations.

The average precision and recall rate of three algorithms are shown in Figure 2.

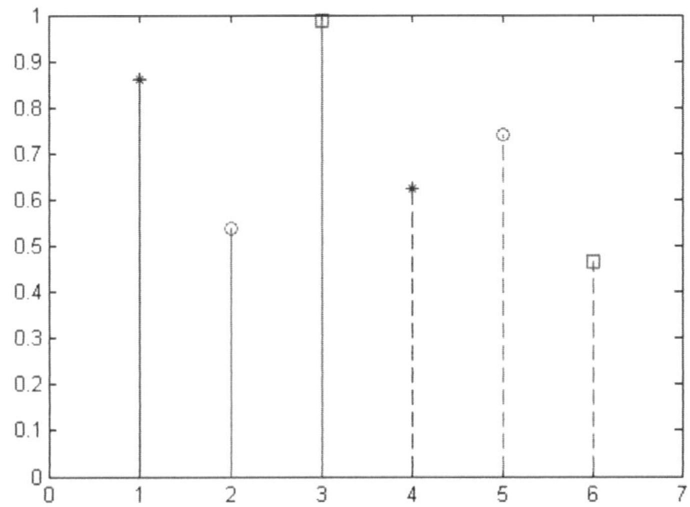

Figure 2. Average accuracy and recall of three algorithms.

In Figure 2, the abscissa "1", "2" and "3" refer to the average precision of GLC algorithm, GSP algorithm and ScSIFT algorithm, and are drawn by solid line. "4", "5" and "6" refers to the average recall rate of the three algorithms, drawn in dashed lines.

In contrast to Table 1, Figure 1 and Figure 2, it shows that the similarity of near-duplicate video detection based on the GLC feature is similar to the ScSIFT feature for most editing transformations, and is only obviously lower than ScSIFT feature for the transform like gray scale increase of 64. The average precision rate of near-duplicate video detection based on GLC feature is more than 85%. At the same time, the precision rate based on GLC feature is slightly lower than GSP, with an average of 62.5%, which is significantly higher than that of ScSIFT, which is 46.5%. As to the efficiency, the average run time of GLC algorithm is 26.48 seconds, 4.32 seconds slower than the GSP algorithm, but significantly better than ScSIFT feature which is 159.08 seconds. Therefore, GLC algorithm can achieve a higher level of accuracy with an average of more than 85%;recall rate is slightly less, but still around 62%. The efficiency of GLC algorithm is not the best, but about 25 seconds of running time is still in acceptable range. In summary, the GLC algorithm can ensure recall and precision at the same time, while ensures running fast. Therefore, near-duplicate detection algorithm based on the global GSP feature and local ScSIFT feature fusion is more effective than the other two detection algorithms based on single feature.

4. Conclusions

Aiming at the problem that near-duplicate video detection method based on global feature is fast but inaccurate and the detection method based on local feature is accurate but time-consuming, a near-duplicate video detection algorithm combining global GSP feature and local ScSIFT feature is proposed. The algorithm combines the high efficiency of global GSP features with the high accuracy of local ScSIFT features. Experiments show that the proposed algorithm has improved the detection algorithm of using only single feature. In the future, we will further study near-duplicate video detection based on video clips to improve the detection efficiency.

Acknowledgments

This work has been supported by the National Natural Science Foundation of China (61571453), Hunan Provincial Department of Education Science and Technology Program (15A020) and Changsha Science and Technology Project (ZD1601014).

IOP Conf. Series: Journal of Physics: Conf. Series **960** (2018) 012033 doi:10.1088/1742-6596/960/1/012033

References

[1] Kruitbosch G and Nack F 2008 *Proc. 3rd ACM Int. Workshop on Human-Centered Computing (Vancouver)* (ACM) pp 7-10

[2] Wu X, Ngo C W, Hauptmann A G and Tan H K 2009 *IEEE Trans. Multimedia* **11** 196

[3] Shen H T, Zhou X, Huang Z, Shao J and Zhou X 2007 *Proc. Int. Conf. on Very Large Data Bases (Vienna)* (the United States: VLDB Endowment) pp 1374-1377

[4] Wu P H, Thaipanich T and Kuo C J 2009 *Proc. Int. Conf. on Image Processing (Cairo)* (New York : IEEE) pp 237-240

[5] Huang Z, Shen H T, Shao J, Cui B and Zhou X 2010 *IEEE Trans. Multimedia* **12** 386

[6] Bertini M, Bimbo A D and Nunziati W 2006 *Proc. Int. Conf. on Image and Video Retrieval (Tempe)* vol 4071 (Berlin: SPRINGER-VERLAG BERLIN) pp 133-142

[7] Buisson J O and Frélicot C 2007 *IEEE Trans. Multimedia* **9** 293

[8] Sarkar, Singh V, Ghosh P, Manjunath B S and Singh A 2010 *IEEE Trans. Circuits and Syst. Video Technol.* **20** 870

[9] Xiao Niu, Yuxiang Xie, Chen Li and Xidao Luan 2016 *Int. Conf. on Signal and Image Processing (Beijing)* (New York : IEEE) pp 103-107

Study on the Rule of Super Strata Movement and Subsidence

YAO Shunli[1,*], YUAN Hongyong[1], JIANG Fuxing[2], CHEN Tao[1] and WU Peng[3]

[1] Institute of Public Safety Research, Department of Engineering Physics, Tsinghua University, Beijing 100084, China;
[2] School of Civil and Environmental Engineering, University of Science and Technology Beijing, Beijing 100083, China;
[3] Beijing Global Safety Technology Co., Ltd., Beijing 100093, China.

*yaoshunli1@126.com

Abstract. The movement of key strata is related to the safety of the whole earth's surface for coal mining under super strata. Based on the key strata theory, the paper comprehensively analyzes the characteristics of the subsidence before and after the instability of the super strata by studing through FLAC3D and microseismic dynamic monitoring of the surface rock movement observation. The stability of the super strata movement is analyzed according to the characteristic value of the subsidence. The subsidence law and quantitative indexes under the control of the super rock strata that provides basis for the prevention and control of surface risk, optimize mining area and face layout and reasonably set mining boundary around mining area. It provides basis for the even growth of mine safety production and regional public safety.

1. Introduction

The coal seam mining may cause overlaying strata movement and surface subsidence (it is the result of bottom-up strata fracture which passes to the earth's surface [1], [2], [3]. With the development of towns [4], surface subsidence in surrounding areas which is caused by coal mining has become a risk of public safety [5], [6].

Because of the ambiguity and complexity of geological conditions of coal mine and the influence of mining factors [7], the rule of overlaying strata movement and characteristics of surface subsidence within the practical mining process are complex and changeable [8]. If there is a super thick and hard stratum in the overlaying strata, surface subsidence has the feature of mutability. With the increase of the mining area, surface subsidence risk will change accordingly. We should do dynamic analysis of the subsidence rule during the mining process and strengthen the protection of surrounding risk bearing bodies.

Taking the real example of super strata-oriented deep mining as its background and on the basis of key strata theory [9], overlaying strata spatial structure theory and mine pressure theory, this article applies FLAC3D numerical simulation analysis and adopts microseismic monitoring and surface rock movement observation to analyze the fracture development trajectory of the overlaying strata before and after the super strata is stable and the subsidence characteristics of different mining ranges through the monitoring of more than four consecutive years. Through the dynamic assessment of surface risk, we will give the surface crisis time and affected range under that strata condition which can serve as the decision-making reference for the classification and prevention of subsidence risk and the safe mining in the surrounding area of towns.

2. Project Background

The boundary of Tengdong Coal Mine in Shandong Province is near to five towns and villages, the Beijing-Kowloon Railway and the G104 National Highway. Thus the pressure of risk prevention in surrounding area is high and the mining boundary should be clarified further according to practical surface subsidence characteristics.

The initial mining area of that mine is the Mining Area 1 and its available coal seam is 3 Coal Seam (pure coal thickness: 3.5~8.0m; average thickness: 5.27m; coal seam dip: NE; dip angle: 2~7° ; average mining depth of the mining area: 945m). The distribution of the roof strata is shown in Table 1: there are 70m-thick super hard conglomerate strata and the average distance between them and the coal seam is 90m. The initial working face is No. 3107 (width: 130m) of the Mining Area 1; No. 3105 (width: 120m) and No. 3109 (width: 150m) are the second and third successive working face respectively. Details of the layout are displayed in Figure 1.

Table 1. Roof key strata distribution in the Mining Area 1

Rock Name	Average Thickness/m	Distance to the Roof of Coal Seam/m	Influence on Surface Subsidence
To the Surface	790	160	Limited direct influence
Conglomerate	70	90	Pallet effect
Fine-Medium Sandstone	17.48	38.5	Limited influence
Fine Sandstone	14.76	7.3	Limited influence
Coal Seam	5.27		

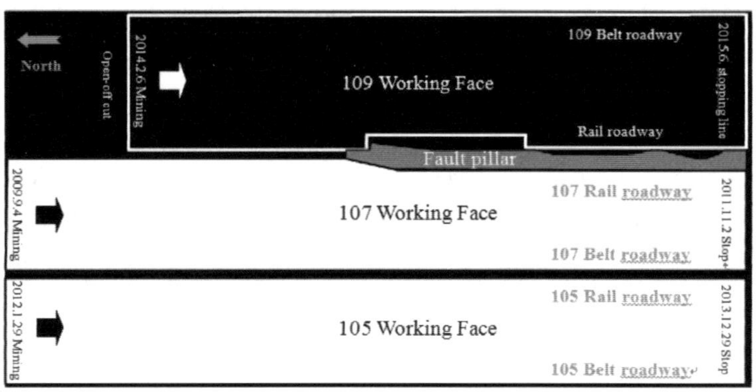

Figure 1. Layout of working face in the Mining Area 1

According to the key strata theory, the 70m-thick conglomerate strata in the bend and subsidence belt is the main key strata of that mine, and its stability and movement features will directly influence the pressure distribution and the surface subsidence characteristics of the mining area.

This article adopts theories to analyze the evolution process of overlaying strata spatial structure and the development process of super thick conglomerate strata instabilityinstable and it provides theoretical references for the clarification of onsite research targets and research methods.

Applying finite-difference numerical simulation, we analyze the development and change trend of the super thick conglomerate strata's plasticity during mining and clarify its movement rules to establish a basis for the monitoring of overlaying strata fracture and surface subsidence.

Through the microseismic monitoring of the roof strata's fracture process, we analyse the stability of super strata; we also apply surface rock movement observation to analyze the characteristics of surface consecutive subsidence. The Table 2 shows detailed monitoring time.

Table 2. Summary table of real-time monitoring of mining process in the Mining Area 1

Working Face Name	Strike Length/m	Mining Time	Microseismic Monitoring Time	Surface Rock Movement Observation Time
107 Working Face	1072	2009.9.4～2011.11.2	2010.5.10～2011.11.2	2009.7～2011.12
105 Working Face	1031	2012.1.29～2013.12.29	2012.1.29～2013.12.29	2012.1～2013.12
109 Working Face	686	2014.4.1～2015.8	2014.2.6～2015.8	2014.10～2015.8

3. Analysis of Super Strata Movement Rules

3.1. Analysis of Super Strata Movement Rules through Numerical Simulation

We uses FLAC3D numerical simulation to analyze the plasticity change process of roof strata and fracture characteristics of super thick key strata during the thick coal seam mining. According to the range of the Mining Area 1, the size of model is 800m×227m and the model adopts Mohr-Coulomb failure criterion. We make the calculation balanced (the calculation of a set of parameters is fully completed) every 10m of propelling and totally we propelled for 500m.

The mine is dug according to the model and we analyze the process of plasticity area of the roof strata in different stages. We start to dig from the location where is 100m left to the model. When the digging width of the initial working face is 130m, the plasticity failure changes of roof strata after the calculation is balanced are displayed in Figure 2 (a). The plasticity failure of overlaying strata is in the area where is 50~60m away from the coal seam and the main failure is tensile fracture with partial shear failure in goaf boundary.

When we continue digging to the second working face with the width of 250m, the plasticity failure changes of roof strata after the calculation is balanced are displayed in Figure 2 (b). The plasticity failure of overlaying strata is expanded to the area where is 90m to the top of coal seam and the main failure is tensile fracture with expanding shear failure range in goaf boundary. In this stage, the damage area of roof strata has the distribution feature that super strata baseplate is the boundary, which means that the overlaying strata fracture in the mining of the second working face has developed to the super thick conglomerate. Because the conglomerate is thick and its lithology is strong, there is no plastic failure, but there are partial lamellar tensile fracture areas above the conglomerate, which means that slow subsidence of super thick conglomerate results in the subsidence of upper weak strata and causes damages.

When the mining width of three working faces is 400m, the plasticity failure changes of roof strata after the calculation is balanced are displayed in Figure 2 (c). There are tensile fractures in the upper area of the conglomerate which are in the two sides of the goaf. With the propelling of working faces, the plastic failure range expands gradually, and the shear failure range in the goaf strata under the conglomerate expands gradually in consequence.

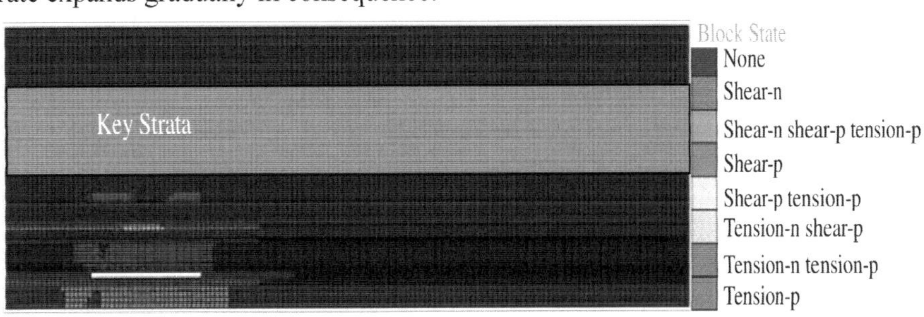

(a) Coal mining width 130m

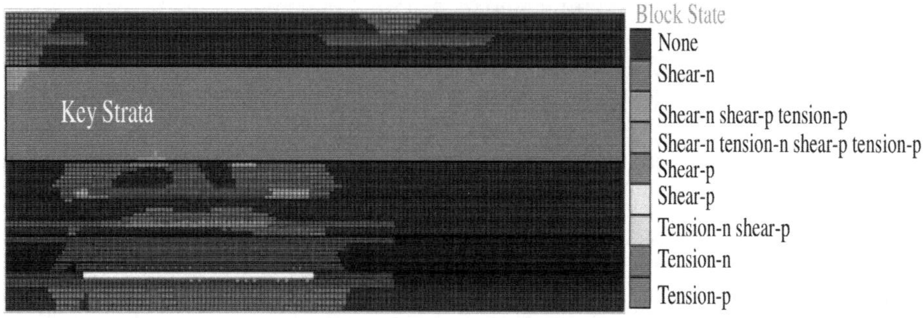

(b) Coal mining width 250m

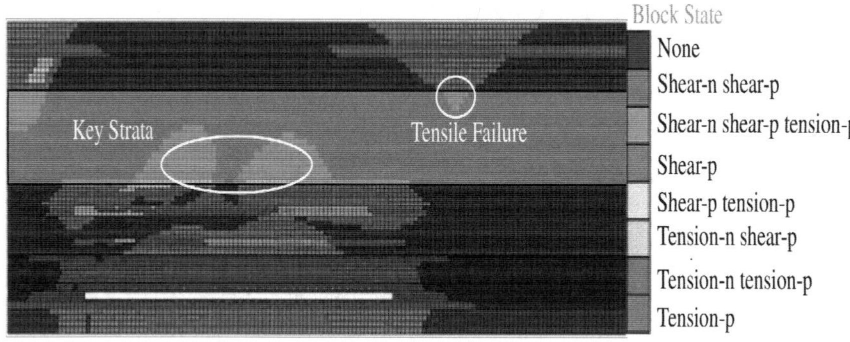

(c) Coal mining width 400m

Figure 2. The plastic failure development model of surrounding rock in coal seam mining

We know from the simulation analysis that with the increase of mining range, locations of overlaying strata fracture becomes higher and lower strata fracture causes gradual bend and subsidence of higher key strata. The maximum failure range of upper strata of the goaf is in "square" stage. Under the influence of key strata failure, its lower coal rock mass and partial overlaying strata plastic area expand with the expansion of mining area, which indicates that super thick conglomerate controls the stage development of the overlaying strata plastic area.

According to the plastic change trend of roof strata and the fracture stage of super thick conglomerate, we know that the roof strata movement range of Mining Area 1 shows stage features when the two working faces are finished and the mining width of the third working face is 400m.

3.2. Super Strata Fracture Monitoring Analysis

Because the geological condition of mine is complex and it's hard to judge the strata nature, structural joint fracture and other elements, under the influence of mining, super strata movement rules still need dynamic process analysis by the onsite monitoring system. This article adopts high-precision microseismic system to monitor the whole process and further analyze movement characteristics of super strata. According to theories and simulation analysis, we categorize the super strata monitoring into before-instable stage and instable stage to do focused monitoring and analysis.

3.2.1. Fracture Field Analysis in Before-instable Stage. No. 3107 working face is the initial working face and the monitoring time is May 10th 2010 ~ November 2nd 2011 when the working face has entered "square" periodical circulation stage. We take microseismic monitoring data in April 2011 as the representative to analyze overlaying strata movement rules in mining (Figure 3 shows the details). According to distribution characteristics of microseismic events, we see that the position of roof strata movement of the working face is stable after entering stable mining, and the average movement height of overlaying strata is within 60m during mining.

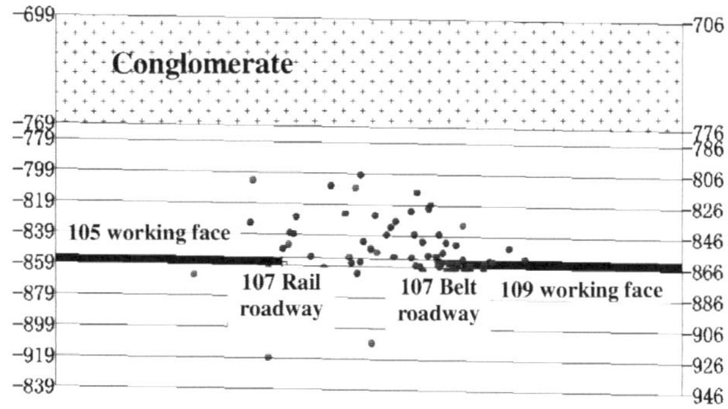

Figure 3. The dip section of microseismic events in No.107 working face "single square"

No. 3105 working face is the second working face. According to analysis results, we know that the overlaying strata spatial structure is in the evolution process when the working face is mined to "double square" stage and surrounding rocks move violently. Thus we analyze the microseismic monitoring data in March 25th 2012 ~ June 2nd 2012 (footage: 110~257m) with attention and analyze the overlaying strata evolution rules. From Figure 4, we know that mining to about 250m, it will enter "double square" stage. The roof strata fracture continues developing to higher positions and the overlaying strata movement height is within 90m domain where is close to the bottom of conglomerate; under the influence of super thick conglomerate protection, there is less concentration of microseismic event of the lower area of conglomerate in the later period.

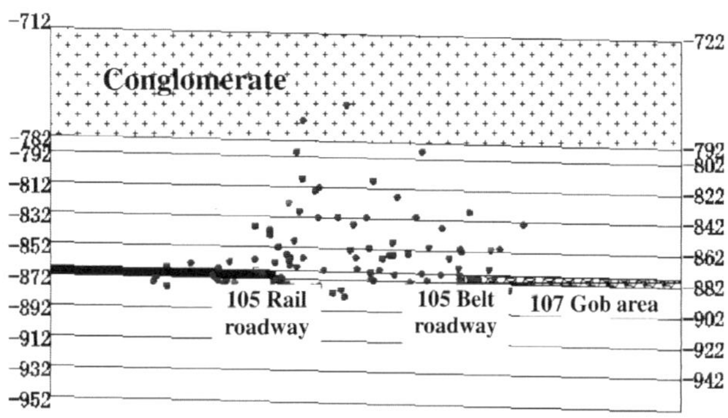

Figure 4. The dip section of microseismic events in No.105 working face "double square"

In the stable situation of super thick conglomerate, microseismic monitoring disclose the overlaying strata fracture range and development trend in the mining of No. 3107 and No. 3105 working face; when No. 3105 working face enters "double square" stage, the overlaying strata movement range of two working faces reaches the bottom of super thick conglomerate and the vibration field develops forward periodically with super thick conglomerate as the roof in the later mining period.

3.2.2. Fracture Field Analysis in Instable Stage. According to the analysis, we can see that when the No. 3109 working face is mined to the position before "double square" (before mining 250m), the roof strata fracture develops to the position of super thick conglomerate. When the working faces are mined to "square" (around 400m) of three goafs, the overlaying strata spatial structure develops to the maximum span status, according to which we can know that there may be instable movements of the super thick conglomerate in the period between "double square" to "triple square".

From Figure 5,through the dip section analysis of microseismic events, we find that the height of roof strata fracture develops from low position to high position during mining, mainly in 135m domain, and microseismic events which reach high-position conglomerate are mainly on sides of the track tunnel fault coal pillar (width: 35m).

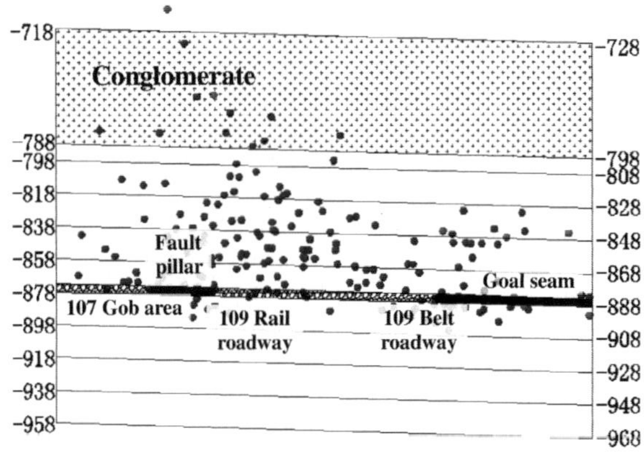

Figure 5. The dip section of microseismic events in No.109 working face "triple square"

Meanwhile, during the mining period of February 1st 2015 ~March 15th 2015 (footage: 307~380m), the plane distribution range of microseismic events in the No. 3109 working face expand gradually and the microseismic energy increases dramatically in this stage; thus we judge that super thick conglomerate strata is in the instable stage.

When No. 3109 working face is mined to around 300m, the vibration field gradually develops to high-position super strata. In the super thick conglomerate instable movement, there is fracture development in the lower area of the conglomerate beside No. 3107 goaf, and the microseismic energy increases dramatically and the range expands. From this we can see that the super thick conglomerate has the feature of continuous fracture movement.

4. Relation between Super Strata Stability and Surface Subsidence

Strata movement observation system is deployed on the surface of the Mining Area 1 to analyze the subsidence figure and subsidence speed in the mining process in real time and comprehensively analyze the relation between super strata movement and surface subsidence.

In Figure 6, an east-west (along approximate dip direction) observation line is set on the surface of the Mining Area 1 and there are 36 observation points (serial numbers are H1 to H36). Both No. 3107 and No. 3105 working face have one observation line on their surface respectively and two observation lines are set along approximate strike direction (serial numbers are Z1 to Z70 and T1 to T50 respectively). After mining No. 3109 working face, an observation line is set on the surface (serial numbers are Z1—Z47). Figure 2 shows the detailed monitoring time.

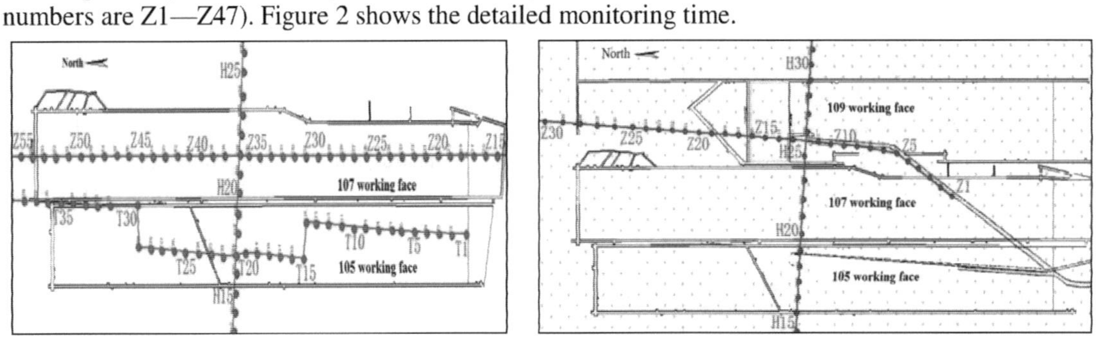

(a)The deployment of surface rock movement observation point of No. 107 and No. 3105 working face;

(b)The deployment of surface rock movement observation point of No. 109 working face

Figure 6. Surface rock movement map in the Mining Area 1

According to simulation calculation and microseismic monitoring analysis, we find that the stability of super strata influences the process of overlaying strata fracture. Thus we divide the surface subsidence monitoring into before-instable stage and instable stage of the super strata to do analysis.

4.1. Surface Subsidence Analysis in Before-instable Stage of Super Strata

Through analyzing the surface rock movement observation data of No. 3107 working face after the mining is stable in March 2012, we acquire that the maximum subsidence of strike observation point is 0.211m (in Z46 observation point); the maximum subsidence of dip observation point is 0.125m (in H24 observation point). The dip subsidence section line is displayed in Figure 7. The subsidence of No. 3107 working face is slow before and after mining and the maximum subsidence speed is 0.2mm/d.

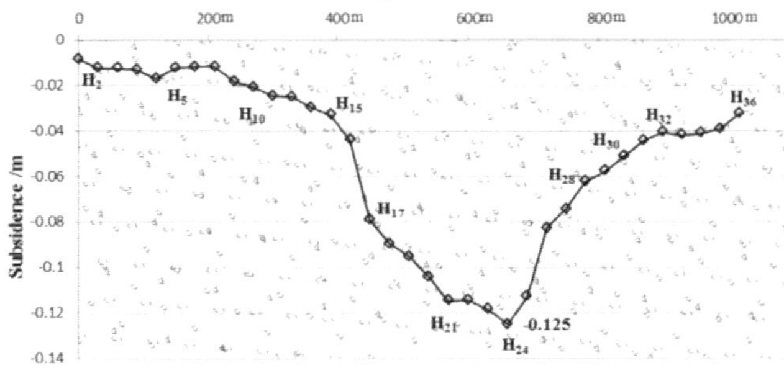

Figure 7. The dip section of subsidence analysis chart after mining No.107 working face

Through analyzing the surface rock movement observation data of No. 3105 working face after the mining is stable in December 2013, we acquire that the maximum subsidence of strike observation point is 0.278m (in Z45 observation point which is above No. 3107 working face); the maximum subsidence of dip observation point is 0.219m (in H24 observation point). The dip subsidence section line is displayed in Figure 8. Compared with No. 3107 working face, the surface subsidence process of No. 3105 working face before and after mining doesn't have obvious difference and the maximum subsidence speed is not higher than 0.3mm/d.

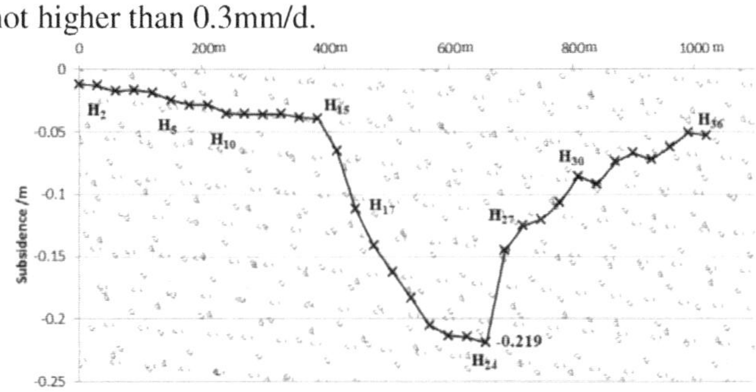

Figure 8. The dip section of subsidence analysis chart after mining No.105 working face

The two working faces are restricted by mining range and super thick conglomerate, and the roof strata are not mined adequately. In the whole process of surface movement, the subsidence speed of any point on the surface is: it is slow in the beginning; then it increase gradually to the maximum; finally it becomes slow gradually, tending to stop. Generally speaking, the subsidence speed change of point is continuous and gradual in terms of time and space, but because the subsidence reduction effect of key strata is obvious in the mining process, each point's subsidence speed is very slow.

Before the super strata are fractured, there is no mutation phenomenon in subsidence and no collapse pit on the surface. In addition, no fracture is observed and the whole subsidence basin looks like a pan. Because of the limitation of working face mining range, there is no flat bottom phenomenon on the surface; that is to say, the mining is inadequate. But the flat bottom phenomenon shows a little in the strike observation line.

4.2. *Surface Subsidence Analysis of the Super Strata in the Instable Stage*

According to the development rule of super strata unstability and surface subsidence characteristics when are mentioned above, we comprehensively analyze the relation between roof strata stability and surface subsidence, and the mining range of three working faces and surface subsidence curves of super strata in instable stage are displayed in Figure 9.

In the mining of 3109 working face, analyzing surface rock movement observation data during October 29th 2014 ~ March 13th 2015, we can find that the maximum subsidence speed of the surface point increases obviously when the working face propelling distance approaches to 300m. According to the calculation of observation results during January 18th 2015 to February 2nd 2015, we get that the subsidence of H23 is 0.052m within 14 days and its speed is 3.5mm/d. Compared with the first two mining, the subsidence speed and volume of that stage increase greatly and the maximum subsidence location is in the middle area of No. 3107 working face.

The surface subsidence monitoring and microseismic monitoring analysis both manifest that there is the instable movement of the super strata. But there no flat bottom phenomenon of the surface subsidence curve, which means the fracture of roof strata is inadequately and the super strata will still fracture and subside while influence the safety of the mining area, which will provide decision-making basis for safe production.

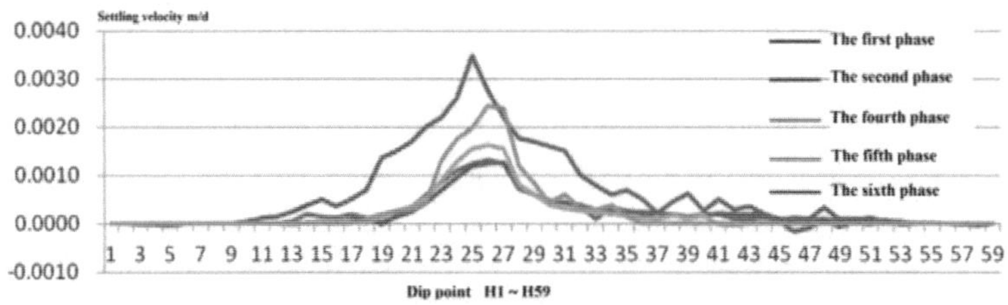

(a) The chart of subsidence speed of dip observation after the mining of No. 109 working face compared with the first period

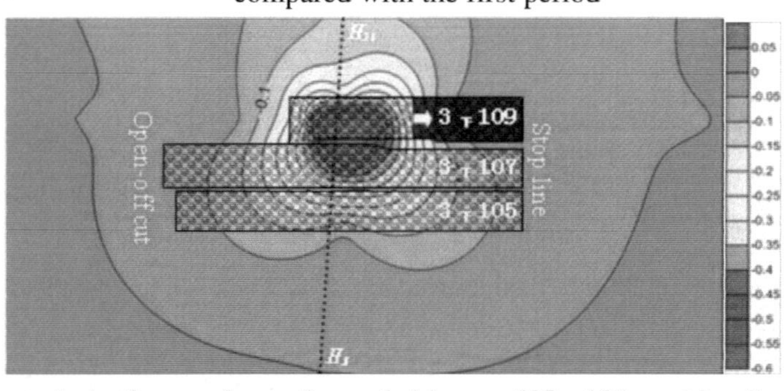

(d) The plane analysis diagram for surface subsidence of No. 109 working face after mining

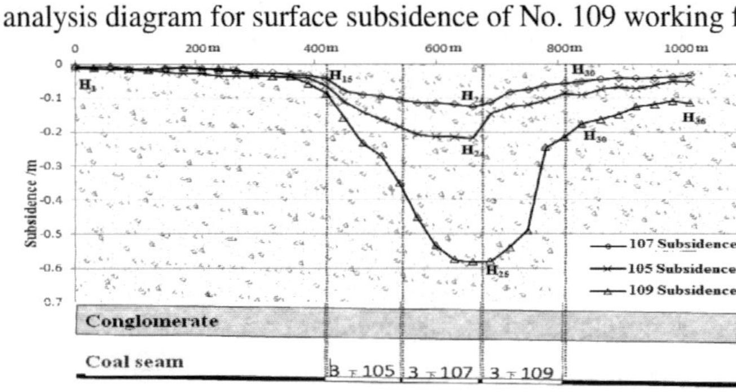

(e) The dip section of surface subsidence of the mining of three working faces

Figure 9. The subsidence analysis diagram of the instable stage of the super strata

The topsoil average thickness of the mine is about 18m; the average depth of the three working faces in

No. 3 is 945m; the width of goaf is 400m; the breadth depth ratio is 0.42; the average thickness of coal seam is 5.27m. Through long-term observation of the surface rock movement observation station, the maximum subsidence of March 13th 2015 is 582mm.

According to characteristics of super strata-controlled surface subsidence, we can clearly find that the biggest risk of surface subsidence is in the instable stage, which can be used for guiding surface risk prevention time and controlled range; according to the distance between the mining boundary and surrounding towns, villages and railways, we can propose proper monitoring plans to provide technical support for optimizing the mining boundary and mining succession, and they can be used as the basis for the safe production of mines and the balanced development of regional public safety.

5. Conclusion

According to numerical simulation analysis, microseismic monitoring and surface rock movement observation, we find the rules of super strata movement and characteristics of surface subsidence: before becoming instable, the surface subsidence is slow and steady; during the instable period, the surface subsidence is swift and its range increases dramatically; meanwhile, surface subsidence rules manifest the stability of super strata movement.

On the basis of monitoring, we did quantitative analysis on the subsidence stage risk degree and affected range of the super strata-controlled surface subsidence while clarifying the prevention time and controlled range of surface risks and optimizing the mining boundary and mining succession to provide references for the safe production of mines and the balanced development of regional public safety.

Acknowledgments
This thesis is supported by the Science and technology project: The technical equipment r&d and application demonstration of the risk assessment and warning for urban disasters and accidents (No.2015BAK10B00)

References

[1] JIANG Chenguang,HE Yong,LIANG Anbo,et al.(2003). New knowledge on law of ground sinking in mining area. Chinese Journal of Rock Mechanics and Engineering, 22,1,162-165.
[2] PENG Linjun,ZHAO Xiaodong,LI Shucai. (2011).Simulating research on rules of surface subsidence due to deep mining. Rock and Soil Mechanics, 32, 6,1910-1914.
[3] M.I.Álvarez Fernández, C.González Nicieza. (2005).Generalization of the n-k influence function to predict mining subsidence.EngineeringGeology, 80, 2,1-36.
[4] LI Shuzhi,DIAO Naiqin. (2015).Mining cities of subsidence area wetland ecological construction planning and construction of technology research and application . MINE Surveying, 43,3,65-69.
[5] FANG Jun.(2016). This paper discusses the classification system of land reclamation unit based on suitability evaluation . MINE Surveying, 44,3,61-64.
[6] ZHANG Changmin.(2009). Study on the characteristics and hazard prediction of coal mine mining. Beijing:Geological institute of China seismological bureau,31-37.
[7] XU Jialin,QIAN Minggao,ZHU Weibing. (2005).Study on influences of primary key stratum on surface dynamic subsidence. Chinese Journal of Rock Mechanics and Engineering, 24,5,787-792.
[8] ZHU Weibing,XU Jialin,SHI Xishu. (2009).In situ test study of the influence of the key strata movement on the surface subsidence . Chinese Journal of Rock Mechanics and Engineering,28,2,403-409.
[9] QIAN Minggao,MIAO Xiexing,XU Jialin,et al. (2000).Key strata theory. Xuzhou:China University of Mining and Technology Press,54-69.

Design and Implementation of Pointer-Type Multi Meters Intelligent Recognition Device Based on ARM Platform

Yang Cui[1], Wang Luo[1,*], Qiang Fan[1], Qiwei Peng[1], Yiting Cai[2], Yiyang Yao[3] and Changfu Xu[4]

[1]NARI Group Corporation (State Grid Electric Power Research Institute), Nanjing, China.
[2]State Grid Wenzhou Electric Power Company, Wenzhou, China.
[3]State Grid Zhejiang Electric Power Company Information and Communication Technology Branch, Hangzhou, China.
[4]State Grid Jiangsu Electric Power Research Institue, Nanjing, China.

*Email address: luowang@sgepri.sgcc.com.cn

Abstract. This paper adopts a low power consumption ARM Hisilicon mobile processing platform and OV4689 camera, combined with a new skeleton extraction based on distance transform algorithm and the improved Hough algorithm for multi meters real-time reading. The design and implementation of the device were completed. Experimental results shows that The average error of measurement was 0.005MPa, and the average reading time was 5s. The device had good stability and high accuracy which meets the needs of practical application.

1. Introduction

Tap water is an important part to keep society running normally. With the development of water industry and information technology, the high-standard condition monitoring of secondary water-supplying equipment is truly needed. Due to the high reliability and accuracy of mechanical pressure gauges, mechanical pointer meter is equipped in many secondary water-supplying equipment [1-3], such as water inlet and outlet pressure meter. At present, the inspection and repair work of secondary water-supplying pump stations mainly depend on manual, so only remedial measures can be taken when failed, its low accuracy, long spacing interval, high cost, poor reliability and false or missed detection easily to cause huge economic losses. Water industry, therefore, urgently need more intelligent development. The operation parameters of the secondary water-supplying equipment through remote intelligent monitoring to achieve precise control, it can early detection of equipment failure thus reduce the loss, besides, it also accurately grasp the real-time water situation of residents so it improve the quality of water supply. Therefore, it is of great significant to develop an efficient and accurate Pointer-type meter recognition device for secondary water-supplying equipment.

2. Research actuality

At present, there are many latest advances algorithms for Pointer-type meter intelligent recognition [4-6]. Course this study adopts low-power processing chip based ARM, it has high-accuracy and low-complexity. Literature [7] was proposed to divide and extract the meter by using the hue and saturation value of HSV model, but the interference of outdoor light and the deterioration of meter color may increase the detection error. Literature [8] proposed to extract sub-images of the matching meter dial area by SIFT. The SIFT algorithm has the robustness and high speed of rotation scale without deformation, but it takes a long time to calculate descriptors. It has limitations in the real-time control of the robot. In the case of pointer identification, it usually obtained indicators of the meter through various image processing algorithms after got subregion of the meter. At present, pointer identification algorithm identified pointers and scales by Hough, but it is not suitable in identification the meter of multi-symbol, and need to marked and added all valid pixels, lead to great amount of calculation. Literature [9] proposed Silhouette method to obtain the meter showing and then read the accurate position by linear fitting. It filtered noise by constrained hough-transfer and iterative algorithm of central projection analysis, but will affect the final reading of the meter because of the angle difference between the two dials. Literature [10] proposed a kind of design method to read long-distance pointer-type chemical meter. Firstly, it obtained the place of meter by Silhouette method, after that, calculated the angle of pointer in Hough. It is suitable to detect meter pointer under constant lighting condition.

Water industry pump stations usually in the basement or suburbs. Because many old pump stations don't have conditions to join Broadband Networks or bandwidth be limited, it can't provide high quality stable access. If the secondary water-supplying equipment meter image data collected by image capture device is transmitted uniformly back to the background for identification and analysis. It leads to high network bandwidth be occupied in wired networks and incur a large amount of communication costs in 4G-LTE. How to discern the number in secondary water-supplying equipment meter with high accuracy, long time and stable? We combined the morphological characteristics of water-supplying equipment meter, adopted the image denoising, binarization, skeleton extraction and Hough transform based on ARM image processing platform, proposed a high-reliability and high-precision pointer-type multi meter reading recognition device. It can transmitted the result to the remote control center in real-time and occupied low network bandwidth, besides, transmitted the video and images when needed. It implemented the long-distance, accurate and real-time monitoring in secondary water-supplying equipment meter.

3. Proposed method

3.1. Equipment Overall Design

According to the design concept of the instrument reading device, the hardware device is designed as shown in Figure 1. The system consists of a processor unit, an image acquisition unit, an LED light source, a power supply, and some drivers.

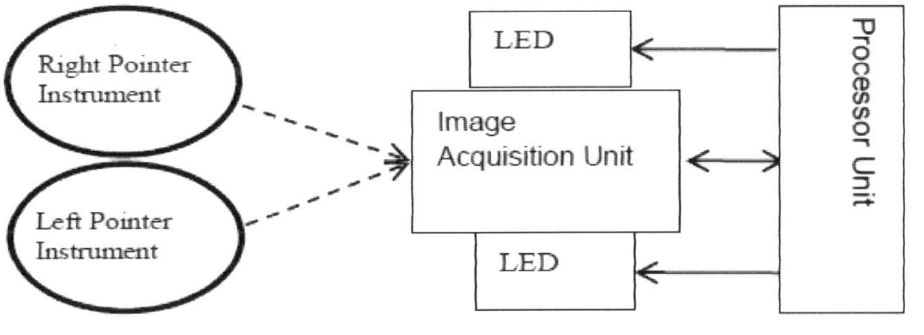

Figure 1. System structure diagram

3.2. Hardware Design

This device uses the processor unit based on H.265/HEVC standard of the industry's first high-definition network camera based on Hisilicon Hi3516A processor. The chip uses high performance ARM Cortex-A9 processor with the co processing unit IVE2.0 of the second generation smart analysis of Hisilicon can provide excellent image quality. The image acquisition unit uses 4 million pixel, wide dynamic, low illumination and 2.8-12mm automatic zoom CMOS sensor OV4689. The device adopts DC12V power supply, and the typical power consumption is 3W. Size is 38mm*38mm double plate structure, as shown in Figure 2 below.

Figure 2. System Hardware

3.3. Algorithm Flow

First, Pointer-type multi meter recognition should extracted instrument scales and pointers, as shown in figure 3, and then through the cross compiler tool algorithm of Hisilicon to downloaded to ARM platform.All the algorithm developed by Visual Studio 2012 platform.

Figure 3. Pointer-type multi meter

A common 1.6 stage high precision pressure gauge in figure 3 is taken as an example to read the pointer instrument, and the algorithm flow is shown in figure 4.

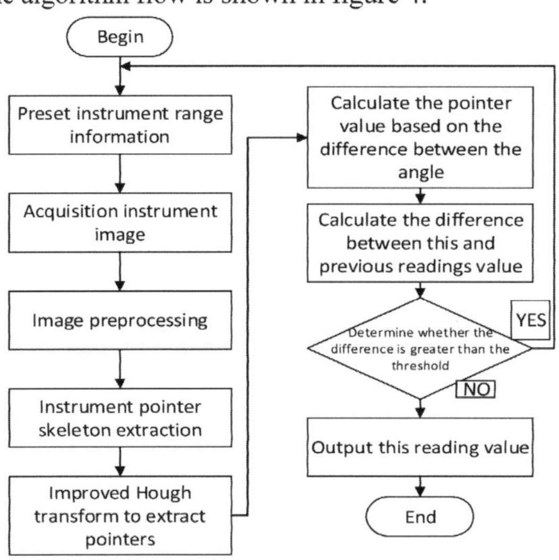

Figure 4. Pointer recognition

The device realizes the two pointer of the meter reading at the same time. And because the relative position of the actual equipment and instrument in the process remain unchanged, considered the recognition accuracy and efficiency of the algorithm, First of all, you need to use self-developed tools for the division of the left and right instrument area, and the preset of the center of the dial and the starting point of the scale, only need to preset once.

The two meter image were acquisited by image acquisition unit, in order to ensure that all day long 24 hours can be normal acquisition, this paper presents a design of LED lights, and the instrument and the lens distance is about 15cm, the sampling frequency was every 5 seconds per pictures, the resolution was CIF, the format was .jpg.

- For image preprocessing, image acquisition will produce light are high value noise due to light intensity changes of light pollution and other issues, the pointer to read before using 3 * 3 Gauss template for image filtering, eliminate noise, this paper use Gauss blur to eliminate noise effects twice. Then, the image to be measured is converted into gray image by means of (1):

$$g = 0.299R + 0.587G + 0.114B \tag{1}$$

- After the image acquisition instrument, in order to facilitate the subsequent pointer extraction, needs to be separated from the background and some dark spots, because the water pump industry pointer instrument is generally installed in the indoor environment, the light changes will cause changes in the dial gray value of the segmentation effect is not good by use fixed threshold method, so this paper using the Otsu method for image segmentation. Figure 5 shows in different light conditions after image binarization meter counter, seen from the experimental images, produced even in the shadow of the pointer around, Otsu method still can better segment the area pointer.

Figure 5. Instrument image under different light intensity and binary image from Otsu thresholding method.

- The pointer which processed by binarization has a certain width, if directly using the classical Hough transform to extract line pointer in this case ,pointer skeleton will appear extracted was not in the state where the straight line overlap, so at the first ,we did pointer to refine and extracted skeleton of the pointer, and then we got the pointer angle. In this paper, a new skeleton extraction algorithm based on Euclidean distance transform ([9]) is used to refine the pointer. Based on Euclidean distance transform, and through the local comparison, the candidate points of skeleton points are rapidly determined, and then the skeleton points are obtained through certain rules, which can ensure the connectivity of the skeleton as well as the real-time. As shown in figure 6.
- In the aspect of extracting pointer, this paper adopts the method of extracting straight line based on Hough transform, The classic Hough change is a straight line of information passing through a point (x, y) in Cartesian coordinate system,and converts the normal line to information in polar coordinates (ρ, θ), formula is as follows:

$$\rho = x\cos\theta + y\sin\theta \tag{2}$$

Then, the cumulative mark of the full line of the effective pixel points was obtained, and get information about each line segment, Since the line of the pointer passes through the center of the circle, it keeps only the line from the center of the circle to the line where the line is less than 5 pixels, Compare all line segments that were qualified and selected the longest line as the pointer. As shown in figure 7.

Figure 6. Results after skeleton extraction

Figure 7. Pointer to be found

- According to the angle difference between the starting pointer and the current pointer position, the number indicated by the current pointer was calculated.

4. Experimental Analysis

In order to verify the effectiveness of the proposed algorithm and the performance of the device, the algorithm's running speed and recognition accuracy were verified in this section. Two block Pointer-type pressure gauge, dial was installed in a vertical distance image acquisition unit distance of about 15cm, and the image acquisition unit was on the table in the astrotech vertical axis. The device was designed to carry out uninterrupted measurement for 5 minutes under strong light, weak light and normal illumination. An image was taken every five seconds for reading and preserving the image.

4.1. Algorithm Accuracy

Table 1 shows the test error of the two meter under normal illumination for 1 minute under different conditions. The actual value of the pressure gauge is equal to the average value of the 3 readings.

Table 1. Algorithm results

Left Instrument			Right Instrument		
Measured value	True value	Absolute error	Measured value	True value	Absolute error
0.001	0	0.006	0.002	0	0.002
0.106	0.105	0.004	0.114	0.111	0.003
0.162	0.164	0.002	0.208	0.213	0.005
0.256	0.254	0.002	0.338	0.339	0.007
0.311	0.315	0.004	0.478	0.475	0.002
0.389	0.388	0.003	0.501	0.500	0.004
0.451	0.450	0.005	0.589	0.588	0.002
0.601	0.609	0.008	0.643	0.654	0.009
0.654	0.656	0.002	0.701	0.700	0.006
0.725	0.722	0.003	0.887	0.886	0.005

The experimental data obtained from table 1 shows that the reading accuracy of this algorithm is basically consistent with that of manual readings, and it can fully meet the requirements of practical reading of instruments.

Table 2 shows the average absolute error and the maximum absolute error of 5 minutes under different light intensities.

Table 2. Algorithm results in different illumination conditions

Illumination Conditions	Instrument Position	Mean Absolute Error	Max Absolute Error
Strong light	Left instrument	0.005	0.011
	Right instrument	0.006	0.009
Normal light	Left instrument	0.003	0.008
	Right instrument	0.004	0.008
Weak light	Left instrument	0.003	0.010
	Right instrument	0.005	0.009

As can be seen from table 2, the error of the algorithm under different illumination conditions can meet the requirements of practical application.

4.2. Device Performance

In order to verify the speed of the device, the average reading time in the 5 minute test under the strong light, weak light and normal light are tested in the experiment, as shown in table 3.

Table 3. Average time of instrument reading under different illumination conditions

Illumination Conditions	Average Reading Time(s)
Strong light	5.41
Normal light	4.98
Weak light	5.55

As can be seen from table 3, the recognition time of this algorithm is about 5 seconds, which meets the requirements of practical application of the device.

4.3. Device Stability

In order to verify the stability of device, test the device during the day and night 8 o'clock to 16 O'clock , 20 o'clock to 4 o'clock for 8 consecutive hours of work, the average reading time, the average absolute error and maximum absolute error as shown in table 4.

Table 4. Device stability

Time Slot	Average Reading Times(s)	Average Absolute Error	Max Absolute Error
8:00-16:00	5.12	0.004	0.011
20:00-04:00	5.45	0.005	0.012

As can be seen from table 4, the device in this paper has good stability and no jump in the reading results, thus meeting the requirements of continuous monitoring in practical applications.

5. Conclusion

This paper adopts low power consumption ARM Hisilicon mobile processing platform and OV4689 camera, combined with a new skeleton extraction distance transform algorithm and the improved Hough algorithm for multi meter real-time reading. The design and implementation of the device were completed. The experimental results show that the accuracy, stability and processing time of the equipment meet the requirements of instrument recognition in practical applications. It shows that this instrument is a feasible method and has a good application prospect. At present, the purpose of this paper is to focus on single pointer and dial scale uniform pointer instrument. In the future work, we will pay more attention to the problem of dual pointer instrument and non-uniform distribution instrument.

Acknowledgment

This research was supported by the project as follows:
State Grid Corporation Science and Technology Project (named the pilot application on network access security for patrol data captured by unmanned planes and robots and intelligent recognition based on big data platform).

References

[1] Lihui Zhou, Yongsheng Zhang,Yong Sun. Development and Application of Intelligent Substation Inspection Robot[J]. Power System Automation, 2011, 35(19):85-88.
[2] Wenchao Cai, Qian Yu, Hong Wang. Detection and Reading of Circular Instrument Based on Fast Hough Transform [C]. National academic annual meeting of Chinese society of artificial intelligence. 2003.
[3] Ye Fang, Zhiqiang Min, Yunfeng Zhou.An instrument identification algorithm for substation inspection robot [J]. Automation and instrumentation, 2013, 28(5):10-14.
[4] Chao Zheng, Shaorong Wang, Yihan Zhang, Pengxiang Zhang, Yong Zhao, "A Robust and Automatic Recognition System of Analog Instruments in Power System by Using Computer Vision", Measurement, pp. , 2016, ISSN 02632241.
[5] L. Fang, J. Wang and R. Xiong, "A coarse-to-fine approach for industrial meter detection and its application," 2016 IEEE Workshop on Advanced Robotics and its Social Impacts (ARSO), Shanghai, 2016, pp. 214-220.doi: 10.1109/ARSO.2016.7736284
[6] Z. Yang, W. Niu, X. Peng, Y. Gao, Y. Qiao and Y. Dai, "An image-based intelligent system for pointer instrument reading," 2014 4th IEEE International Conference on Information Science and Technology, Shenzhen, 2014, pp. 780-783.doi: 10.1109/ICIST.2014.6920593
[7] Liangui Wang. Automatic Inspection Technology of Automobile Instrument Based on Visual Automatic Identification Function[J], Automation and instrumentation.2015(11) :82－83.
[8] Zhang T Y, Suen C Y. A fast parallel algorithm for thinning digital patterns[M]. ACM, 1984.
[9] Guo Z, Hall R W. Parallel thinning with two-subiteration algorithms[J]. Commun. ACM; (United States), 1989, 32:3(32):359-373.
[10] Chao Xu, Xiao Xiao, Yan Luo, Huijie Hu,Min Liu. A New Skeleton Extraction Method Based on Distance Transform [J]. Journal of instrument and instrument ,2012,(12):2851-2856.

Experiment of Azimuth-invariant Bistatic UHF UWB SAR

Hongtu Xie[1,2,3,4,*], Shaoying Shi[3], Junfa Mao[1], Fuhai Li[2], Daoxiang An[4,*], Zhimin Zhou[4] and Guoqian Wang[5,*]

[1]Key Laboratory of Ministry of Education of Design and Electromagnetic Compatibility of High Speed Electronic Systems, Shanghai Jiao Tong University, Shanghai, China
[2]College of Electrical and Information Engineering, Hunan University, Changsha
[3]Air Force Early Warning Academy, Wuhan, China
[4]College of Electronic Science, National University of Defense Technology, Changsha
[5]Hunan Institute of Traditional Chinese Medicine, Changsha, China
[*]Hongtu Xie and Guoqian Wang are the co-first authors, and Daoxiang An is the corresponding author.

*e-mail: xht20041623@163.com.

Abstract. Bistatic ultrahigh frequency ultrawideband synthetic aperture radar (UHF UWB SAR) has the well ability of the penetrating the foliage, high-resolution imaging and providing the increased information. In the paper, an imaging experiment of the azimuth-invariant bistatic UHF UWB SAR is described and the result is proposed. In August 2015, an along-track bistatic UHF UWB SAR experiment was conducted in China, and the raw data was collected. In this bistatic SAR system, the transmitter and receiver are both carried by a vehicle and separated by an invariable distance. The aim was to investigate the imaging property of the bistatic UHF UWB SAR system. Bistatic image was obtained using the subaperture spectrum-equilibrium method integrated with the fast factorized back projection algorithm (FFBPA). Experiment results prove the validity of the bistatic UHF UWB SAR experiment.

1. Introduction

Presently, a number of the remote sensing and geoscience applications are developed on the synthetic aperture radar (SAR) [1]-[3]. In recent years, the bistatic SAR imaging experiment was carried out by several countries in the military and civilian fields, and some experiment results were obtained [4]-[6].

Compared with the monostatic SAR, bistatic SAR has the complex configurations since it can be designed and deployed flexibly to satisfy different applications [7]-[9]. For the same scene, the bistatic SAR has the greater anti-jamming ability than the monostatic case. If the bistatic SAR system includes more than one transmitters or receivers, the imaging scene can be illuminated at the different angles, which can acquire the added information of scenes. Along-track bistatic SAR is an azimuth-invariant bistatic SAR system, whose transmitter and receiver are both carried by one platform and separated by an invariable distance. Along-track bistatic SAR not only inherits advantages of the general bistatic SAR, but also is easy to implement imaging, thus it has attracted the much attention.

Ultrahigh frequency (UHF) ultrawideband (UWB) SAR system operates in the UHF UWB signal with a large fractional bandwidth and the wide antenna beamwidth associated with a large integration angle, so it can provide the higher resolution SAR image [10]-[12]. Hence, it is a significant advantage that the along-track bistatic SAR operates in the UHF UWB mode. Some countries have conducted the bistatic UHF UWB SAR experiments, and several results were obtained [13] [14].

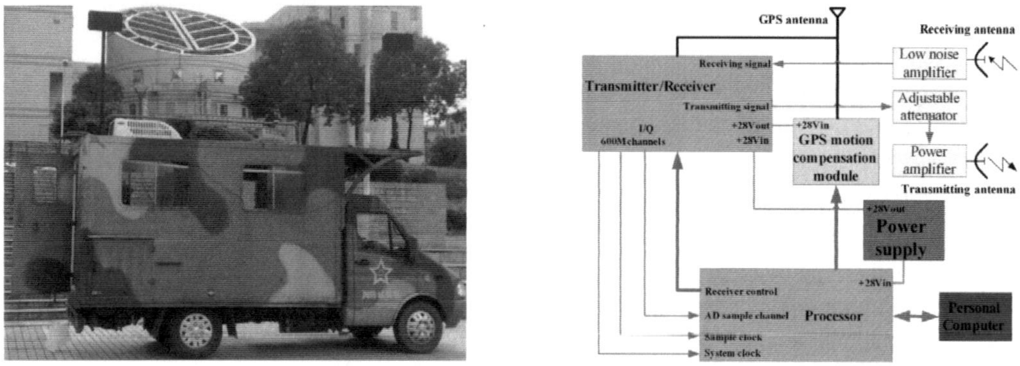

(a) (b)

Figure 1. (a) Along-track bistatic UHF UWB SAR system; (b) Its block diagram.

However, the along-track bistatic UHF UWB SAR experiment also poses the various challenges, such as the synchronization between the transmitter and receiver, position measurement of the radars, polarimetric calibration, collection geometry planning and data processing. Most important challenge is the successful operation of the along-track bistatic UHF UWB SAR experiment.

The paper presents the experiment of the along-track bistatic UHF UWB SAR. In Section 2, the along-track bistatic UHF UWB SAR system is given, focusing on the collection ability of the bistatic UHF UWB SAR data. The along-track bistatic UHF UWB SAR imaging experiment and the data processing are discussed in Section 3. The experimental results are presented in Section 4.

2. Development of the along-track bistatic UHF UWB SAR

Monostatic UHF UWB SAR system was developed by National University of Defense Technology (NUDT) in 2008, operating at UHF-band with a fractional bandwidth of about 0.25 [3]. In 2011, a programme of the upgrades to add a bistatic collection ability to the monostatic UHF UWB SAR was commenced using the hardware and software of the monostatic UHF UWB SAR. In August 2015, the experiment of the along-track bistatic UHF UWB SAR was conducted in campus in Changsha, China. Figure 1(a) gives the along-track bistatic UHF UWB SAR system, which is fixed on the vehicle, and the transmitting and receiving antennas are mounted at the front and end of the top of the vehicle. The transmitter transmits the chirp signal to the scene, while the receiver illuminates the same region and receives the scattered signal of the scene. Figure 1(b) gives the block diagram of this bistatic SAR, which contains the transmitting and receiving antennas, global positioning system (GPS) antenna, transmitter/receiver, GPS motion compensation module, processor, power supply, personal computer (PC), etc. To void the antenna's coupling and interfere for the near-field imaging, the caliber coupling multilayer patch antennas used for transmitting the radar signal and receiving the scattered signal are mounted on the vehicle separately. The transmitting and receiving antennas have the 3 dB azimuth beamwidth from approximately 30° to 55° and 3 dB elevation beamwidth from approximately 70° to 90°. The transmitter and receiver have the center controller, GPS synchronization module, frequency synthesizer, wave generator, and so on. One pulse per second (1PPS) signal from the GPS receiver is transferred into the center controller to control the system timing reference, and GPS synchronization module provides the coherence frequency reference for the SAR system, and then it is used to drive a frequency synthesizer to produce various clock signals for different parts of this SAR system. Under the control of the center controller, the wave generator generates the chirp signal. It is then modulated into the radio frequency signal, attenuated by the adjustable attenuator and amplified by the power amplifier. Finally, it is transmitted by the transmitting antenna. Simultaneity, the receiving antenna receives the scattered signal of the scene. It is then transferred into the low noise amplifier, down-converted into the intermediate frequency signal, filtered and amplified by the power amplifier, and demodulated into the in-phase and quadrature signals. Finally, the in-phase and quadrature signals are sampled by a high-speed sampler, converted by the analogue-to-digital converter, and recorded on the solid state electronic array and fed to PC. This bistatic SAR system's position is obtained by the GPS motion compensation module and recorded on the electronic array to offer the motion compensation.

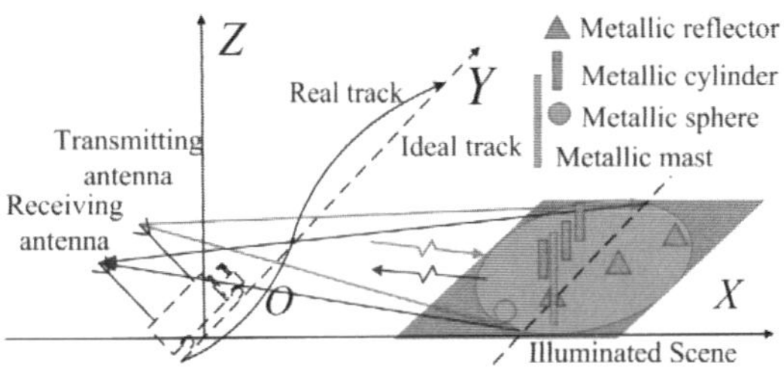

Figure 2. Geometry of the bistatic experiment.

Figure 3. Optical image of the imaging scene.

Table 1. Parameters of the along-track bistatic UHF UWB SAR system.

Parameter	Value	Parameter	Value
Signal frequency	UHF	Signal bandwidth	200MHz
Sampling frequency	220Hz	Pulse duration	1us
Pulse repetition frequency	500Hz	Incidence angle	82.3°

3. Imaging experiment of the along-track bistatic UHF UWB SAR

3.1. *Vehicle experiment*

The bistatic data collection ability of the UHF UWB SAR system was exercised in August 2015. The along-track bistatic UHF UWB SAR system was performed well, simultaneously producing a set of collected bistatic scattered signal. Figure 2 shows the geometry of the bistatic experiment. The dashed line is the vehicle nominal track, and the solid curve is its actual track. The transmitting antenna continually illuminates the scene including several targets, and then the receiving antenna receives the bistatic scattered signal of the scene. Parameters of the along-track bistatic UHF UWB SAR system is listed in Table 1. The vehicle moves along the Y axis direction with the speed about 3.5m/s. Initial positions of the transmitting and receiving antennas are about (0, 2, 4)m and (0, -2, 4)m at zeros time.

To collect data over a range of the incidence angle, the distance of the scene center was varied: the typical distance ranged from 35m to 55m, which produces the incidence angle transmitting/receiving antenna from 80.3° to 80.8°. Finally, one data site was selected for the data processing: the scene was a flat area, which has the size about 50m in the X axis and 80m in the Y axis. The scene center is (45, 40, 0)m. Figure 3 shows the optical image of imaging scene, which includes some various targets, such as the metallic reflectors, metallic cylinders, as well as metallic sphere and metallic mast, were deployed at the scene prior to the imaging experiment, in order to validate the imaging property of the bistatic UHF UWB SAR system and scattering characteristics of various targets. The positions of the metallic reflectors are about (37.8, 38.8, 0)m, (48.2, 49.2, 0)m and (58.6, 59.6, 0)m, respectively. The positions of the metallic cylinders are about (32.7, 46.6, 0)m, (32.7, 49.2, 0)m and (32.7, 51.8, 0)m, respectively. The position of the metallic mast is about (40, 31, 0)m.

3.2. *Data processing*

Along-track bistatic UHF UWB SAR system is subject to some difficulties in the data processing, i.e., the huge amount of the scattered signal, serious range-azimuth coupling and complex motion errors, as well as the serious radio frequency interfere (RFI), which makes the data processing more complicated. The accurate disposal of these problems is an intractable issue for the frequency domain algorithms, but it can be precisely managed in the time domain back projection algorithm (BPA) [7]. To reduce its high computation, the efficient implementation of the BPA has been applied for the bistatic SAR imaging, i.e., the fast factorized BPA (FFBPA) [15], which can be used for this bistatic SAR imaging.

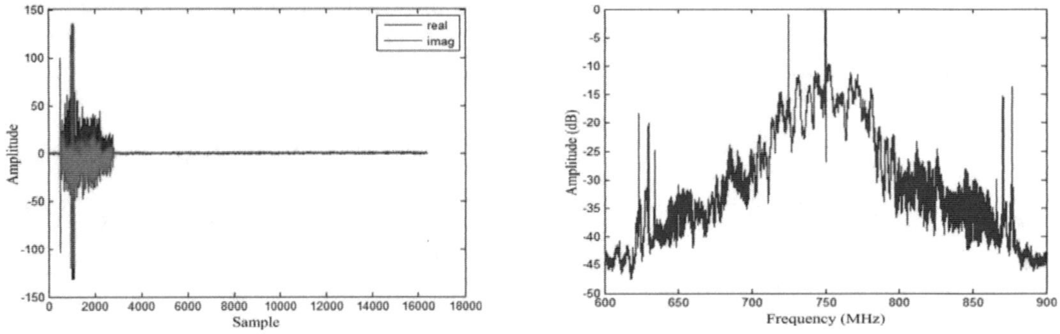

Figure 4. Bistatic scattered data of the illuminated scene. Left column: Bistatic scattered data from one transmitted radar pulse, including the real (blue line) and imaginary (red line) components; Right column: Sum of the frequency spectrum of the bistatic scattered data from 100 radar pulses.

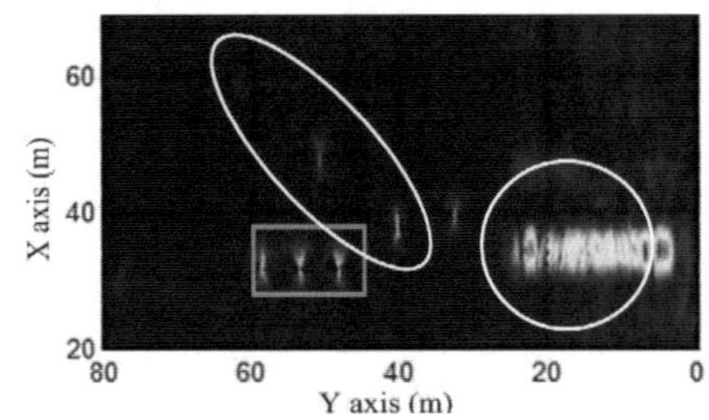

Figure 5. Bistatic UHF UWB SAR image obtained by the FFBPA

Besides, there is a strong radio frequency interference (RFI) saturated by the receiving antenna working in the UWB band. The influence of RFI signal may be extremely significant since their power is much higher than the reflected power of the bistatic UHF UWB SAR. To remove its influence, a subaperture spectrum equilibrium method [16] integrated with a FFBPA is used in the data processing.

Finally, before the RFI suppression is performed, the bistatic UHF UWB SAR data (including the GPS data) should be preprocessed. First, the GPS data should be upsampled and filtered, and then transferred into the Cartesian coordinate system to obtain the local position of the transmitting and receiving antennas. Then, the scattered data of the scene should be downsampled and filtered.

4. Experiment results

Figure 4 gives the bistatic scattered data of the scene in time-domain and frequency-domain. In the left of Figure 4, the real (blue line) and imaginary (red line) components of the bistatic scattered data is complicated due to the bistatic geometry. And, it can be seen that the scattered data of targets in the frequency domain in the right of Figure 4, since targets are deployed at the flat and open square, and the power of their scattered data is strong.

Figure 5 shows the bistatic UHF UWB SAR image obtained by the FFBPA, which is shown that the scene is well reconstructed. First, it is seen that the targets (the metallic reflectors, cylinders, mast and sphere) are well focused. Second, it can be found that the relative scattered power of the metallic cylinder (as well as metallic mast) in the yellow ellipse will become strong compared with the metallic reflectors in the red rectangle. Finally, we find that only a metallic sphere in the white circle is seen.

In order to quantitatively evaluate the focusing quality of the bistatic UHF UWB SAR image in Figure 5, the amplitude profiles of the imaging result of one metallic reflector in the X and Y axes are extracted from Figure 5. Moreover, the resolutions of the selected focused metallic reflector in the X and Y axes can be measured based on the width of the amplitude profiles at -3dB. The measured resolutions of the selected metallic reflector in the X and Y axes are 0.21m and 0.23m, respectively.

5. Conclusion

Imaging experiment of the azimuth-invariant bistatic UHF UWB SAR has been performed using the monostatic UHF UWB SAR system. After the processing of the acquired bistatic data, the high quality bistatic UHF UWB SAR image was obtained, which validates the imaging property of the bistatic UHF UWB SAR system. A more extensive of the bistatic UHF UWB SAR experiments with different operation modes will be carried out in the future.

Acknowledgments

This work was supported in part by the National Natural Science Foundation of China (No. 61571447), in full by Scientific Research Project of Education Department of Hunan (No. 16C1233), and in part by the Domain Foundation of Equipment Advance Research (No. 6140410301 and No. 6140410304).

References

[1] Cumming I and Wong F 2005 *Digital Processing of Synthetic Aperture Radar Data: Algorithm and Implementation* (Norwood, MA, USA)
[2] Xie H, An D, Huang X and Zhou Z 2016 Spatial resolution analysis of low frequency ultrawidebeam–ultrawideband synthetic aperture radar based on wavenumber domain support of echo data *J. Appl. Remote Sens.* **095033** 1-20.
[3] Xie H, Shi S, Xiao H, Xie C, Wang F and Fang Q 2016 Efficient time-domain imaging processing for one-stationary bistatic forward-looking SAR including motion error *Sensors* **16** 1-27.
[4] Walterscheid I et al 2010 Bistatic SAR experiments with PAMIR and TerraSAR-X-Setup, processing, and image results. *IEEE Trans. Geosci. Remote Sens.* **48** 3268-3279.
[5] Rodriguez-Cassola M et al 2012 First bistatic spaceborne SAR experiments with TanDEM-X *IEEE Geosci. Remote Sens. Lett.* **9** 33-37.
[6] Baqué R et al 2010 LORAMbis A bistatic VHF/UHF SAR experiment for FOPEN. *Proc. IEEE Radar Conf.* (Washington, DC, USA) pp 832-837.
[7] Xie H, An D, Huang X and Zhou Z 2015 Fast time-domain imaging in elliptical polar coordinate for general bistatic VHF/UHF ultra-wideband SAR with arbitrary motion *IEEE J. Sel. Topics Appl. Earth Observ.* **8** 879-895.
[8] Xie H, An D, Huang X and Zhou Z 2016 Efficient raw signal generation based on equivalent scatterer and subaperture processing for one-stationary bistatic SAR including motion errors *IEEE Trans. Geosci. Remote Sens.* **54** 3360-3377.
[9] Xie H, An D, Huang X and Zhou Z 2015 Research on spatial resolution of one-stationary bistatic ultrahigh frequency ultrawidebeam-ultrawideband SAR based on scattering target wavenumber domain support *IEEE J. Sel. Topics Appl. Earth Observ.* **8** 1782-1798.
[10] Davis M 2011 *Foliage Penetration Radar: Detection and Characterization of Objects Under Trees* (New York, NY, USA).
[11] An D, Li Y, Huang X, Li X and Zhou Z 2014 Performance evaluation of frequency-domain algorithms for chirped low frequency UWB SAR data processing *IEEE J. Sel. Topics Appl. Earth Observ.* **7** 678-690.
[12] Xie H et al 2017 Fast factorized backprojection algorithm for one-stationary bistatic spotlight circular SAR image formation *IEEE J. Sel. Topics Appl. Earth Observ.* **10** 1494 -1510.
[13] Ulander L, Frölind P, Gustavsson A, Jonsson T, Ragnarsson R, Stenström G 2014 Development of CARINA bistatic VHF-band SAR *Proc. Int. Radar Conf.* (Lille, France) pp 601-606.
[14] Ulander L, Frölind P, Gustavsson A, Ragnarsson R and Stenström G 2015 VHF/UHF bistatic and passive SAR ground imaging *Proc. IEEE Int. Radar Conf.* (Arlington, USA) pp 669-673.
[15] Ulander L, Frölind P, Gustavsson A, Murdin D and Stenström G 2010 Fast factorized back-projection for bistatic SAR processing *Proc. EUSAR* (Aachen, Germany) pp1002-1005.
[16] Dong Z, Liang D and Huang X 2008 A RFI suppression algorithm based channel equalization for the VHF/UHF UWB SAR *J. Elec. Inf. Tech.* **30** 550-553.

CCISP 2017 IOP Publishing
IOP Conf. Series: Journal of Physics: Conf. Series **960** (2018) 012038 doi:10.1088/1742-6596/960/1/012038

Text Line Detection from Rectangle Traffic Panels of Natural Scene

Shiyuan Wang[1,b], Linlin Huang[1,a] and Jian Hu[1,c]

[1]School of Electronics and Information Engineering, Beijing Jiaotong University, 100044, Beijing, China

E-mail: [a]huangll@bjtu.edu.cn; [b]15120030@bjtu.edu.cn; [c]jhu@bjtu.edu.cn

Abstract. Traffic sign detection and recognition is very important for Intelligent Transportation. Among traffic signs, traffic panel contains rich information. However, due to low resolution and blur in the rectangular traffic panel, it is difficult to extract the character and symbols. In this paper, we propose a coarse-to-fine method to detect the Chinese character on traffic panels from natural scenes. Given a traffic panel Color Quantization is applied to extract candidate regions of Chinese characters. Second, a multi-stage filter based on learning is applied to discard the non-character regions. Third, we aggregate the characters for text lines by Distance Metric Learning method. Experimental results on real traffic images from Baidu Street View demonstrate the effectiveness of the proposed method.

1. Introduction

Traffic sign detection and recognition system is an important part of intelligent transportation systems. It plays an important role in drivers and pedestrian safety. While the detection and recognition of traffic sign are mainly focused on symbol, the detection of text within rectangular traffic panel is far less developed. However, text contained within rectangular traffic panel images can be of great semantic value, and so is an important step to-wards information retrieval. Text detection within rectangular traffic panel still remains a challenge in computer vision and has particularly important application value in the future transportation systems.

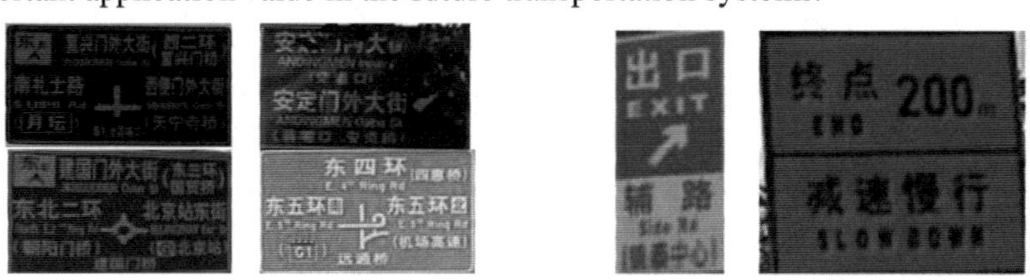

(a) (b)

Figure 1. Rectangular traffic panel (a) illumination effects (b) different of color, size, distribution etc

In the real situation, we get a variety of traffic panels that are taken in uncontrolled outdoor scenes. The various conditions such as illumination effects or weather conditions in the outdoor bring a huge challenge to our text detection. Traffic panels have a wide range of changes in color and size. Therefore many traditional methods have no effect on all traffic panels. Some samples with the small size and low resolution cause some text to blur. It is a huge obstacle in detecting text. In addition, the size and color of character is different, the distribution of characters in the traffic panel is not rule and the distance between characters is also variation. These factors have a direct impact on text aggregation results. Figure 1 shows sample from our database.

2. Related work

The work of text detection has been studied by many scholars in recently years. It is mainly divided into texture-based methods [1], region-based methods [2] and edge-based methods [3]. Zhong *et al.* [4] used the spatial variance of horizontal direction to obtain the approximate position of the text, and then extracted the text by color segmentation in the color image. Anh-Nga Lai *et al.* [5] used the local Color quantization and K-means clustering to detection Korean text in the natural scene. The biggest problem of text detection using these methods is that the computational efficiency is low, and the method is reduced in the practical application.

There are some scholars who are studying the text within traffic panel. However English text is the main research language rather than Chinese text. Greenhalgh *et al.* [6] located many candidate regions by MSERs and saturation, hue and value color. Individual characters are grouped into text lines. X. Rong *et al.* [7] established a top-down network framework that was able to locate and identify text contained traffic panel in the natural scene. The authors build a two-stage cascade framework inspired by the You Only Look Once (YOLO) detector [8]. Yet the method of CNN has relatively high requirements to hardware and datasets.

3. Text line detection based on multi-stages

The diagram of the proposal method is shown in Figure 2. The region of interest (ROI) is character candidate region in Figure 2. We use multi-stage detection method in order to improve the detection speed as well as accuracy. First, we employ Color quantification (CQ) method to detect the character candidate regions, and discard the non-character candidate region by machine learning method. Second, we recall the lost character region. Finally, the candidate regions are grouped into text line by the Distance Metric Learning method [9].

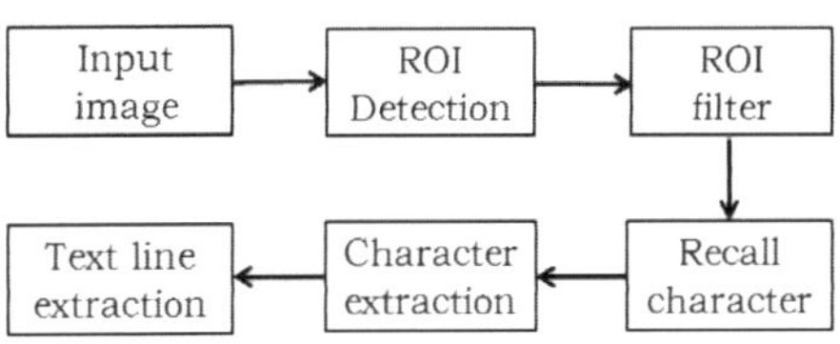

Figure 2. Text line extraction

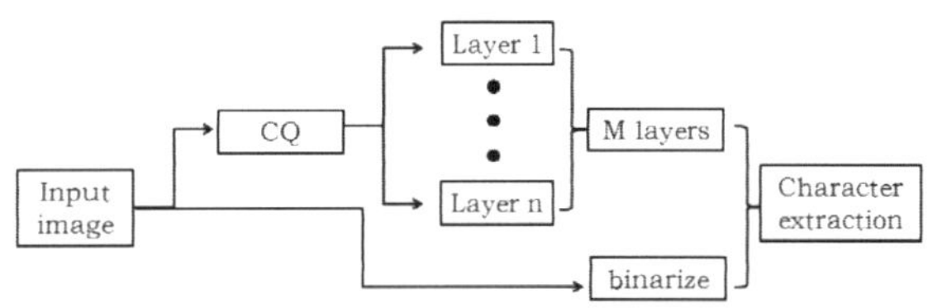

Figure 3. Character candidate region detection

3.1. Character candidate region detection

Due to the image with blurred and low resolution in our database, our proposed detection method must be able to solve these problems. Colour quantification (CQ) can solve the trouble of murky image and find the edge of the character. In order to save execution time and less memory, Octree structure is used. The diagram of character candidate region detection is shown in Figure 3.

CQ algorithm chooses K different colors as the initial color in the octree, and then sequentially reads the image data. When encountering a new color, the color is merged and the number of this color is stored. Finally, all the pixels are divided and have their own quantified categories. We get the different image layer according to the value of quantization. As shown in equation (1).

$$g_k(x,y) = \begin{cases} 255 & if\ I(x,y) == k \\ 0 & otherwise \end{cases} \tag{1}$$

$g_k(x,y)$ is the pixel value of x, y location in kth layer. k is the kinds of quantization, $k = 1, \dots K$. $I(x,y)$ represents the pixel value of x, y location in quantized image. We get many binary images in different layer. However, some layers of images do not contain characters; these layers will be abandoned by enacted rule. The rule is shown in equation (2).

$$r_k = p_k/p \tag{2}$$

r_k is the ratio in kth layer, p_k is the number of foreground point, p is the number of all pixels. When $r_k <$ threshold, the kth layer is no longer used in the follow process. The multi-layers are shown in Figure 4. Surely, only partial layers are shown, these binary images are obtained according to equation (1). From the comparison of four images, we know each image contains the unique information and other images don't have. In (a) and (c) graphics, we see that areas surrounded by these red boxes are special because they contain relatively clear character to compare to other layers in the same regions. Obviously characters in (b) are clearer than other layers. No characters exist in (d), so the binary layer will be discarded through the equation (2). We consider that the whole image is composed of different binary layers. In additional, the binary multi-layers are clearer and easier to detect character region. Finally, we employ connected components (CCs) method to extract character using all binary images.

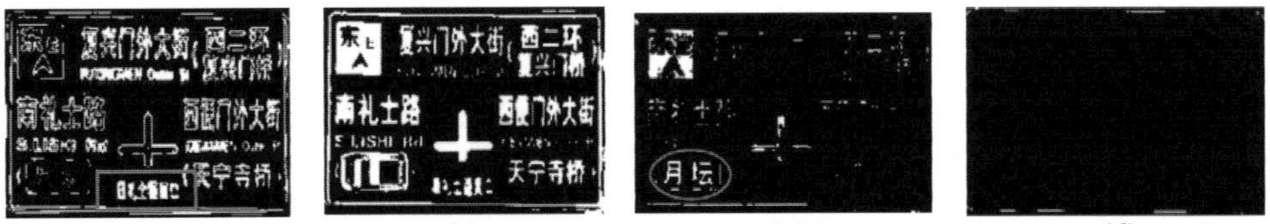

(a) (b) (c) (d)

Figure 4. Partial binary layers after quantization (a),(b),(c),(d) represent different layers

3.2. Multi-stage Filter

Candidate regions contain character, number and sign etc. but only the character is our detection target. We design a two-stage filter to discard the non-character candidate region. From Table 1, we choose the different feature and the classifier is Adaboost in the first filter, the result is shown. In order to extract more character regions, we apply the geometric features in the first filter. In first stage, we extract four parameters as the geometric features and use Adaboost classifier to classify. They are the ratio of width and height in the candidate boxes, the ratio of width in the candidate boxes and width in the panel images, the ratio of height in the candidate boxes and height in the panel images, the ratio of area in the candidate boxes and area in the panel images. We find this coarse filter is very efficient and the time-consuming is low. The obvious non-character areas are thrown away. But some number or the candidate regions that are similar to character region are still saved. The next stage will remove these candidate boxes.

Table 1. Different feature result in the first filter stage

Feature	Recall (%)	Precision (%)
Geometric features	90.3	30.1
HOG	89.5	36.0
Gabor	86.1	34.5
LBP	84.2	38.7

Form Table 2, we know HOG feature has the good performance in recall rate. So we use the HOG feature to extract descriptors in the candidate regions and give the judgment by Adaboost classifier in second stage. As we all know, HOG feature can maintain a good invariance in the geometric and optical deformation of the image. The shape of character, number and sign is different; the main characteristic is employed to distinguish character and non-character. HOG feature shows superior performance in this stage.

Table 2. Different feature result in the second filter stage

Feature	Recall (%)	Precision (%)
HOG	86.5	68.7
Gabor	84.6	65.3
LBP	82.8	70.6

3.3. Recall character regions

In the front of filter, some character regions must be misjudged as the non-character and be lost. So these character regions should be recalled to enhance the result in this situation. After filtering, each candidate boxes get the confidence. We employ the value of confidence to find lost character regions. Through our statistics and observation, characters are usually coming in the form of text lines. That is to say, we can find the character by this relation between characters. The diagram of recall character is shown in Figure 5.

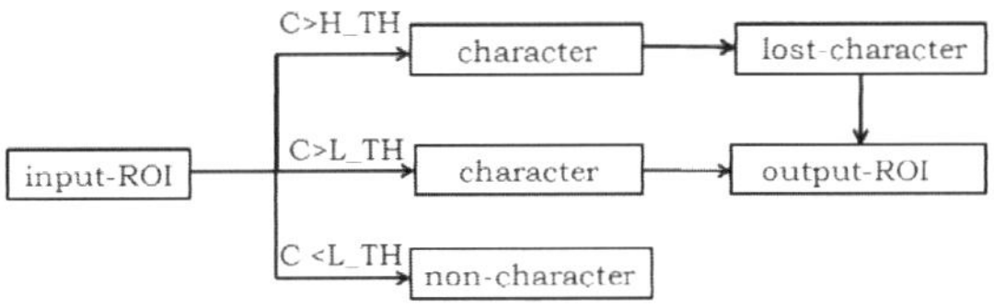

Figure 5. Recall Character regions

The diagram shows the process of recalling character regions. C represents the confidence of character candidate regions. H_TH is the high threshold of confidence, L_TH is the low threshold of confidence. Region of interest (ROI) is also character candidate region. First, the confidence of ROI compares withL_TH, if $C < L\_TH$, ROI is considered as non-character regions. If $C > L\_TH$ and $C < H\_TH$, we regard the ROI as character region, ROI will directly output. Surely, if $C > H\_TH$, we think ROI that is called seed region must be character region, and find the other lost character candidate regions surrounding it. If there are some lost ROIs in the left or right of seed regions, we judge the height between ROIs and the seed regions; if their height is almost the same, we think these ROIs are character regions and output to the next stage.

3.4. Text extraction on traffic panel

Through the previous steps, we have extracted character regions in the rectangular traffic panel. In the next steps, the extracted characters are merged into text lines. We apply the Distance Metric Learning method to set the condition of the aggregation, like the combined threshold between the characters and different row. The function of distance metric learning is defined as:

$$d(u, v; w) = w^T \boldsymbol{x}_{\boldsymbol{u},\boldsymbol{v}} \tag{3}$$

u, v represent the character regions, d is the distance between u and v. w is weight from the logistic regression. $\boldsymbol{x}$ is the feature vector and contains the relationship between u and v. As follows:

$$d_x = \begin{cases} abs(x_v - x_u - w_u)/\max(w_v, w_v) & if\ x_v < x_u \\ abs(x_u - x_v - w_u)/\max(w_v, w_v) & otherwise \end{cases} \tag{4}$$

$$\begin{aligned} d_{top} &= abs(y_u - y_v)/\max(h_u, h_v) \\ d_{down} &= abs(y_u - y_v + h_u - h_v)/\max(h_u, h_v) \end{aligned} \tag{5}$$

d_x is the adjacent distance, d_{top}, d_{down} are the different of alignment, d_w ,d_h which are the difference of the width and height are obtained by equation (5) but the input is the width and height of u, v. We use the weight of feature vector to obtain the $d(u, v; w)$. In order to get the parameter of w in the logistic regression, we choose the maximum distance in the same row and the smallest distance in the adjacent row. If $d(u, v; w)$ is less than zero, the u and v must be merged. The merger should be stop until no character needs to be merged. The result is shown in Figure 6.

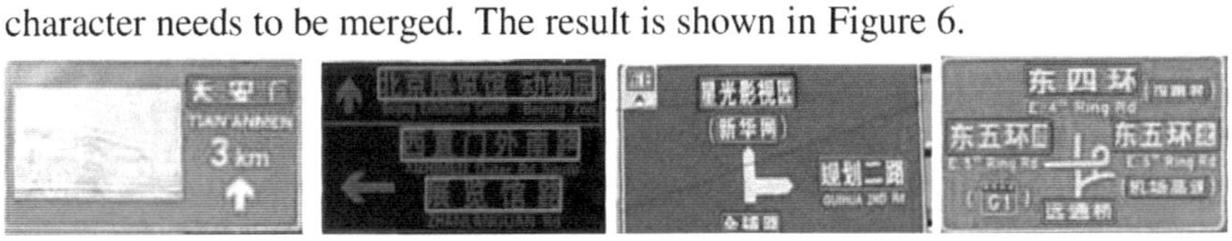

Figure 6. The result of text line extraction

4. Experimental results

We evaluated our proposed method and faster R-CNN[10] of the text detection on our dataset. The experiments were performed on computer with cpu i7 and GPU 1080Ti. Our dataset contains 1073 training images and 436 test images. The size of the rectangular traffic panel varies between 50× 27 and 330 × 163 pixels. These images contain 11962 training characters and 5114 test characters, 4148 training strings and 1696 test strings.

First, we estimate Chinese character extraction. The result of Chinese character detection depends on the Intersection over Union(IoU) between our detection region and ground truths. We set the Threshold of overlap is 0.7. Chinese character detection results are given below.

Table 3. Chinese character detection results

Method	Recall (%)	Precision (%)
Our method	88.63	95.81
Faster R-CNN	76.31	86.47

From Table 3, we know that we acquire the relatively good result in the Chinese character detection. The precision reaachs up to 95.81% and the recall is 88.63%. At the same time, we also apply IoU to evaluate Chinese text line extraction. The IoU is also 0.7. the Chinese text line extraction results is shown as Table 4. the result also proves that our proposed method is efficient.

Table 4. Chinese text line detection results

	Recall (%)	Precision (%)
Text line	89.83	93.24

5. Conclusion

In this paper, we propose an effective approach for detecting text in Chinese rectangular traffic panel image using multi-layered detection method based on Octree Quantification on our database; the proposed method has a good performance for rectangular traffic panel text extraction, which obtains a higher precision 93.24%. Compared to Faster R-CNN, the database is relatively small, so the deep learning result is not particularly good. But from the small dataset, the experimental results show the feasibility of our proposed algorithm.

Acknowledgements

This work was supported in part by the National Natural Science Foundation of China (NSFC) Grant 61271306.

References

[1] Yin, X. C., et al. Robust Text Detection in Natural Scene Images. IEEE Transactions on Pattern Analysis & Machine Intelligence, 2013, 36 pp 1-1.

[2] Yi, C., Tian, Y., Arditi, A.: Portable camera-based assistive text and product label reading from hand-held objects for blind persons. IEEE/ASME Trans on Mechatronics. 2014, 19 pp 808-817.

[3] C. Yao, X. Bai, W. Liu, Y. Ma and Z. Tu. Detecting texts of arbitrary orientations in natural images. Computer Vision and Pattern Recognition, 2012, 157 pp 1083-1090.

[4] Zhong Y, Karu K, Jain A K. Locating text in complex color images. Proceedings of International Conference on Document Analysis and Recognition (ICDAR), 1995, 1 pp 146-149.

[5] Lai, Anh Nga, et al. Korean Text Extraction by Local Color Quantization and K-means Clustering in Natural Scene. Intelligent Information and Database Systems, 2009, 200 pp 138-143.

[6] Greenhalgh J, Mirmehdi M. Recognizing Text-Based Traffic Signs. IEEE Transactions on Intelligent Transportation Systems, 2015, 16 pp 1360-1369.

[7] Rong, Xuejian, C. Yi, and Y. Tian. Recognizing Text-Based Traffic Guide Panels with Cascaded Localization Network. European Conference on Computer Vision, 2016, pp 109-121.

[8] Redmon, J., Divvala, S., Girshick, R., Farhadi, A. You Only Look Once: Unified, Real-Time Object Detection. CVPR, 2015 pp 779-788.

[9] Nguyen, Bac, C. Morell, and B. D. Baets. Large-scale distance metric learning for k -nearest neighbors regression. Neurocomputing, 2016 214 pp 805-814.

[10] Ren, Shaoqing, et al. Faster R-CNN: Towards Real-Time Object Detection with Region Proposal Networks. IEEE Transactions on Pattern Analysis & Machine Intelligence, 2017 39 pp 1137.

A Direction Finding Method with A 3-D Array Based on Aperture Synthesis

Shiwen Li[1], Liangbing Chen[2,*], Zhaozhao Gao[1] and Wenfeng Ma[2]

[1] Science and Technology on Electronic Information Control Laboratory, Chengdu, 610036, China
[2]School of Information Engineering, Nanchang University, Nanchang, 330031, China

*Liangbingchen@gmail.com

Abstract. Direction finding for electronic warfare application should provide a wider field of view as possible. But the maximum unambiguous field of view for conventional direction finding methods is a hemisphere. It cannot distinguish the direction of arrival of the signals from the back lobe of the array. In this paper, a full 3-D direction finding method based on aperture synthesis radiometry is proposed. The model of the direction finding system is illustrated, and the fundamentals are presented. The relationship between the outputs of the measurements of a 3-D array and the 3-D power distribution of the point sources can be represented by a 3-D Fourier transform, and then the 3-D power distribution of the point sources can be reconstructed by an inverse 3-D Fourier transform. And in order to display the 3-D power distribution of the point sources conveniently, the whole spherical distribution is represented by two 2-D circular distribution images, one of which is for the upper hemisphere, and the other is for the lower hemisphere. Then a numeric simulation is designed and conducted to demonstrate the feasibility of the method. The results show that the method can estimate the arbitrary direction of arrival of the signals in the 3-D space correctly.

1. Introduction

In electronic warfare field, direction finding of the signals in the space is one of the most important aspects. Many direction finding methods have been developed, such as interferometer direction finding [1], amplitude comparison direction finding [2], TDOA-based direction finding [3], MUSIC based on the array signal processing [4, 5], direction based on sparsity [6], and so on.

For electronic warfare application, direction finding should provide the field of view as wider as possible. It's very important for some special application. But conventional direction finding methods are based on a linear array or a planar array, which causes ambiguity in the practical application. When the minimum antenna spacing is half a wavelength, the methods can estimate the direction of arrivals of the signals in the hemisphere. Under this condition, they can obtain the maximum field of view, equal to 2π steradian, which is the solid angle of a hemisphere. But the signals exist in the 4π steradian (4π is the solid angle of the whole sphere), these methods cannot distinguish the direction of arrival of the signals from the back lobe of the antennas, i.e., there is ambiguity.

In this paper, a direction finding method based on aperture synthesis (DFAS) radiometry with a 3-D array is proposed. It utilizes a 3-D array to receive the signals, and an image including the spatial distribution of the sources is outputted. Firstly, the model of the direction finding system and the fundamentals are presented. Then in order to demonstrate the feasibility of the method, numeric simulation is conducted and analyzed.

2. The fundamentals of the DFAS

The illustration of a DFAS is shown as figure 1. The system contains a 3-D array, receivers, and digital correlators. The array receives the signals emitted by the point sources. The signals pass through the receivers and are correlated between any pair of antennas in the digital correlators. The output of digital correlators is called visibility function. Then visibility function is processed by 3-D Fourier transform to obtain a 3-D image including the spatial distribution of the point sources. Thus, the direction of arrivals of these point sources can be obtained from the image

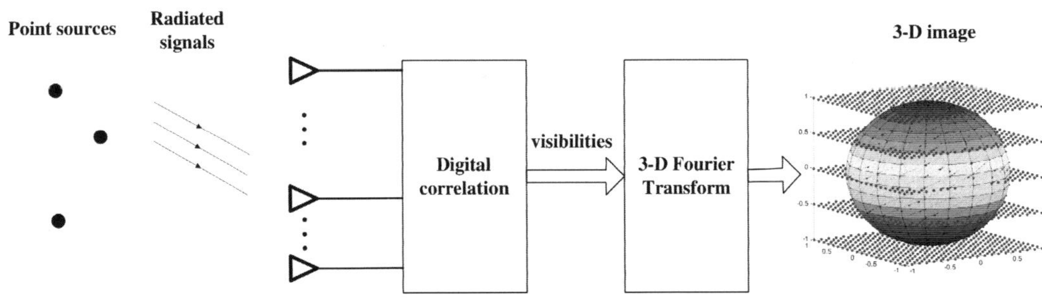

Figure 1. The illustration of the model of a DFAS system

According to the principle of the aperture synthesis radiometry in astronomy or Earth remote sensing [7-9], the correlation between the signals received by any pair of antennas is called visibility function. For a pair of antennas a_i and a_j with the coordinates (x_i, y_i, z_i) and (x_k, y_k, z_k) normalized respect to the wavelength, the relationship between the visibility function and the distribution of point sources is expressed as follow:

$$V(u,v,w)=\iint P(\xi,\eta)e^{-j2\pi(u\xi+v\eta+w\cos\theta)}d\xi d\eta \tag{1}$$

Where (u,v,w)=(x_k —x_i, y_k —y_i, z_k —z_i) are the spatial frequencies, $(\xi,\eta)=(\sin\theta\cos\varphi,\sin\theta\sin\varphi)$ are the directional cosines. $P(\xi,\eta)$ is the power of the point source. For different pair of antennas, one obtains the different samples of the visibility function. And a 3-D array can provide a 3-D spatial frequency distribution and the corresponding 3-D samples of the visibility function.

It can not reconstruct $P(\xi,\eta)$ from the 3-D visibility samples directly. But (1) can be rewritten in the form of a 3-D Fourier transform involving the third direction cosine n defined with respect to w axis.

$$V(u,v,w)=\iiint P(\xi,\eta,\sigma)\delta(\cos\theta-\sigma)e^{-j2\pi(u\xi+v\eta+w\sigma)}d\xi d\eta d\sigma \tag{2}$$

The delta function $\delta(\cos\theta-\sigma)$ maintains the condition $\sigma=\cos\theta$. And the spatial distribution is rewritten as $P(\xi,\eta,\sigma)$. When σ is greater than zero, it indicates the point source locates at the upper semisphere. When σ is smaller than zero, it indicates the point source locates at the lower semisphere. Thus, it can distinguish the point source in the 4π space. The spatial distribution of the point sources can be reconstructed by 3-D inverse Fourier transform:

$$P(\xi,\eta,\sigma=\cos\theta)=\iiint V(u,v,w)e^{+j2\pi(u\xi+v\eta+w\sigma)}dudvdw \tag{3}$$

For (ξ,η,σ), they satisfy the condition:

$$\xi^2+\eta^2+\sigma^2=1 \tag{4}$$

Obviously, it is the function of the sphere in the 3-D coordinates. For actual application, only some samples of the 3-D visibility function can be measured. Thus, the spatial distribution image of the point sources can be estimated by discrete Fourier transform:

$$P(\xi_i,\eta_j,\sigma_k=\cos\theta)=\Delta u\Delta v\Delta w\sum\sum\sum V(u_l,v_m,w_n)e^{+j2\pi(u_l\xi_j+v_m\eta_j+w_n\sigma_k)} \tag{5}$$

Generally speaking, this method provides a mapping from a 3-D spatial frequency domain to a sphere spatial domain. But it is not convenient to display a power distribution of the point sources in a 3-D viewpoint. It should be simplified further. According to (4), σ depends on ξ and η. So the whole sphere can be projected to two unit circle. One of the circle with $\sigma>0$ denotes the upper hemisphere, and the other with $\sigma<0$ denotes the lower hemisphere. The illustration is shown in Figure 2. Then the direction of the arrivals of the point sources can be obtained from the reconstructed 2-D image.

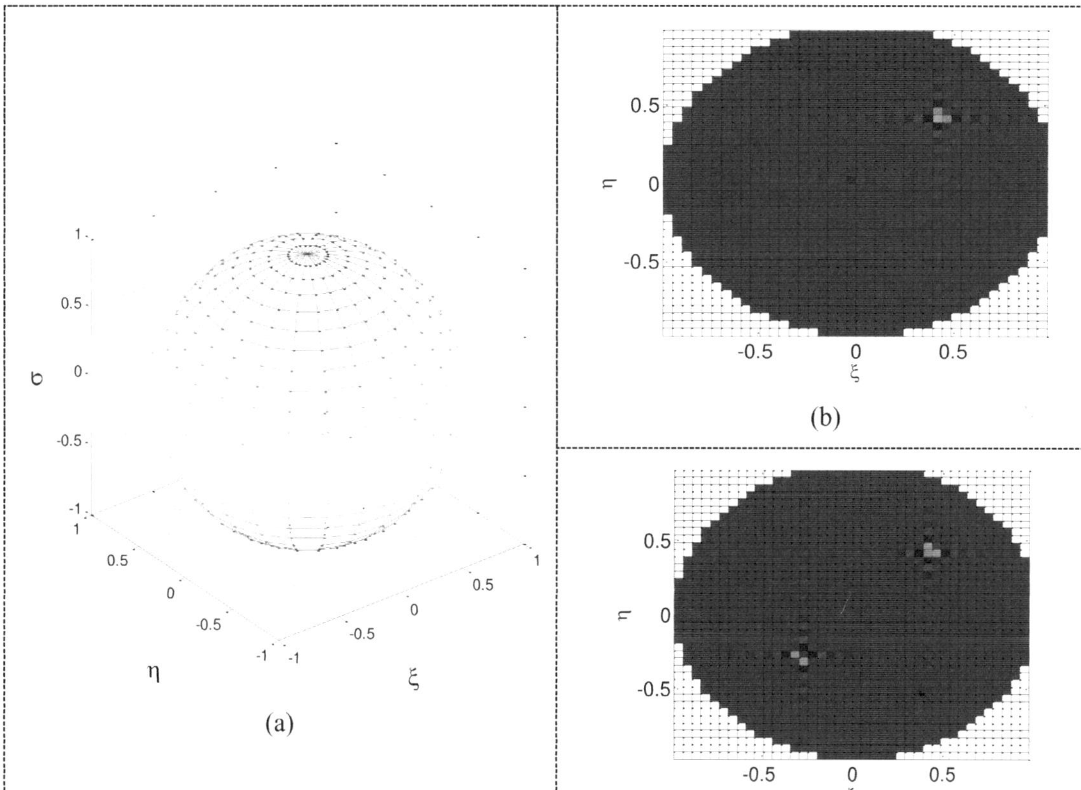

Figure 2. The illustration of the whole sphere projected to two unit circles. (a) The whole sphere. (b) The unit circle with $\sigma>0$. (c) the unit circle with $\sigma<0$

3. Numerical Simulation for DFAS

The distribution of the 3-D spatial frequencies depends on the 3-D array arrangement. Thus, how to design an optimal 3-D array is an important problem. Here, the antenna array is not referred, and the distribution of 3-D spatial frequencies is directly provided. Considering a set of spatial frequency samples $\{(u_l,v_m,w_n)|-20\le l\le 20,-20\le m\le 20,-5\le w\le 5\}$, The spatial frequency samples are uniformly distributed with the spacing of 0.5. Three point sources locate at $(\theta_1,\varphi_1)=(\pi/3,\pi/3)$, $(\theta_2,\varphi_2)=(2\pi/3,\pi/3)$ and $(\theta_3,\varphi_3)=(\pi/4,3\pi/4)$ respectively. The corresponding directional cosines are about $(\xi_1,\eta_1,\sigma_1)=(0.43,0.75,0.5)$, $(\xi_2,\eta_2,\sigma_2)=(0.43,0.75,-0.5)$, and $(\xi_3,\eta_3,\sigma_3)=(-0.5, 0.5,0.71)$ respectively. The first point source and the third one locate on the upper hemisphere, and the second one locate on the lower hemisphere.

For a 2-D array, the point source with $(\theta_1,\varphi_1)=(\pi/3,\pi/3)$ and the point source with $(\theta_2,\varphi_2)=(2\pi/3,\pi/3)$ can not be distinguished, and the point source with $(\theta_2,\varphi_2)=(2\pi/3,\pi/3)$ will be estimated with the direction of the arrival of $(\theta_1,\varphi_1)=(\pi/3,\pi/3)$. It is because the directional cosines of the two point sources for a 2-D array are the same, i.e., there is ambiguity.

For a 3-D array above, the reconstructed distribution image of the point sources are shown in figure 3. Figure 3(a) is the reconstructed 2-D image corresponding to the projection of the 3-D upper hemisphere, and figure 3(b) is the reconstructed 2-D image corresponding to the projection of the 3-D lower hemisphere. It can be found there are two point sources in figure 3(a), and the corresponding directional cosines are $(\xi_1^{'},\eta_1^{'})=(0.44,0.73)$ and $(\xi_3^{'},\eta_3^{'})=(-0.49,0.49)$ respectively, which is approximate to the actual directional cosines of the first point source and the third one. It means that the two point sources located on the upper hemisphere can be correctly reconstructed. In figure 3(b), there is only one point source, and the corresponding directional cosine is $(\xi_2^{'},\eta_2^{'})=(0.44,0.73)$, which is approximate to the actual directional cosine of the second point source, which locates on the lower hemisphere. It means that DFAS can correctly reconstruct the distribution of the point sources locating on the whole sphere, and the direction of arrivals can be correctly estimated.

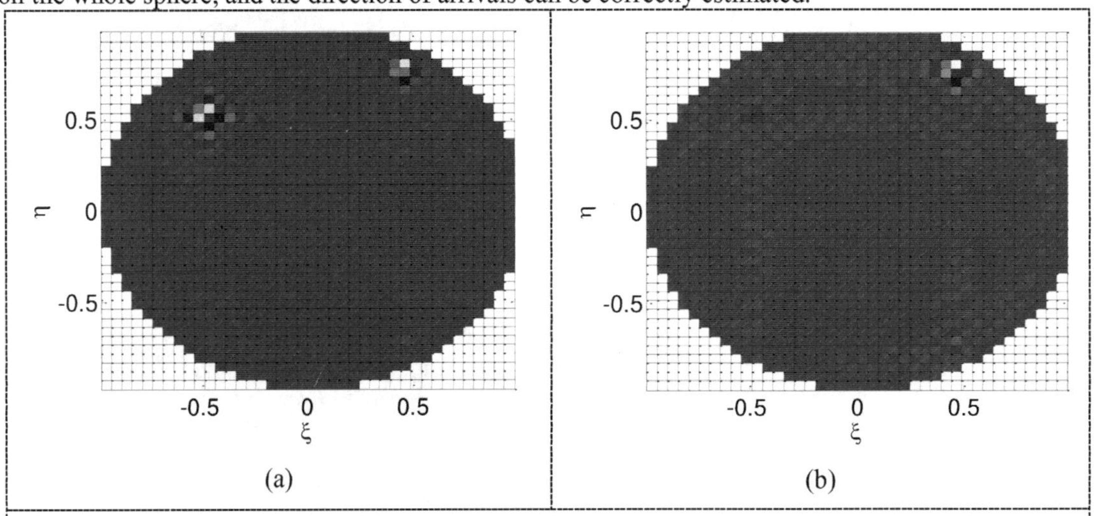

Figure 3. The simulation results of DFAS. (a) the reconstructed 2-D distribution image corresponding to the projection of the 3-D upper hemisphere. (b) the reconstructed 2-D distribution image corresponding to the projection of the 3-D lower hemisphere

4. Conclusion

The aperture synthesis radiometry in astronomy and Earth remote sensing is introduced to estimate the direction of arrivals from the whole sphere. The simulation results show that it can handle the ambiguity caused by the planar array. In this paper, there is only a brief presentation of DFAS, and more problems should be discussed further.

Acknowledgments

This work was supported in part by the National Natural Science Foundation of China (NSF41505015), Science Foundation of the Department of Science & Technology of Jiangxi Province (20151BAB213034), and Foundation of Science and Technology on Electronic Information Control Laboratory.

References

[1] Lee J H and Woo J M 2015 *IEEE Antennas Wireless Propag. Lett.* **14** 719
[2] Orduyilmaz A, Serin M and Yildirim A 2015 *Proc. Signal Processing and Communications Applications Conference* (IEEE) p 109
[3] Cui X, Yu K and Lu S 2015 *IEEE Transactions on Instrumentation & Measurement* **64** 2347
[4] Guo Z, Wang X and Wei H 2017 *IEEE Transactions on Wireless Communications* **16** 5384
[5] Ciuonzo D and Rossi P S 2017 *IEEE Signal Processing Letters* **24** 397
[6] Liu Z M, Huang Z T, Zhou Y Y and Liu J 2012 *IEEE Transactions on Aerospace & Electronic Systems s* **48** 2690
[7] Zhang C, Liu H, Wu J, Zhang S, Yan J, Niu L, Sun W and Li H 2014 *IEEE Transactions on Geoscience & Remote Sensing* **53** 207
[8] Ruf C S, Swift C T, Tanner A B and Vine D M L 1998 *IEEE Transactions on Geoscience & Remote Sensing* **26** 597
[9] Thompson A R, Morgan J M and Swenson G W 2001 *Interferometry and Synthesis in Radio Astronomy* (New York: Wiley–Interscience)

Three-Component Decomposition of Polarimetric SAR Data Integrating Eigen-Decomposition Results

Da Lu[1,*], Zhihua He[1] and Huan Zhang[1]

[1]AVIC Leihua Electronic Technology Institute, LiangXi Road #796, Wuxi, China

*lartely@163.com

Abstract. This paper presents a novel three-component scattering power decomposition of polarimetric SAR data. There are two problems in three-component decomposition method: volume scattering component overestimation in urban areas and artificially set parameter to be a fixed value. Though volume scattering component overestimation can be partly solved by deorientation process, volume scattering still dominants some oriented urban areas. The speckle-like decomposition results introduced by artificially setting value are not conducive to further image interpretation. This paper integrates the results of eigen-decomposition to solve the aforementioned problems. Two principal eigenvectors are used to substitute the surface scattering model and the double bounce scattering model. The decomposed scattering powers are obtained using a constrained linear least-squares method. The proposed method has been verified using an ESAR PolSAR image, and the results show that the proposed method has better performance in urban area.

1. Introduction

Polarimetric synthetic aperture radar (PolSAR) is sensitive to the geometric structure, orientation and physical characteristics of scattering targets. Polarimetric target decomposition (PTD) is a useful tool to identify and separate scattering mechanisms [1]-[4]. One of the most used PTD methods, Freeman and Durden0 three-component decomposition method, decomposes PolSAR data into three power categories: surface, double-bounce, and volume scattering power. When decomposing surface scattering power and double-bounce power, some artificial settings were applied in the three-component decomposition due to the number of unknowns is larger than it of the observation. This approach is best when either double bounce contribution or surface contribution is close to zero. In real case, especially in double-bounce and surface contribution mixed place, speckle-like decomposition results will be introduced due to artificial settings. Another problem of three-component decomposition method is that the volume scattering power is usually overestimated in oriented urban areas. Though it can be partly solved by deorientation process [5], it still exists in some highly oriented urban areas [6]. This paper integrates the results of eigen-decomposition to solve these two problems. As discussed in [7], it is needless to artificially set parameters to be fix values, and the eigen-decomposition results can directly provide parameters. Based on this, the surface and double bounce scattering models can be calculated by eigen-decomposition results. Finally, the decomposed scattering powers are obtained using a constrained linear lest-square method.

A brief description of the decomposition algorithm is presented in Section 2. Section 3 provides the three-component scattering decomposition algorithm integrating eigen-decomposition. In Section 4 the algorithm is illustrated using ESAR PolSAR data. Finally, conclusion is given in Section 5.

2. Decomposition algorithms

2.1. Three-Component Decomposition method

Fully polarimetric SAR system measures the 2×2 complex scattering matrix $[S]$ of each cell in image. The Pauli vector can be defined as

$$\vec{k}_P = \frac{1}{\sqrt{2}}\left[S_{hh}+S_{vv}, S_{hh}-S_{vv}, 2S_{hv}\right]^T \tag{1}$$

Where S_{hh}, S_{hv} and S_{vv} are elements of $[S]$.

The coherence matrix is defined as

$$[T] = \left\langle \vec{k}_P \vec{k}_P^H \right\rangle = \begin{bmatrix} T_{11} & T_{12} & T_{13} \\ T_{21} & T_{22} & T_{23} \\ T_{31} & T_{32} & T_{33} \end{bmatrix} \tag{2}$$

The rotated coherency matrix with rotation angle θ can be obtained as

$$\begin{aligned}\left[T(\theta)\right] &= \begin{bmatrix} T_{11}(\theta) & T_{12}(\theta) & T_{13}(\theta) \\ T_{21}(\theta) & T_{22}(\theta) & T_{23}(\theta) \\ T_{31}(\theta) & T_{32}(\theta) & T_{33}(\theta) \end{bmatrix} \\ &= \left[R_P(\theta)\right][T]\left[R_P(\theta)\right]^H \end{aligned} \tag{3}$$

Where $\left[R_P(\theta)\right] = \begin{bmatrix} 1 & 0 & 0 \\ 0 & \cos 2\theta & \sin 2\theta \\ 0 & -\sin 2\theta & \cos 2\theta \end{bmatrix}$

The angle θ can be obtained as [5]

$$\theta = \frac{1}{4}\left(\tan^{-1}\left[\frac{2\operatorname{Re}(T_{23})}{T_{22}-T_{33}}\right] \pm n\pi\right) \quad n = 0,1 \tag{4}$$

Therefore, the rotated coherency matrix can be expressed as the sum of three matrices including surface, double-bounce, and volume scattering matrices.

$$\left[T(\theta)\right] = f_s[T]_s + f_d[T]_d + f_v[T]_v \tag{5}$$

Where f_s, f_d and f_v are the coefficients corresponding to the surface scattering coherency matrix $[T]_s$, the double-bounce scattering coherency matrix $[T]_d$ and the volume scattering coherency matrix $[T]_v$ respectively. $[T]_s$ or $[T]_d$ will be set to be a fixed value depending on which scattering is dominant.

2.2. Eigen-decomposition

Eigen-decomposition of the coherence matrix can be expressed as

$$[T] = \sum_{i=1}^{3} \lambda_i \vec{e}_i \vec{e}_i^H \tag{6}$$

Where λ_1, λ_2 and λ_3 are the eigenvalues, $\vec{e}_i$ represents the normalized eigenvector corresponded with λ_i, e_{im} is the $m-th$ element of the $\vec{e}_i\ (m=1,2,3\ \ i=1,2,3)$.

The two principle eigenvectors $\vec{e}_1$ and $\vec{e}_2$ can be viewed as the same as the surface and double-bounce scattering mechanism[7]. This means that the eigen-decomposition results can be directly used as prior information to fix unknown values in three-component decomposition method.

3. Modified three-component scattering decomposition

The three-component scattering decomposition suffers from scattering mechanism ambiguity in some oriented urban areas. Chen *et al.*[8] proposed a general model-based decomposition, which by separating double and odd bounce orientation angles, to solve the volume contribution overestimation problem. However, this approach used a nonlinear least squares optimization technique that will introduce heavy computation burden.

In this paper, the eigen-decomposition method are integrated with model-based decomposition. The $[T]_s$ and $[T]_d$ can be determined directly by $\vec{e}_1$ and $\vec{e}_2$,

$$[T]_s = \vec{e}_s\vec{e}_s^{\,H}, [T]_d = \vec{e}_d\vec{e}_d^{\,H} \tag{7}$$

Where

$$\begin{cases} \vec{e}_s = \vec{e}_1, \vec{e}_d = \vec{e}_2 & \mathrm{Re}\{e_{11}/e_{12}\} < \mathrm{Re}\{e_{21}/e_{22}\} \\ \vec{e}_s = \vec{e}_2, \vec{e}_d = \vec{e}_1 & \mathrm{Re}\{e_{11}/e_{12}\} \geq \mathrm{Re}\{e_{21}/e_{22}\} \end{cases} \tag{8}$$

The proposed decomposition can be written as

$$[T] = f_s[T]_s + f_d[T]_d + f_v[T]_v + [T]_{\mathrm{residual}} \tag{9}$$

The residual matrix $[T]_{\mathrm{residual}}$ is used to measure the performance of the decomposition results. The criterion is to minimize $\left\|[T]_{\mathrm{residual}}\right\|_2$ (L2-norm of $[T]_{\mathrm{residual}}$).

$$\begin{aligned} &\text{minimize} \quad \left\|[T]_{\mathrm{residual}}\right\|_2 \\ &\text{subject to} \quad [T]_{\mathrm{residual}} = [T] - f_s[T]_s + f_d[T]_d + f_v[T]_v \end{aligned} \tag{10}$$

Note that $[T]_s$, $[T]_d$ and $[T]_v$ are all known, f_s, f_d and f_v should be no less than zero. The decomposition turns to be a linearly constrained least-squares problem[9],

$$\min_{\vec{x}} \frac{1}{2} \left\| \begin{bmatrix} T_{s11} & T_{d11} & T_{v11} \\ \mathrm{Re}\{T_{s12}\} & \mathrm{Re}\{T_{d12}\} & \mathrm{Re}\{T_{v12}\} \\ \mathrm{Im}\{T_{s12}\} & \mathrm{Im}\{T_{d12}\} & \mathrm{Im}\{T_{v12}\} \\ \mathrm{Re}\{T_{s13}\} & \mathrm{Re}\{T_{d13}\} & \mathrm{Re}\{T_{v13}\} \\ \mathrm{Im}\{T_{s13}\} & \mathrm{Im}\{T_{d13}\} & \mathrm{Im}\{T_{v13}\} \\ T_{s22} & T_{d22} & T_{v22} \\ \mathrm{Re}\{T_{s23}\} & \mathrm{Re}\{T_{d23}\} & \mathrm{Re}\{T_{v23}\} \\ \mathrm{Im}\{T_{s23}\} & \mathrm{Im}\{T_{d23}\} & \mathrm{Im}\{T_{v23}\} \\ T_{s33} & T_{d33} & T_{v33} \end{bmatrix} \cdot \vec{x} - \begin{bmatrix} T_{11} \\ \mathrm{Re}\{T_{12}\} \\ \mathrm{Im}\{T_{12}\} \\ \mathrm{Re}\{T_{13}\} \\ \mathrm{Im}\{T_{13}\} \\ T_{22} \\ \mathrm{Re}\{T_{23}\} \\ \mathrm{Im}\{T_{23}\} \\ T_{33} \end{bmatrix} \right\|_2^2 \quad \text{such that} \begin{cases} 0 \leq x_1 \\ 0 \leq x_2 \\ 0 \leq x_3 \end{cases} \tag{11}$$

Where $\vec{x} = [x_1, x_2, x_3]^T = [f_s, f_d, f_v]^T$.

Since the trace of $[T]_s$, $[T]_d$ and $[T]_v$ are all equal to one, f_s, f_d and f_v can be regarded the contribution P_s, P_d and P_v of each scattering mechanism to the span.

4. Experimental results and discussion

The ESAR L-band fully polarized images were used to validate the modified three-component scattering power decomposition model.

The results of applying the original Freeman decomposition (F3O)[2], three decomposition with deorientation (F3R)[5] and the proposed method (E3) to the three patches (forest, oriented urban and orthogonal urban) of ESAR image are shown in Figure 1. The RGB represents double-bounce, volume and surface scattering components respectively.

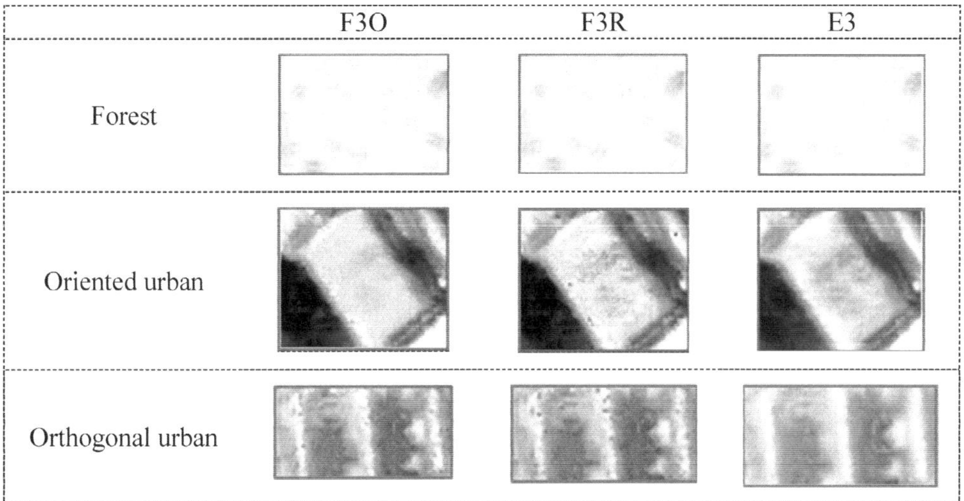

Figure 1. Decomposition results of patches (forest, oriented urban and orthogonal urban) of the ESAR image.

As can be seen in Figure 1, in forest area, all of the three methods have the similar decomposition results. In oriented urban area, volume scattering is the dominant scattering in the result of F3O. This is because oriented area will induce significant cross-polarization power. Though in F3R, orientation angle compensation can reduce cross-polarization power and the urban area appears to be not that green as F3O results, the dominant scattering is still volume scattering. This is because the orientation angle of oriented urban is very high. The proposed method, however, can reduce volume scattering contribution significantly. In orthogonal urban area, which is the orthogonal urban area, the dominant scattering powers are all double bounce contribution. The results of this area are also similar. However, speckle-like decomposition results are shown in F3O and F3R results, this is because of the artificial settings. In E3 result, the decomposition results are smooth, which would be helpful for further image interpretation.

In order to examine results quantitatively, the decomposed power contributions of the three methods are shown in Table 1.

Table 1. Scattering power contribution (%).

Patch Name	Method	P_s	P_d	P_v
	F3O	17.45	20.19	62.36
Oriented urban	F3R	23.50	28.78	47.72
	E3	41.68	25.86	32.46

The contribution of volume scattering in oriented urban area is suppressed to 32.46% by proposed method. Both the double bounce and surface scattering contributions are enhanced accounting to 41.68% and 25.86% respectively. Therefore, the proposed method improves the decomposition results and can be more consistent with the real situation of oriented urban areas.

5. Conclusion

This paper integrates the results of eigen-decomposition to solve the volume scattering overestimation in some oriented urban areas and artificial parameter setting problems. Two principal eigenvectors are used to substitute the surface scattering and double bounce scattering models. The decomposed scattering powers are obtained using a constrained linear least-squares method. The proposed method has been verified using an ESAR PolSAR image, and the results show that the proposed method has better performance in oriented urban area.

References

[1] Cloude S R and Pottier E 1996 A review of target decomposition theorems in radar polarimetry *IEEE Trans. Geosci. Remote Sens.* **34** 498–518

[2] Freeman A and Durden S L 1998 A three-component scattering model for polarimetric SAR data *IEEE Trans. Geosci. Remote Sens.* **36** 963–73

[3] Yamaguchi Y, Moriyama T and Ishido M 2005 Four-component scattering model for polarimetric SAR image decomposition *IEEE Trans. Geosci. Remote Sens.* **43** 1699–706

[4] Zhang L, Zou B and Cai H 2008 Multiple-component scattering model for polarimetric SAR image decomposition *IEEE Geosci. Remote Sens. Lett.* **5** 603–7

[5] An W, Cui Y and Yang J 2010 Three-component model-based decomposition for polarimetric SAR data *IEEE Trans. Geosci. Remote Sens.* **48** 2732–9

[6] Chen S, Ohki M and Shimada M 2013 Deorientation effect investigation for model-based decomposition over oriented built-up areas *IEEE Geosci. Remote Sens. Lett.* **10** 273–7

[7] Van Zyl J J, Arii M and Kim Y 2011 Model-based decomposition of polarimetric SAR covariance matrices constrained for nonnegative eigenvalues *IEEE Trans. Geosci. Remote Sens.* **49** 3452–9

[8] Chen S, Wang X and Xiao S 2014 General polarimetric model-based decomposition for coherency matrix *IEEE Trans. Geosci. Remote Sens.* **52** 1843–55

[9] Coleman T F and Li Y 1996 A reflective Newton method for minimizing a quadratic function subject to bounds on some of the variables *SIAM J. Optimization* **6** 1040–58

Visual Perception Based Rate Control Algorithm for HEVC

Zeqi Feng[1], PengYu Liu[1,*] and Kebin Jia[1]

[1]Beijing University of Technology, Beijing, China

* liupengyu@bjut.edu.cn

Abstract. For HEVC, rate control is an indispensably important video coding technology to alleviate the contradiction between video quality and the limited encoding resources during video communication. However, the rate control benchmark algorithm of HEVC ignores subjective visual perception. For key focus regions, bit allocation of LCU is not ideal and subjective quality is unsatisfied. In this paper, a visual perception based rate control algorithm for HEVC is proposed. First bit allocation weight of LCU level is optimized based on the visual perception of luminance and motion to ameliorate video subjective quality. Then λ and QP are adjusted in combination with the bit allocation weight to improve rate distortion performance. Experimental results show that the proposed algorithm reduces average 0.5% BD-BR and maximum 1.09% BD-BR at no cost in bitrate accuracy compared with HEVC (HM15.0). The proposed algorithm devotes to improving video subjective quality under various video applications.

1. Introduction

Since the development of multimedia technology, the categories of video services have rapidly grown in recent years. To meet the growing demand for video applications, Video Coding Experts Group (ITU-T VCEG) and Moving Picture Experts Group (ISO/IEC MPEG), the two major standardization organizations, forming Joint Collaborative Team on Video Coding (JCT-VC), jointly develop the next generation of video coding standard—high efficiency video coding (HEVC) standard [1]. As the inheritance and development of H.264/AVC [2], HEVC pays more attention to high-resolution videos. Under the same subjective visual quality, HEVC saves nearly 50% bitrate [3] in comparison with H.264/AVC.

Rate control is an indispensably important video coding technology to cope with serious bandwidth undulation and the limitation of transmission delay during video communication. It aims to allocate bits reasonably and keep video quality stable. At present, HEVC adopts the rate control algorithm based on R-λ model, which proposed by B Li et al. [4]. This algorithm effectively improves bitrate accuracy as well as the rate distortion performance of reconstructed video and it is added to the test model HM10.0 [5]. Until HM15.0 [6] it still uses the rate control algorithm based on R-λ model. However, it is found that target bit allocation of large coding unit (LCU) level is based on the mean absolute difference (MAD) of each LCU in the above rate control algorithm. MAD does not fully reflect image content characteristics and ignores subjective vision-experience. For key focus regions, bit allocation is not ideal and subjective quality is unsatisfied. It is difficult to meet users' needs. Therefore, various research studies have been carried out to improve video subjective quality. Firstly, some rate control algorithms focused on image structure characteristics. Zhu et al. [7]-[9] joined structural similarity index (SSIM) into bit allocation of LCU level to obtain better rate distortion performance. Nevertheless, SSIM only reflects the structure of LCUs, and still does not consider subjective visual perception. Then, some other rate control algorithms were optimized by overcoming subjective distortion. Kim J [10] introduced compliant perceptual video coding scheme based on just noticeable distortion (JND) model in both transform and pixel domains to improve subjective visual perception. Wei [11] made use of saliency map to calculate average saliency value, and further adaptively adjusted quantization parameter (QP) to determine bits of each LCU. Table 1 shows the rate distortion and time complexity of the above different algorithms by setting HEVC test video sequence "Cactus" as an example, where the rate distortion performance is measured by Bjøntegaard delta bit rate (BD-BR) [12]. As shown in Table 1, the subsequent algorithms spend more time, but demonstrate better rate distortion performance compared with the aforementioned rate control algorithm based on SSIM. Thus the rate control algorithms guided by overcoming subjective distortion deserve to be further studied. Throughout above algorithms, they try to achieve better performance according to the analysis of visual perception characteristics. However, some parameters involved in the analysis process increase computational complexity and inevitably consume extra time in encoding. Hence it is not conducive to real-time applications.

Table 1. Rate distortion performance and time complexity for different algorithms

	[7] vs HM13.0	[10] vs HM13.0	[11] vs HM11.0
BD-BR	-6.1%	-14.4%	-19.9%
ΔT	+3%	+11.25%	-

Therefore, how to effectively use the encoded information to analyse human visual perception characteristics and fully integrate it into rate control of LCU level for HEVC, has become a key issue to improve subjective visual perception in the case of serious bandwidth undulation and the limitation of transmission delay.

The rest of this paper is organized as follows. Section 2 reviews the rate control benchmark scheme in HEVC. Section 3 describes the proposed visual perception based rate control algorithm. In Section 3.1 and 3.2, the luminance weight and the motion weight are defined respectively. In Section 3.3, the bit allocation weight is calculated by the luminance weight and the motion weight. In Section 3.4, a modified method of λ and QP is provided. Finally, extensive experimental results are illustrated and discussed in Section 4 to demonstrate the performance of the proposed rate control algorithm, followed by the conclusion in Section 5.

2. LCU level rate control for HEVC

LCU level rate control has two phases: allocate target bits to each LCU and determine QP by $R-\lambda$ model. Firstly, the target bits of each LCU depend on the target bits of current encoding frame as follows:

$$T_{curLCU} = \frac{T_{curPic} - Bit_{head} - Code_{LCU}}{\sum\limits_{NotCodedLCUs} \omega_i} \times \omega_{curLCU} \tag{1}$$

where T_{curPic} denotes the target bits of current encoding frame. Bit_{head} is the bits of header estimated from the encoding frame. $Code_{LCU}$ is the total bits for the encoded LCUs of the current frame. ω_i and ω_{curLCU} denote the bit allocation weights of the i th LCU and current encoding LCU respectively. ω_{curLCU} is carried out as follow equations (2) (3):

$$MAD = \frac{1}{N_{\text{pixels}}} \sum_{AllPixelsInCurLCU} \left| P_{cur} - P_{pred} \right| \tag{2}$$

where N_{pixels} denotes the number of pixels in the current LCU. P_{cur} and P_{pred} are original pixel luminance value in the current LCU and predicted pixel luminance value in the corresponding location LCU of the reference frame respectively.

$$\omega_{curLCU} = MAD^2 \tag{3}$$

Then, according to $R-\lambda$ model, λ can be given by

$$\lambda = \alpha \times bpp^{\beta} \tag{4}$$

where α and β are related to video content. bpp denotes the target bits per pixel of the current LCU.

Through the relationship between QP and $\ln\lambda$, QP is carried out:

$$QP = 4.2005\ln\lambda + 13.7122 \tag{5}$$

In the above bit allocation process, the bit allocation weight per LCU is calculated on the basis of MAD, but MAD only reflects the luminance difference between pixels without taking human visual characteristics into account, as shown in equations (2) (3). Thus, a visual perception based rate control algorithm is proposed to improve video subjective quality in the following.

3. Proposed LCU level rate control algorithm

Previous studies have shown that human visual system (HVS) is more sensitive to luminance than colour, and motion regions are usually the focus of videos, whose intensity of motion also affects human visual perception [13]. Therefore, according to the perception of luminance and motion based on HVS, the luminance weight and the motion weight are defined, and then they are used to guide LCU level rate control for optimizing it. The flow chart of the proposed LCU level rate control algorithm is illustrated in figure 1. It is noted that the parameters of the algorithm come from encoded information without taking any extra time.

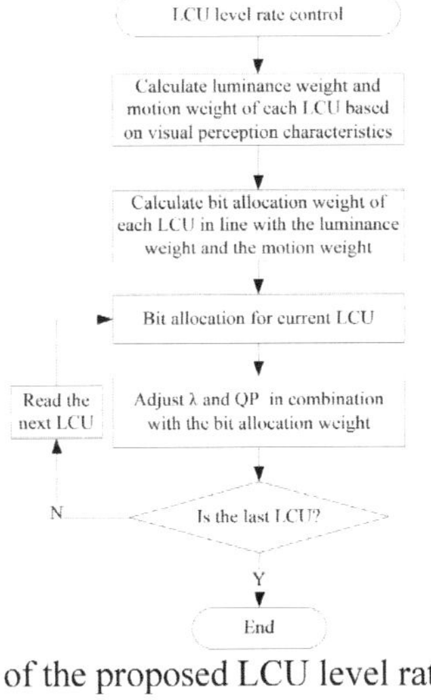

Figure 1. Flow chart of the proposed LCU level rate control algorithm

3.1 The luminance weight

Some researchers found that luminance affects human visual perception: at higher or lower luminance, human visual perception is lower, but in a specific range of luminance, human visual perception is higher. Background luminance sensitivity threshold curve [14] reflects the sensitivity of eyes to different luminance, as shown in figure 2. The X-axis denotes background luminance (the range is 8-bit gray value 0-255). The Y-axis denotes the eyes sensitivity threshold: the smaller the threshold, the higher the sensitivity in the corresponding luminance, and vice versa.

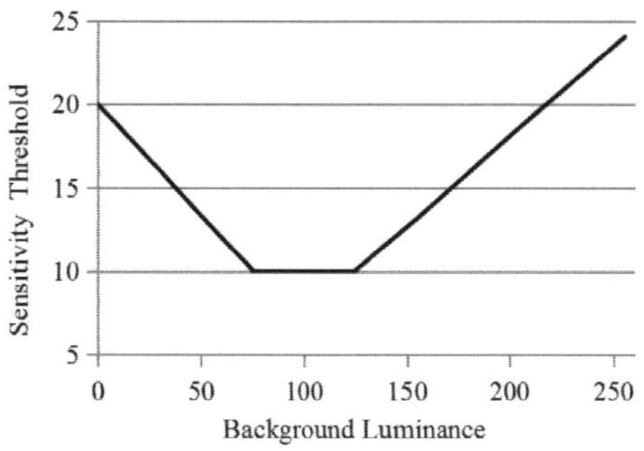

Figure 2. Background luminance sensitivity threshold curve

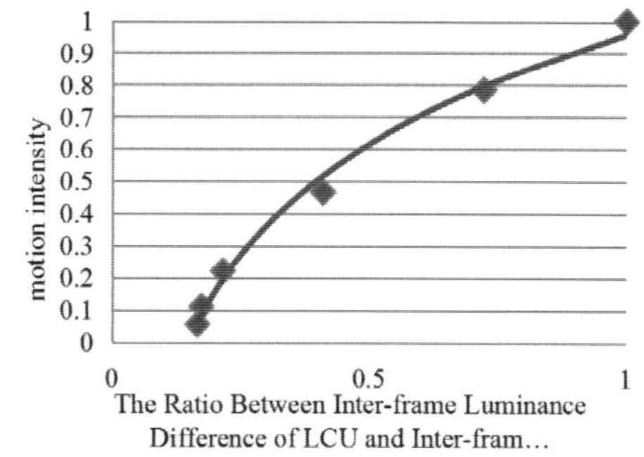

Figure 3. The relationship between motion intensity and inter-frame luminance difference

With reference to the above curve, the luminance weight is set to 1.0 in the luminance from 75 to 125, where human visual perception is higher. And based on this, the luminance weight of the remaining luminance is calculated. The luminance weight of the i th LCU is specified by

$$L_i = \begin{cases} \dfrac{1}{2.0 - 0.0133 \times \overline{Y_i}} & 0 \le \overline{Y_i} < 75 \\ \dfrac{1}{0.0108 \times \overline{Y_i} - 0.3462} & 125 < \overline{Y_i} \le 255 \\ 1.0 & 75 \le \overline{Y_i} \le 125 \end{cases} \tag{6}$$

where $\overline{Y_i}$ specifics the luminance mean of the i th LCU in the current encoding frame. The range of L_i is (0, 1).

3.2 The motion weight

Motion regions of video sequences tend to be more noticeable. In order to describe the intensity of motion in the motion areas, the motion intensity is defined by the ratio between the motion vector magnitude of the current LCU and the motion vector magnitude sum of all LCUs in the current encoding frame, where magnitude of unit motion vector is 64 pixels. Since the motion vector cannot be derived before encoding, it is necessary to find a parameter which could also represent motion information and be deduced before encoding. The inter-frame difference of luminance just satisfies the above two requirements at the same time and it has achieved good performance in moving target detection. Based on this, the relationship is explored between the motion intensity and the inter-frame difference of luminance. Experimental results show that the relationship between the motion intensity F_i and the inter-frame luminance difference of the i th LCU is as follows:

$$F_i \approx \ln \frac{D_i}{D_s} \tag{7}$$

where D_i and D_s are inter-frame luminance difference of the i th LCU and current encoding frame respectively.

As shown in figure 3, when the ratio is greater and the motion is more active, visual attention will be higher. Accordingly, the motion weight M_i is defined as follow equations (8) (9):

$$M_i' = \ln \frac{D_i}{D_s} \tag{8}$$

Normalizing M_i' as:

$$M_i = \frac{1}{1+e^{-M_i'}} = \frac{1}{1+\frac{D_s}{D_i}} \tag{9}$$

where the range of M_i is (0, 1).

3.3 The bit allocation weight

To sum up, the visual perception weight is affected by the luminance weight and the motion weight respectively. Besides, the two weights are also mutually constrained. It is just as interaction effect in statistics. For example, when the background luminance is not within the visual perception range, the motion intensity will no longer affect visual perception. In other words, the luminance weight restricts the influence of the motion weight on the visual perception weight. Based on this, the interaction term between the motion weight M and the luminance weight L is taken as the basis for bit allocation. The interaction term is the product of the two weights. The bit allocation weight of the i th LCU is carried out:

$$\omega_i = L_i \times M_i \tag{10}$$

Normalizing ω_i as:

$$\omega_i' = \frac{\omega_i}{\sum_{i=1}^{N} \omega_i} \tag{11}$$

where N is the number of LCUs in the current encoding frame. The range of ω_i' is (0, 1).

3.4 Adjust λ and QP

For HEVC, λ not only determines QP, but also affects other parameters which being closely related to rate distortion performance, such as coding unit (CU) partition mode, transformation unit (TU) mode and prediction unit (PU) mode [4]. To improve rate distortion performance, λ of the current LCU is modified in combination with the bit allocation weight as follow:

$$\lambda_{new} = \lambda \times K_\lambda \tag{12}$$

The K_λ used by

$$K_\lambda = \begin{cases} 1.0 & I\ frame \\ 1.2 \times \omega_i' + 0.3 & others \end{cases} \tag{13}$$

I frame is performed to intra-frame coding as a reference frame for subsequent encoding frames, which is not processed.

Due to λ adjusted, QP and bitrate accuracy inevitably changes with λ in LCU level. If also adopts $R-\lambda$ model in HEVC, QP_{HEVC} will be calculated as:

$$QP_{HEVC} = 4.2005 \ln \lambda_{new} + 13.7122 = QP_{Target} + 4.2005 \ln K_\lambda \tag{14}$$

where QP_{Target} is the quantization parameter corresponding to target bitrates of the current LCU, which is adopted as final QP to ensure bitrate accuracy.

4. Experimental result

To verify the performance of the proposed algorithm, it is tested on HEVC test model version 15.0 (HM15.0). The performance is measured in four aspects: rate distortion performance, bitrate accuracy, bitrate fluctuation and video subjective quality. Test video sequences include "BQSquare", "BlowingBubbles", "BasketballDrill", "PartyScene", "FourPeople". For each sequence, the specific encoder configurations are as follow: four different fixed QPs as 25, 30, 35 and 40, low delay (LD) condition and LCU size as 64×64. Other configurations reference the common test conditions of HEVC standardization.

IOP Conf. Series: Journal of Physics: Conf. Series **960** (2018) 012041 doi:10.1088/1742-6596/960/1/012041

4.1 Rate distortion performance and bitrate accuracy

The BD-BR is used to evaluate rate distortion performance of the proposed algorithm and bitrate accuracy is measured by

$$E = \frac{\left| bits_{RC} - bits_{Target} \right|}{bits_{Target}} \times 100\% \quad (15)$$

where E denotes bitrate error, $bits_{RC}$ is actual bits, $bits_{Target}$ is the target bits.

Table 2 shows the comparison between the proposed algorithm and HM15.0 rate control algorithm in rate distortion performance and bitrate accuracy. As shown in table 2, for various resolution sequences, the proposed algorithm reduces 0.5% BD-BR with 0.06% bitrate accuracy drop. For the sequences with slightly undulation luminance and motion object distinguishing from background, the encoder performance is significantly improved. To further evaluate the performance of the proposed algorithm, figure 4 gives rate distortion curves of "BasketballDrill". At the same bitrate, the peak signal to noise ratio (PSNR) of the proposed algorithm is higher than the PSNR of HM15.0 rate control algorithm. Based on the analysis above, compared with rate control benchmark algorithm in HM15.0, the proposed algorithm could achieve better rate distortion performance at no cost in bitrate accuracy.

Table 2. Rate distortion performance and bitrate accuracy between the proposed algorithm and HM15.0 rate control algorithm

Sequence	ΔKbps(HM15.0)	ΔKbps(Proposed)	BD-BR
BQSquare	0.14%	0.13%	-0.80%
BlowingBubbles	0.09%	0.12%	-0.44%
BasketballDrill	0.01%	0.01%	-1.09%
PartyScene	0.16%	0.17%	-0.51%
FourPeople	4.30%	4.59%	0.32%
Average	0.94%	1.00%	-0.50%

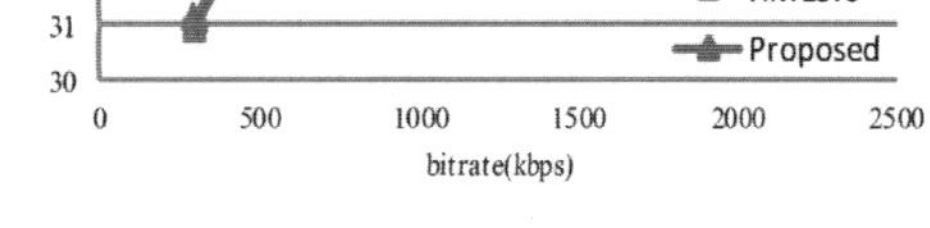

Figure 4. Rate distortion curves of "BasketballDrill"

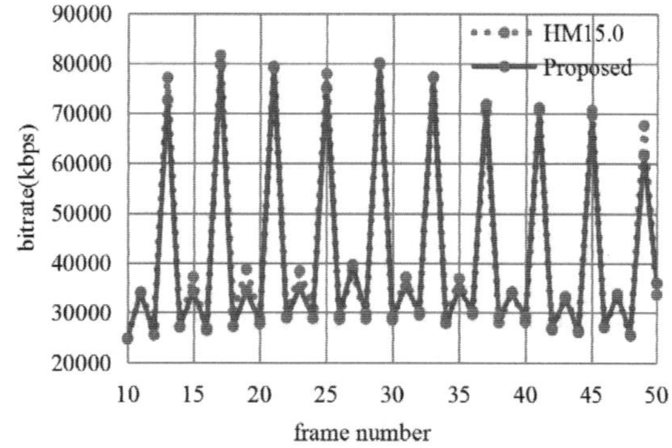

Figure 5. Bitrate fluctuation of "BasketballDrill" (QP=25)

4.2 Bitrate fluctuation

To validate the effectiveness of the proposed algorithm, bitrate fluctuation is measured in HM15.0 rate control algorithm and the proposed algorithm. Figure 5 shows the bitrate curves of different algorithms by setting "BasketballDrill" (QP=25) as an example. Compared with HM15.0 rate control algorithm, the bitrate curve of the proposed algorithm is smoother. It is beneficial to real-time video transmission.

4.3 Video subjective quality

To evaluate the performance of the proposed algorithm, table 3 shows the reconstructed video subjective quality of the two rate control algorithms for the 15th frame of "BasketballDrill" (QP=25) and the 17th frame of "PartyScene" (QP=30). The two algorithms have little difference in PSNR of these frames. As shown in table 3, compared with HM15.0 rate control algorithm, the details of the reconstructed frame are richer. For the sequence "BasketballDrill", the number on the athlete's clothing is closer to original frame at the proposed algorithm. And for the sequence "PartyScene", the girl's facial contours are clearer at the proposed algorithm. On top of that, the bit allocation of the proposed algorithm is more consistent with visual characteristics. Therefore, the proposed algorithm could achieve better video subjective quality compared with HM15.0 rate control algorithm.

Table 3. Video subjective quality between the proposed algorithm and HM15.0 rate control algorithm

	Sequence	Original Frame	HM15.0	Proposed
Entire Frame and Bit Allocation	BasketballDrill		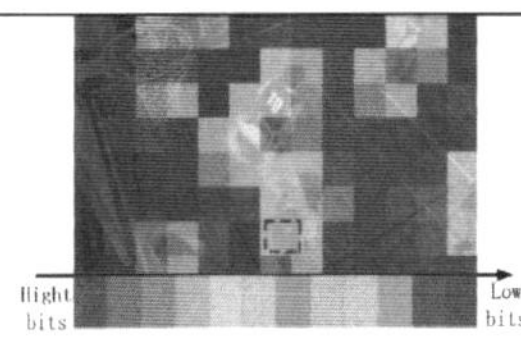 PSNR 38.6613dB	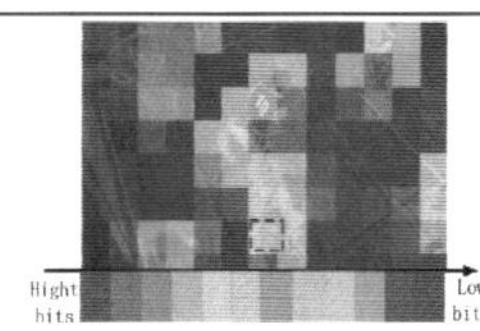 PSNR 38.6716dB
Local Amplification		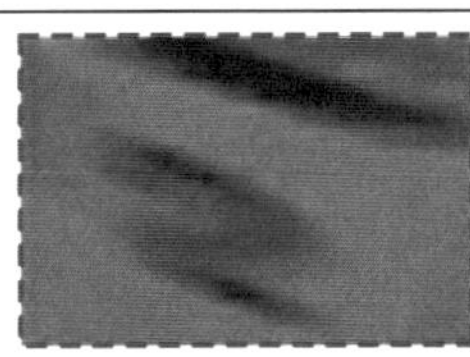	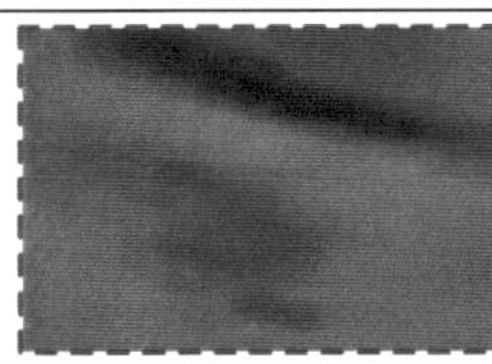	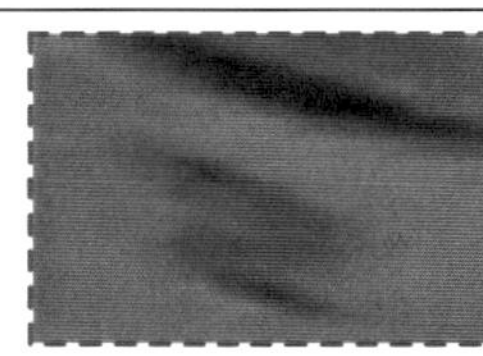
Entire Frame and Bit Allocation	PartyScene		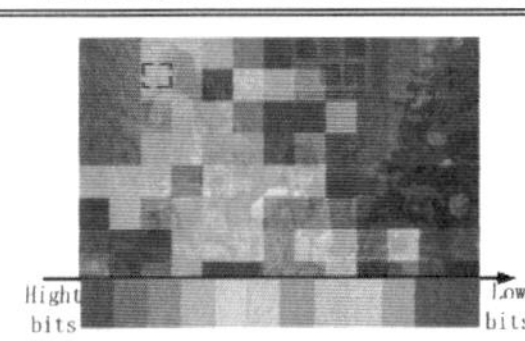 PSNR 31.1830dB	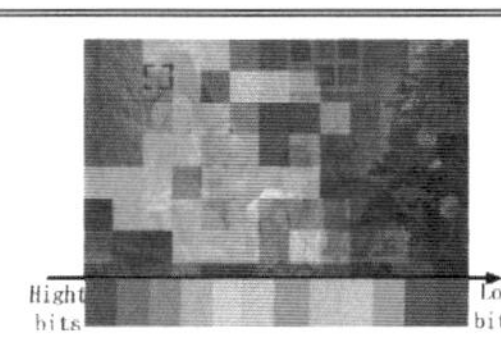 PSNR 31.1928dB
Local Amplification		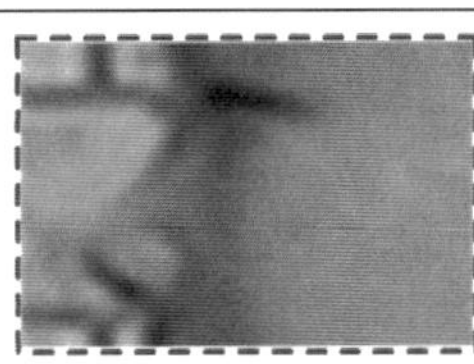	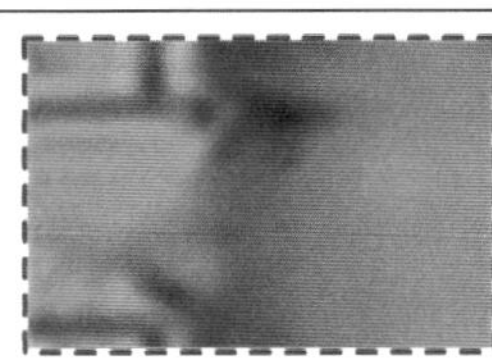	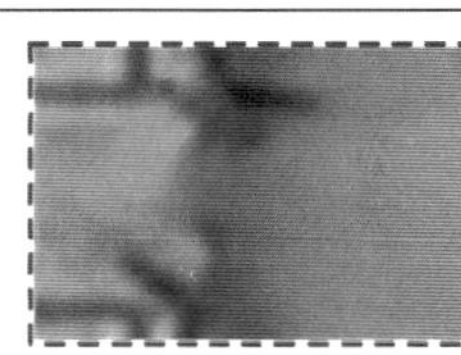

5. Conclusions

In this paper, a visual perception based rate control algorithm for HEVC is proposed. In order to ameliorate video subjective quality, first luminance weight and motion weight are defined based on the visual perception of luminance and motion, and then bit allocation weight of LCU level is calculated in line with luminance weight and motion weight. To improve rate distortion performance, λ and QP are adjusted based on the bit allocation weight. Experimental results demonstrate that the proposed algorithm improves rate distortion performance at no cost in bitrate accuracy. The proposed algorithm yields 0.5% average reduction and 1.09% maximum reduction in BD-BR, which alleviates the contradiction between the reconstructed video quality and the encoding resources. The experiments also show that the bit allocation of the proposed algorithm is more consistent with visual characteristics, which polishes up the reconstructed video subjective quality. Note that the parameters involved in the proposed algorithm are derived from encoded information without introducing additional computation. The proposed algorithm effectively ameliorates the performance of HEVC encoder.

Acknowledgments
This paper is supported by the Project for the National Natural Science Foundation of China under Grant No. 61672064, the Beijing Natural Science Foundation under Grant No. KZ201610005007, 4172001, the China Postdoctoral Science Foundation under Grants No. 2015M580029, 2016T90022, the Beijing Postdoctoral Research Foundation under Grant No. 2015ZZ-23 and Project supported by Chaoyang District of Beijing Postdoctoral Research Foundation under Grant No. 2016ZZ- 01-15.

References

[1] Sullivan G J, Ohm J R, Han W J, et al. Overview of the high efficiency video coding (HEVC) standard. *Circuits and Systems for Video Technology, IEEE Transactions on,* 2012, **22**(12): 1649-1668.

[2] Wiegand T, Sullivan G J, Bjøntegaard G, et al. Overview of the H.264/AVC video coding standard. *Circuits and Systems for Video Technology, IEEE Transactions on,* 2003, **13**(7): 560-576.

[3] Li B, Sullivan G J, Xu J. Comparison of compression performance of HEVC draft 7 with AVC high profile. *JCTVC-J0236 r1, 10th JCT-VC Meeting*. 2012.

[4] Li B, Li H, Li L, et al. Rate control by R-lambda model for HEVC. *Jt. Collab. Team Video Coding (JCT-VC) of ITU-T SG, K0103*, 2012.

[5] "High Efficiency Video Coding (HEVC) Test Model 10 (HM10) Encoder Description" *document Joint Collaborative Team on Video Coding of ISO/IEC and ITU-T, JCTVC-L1002,* Geneva, CH, Jan. 2013.

[6] "High Efficiency Video Coding (HEVC) Test Model 15 (HM15) Encoder Description" *document Joint Collaborative Team on Video Coding of ISO/IEC and ITU-T, JCTVC- Q1002*, Valencia, ES, 27 Mar. 2014.

[7] Zhu T Z, Yu M, Jiang G Y, et al. SSIM-based rate distortion optimization for HEVC intra coding. *Journal of Optoelectronics Laser*, 2014, **25**(12):2362-2370.

[8] Zhao H, Xie W, Zhang Y, et al. An SSIM-motivated LCU-level rate control algorithm for HEVC. *Picture Coding Symposium. IEEE*, 2013:85-88.

[9] Gao W, Kwong S, Zhou Y, et al. SSIM-Based Game Theory Approach for Rate-Distortion Optimized Intra Frame CTU-Level Bit Allocation. *IEEE Transactions on Multimedia*, 2016, **18**(6):988-999.

[10] Kim J, Bae S H, Kim M. An HEVC-Compliant Perceptual Video Coding Scheme Based on JND Models for Variable Block-Sized Transform Kernels. *IEEE Transactions on Circuits & Systems for Video Technology*, 2015, **25**(11):1786-1800.

[11] Wei H, Zhou X, Zhou W, et al. Visual saliency based perceptual video coding in HEVC. *IEEE International Symposium on Circuits and Systems*. IEEE, 2016:2547-2550.

[12] Bjontegaard, Gisle. "Calculation of average PSNR difference between RD-curves." 2001.

[13] Fei M Y. Peng Z J, Li C H, Chen F, Yu M, Jiang G Y. Human perception features based rate distortion optimization for HEVC. *Journal of Image and Graphics*, 2015, **20**(7): 857-864.

[14] Yu Like, Dai Feng, Zhang Yongdong. Novel rate distortion optimization strategy based on perceptual properties of texture and luminance. *Journal of Image and Graphics*, 2012, **17**(1):54-61.

A Method of Character Detection and Segmentation for Highway Guide Signs

Jiawei Xu[1,*] and Chongyang Zhang[1]

[1]School of Computer Science and Engineering, Nanjing University of Science and Technology, Nanjing, China

*2433041042@qq.com

Abstract. In this paper, a method of character detection and segmentation for highway signs in China is proposed. It consists of four steps. Firstly,the highway sign area is detectedby colour and geometric features, andthe possible character region is obtained by multi-level projection strategy. Secondly, pseudo target character region is removed by local binary patterns (LBP) feature.Thirdly,convolutional neural network (CNN)is used to classify target regions. Finally,adaptive projection strategies are used to segment characters strings. Experimental results indicate that the proposed method achieves new state-of-the-art results.

1. Introduction

The acquisition of visual information of the road is a part of the study of unmanned vehicles. The method proposed in this paper is to detect and segment the characters of Chinese highway signs. Domestic and foreign researchers have proposed some methods for the acquisition of visual information of the road. The extraction of the traffic sign area of interest can be achieved by a combination of colour features, shape features and texture features [1,2,3,4]. For the acquisition of character information in natural scene, there are methods based on connected component [5], texture feature [6] and both of them [7]. Some articles have proposed methods for locating characters on traffic signs in Chinese road scenes. Gu et al. [4] propose a method to obtain rectangular traffic sign regions by colour and shape features, and locate Chinese characters by one vertical projection and one horizontal projection on the binary image. Liu et al. [8] propose a method to get the binary image of the road scene by improving the SCW algorithm. By analyzing the connected components, the rules are defined, and the signs and characters are located. Ni et al. [9] consider that the traffic sign area is obtained by combining shape feature and classifier. Then the image is binarized by K-means colour clustering, and the connected components are analyzed to segment characters. Chen et al. [10] use the colour feature and the HOG feature combined with the support vector machine to obtain the traffic area of interest, and obtains the characters and symbols by analyzing the geometric and location information of the connected components. For Chinese traffic signs with characters, a Chinese character may consist of multiple connected components, and the information of the layout is complicated. For these characteristics, the above mentioned methods have some limitations. In this paper, our method aims to overcome the difficulties above and obtain the character information of the highway signs in China. The character areas we want to detect include highway number, distance character, direction symbol, and Chinese place name. Firstly, the areas of highway sign are obtained by colour segmentation and filtering based on geometric features, and the tilted signs need to be corrected. Then a multi-level projection strategy is used to obtain character regions on the binary image. Next, the LBP [11,12] features of character regions are extracted, and are compressed by the model of piecewise function. Target character regions and pseudo target character regions are classified by the AdaBoost classifier [13,14]. Finally, the target character regions are classified by CNN [15,16,17,18,19], and four different character regions such as highway number, distance character, direction symbol and Chinese place name are obtained. Different types of characters are segmented by adaptive projection strategies. The framework of our method is shown in Figure 1. Section 2, 3, 4 introduce the details of the method. Section 5 introduces the experiment. Conclusions are made in section 6.

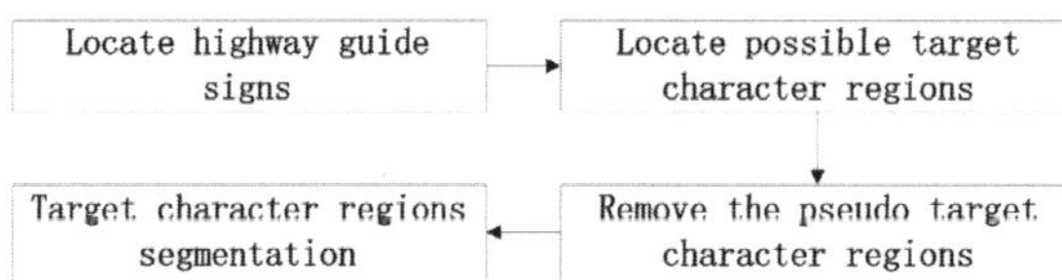

Figure 1. Algorithm flowchart

2. Location of Highway Guide Sign

A highway guide sign is a rectangular panel of dark green, which can be roughly detected by colour segmentation. In order to avoid the influence of light and other factors, a robust colour segmentation method is proposed.

Pixel values of the image in the RGB colour space and the HSV colour space are obtained. Then the brightness of the image is divided into three intervals: v_1 , v_2 , and v_3 . In the three different intervals, the separation thresholds of the colour space are obtained statistically, and the dark green regions are obtained by the thresholds of the two colour spaces. After colour segmentation, part noisesare removed by area feature. Then the operations of morphological dilation, cavity filling, and morphological corrosion are performed. According to the aspectratio of the region and the area ratio of the connected component to the external rectangle, noisesare filtered out. Finally, the regions of highway guide signs can be obtained. Figure 2 shows some test results.

Figure 2. The result of highway guide signs detection in some scene of highway

3. Detection of target character regions

3.1. Multi-level projection strategy

A multi-level projection strategy is proposed to obtain the possible character regions from the highway guide signs. First of all, a binary image of the guide sign is obtained by binarization, and corner points of the sign are found by Hough transform on the binary image. According to the corner points, affine transformation can be used to correct the tilt sign. By analysis of the layout, we can see that the layout is very complex, and a Chinese character may consist of multiple connected components. In order not to destroy the integrity of characters, a plurality of combined projection operations can be carried out, and the complicated layout can be completely cutoff.Each combined projection operation consists of a vertical projection operation and a horizontalprojectionoperation.Projection is done on a binary image. Vertical projection operation means the accumulation of pixels which consists of connected components, and is applied to each row of image. A row of the image, of which projection value is 0, can be cut off. Similarly, Horizontalprojection operation is for image columns. After a combined projection operation, it is necessary to judge whether the obtained regions satisfy the condition for continued projection.If the condition is satisfied, a combined projection operation will continues until complete segmentation is performed. By analysis of a large number of layouts,regions that meet the following three situations need a combined projection operation.

- A region accounts for a larger proportion of the layout.
- Aregion is vertical strip.
- The upper part of the regionconsists of Chinese characters, and the lower part of the regionconsists of English alphabets. The Chinese characters are usually a name of a place, and the English alphabets are Pinyin or English name of the place. This is a fixed pattern.

After cutting, individual components are merged into character regions by character lines and layout rules. As shown in Figure 3, the target regions include highway numbers, distance characters, direction symbols, and Chinese place names in the white font. Others are pseudo target regions.

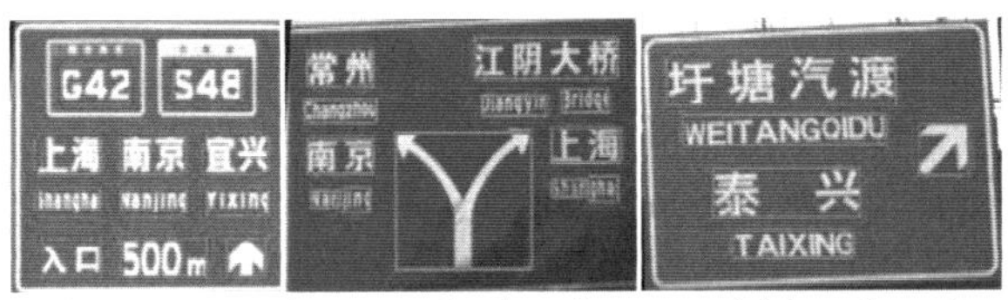

Figure 3. The character regionsareobtainedby a multi-level projection strategy

3.2. LBP feature extraction and feature dimension compression

LBP feature [12] is an operator used to describe the local texture feature of an image, which has the advantages of rotational invariance and grayscale invariance. The original LBP operator defines a 3*3 square window and takes the value of the center pixel of the window as the threshold. Comparing the adjacent pixel values to the threshold, if the pixel value is greater than the threshold, then the pixel position is set to 1, otherwise 0.Thus, you can get an 8 bit binary number and convert it to decimal number, which is the LBP value of the central pixel. As shown in formula (1).

$$LBP(x,y)=\sum_{n}^{n-1}2^{n}S(v_n-v_c),\ S(x)=\begin{cases}1, if(x\geq 0)\\ 0, if(x<0)\end{cases} \tag{1}$$

$S(x,y)$ is a symbolic function, and (x,y) is the central pixel of the window. v_n is the pixel value of neighborhood pixels, and v_c is the center pixel value.A circular neighborhood is used instead of a square neighborhood, and a circular field with a radius of R is allowed to have any number of pixels, which can be adapted to different sizes of texture features.For the central pixel, the points of its neighborhood can be computed by the following formula (2). R is the radius of the circle region, and N is the number of pixels in the neighborhood.

$$\begin{cases}x_n = x_c + R\cos(\frac{2\pi n}{N})\\ y_n = y_c - R\sin(\frac{2\pi n}{N})\end{cases} \tag{2}$$

$$f(x,y)\approx\begin{bmatrix}1-x & x\end{bmatrix}\begin{bmatrix}f(0,0) & f(0,1)\\ f(1,0) & f(1,1)\end{bmatrix}\begin{bmatrix}1-y\\ y\end{bmatrix} \tag{3}$$

If the calculated coordinates and pixel coordinates are biased, the bilinear interpolation is used for approximate processing, such as formula (3).Figure 4 is the image of the character regionsandtheir LBP.

Figure 4. Images of some character regions and their LBP images: G42, Zhongshan East Road, 500m, arrow

The LBP feature of an image can be obtained as follows. The image is divided into some blocks which size is n*n, and the LBP of each block is calculated. Then we calculate the frequency of each value of the LBP, build the histogram ofLBP of theblock, and normalize the histogram.Finally, the histograms of each block are concatenated into a feature vector, which is the LBP texture feature vector of the image.In order to reduce the dimension of the feature, we proposes a LBP feature based on the reduced modelof the piecewise function (RdLBP), as shown in formula (4).

$$f(LBP(x,y)) = \begin{cases} 0, LBP(x,y) \in [0,n), \\ 1, LBP(x,y) \in [n,2n), \\ 2, LBP(x,y) \in [2n,3n), \\ \text{and so on.} \end{cases} \tag{4}$$

3.3. AdaBoost classifier

The AdaBoost algorithm [13] is a way to generate a strong classifier from a series of weak classifiers.Suppose there areafeature vector set X and alabel set Y , then each feature vector corresponds to a label.TheAdaBoost algorithm generates a weak classifier set K after N iterations. Each weak classifier can decide which tag the feature vector belongs to.After the training of $n-1$ iterations, $n-1$ weak classifiers are combined to produce a strong classifier. Then a new strong classifier can be generated by combining the n-th week classifier with the previous strong classifier, such as formula (5). The weight α_n and the weak classifier k_n need to make the strong classifier F_n better than the F_{n-1} classifier.

$$F_n = F_{n-1} + \alpha_n k_n \tag{5}$$

$$F_n = sign(\sum_{n=1}^{N} k_n(x)) \tag{6}$$

The AdaBoost algorithms include Discrete AdaBoost, Real AdaBoost, LogitBoost and Gentle AdaBoost. And the Gentle AdaBoost is used here. The weak classifier k_n is obtained by regression fitting of weight-based weighted least square method from label to feature vector.After normalizing the weights, the final strong classifier is shown in formula (6). Figure 5 is a graph showing the result of removing the pseudo targets in Figure 3.

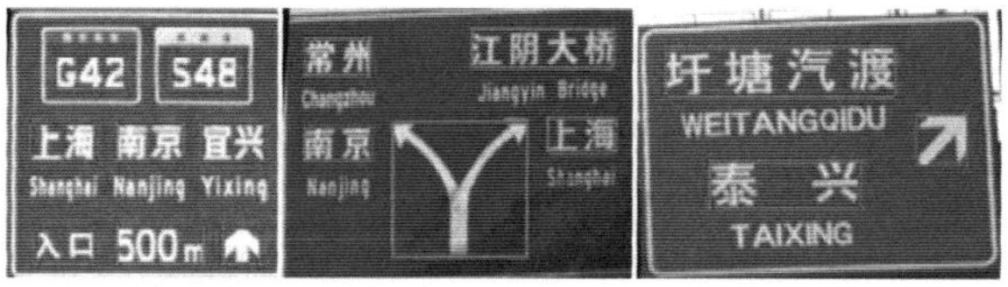

Figure 5. Highway guidesigns after removing pseudo targets

4. Character segmentation

IOP Conf. Series: Journal of Physics: Conf. Series **960** (2018) 012042 doi:10.1088/1742-6596/960/1/012042

There are four types of target character regions: Highway number, distance character, direction symbol and Chinese place name. In order to achieve better cutting effect of characters, the target character regions are classified by CNN, and then adaptive projection strategies are adopted.

Table 1. The network hierarchy of Caffenet

Conv1	Conv2	Conv3	Conv4	Conv5	Fc6	Fc7	Fc8
data	norm1	norm2	conv3	conv4	pool5	fc6	fc7
↓Conv	↓Conv	↓Conv	↓Conv	↓Conv	↓InnerProduct	↓InnerProduct	↓InnerProduct
conv1	conv2	conv3	conv4	conv5	fc6	fc7	fc8
conv1	conv2	conv3	conv4	conv5	fc6	fc7	fc8
↓ReLU	↓ReLU	↓ReLU	↓ReLU	↓ReLU	↓ReLU	↓ReLU	↓Softmax
conv1	conv2	conv3	conv4	conv5	fc6	fc7	prob
conv1	conv2			conv5	fc6	fc7	
↓Pool	↓Pool			↓Pool	↓Dropout	↓Dropout	
pool1	pool2			pool5	fc6	fc7	
pool1	pool2						
↓LRN	↓LRN						
norm1	norm2						

4.1. Classification based on CNN

CNN is a multi-layer artificial neural network [18], which usually consists of feature extraction layer and feature mapping layer. The input of each neuron is connected to the local accepted domain of the previous layer and extracts the feature of the region. Once the local feature is extracted, the positional relationship between it and other features is determined. Each computational layer of the network consists of multiple feature mappings, each of which is a plane, and the weights of all the neurons on the plane are equal. Each convolutional layer follows a computational layer for local averaging and secondary extraction, and this unique feature extraction structure reduces the feature resolution. Caffenet network structure [15] is adopted here. It consists of eight weighted layers. The first five layers are convolutional layer, and the latter three are inner product layer. The structure of the network is shown in Table 1.

Convolution is an operation of analytic mathematics. Assuming that $f(x)$ and $g(x)$ are integrable functions [17] on the set R, the formula (7) defines the convolution of function $f(x)$ and $g(x)$.

$$F = \int_{-\infty}^{+\infty} f(a) \cdot g(x-a)da \tag{7}$$

Convolution of different positions of image can obtain different eigenvalues. The convolutional layer is the core layer of convolutional neural network, and convolution operation is performed by parameter setting. The pooling layer is basically the same as the convolutional layer, with the aim of reducing the data dimension, similar to downsampling. The LRN layer normalizes the local region of an input and achieves the effect of lateral inhibition. The activation layer activates the input elements, using the ReLU activation function [19].

$$F(x_i) = \begin{bmatrix} p(y_1 \mid x_i) \\ p(y_2 \mid x_i) \\ \cdots \\ p(y_k \mid x_i) \end{bmatrix} = \frac{1}{\sum_{j=1}^{k} e^{w_j^T x_i}} \begin{bmatrix} e^{w_1^T x_i} \\ e^{w_2^T x_i} \\ \cdots \\ e^{w_k^T x_i} \end{bmatrix} \tag{8}$$

The innerproductlayer is actually also a convolutional layer, but its convolutional kernel is the same size as the original data.Dropout is a strategyof preventing over-fitting, which allows the network to hide the weight of the nodes which belong to default hidden layer randomly.The ultimate goal of the Softmax layer is to obtain the probability likelihood values for each class, and the output function [17] follows the formula (8).

$P(y_i|x_i)$ indicates the probability that x_i belongs to y_i , and w is parameter of the model. $\frac{1}{\sum_{j=1}^{k} e^{w_j^T x_i}}$ normalizes the probability distribution.

The model is obtained bytraining of thenetwork.Firstly, the corresponding values of output and loss function are calculated by forward propagation,and then the gradient of loss is calculated by back propagation, and the new weights and parameters are obtained.Finally, the appropriate model is obtained.In the prediction, the original image is fed into a convolutional neural network, and the category and probability value are obtained.

4.2. Different types of character regions segmentation

After getting the category of character area, adaptive projection strategies are adopted.A Chinese character may contain more than one connected component. When cutting Chinese place names, it is necessary to consider the aspect ratio of Chinese characters, and cutting is based on the horizontal projection strategy. Generally speaking, the width to height ratio of Chinese characters is about equal to 1.Distance characters and highway numbers contain both English letters and numbers, in which a single character contains only one connected component, and the aspect ratio of different characters fluctuates considerably. We can directly cut each character by horizontal projection.A direction symbol contains only one symbol, and does not need to be cut again.After cutting all the characters,the non-character parts of the region above and below are removed by vertical projection. The horizontal projection and vertical projection here are the same as those in front.Figure6 shows an instance.

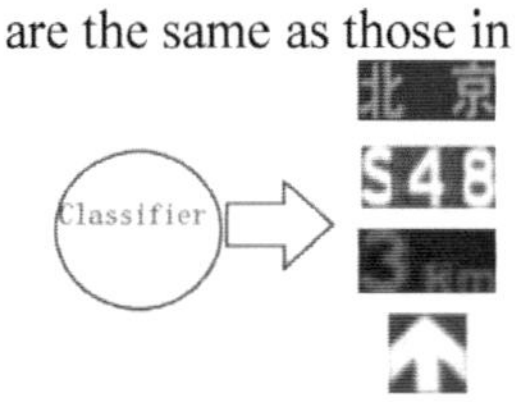

Figure 6. An instance of cutting different character regions: Beijing, S48, 3km, arrow

5. Experiment

5.1. Experimental condition

The data used in the experiment is photos of actual highway scene collected by unmanned vehicles, and the resolutions of these photos are 2448*2050.Pictures are collected in the morning and afternoon. In the morning, from Nanjing to Wuxi, the weather is fine. In the afternoon, from Wuxi to Nanjing, the weather is cloudy.Throughout the process, the speed is about 90km/h, and the direction and intensity of illumination vary. The experimental software condition is Windows 7+vs2013+opencv 2.4.10, and the hardware condition is Intel Core i5-2450M CPU and 4GB of RAM.

5.2. Experimental results

In this experiment, Precision andRecall are used to describe the experimental results, as shown in the formula (9).

$$\begin{cases} Precision = \frac{TP}{TP+FP} \\ Recall = \frac{TP}{TP+FN} \end{cases} \tag{9}$$

TP is the correct detected region, FP is the wrong detected region, and FN is the correct region for missed detection. As to the detection of character regions, this method combined with machine learning can detect more character region information than [4,8,9,10]. The correct regions include the highway number regions, the distance character regions, the direction symbol regions and the Chinese place name regions. The wrong regions are some other regions. The experimental results are shown in Table 2. Firstly, the possible character regions are obtained by multi-level projection. Then each region is normalized to 60*60, and each region is divided into some blocks of which size is 10*10, and the LBP feature of each region is obtained statistically. Next the dimension of the LBP feature is reduced, and the size of the dimensionality reduction is 8. The dimension of the original LBP feature is 9216, and the dimension of the LBP feature is 1152 after reduction. Finally, the AdaBoost classifier is used to remove the pseudo target character regions.

Table 2. Experiment of detecting the target character regions

Method	Number of character regions	Accuracy (%)	Recall (%)	Time (ms)
multi-level projection	1498	69.69	97.46	44
multi-level projection +LBP+AdaBoost	1498	96.30	97.19	45
multi-level projection +RdLBP+AdaBoost	1498	96.48	97.19	35

Experimental results show the detection effect is not very good when the pseudo targets are removed only by the geometric feature and not by the machine learning method. For example, some other character regions such as low discrimination regions of Pinyin will affect the detection. It can be seen that the use of LBP combined with AdaBoost is very effective. At the same time, the experimental results before and after dimension reduction are similar, but the time of execution is obviously shortened after dimension reduction.

For the experiment of character segmentation, the research[20] proposed a method based on connected component and a method based on projection. Compared with the segmentation method in this paper, the experimental results are shown in Table 3.

In the experiment, if each character in character region is correctly segmented, the region is considered to be correctly segmented. The experimental results show that the segmentation method proposed in this paper is more accurate than the other two methods.

Table 3. Experiment of character segmentation

Method	Accuracy (%)
A segmentation method based on connected component	81.02
A segmentation method based on projection	94.69
The segmentation method of this paper	97.76

6. Conclusion

In this paper, the method of detecting and segmenting the character for highway guide sign is effective. We can detect the useful characters on the sign, including the place name, distance, direction and so on.This provides visual information for unmanned vehicles and supports double validation of visual information and satellite navigation information on expressways. However, there are many factors affecting the quality of images, and the environment of the road changes drastically. Further research is needed to do to improve the adaptability and robustness of the algorithm.And after getting the characters, we need to recognize the characters and understand the semantics, which needs further research.

Acknowledgements

We acknowledge the support of the National Science and Technology Major Project of China, under

Contracts 2015ZX01041101.

References

[1] S. Salti, A.Petrelli, F. Tombari, N. Fioraio, and L. D. Stefano. Traffic sign detection via interest region extraction. *Pattern Recognition*, 48(4):1039-1049, 2015.
[2] W. Chen, C. Li, and Z. Wang. Road Traffic Sign Detection Using Color and shape. *Journal of Xiamen University(Natural Science)*, 46(5):635-640, 2007.
[3] Y. Chen, F. Ye, F. Xiao, and Q. Li. Detection and Recognition of Traffic Signs Based on HSV Vision Model and Shape Features. *J. Jianghan Univ.(Nat. Sci. Ed.)*, 44(2):119-125, 2016.
[4] M. Gu, Z.Cai, and X. Ren. Detection algorithm of traffic sign with character in nature scenes. *Journal of Central South University(Science and Technology)*, 44(5): 120-127, 2013.
[5] S. Tian, S. Lu, B. Su, and C. Tan. Scene Text Segmentation with Multi-level Maximally Stable Extremal Regions. *In Proceedings of the 22nd International Conference on Pattern Recognition*, pages 2703-2708, 2014.
[6] J. Fabrizio, B.Marcotegui, and M. Cord. Text detection in street level images. *Pattern Analysis and Applications*,16(4):519-533, 2014.
[7] G. Zhou, Y. Liu, Z. Tian, and Y. Su. A New Hybrid Method to Detect Text in Natural Scene. *In Proceedings of the 18th IEEE International Conference on Image Processing*, pages 2605-2608, 2011.
[8] F. Liu, and H. Sha. An Algorithm for the Detection of Text on Road Signs Based on Improved SCW. *Journal of Image and Graphics*, 12(10):1869-1872, 2007.
[9] C. Ni. Study on Traffic Sign Text Detection and Recognition Based on Video. *Beijing Jiaotong University*, 2015.
[10] Y. Chen. Study on Traffic Panel Layout Understanding from Natural Scenes. *Beijing Jiaotong University*, 2016.
[11] J. Ren, X. Jiang, and J. Yuan. Learning LBP structure by maximizing the conditional mutual information. *Pattern Recognition*, 48(10): 3180-3190, 2015.
[12] M.Ghorbani, A. T.Targhi, and M. M.Dehshibi. HOG and LBP: Towards a robust face recognition system. *InInternational Conference on Digital Information Management*, pages 138-141, 2016.
[13] C. Yan, Y. Wang, J. Li, and Z. Zhang. Theory deduction of AdaBoost classification. *Journal of Southeast University(Natural Science Edition)*, 41(4):701-705, 2011.
[14] D. Yang, Z. Zhao, and Q. Bai. Face detection based on AdaBoost and Space Support Vector Domain classifier. *Journal of Circuits and Systems*, 18(2):321-325, 2013.
[15] Y. Jia, E. Shelhamber, J. Donahue, S. Karayev, J. Long, R. Girshick, et al. Caffe: Convolutional architecture for fast feature embedding. *In EprintArxiv,* pages 675-678, 2014.
[16] R. Girshick. Fast R-CNN. *In IEEE International Conference on Computer Vision*, pages 1440-1448, 2015.
[17] F. Li. Convolutional Neural Network and Its Applications. *Dalian University of Technology*, 2014.
[18] H. Lu, and Q. Zhang. Applications of Deep Convolutional Neural Network in Computer Vision. *Journal of Data Acquisition and Processing*, 31(1):1-17, 2016.
[19] A.Krizhevsky, I.Sutskever, and G. E. Hinton. ImageNet classification with deep convolutional neural networks.International Conference on Neural Information Processing Systems,25(2):1097-1105, 2012.
[20] Q. Weng. Research on Optimized Preprocessing and Character Segmentation of Irregular Vehicle Plate. *Zhejiang University*, 2014.

Robust Frequency Invariant Beamforming with Low Sidelobe for Speech Enhancement

Yiting Zhu[1,*] and Xiang Pan[1]

[1]College of Information Science and Electronic Engineering, Zhejiang University, Zheda Road, No.38,310000, Hangzhou, People's Republic of China

*Email: zhuyiting@zju.edu.cn

Abstract. Frequency invariant beamformers (FIBs) are widely used in speech enhancement and source localization. There are two traditional optimization methods for FIB design. The first one is convex optimization, which is simple but the frequency invariant characteristic of the beam pattern is poor with respect to frequency band of five octaves. The least squares (LS) approach using spatial response variation (SRV) constraint is another optimization method. Although, it can provide good frequency invariant property, it usually couldn't be used in speech enhancement for its lack of weight norm constraint which is related to the robustness of a beamformer. In this paper, a robust wideband beamforming method with a constant beamwidth is proposed. The frequency invariant beam pattern is achieved by resolving an optimization problem of the SRV constraint to cover speech frequency band. With the control of sidelobe level, it is available for the frequency invariant beamformer (FIB) to prevent distortion of interference from the undesirable direction. The approach is completed in time-domain by placing tapped delay lines(TDL) and finite impulse response (FIR) filter at the output of each sensor which is more convenient than the Frost processor. By invoking the weight norm constraint, the robustness of the beamformer is further improved against random errors. Experiment results show that the proposed method has a constant beamwidth and almost the same white noise gain as traditional delay-and-sum (DAS) beamformer.

1. Introduction

Microphone arrays combined with FIBs are widely utilized in speech enhancement [1-3]. Compared to the frequency-domain broadband beamformer, the FIB in time-domain is attractive for its low computational complexity and constant frequency response with uniform beamwidth to avoid the effect of low pass filtering.

Harmonic nesting based on discrete Fourier transform is a traditional approach which achieves the frequency-invariant property between octaves [4]. However this method operates different subarrays at different frequency bands so that the beam pattern within one octave is still frequency dependent. In [5], harmonic nesting and filter-and-sum beamforming are combined to improve the frequency-invariant property within an octave. Instead of focusing on linear arrays like the above works, a simple FIB design utilizing multi-dimensional inverse Fourier transform is investigated in[6-7], which can be used for two or three-dimensional arrays. Recently, the convex optimization is utilized in the FIB design as a direct optimization approach [8]. However such beamformer using Frost processor needs pre-steering delays which are not an integer multiple of the sampling period, leading to some difficulties in implementation. In [9], the pre-delays are avoided by tapped delay lines (TDL) structure and complex weighted method. Another drawback of many convex optimization approaches is that they cover only one or two octaves. To get the closed-form array weights of the FIB[10] and further improve the frequency invariant, the least squares (LS) cost function combined with the SRV constraint is proposed, whose frequency band covers five octaves (speech frequencies). However, direct implementation of the least square cost function neglects the robustness of the beamformer so that the output signal has bad intelligibility and can only be used in source localization.

In this paper, we combine the spatial response constraint and the sidelobe control together as the cost function, and use the convex optimization method to solve the beam pattern synthesis problem to achieve a frequency-invariant response. To improve the robustness of the FIB, norm constraint of the steering weights are imposed, so that the proposed beamformer has good speech intelligibility just like conventional beamformer. Moreover, we use TDL to implement integer pre-delays [11] before FIR filter with real coefficient to avoid the complex structure of Frost processor [12].

2. Improved FIB

Consider an M-element array with a known arbitrary geometry, the sensors spatially sample the signal field at the locations $\boldsymbol{p}_m, m=1,2,...,M$ .The input is a plane wave with frequency region $[F_l, F_u]$ propagating in the direction θ, $\boldsymbol{u}(\theta)$ represents a unit vector that can be expressed as

$$\boldsymbol{u}(\theta)=\begin{bmatrix}-\cos\theta \\ -\sin\theta\end{bmatrix} \quad \text{* MERGEFORMAT (1)}$$

The time delay based on coordinate origin of every sensor is defined as

$$\tau_m(\theta)=\frac{\boldsymbol{u}(\theta)^T \boldsymbol{p}_m}{c} \quad \text{* MERGEFORMAT (2)}$$

where c is the velocity of propagation in the medium, and the superscript T denotes the transpose.

Figure.1 shows the modified time-domain beamformer structure. The processor has M sensors and each sensor is followed by a TDL and a FIR filter of length L. The beamforming output is the sum of the ML weighted tap signals.

Let T_s represents the sampling period. T_m=int$[-\tau_m(\theta_0)/T_s]\times T_s, m=1,2,...,M$ is the integer pre-delay attached after each sensor, where int$[\cdot]$ denotes the nearest integer so that the pre-delays are more convenient for practical implementation by using TDL.

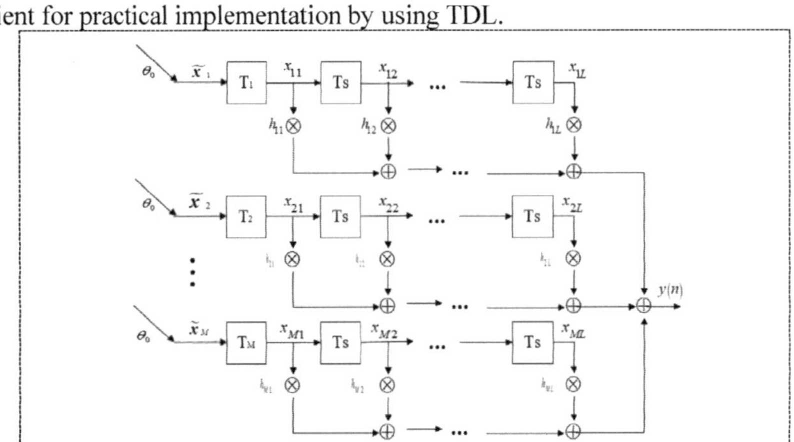

Figure 1. Time-domain beamformer with TDL and FIR filters

The sampled data from the mth sensor is $\tilde{x}_m(n)$. Thus, the data after integer pre-delay is defined as $x_m(n)=\tilde{x}_m(n-T_m)$. And the data of of every FIR filter tap is given by

$$x_{ml}(n)=x_m(n-(l-1)\times T_s), l=1,...,L \quad \text{* MERGEFORMAT (3)}$$

Thus, complex frequency response of the mth sensor FIR filter at frequency f_k is

$$H_m(f_k)=\sum_{l=1}^{L}h_{ml}e^{-j(l-1)2\pi f_k T_s}=\boldsymbol{e}^T(f_k)\boldsymbol{h}_m \quad \text{* MERGEFORMAT (4)}$$

where $\boldsymbol{e}(f_k)=\left[1,e^{-j2\pi f_k T_s},\ldots,e^{-j(L-1)2\pi f_k T_s}\right]^T$, and $\boldsymbol{h}_m=\left[h_{m1},h_{m2},\ldots,h_{mL}\right]^T$ are the adjustable real weights which follow behind the mth sensor, f_k is the frequency grid of the frequency region $[F_l,F_u]$.

The weight vector of mth sensor at frequency f_k is caused by the combined effects of pre-delay and FIR filter, which can be expressed as

$$w_m(f_k)=\boldsymbol{e}^T(f_k)\boldsymbol{h}_m g_m(f_k), m=1,2,\ldots,M \quad \text{* MERGEFORMAT (5)}$$

where $g_m(f_k)=e^{-j2\pi f_k T_m}$ is the phase shift caused by pre-delay. Finally the weight vector of frequency f_k is given by

$$\boldsymbol{w}(f_k)=\left[\boldsymbol{e}^T(f_k)\boldsymbol{h}_1 g_1(f_k),\boldsymbol{e}^T(f_k)\boldsymbol{h}_2 g_2(f_k),\ldots,\boldsymbol{e}^T(f_k)\boldsymbol{h}_M g_M(f_k)\right]^T \quad \text{* MERGEFORMAT (6)}$$

For every frequency grid f_k, the two-dimensional array response of the beamformer is

$$p(\theta,f_k)=\boldsymbol{w}^T(f_k)\boldsymbol{v}(\theta,f_k), k=1,2,\ldots,K \quad \text{* MERGEFORMAT (7)}$$

where $\boldsymbol{v}(\theta,f_k)$ is the array manifold vector given by

$$\boldsymbol{v}(\theta,f_k)=[e^{-j2\pi f_k\tau_1(\theta)},e^{-j2\pi f_k\tau_2(\theta)},\ldots,e^{-j2\pi f_k\tau_M(\theta)}]^T \quad \text{* MERGEFORMAT (8)}$$

$\tau_m(\theta), m=1,2,\ldots,M$ is the propagation delay of the mth array element form direction θ. Combining Eqs.(6) and Eqs.(8) into Eqs.(7) gives

$$p(\theta,f_k)=\boldsymbol{h}^T[\boldsymbol{v}(\theta,f_k)\bullet\boldsymbol{g}(f_k)\otimes\boldsymbol{e}(f_k)], k=1,2,\ldots,K \quad \text{* MERGEFORMAT (9)}$$

$\boldsymbol{g}(f_k)=\left[g_1(f_k),g_2(f_k),\ldots,g_M(f_k)\right]^T$, $\bullet$ represents the Hadamard product operator, and $\otimes$ represents Kronecker product operator, $\boldsymbol{h}=[h_{11},h_{12},\ldots,h_{1L},\ldots,h_{M1},h_{M2},\ldots,h_{ML}]^T$ is the $ML\times 1$ real-valued weight vector.

To get frequency invariant beam pattern, the SRV constraint is defined as:

$$SRV=\sum_{\theta_m\in\Theta}\sum_{f_k\in[F_l,F_u]}|\boldsymbol{h}^T\boldsymbol{a}(\theta_m,f_k)-\boldsymbol{h}^T\boldsymbol{a}(\theta_m,f_0)|^2 \quad \text{* MERGEFORMAT (10)}$$

where f_0 is reference frequency, $\boldsymbol{a}(\theta,f)=\boldsymbol{v}(\theta,f)\bullet\boldsymbol{g}(f)\otimes\boldsymbol{e}(f)$,and θ_m is the angular grid of the whole angle region Θ, which can be same for different frequency. SRV measures the Euclidean distance of beam response between the reference frequency and other operating frequencies. The smaller the value of SRV is, the better performance of the beamformer has. Obviously, when SRV is fixed to zero, the beamformer has a frequency-invariant response.

In order to suppress interference from undesirable direction, the control of sidelobe level is obtained by adding a constraint for the sidelobe response at reference frequency f_0, which is defined as

$$J_{SL}=\sum_{\theta_i\in\Theta_{SL}}|\boldsymbol{h}^T\boldsymbol{a}(\theta_i,f_0)|^2 \quad \text{* MERGEFORMAT (11)}$$

where Θ_{SL} is the sidelobe region. Combining Eq.(10) and Eq.(11), the cost function can finally be expressed as

$$J=\sum_{\theta_m\in\Theta}\sum_{f_k\in[F_l,F_u]}|\boldsymbol{h}^T\boldsymbol{a}(\theta_m,f_k)-\boldsymbol{h}^T\boldsymbol{a}(\theta_m,f_0)|^2+\alpha\sum_{\theta_i\in\Theta_{SL}}|\boldsymbol{h}^T\boldsymbol{a}(\theta_i,f_0)|^2=\boldsymbol{h}^T\boldsymbol{C}\boldsymbol{h} \quad \text{* MERGEFORMAT (12)}$$

where $\alpha(\alpha>0)$ is a compromise parameter between the sidelobe level and the frequency-invariant property, and

$$C=\sum_{\theta_m\in\Theta}\sum_{f_k\in[F_l,F_u]}[\boldsymbol{a}(\theta_m,f_k)-\boldsymbol{a}(\theta_m,f_0)]\times[\boldsymbol{a}(\theta_m,f_k)-\boldsymbol{a}(\theta_m,f_0)]^H+\alpha\sum_{\theta_i\in\Theta_{SL}}\boldsymbol{a}(\theta_i,f_0)\boldsymbol{a}^H(\theta_i,f_0) \quad \text{* MERGEFORMAT (13)}$$

Let $\boldsymbol{C}=\boldsymbol{U}^T\boldsymbol{U}$ be the Cholesky factorization, we have

$$h^T C h = (U^* h)^T (U^* h)^* = \| U^* h \|^2 \quad \backslash\text{* MERGEFORMAT (14)}$$

Thus, minimizing $h^T C h$ is equivalent to minimizing $\| U^* h \|$. Consequently, the FIB design problem is formulated as follows:

$$\begin{aligned} &\min_{h} \ \varepsilon \\ &\text{subject to } \ h^T a(\theta_0, f_0) = 1 \\ &\| U^* h \| \le \varepsilon \\ &\| h \| \le \Delta \end{aligned} \quad \backslash\text{* MERGEFORMAT (15)}$$

where $h^T a(\theta_0, f_0) = 1$ denotes that the beamformer response at the reference frequency in the look direction achieves the unity response. ε is an auxiliary variable to control the cost function. E.q(15) is different from the constraints in Ref.[10] , because Δ is used to improve the robustness of the FIB in the improved method.

3. Simulation and experimental result

A speech enhancement experiment with a16-element linear microphone array with element spacing of 0.05m is used for evaluating the performance of the improved FIB. Table.1 shows the specific experimental parameters. A speech source is placed at a distance of 10 meters and at the normal direction of the microphone array.

Table 1. Parameters of improved FIB.

experimental parameters	value
Steering direction	0°
Number of microphones	16
Working band	[300,3400]Hz
Sampling frequency	8000Hz
Element spacing	0.05m
Reference frequency	2400Hz
The length of FIR filters	30
Compromise parameter	0.5
The norm of FIR filter tap weights	0.5
Subband number	100
The uniform grid of [-90,90]	2°
The side-lobe region	$[-90^o, -50^o] \cup [50^o, 90^o]$

Figure.2 depicts the beam pattern of the improved FIB over the frequency range of 300 to 3400 Hz in simulation. As can be observed, the resulting beam pattern has approximately constant mainlobe and the response in the look direction is close to the unity response. Moreover, 2.5^o difference between the maximum beamwidth and the minimum beamwidth at all frequencies further displays the frequency-invariant property of the broadband beamformer. Due to using the sidelobe control design, the sidelobe level of the improved FIB is under -17 dB.

Speech signals collected in the outdoor environment with the noise level of 53-55 dB are shown in Figure.3. The 6.5 s speech is chosen from the listening materials of CET-6 (College English Test, Level 6). The corresponding spectrograms are achieved by the short-time Fourier transform (STFT).

To prove the improved FIB has the same white noise gain as the traditional DAS beamformer, the results provided by the DAS beamformer are also shown in Figure.3(c). Obviously, after a large part of high frequency noise has been filtered out, the resulting speech signal's clarity and intelligibility in Figure.3(d) have been significantly improved in comparing with the polluted speech signal in a single channel as shown in Figure.3(b).

Furthermore, the performance of the improved FIB is evaluated by the output signal-to-noise ratio (SNR) and the perceptual evaluation of speech quality (PESQ) index [3]. Note that the PESQ value is calculated using the whole data of 6.5 s and the SNR is calculated with the data of 1.5 s. The results are shown in Table 2. For the speech signal with SNR of 4.6136 dB (single channel), the output SNRs of the DAS-beamformer and the improved FIB are 12.5324 dB and 12.8445 dB, respectively. Correspondingly, the PESQ values for speech in the single channel, the DAS-based beamformer output and the FIB output are 1.287, 1.965 and 1.904, respectively. Therefore, the proposed FIB has almost the same white noise gain as the DAS-based beamformer, which has been proved to be robust in practical implementation.

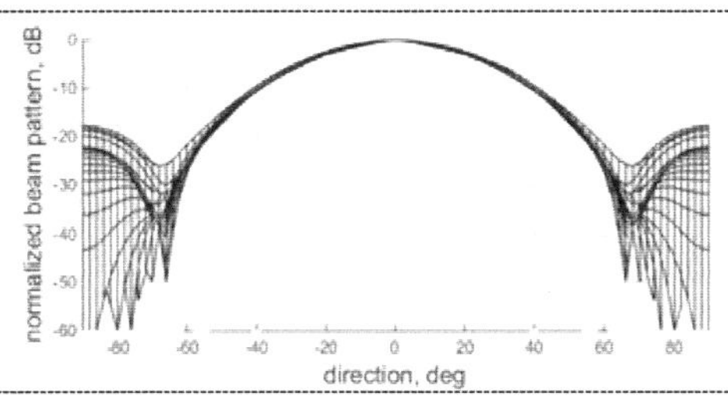

Figure 2. Beam pattern of the improved FIB in simulation for a look direction of 0° over the bandwidth of 300 to 3400Hz using a uniform linear array consisting of 16 microphones.

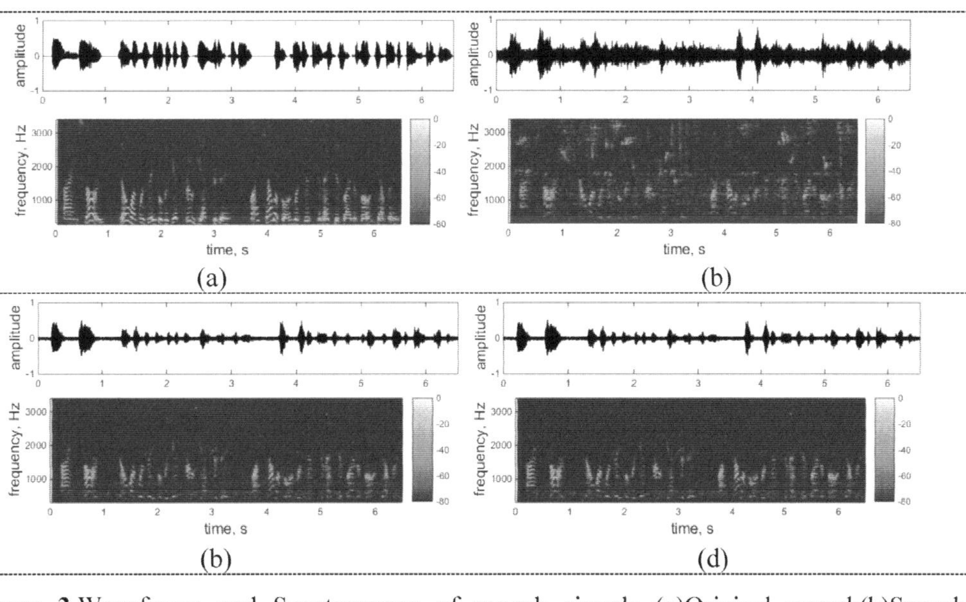

Figure 3. Waveforms and Spectrograms of speech signals (a)Original speech(b)Speech received in channel1(c)DAS-based beamformer output(d)Improved FIB output.

Table 2. The improvements in terms of output SNR and the PESQ index.

Method	SNR	PEAQ
DAS-beamfomer	12.5324	1.965
Improved-FIB	12.8445	1.904

4. Conclusion

A Robust frequency invariant beamforming with low sidelobe is proposed for speech enhancement. Our method minimizes the SRV constraint while maintaining the robust of beamformer using norm constraint on tap weights and providing the sidelobes of array pattern under a given threshold value. Furthermore, the convex optimization-based FIB is suited for wideband signals covering five octaves, and the proposed beamformer is convenient and cheap to implement for using integral tapped pre-delay. This approach can be used in arbitrary geometry array, and does not require significant computational complexity. The outdoor speech trials have verified the effectiveness of the proposed approach in terms of the output SNR and the PESQ measurement.

Acknowledgments

The authors would like to thank the financial support from the National Natural Science Foundation of China(No.6151397 and No.61171148)

References

[1] Mars R, Reju V G, Khong A W H 2014 A frequency-invariant fixed beamformer for speech enhancement *Signal and Information Processing Association Annual Summit and Conference (APSIPA)* pp 1-6
[2] Zou Y X, Li B, Liu W 2010 A broadband speech enhancement technique based on frequency invariant beamforming and GSC *2010 International Conference on Green Circuits and Systems (ICGCS 2010)* pp 175-179
[3] Cho K, Nam K W, Lee J C, Hong S H, Kwon S Y, Han J, Kim D, Lee S, Kim I N 2014 A comparison of frequency-invariant beamforming algorithms for hearing aids-differential microphone-based beamformers and the broadband beamformer *Biomedical Engineering Letters* **4** pp 166-175
[4] Smith R 1970 Constant beamwidth receiving arrays for broadband sonar systems *Acustica* **48** pp 21-26
[5] Chou T 1995 Frequency-independent beamformer with low response error in Proc. IEEE Int. Conf. Acoustics *Speech, Signal Processing* **5** pp2995-2998
[6] Liu W, Weiss S, McWhirter J G, Proudler I K 2007 Frequency invariant beamforming for two-dimensional and three-dimensional arrays *Signal Process* **87** pp 2535-2543
[7] Liu W, Weiss S 2008 Design of frequency invariant beamformers for broadband arrays IEEE *Trans. Signal Processing* **56** pp 855-860
[8] Yan S F, Ma Y L, Hou C H 2007 Optimal array pattern synthesis for broadband arrays *Journal of Acoustic Society of America* **122** pp 2686-2996
[9] Peng, Chen, Yihui, 2008 SRV constraint based FIB design for wideband linear array *Journal of Systems Engineering and Electronics* **21** pp 941-947
[10] Zhao Y, Liu W, Langley R 2011 Application of the least squares approach to fixed beamformer design with frequency-invariant constraintsIET *Signal Processing* **5** pp 281-291
[11] Yan S, Hou C, Ma X 2007 Convex optimization based time-domain broadband beamforming with sidelobe control *Journal of the Acoustical Society of America* **121** pp46-49
[12] Frost O L An algorithm for linearly constrained adaptive array processing 1972 *Proceedings of the IEEE* **60** pp 926-935

A Novel Motion Compensation Method for Random Stepped Frequency Radar with M-sequence

Zhikun Liao[1,*], Jiemin Hu[1], Dawei Lu[1], Jun Zhang[1]
[1]ATR Lab, National University of Defense Technology, Changsha 410073, China

E-mail: [*]lzkun2008@yeah.net

Abstract. The random stepped frequency radar is a new kind of synthetic wideband radar. In the research, it has been found that it possesses a thumbtack-like ambiguity function which is considered to be the ideal one. This also means that only a precise motion compensation could result in the correct high resolution range profile. In this paper, we will introduce the random stepped frequency radar coded by M-sequence firstly and briefly analyse the effect of relative motion between target and radar on the distance imaging, which is called defocusing problem. Then, a novel motion compensation method, named complementary code cancellation, will be put forward to solve this problem. Finally, the simulated experiments will demonstrate its validity and the computational analysis will show up its efficiency.

1. Introduction

The random stepped frequency (RSF) radar is a new kind of synthetic wideband radar which has been widely researched [1-2]. Its transmitted signal waveform looks a bit like the linear stepped frequency (LSF) signal [3]. However, its modulated frequencies of different transmitted pulses are randomly distributed over the given frequency band. The hopping frequencies make it possible to cover ultra-wide frequency band which is necessary for synthesizing the high resolution range profile (HRRP).

The frequency of RSF waveform is modulated by the pseudorandom code, a permutation of unique integer, such as Costas code [4] or Chaos code using Bernoulli mapping [5]. In our work, M-sequence which has been widely applied in communication and radar field to generate quasi-white noise will be employed. Different with the slant ridge ambiguity function (AF) of LSF signal, the AF of RSF signal looks more like a thumbtack. It means there is only one peak in the centre of echo delay and Doppler offset plane and other places are filled with the approximately uniform side-lobes. Meanwhile, it also indicates that, without a precise velocity measurement and motion compensation, the range profile of RSF radar will be destroyed, which is called de-focusing problem. In other words, the RSF signal is a very sensitive to Doppler frequency [6]. This is different with the case of LSF radar system [7], whose synthetic range profile will cyclically migrate and be slowly blurred when target is moving.

Therefore, it is necessary to carry out a precise motion compensation before the distance imaging for RSF radar. In terms of HRRP radar, there are several ways to achieve motion compensation, i.e., minimum waveform entropy (MWE) [8], range velocity iterative alternating project (RV-IAP) [9], cross-correlation between two adjacent pulse-trains [10] and a bank of Doppler filters [3]. They either need a lot of calculations or consume plenty of hardware resources. Besides, the RV-IAP algorithm is proposed to detect range and velocity parameters of multiple targets, however it requires the number of targets as a prior knowledge, which is unknown in most situations.

Taking all of these into account, we propose an effective and efficient motion compensation method, named complementary code cancellation (CCC), for the RSF radar with M-sequence. This paper is organized as follows. The complemented schematic diagram is provided in section 2 to show up the global framework of RSF radar system. In section 3, the defocusing problem is introduced. And in section 4, the motion compensation method based on CCC algorithm is presented. Then in section 5, the performance of the proposed method is characterized with the simulated experiments. Finally, some conclusions are drawn in the last section.

2. Schematic diagram

This section will introduce the brief implemented schematic diagram of RSF radar system based on CCC algorithm in Figure 1.

The pseudorandom code generation module is framed out in the red dash line at right side. Firstly, the M-sequence is generated by linear combination of shift registers. Then, the pseudorandom code could be obtained via decoding the binary flow of this sequence. For the convenience of motion compensation, its complementary code shall be employed in the next coherent processing interval (CPI). It can be seen that this schematic diagram is characterized by easy implementation.

Besides, the bottom part boxed in the red dash line shows the velocity detection method which will be emphatically presented in section 4. The velocity of target should be measured in front of distance imaging, because the range profile is very sensitive to the Doppler frequency, which will be uncovered in the next section.

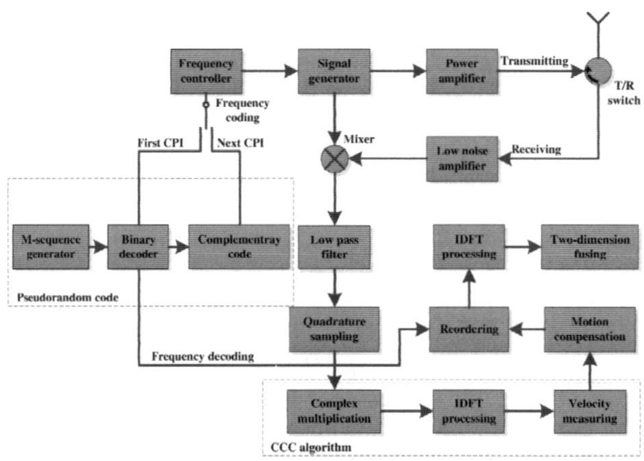

Figure 1. The implemented schematic diagram of RSF radar system based on CCC algorithm.

3. Defocusing problem

In this section, the effect of target movement on the range profile, named defocusing problem, will be briefly discussed. Firstly, the transmitted and received signal model of RSF radar is introduced. Then, the range profile synthesis method is presented as the groundwork of the defocusing problem. Finally, the defocusing problem would be formally explained in the last subsection to indicate the relationship between the target movement and the range profile.

3.1. Signal Model

3.1.1. Transmitted waveform

Consider a RSF radar using a train of N monotone bursts, with a pulse repetition interval (PRI) of T_r seconds and pulse duration of T seconds. And the total synthetic bandwidth is B Hz. In this way, the stepped frequency interval could be obtained as $\Delta f=B/N$. As well as the case of LSF signal [11], the stepped frequency interval and the pulse duration must meet the following tight contrast to present the appearance of grating lobes.

$$\Delta_f \cdot T \leq 1 \tag{1}$$

The complex envelope of RSF signal in a train can be represents as

$$\tilde{s}(t) = \sum_{n=0}^{N-1} \text{rect}\left(\frac{t - nT_r}{T}\right) e^{j2\pi c_n \Delta f (t - nT_r)}. \tag{2}$$

Here, the function rect(x) denotes the unit rectangle function. When $0 \leqslant x \leqslant 1$, it is equal to 1 and 0 otherwise. The frequency step coefficient c_n acquired by the M-sequence is a unique integer randomly chosen from 1 to N.

Let A and f_c denote the transmitting gain and carrier frequency. Then, the transmitted signal can be written as

$$s(t) = A\,\text{Re}\left\{\tilde{s}(t) e^{j2\pi f_c t}\right\}. \tag{3}$$

3.1.2. Echo signal

Consider an ideal point target at the initial distance of R_0 meters with a radial velocity of V meters per second running away from the radar. Then, the echo delay of nth monotone burst can be written as

$$\tau_n = \frac{2(R_0 - nVT_r)}{c}. \tag{4}$$

Here, the parameter c denotes the propagation speed. The RSF echoes are always received in the narrow band. Therefore, based on the stop and jump model, the echo is a delay version of the transmitted signal.

$$s_r(t) = s(t - \tau_n). \tag{5}$$

Substituting (3) into (5) and performing the quadrature digital down conversion processing, the complex echo signal at the sampling range bin where target locates will become

$$s_r\left[n \middle| R_0, V\right] = e^{-j2\pi(f_0 + c_n \Delta f)\tau_n} = e^{-j\frac{4\pi}{c}(f_0 + c_n \Delta f)(R_0 + nVT_r)}. \tag{6}$$

It can be seen that the distance and velocity of target are coupled in the phase of echoes. In the next subsection, the range profile synthesis will introduced to tell how to obtain the distance information of target. However, the velocity is bound to make an impact on the synthesis result, which will be studied in the subsection 3.

3.2. Range profile synthesis method

The echo model of RSF signal can be divided into two parts, i.e., distance item and velocity item. As for the distance item, it is coded by the coefficient c_n. Yet, the coefficient c_n is a kind of unique integer from 1 to N, which means it can be reordered to an ascending sequence. When reordering coefficient c_n, the independent variable in the velocity item will change to the inverse map from n to c_n. We note it as the coefficient $\dot{c}_n$ which is also a random unique integer from 1 to N. In this way, the reordered echo will turn into the following form.

$$\dot{s}_r\left[n \middle| R_0, V\right] = e^{-j\frac{4\pi}{c}(f_0 + n\Delta f)R_0} e^{-j\frac{4\pi}{c}(f_0 + n\Delta f)\dot{c}_n VT_r} \tag{7}$$

After reordering processing, the distance item gets a linear phase. That is to say the range profile can be synthesized by the inverse discrete Fourier transform (IDFT) processing. However, the velocity item remains random. It is conceivable that its effect on the synthesis result depends on the size of velocity. Figure 2 lists the synthetic range profiles of ideal point target at 5.1km under the conditions of different velocities.

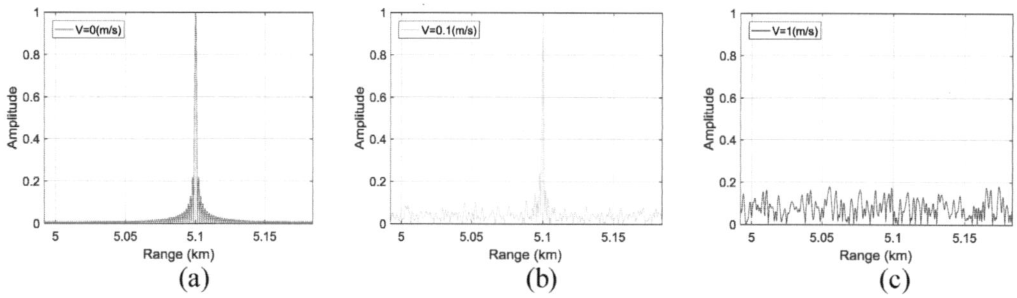

Figure 2. The synthetic range profiles of the ideal point target at 5.1km with the velocity of 0, 0.1 and 1(m/s) for (a), (b) and (c) respectively. The parameters of RSF radar is set as follows: T=1μs, T_r=100μs, f_c=35GHz, B=100MHz, N=128.

3.3. Defocusing problem

The range profiles in figure 2 intuitively illustrate the defocusing problem of RSF radar. Firstly, when the target is stationary relative to the radar, the range profile could be accumulated very well, which has been shown in the red line of figure 2. However, even if the target moves with a small velocity, the peak of range profile will attenuate rapidly, because the carrier frequency f_c is always very huge. When the velocity continues increasing, the peak will disappear completely. In summary, the defocusing problem of RSF radar reflects a more complex coupling relationship between distance and velocity.

Then, we will define a variable to quantify the defocusing problem, named velocity tolerance. It is defined as a velocity threshold that make the peak of range profile of ideal point target attenuate 3dB. It is difficult to directly deduce the velocity tolerance. But it can be seen that it is constrained by the carrier frequency f_c and PRI T_r. Through fitting simulation, we get an approximate result.

$$v_T = \frac{0.24c}{f_0 N T_r} \tag{8}$$

The velocity tolerance indicates the precision threshold of motion compensation. Only if the precision is less than velocity tolerance, the motion compensation method can be regard as being valid.

4. Motion compensation method

4.1. Complementary code cancellation algorithm

In the transmitting stage, the waveform defined by (2) is employed during the first CPI. However, the waveform modulated by the complementary frequency hopping coefficient $\hat{c}_n$ will be used during the next CPI. The frequency hopping coefficients in the adjacent two CPI meet the following relationship

$$\hat{c}_{n+N} + c_n = N, n = 0,1,\cdots,N-1. \tag{9}$$

In the coherent receiver, after the same processing as the first CPI, the sampling echo signal of the second CPI will have the similar form

$$\hat{s}_r\left[n\middle|R_0,V\right] = e^{-j\frac{4\pi}{c}(f_0+\hat{c}_n\Delta f)(R_0+nVT_r+NVT_r)}, n = 0,1,\cdots N-1. \tag{10}$$

Then, multiply each corresponding signals of the both CPI together.

$$\begin{aligned}\rho\left[n\middle|R_0,V\right] &= s_r\left[n\middle|R_0,V\right]\cdot\hat{s}_r\left[n\middle|R_0,V\right] \\ &= e^{-j\frac{4\pi}{c}\left[f_0(2R_0+VNT_r)+N\Delta f(R_0+VNT)_r\right]}\cdot e^{-j\frac{4\pi}{c}(2f_0+N\Delta f)VnT_r}\cdot e^{j\frac{4\pi}{c}c_n\Delta fVNT_r}\end{aligned} \tag{11}$$

It can be seen that the distance item is involved in the first factor which consists of the constant phase part. And the second factor is composed of the linear velocity item which makes up the most important part of velocity measure. At last, the third factor is the disturbing item. But, as long as the velocity satisfies the following relationship, it will not cause a significant interference on the result.

$$\max_{n\in[0,N-1]}\left\{\frac{4\pi}{c}c_n\Delta fVNT_r\right\}\le\frac{\pi}{2}\Rightarrow V\le\frac{c}{8\Delta fN^2T_r} \tag{12}$$

Compared with the velocity tolerance v_T, this is a quite rough condition for the most radar system. Therefore, the third factor can be omitted in the succeeding analysis. As a result, conducting the IDFT processing on the multiplying output, we will obtain the approximate velocity profile

$$\mathrm{P}\left[k\middle|V\right]\approx\left|\frac{\sin\frac{\pi N}{K}\left[\frac{2}{c}(2f_0+N\Delta f)VT_rK-k\right]}{\sin\frac{\pi}{K}\left[\frac{2}{c}(2f_0+N\Delta f)VT_rK-k\right]}\right|. \tag{13}$$

Here, the parameter K denotes the length of IDFT processing. Consequently, searching its maximal value, the measured velocity could be acquired.

$$\bar{k}=\arg\max_{1\le k\le K}\left\{\mathrm{P}\left[k\middle|V\right]\right\}\Rightarrow\bar{V}=\frac{c\bar{k}}{2T_rK\left(2f_0+N\Delta f\right)} \tag{14}$$

Then, the motion compensation factor can be generated via the measured velocity. Compensate the sampling echo in (7). Finally, the range profile could be synthesized by the method mentioned in the previous section. The detailed algorithm procedure has been displayed in the schematic diagram in the Figure 1.

4.2. Algorithm precision analysis

From the velocity profile, the 3dB velocity resolution rate of the presented method can be figured out.

$$\Delta v=\frac{0.442c}{(2f_0+N\Delta f)NT_r} \tag{15}$$

Compare the velocity resolution rate with the velocity tolerance.

$$\Lambda=\frac{\Delta v}{v_T}=\frac{1}{1.086+0.543\,{}^{B}\!/\!_{\Delta f}} \tag{16}$$

It can be seen that the ratio is always less than 1, which means the precision of presented method could meet the demand of the motion compensation.

5. Simulated experiments

The simulated experiments will carried out on the basis of the following parameters: the pulse duration T=1μs, PRI T_r=100μs, the carrier frequency f_c=35GHz, the synthetic bandwidth B=100MHz, the number of bursts in a CPI N=128. The simulated scene is a tank model at the distance of 5.1km with the velocity of 10m/s running away from the radar. The intermediate and final experiment results will be displayed in the Figure 3.

The picture of simulated tank is shown in subfigure (a) of Figure 3. Its velocity profile and range profile are shown by the blue line in subfigure (b) and (c). The red line in subfigure (b) indicates the velocity detecting threshold. It can be seen that the velocity of tank can be well measured. Besides, the red diamond in subfigure (c) also marks out the real RCS values of simulated tank. The peaks of range profile are very close to these values, which means the presented motion compensation method is valid.

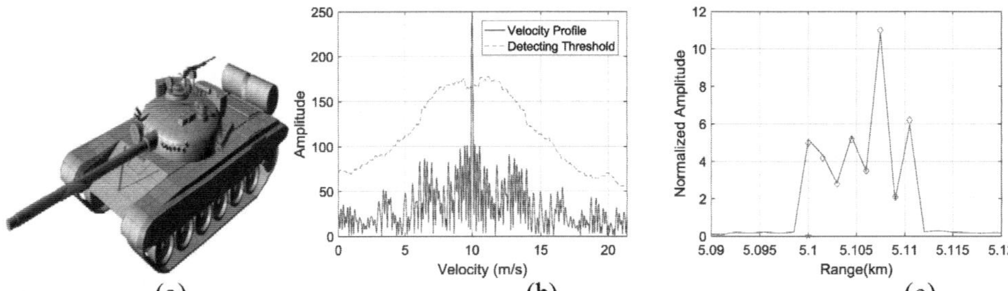

(a) (b) (c)

Figure 3. the intermediate and final results of range profile synthesis processing of a tank (shown in (a)). The velocity profile and range profile is shown in (b) and (c), respectively.

Compared with the existing methods, the motion compensation method presented in the paper features lower computation overhead. The cross-correlation method needs $O(N^2)$ multiplication times. The MWE method needs $O(N^3)$ multiplication times. And the RV-IAP algorithm needs more, due to the iterative search. However, the CCC method just requires $O(N\log_2 N)$ multiplication times because of the fast Fourier transform (FFT). But, the unambiguous velocity interval of CCC method looks not large since the data of two CPI is employed. Thereore, in some application situations, the rough velocity measurement is needed to be performed in advance.

6. Conclusions

The RSF radar can be applied to synthesize the ultra-wide bandwidth to achieve HRRP. However, it suffers from the defocusing problem. In this paper, the schematic diagram of RSF radar based on M-sequence is designed, which benefits from sample structure. Then, the velocity tolerance is imported to indicate the threshold of motion compensation. And the CCC method is proposed to solve the defocusing problem. Finally, the simulated experiments demonstrate that the presented method is novel. It is believed that its effectiveness and efficiency will make it useful in the radar imaging.

References

[1] J. E. Luminati, T. B. Hale, M. A. Temple, M. J. Havrilla and M. E. Oxley, "Doppler aliasing artefact filtering in SAR imagery using randomized stepped-frequency waveforms", Electronics Letter, vol. 4, no. 22, pp. 1447-1448, 2004.

[2] R. J. Axelsson, "Analysis of random step frequency and comparison with experiments", IEEE Trans. Geosci. Remote Sens., vol. 45, no. 4, pp. 890-904, 2007.

[3] N. Levanon, "Stepped-frequency pulse train radar signal", Proc. IEE radar Sonar Navigation, vol. 149, no. 6, pp. 297-309, 2002.

[4] J. P. Costas, "A study of a class of detection waveforms having nearly ideal range-Doppler ambiguity function", IEEE Proc., vol. 72, no. 8, pp. 996-1009, 1984.

[5] Yu Zhang, Wen Hu, Ling Wang and Lu Zhang, "A novel random stepped frequency radar using Chaos", IEEE Radar Conference, 0663-0665, 2014.

[6] T. Huang, Y. Liu, G. Li and X. Wang, "Randomized stepped frequency ISAR imaging", IEEE Radar conference, pp. 553-557, 2012.

[7] J. Hu, J. Zhang, Q. Zhai, R. Zhan and D. Lu, "ISAR imaging using a new stepped frequency signal format", IEEE trans. Geosci. Remote Sens., vol 52, no. 7, pp. 4291-4305. 2014.

[8] P. Song, H. Meng, T. Huang and Y. Liu, "A novel target motion compensation method for randomized stepped frequency ISAR", IEEE Radar conference, pp. 917-921, 2013.

[9] Y. Liu, H. Meng, G. Li and X. Wang, "Range-velocity estimation of multiple targets in randomised stepped-frequency radar", Electronics Letters, vol. 44, no. 17, pp. 1032-1034, 2008.

[10] R. Xu, Z. Cao and Y, Liu, "Motion compensation for ISAR and noise effect," IEEE Trans. Aerospace and Electronic Systems, vol. 5, no. 6, pp. 20-22, 1990.

[11] N. Levanon and E. Mozeson, Radar Signals. Wiley, New York, 2004.

A Divergence Median-based Geometric Detector with A Weighted Averaging Filter

Xiaoqiang Hua[1,*], Yongqiang Cheng[1], Yubo Li[1], Hongqiang Wang[1], Yuliang Qin[1]

[1]School of Electronic Science and Engineering, National University of Defence Technology, Changsha, China

*hxq712@yeah.net

Abstract. To overcome the performance degradation of the classical fast Fourier transform (FFT)-based constant false alarm rate detector with the limited sample data, a divergence median-based geometric detector on the Riemannian manifold of Heimitian positive definite matrices is proposed in this paper. In particular, an autocorrelation matrix is used to model the correlation of sample data. This method of the modeling can avoid the poor Doppler resolution as well as the energy spread of the Doppler filter banks result from the FFT. Moreover, a weighted averaging filter, conceived from the philosophy of the bilateral filtering in image denoising, is proposed and combined within the geometric detection framework. As the weighted averaging filter acts as the clutter suppression, the performance of the geometric detector is improved. Numerical experiments are given to validate the effectiveness of our proposed method.

1. Introduction

Improving the performance of a detector in interference environment is an important subject in the field of signal processing. The classical fast Fourier transform (FFT) based constant false alarm rate (CFAR) detector [1] uses the FFT to obtain the correlation of sample data. Usually, the performance suffers from severe degradation with the limited sample data as the Doppler resolution is poor and the energy of Doppler filter banks spread. To overcome these drawbacks, F. Barbaresco has studied the structure of space of Hermitian positive-definite (HPD) matrix, and has proposed a generalized CFAR technique on a Riemannian manifold of Hermitian positive-definite (HPD) matrices, which was referred to as the Riemannian median-based geometric detector [2]. In the Riemannian median-based geometric detector, the radar received clutter data in each range cell in one CPI is modeled as an HPD matrix. In addition, the Riemannian median detector [3] are derived. This geometric detector has been used to monitor the turbulence of a plane [4], and target detection in coastal X-band and HF surface wave radars [5].

Many metrics can be used to measure the closeness between any two points on the Riemannian manifold of HPD matrices. Different measurements can reflect different structures of this space. Many divergences can be used as measurements. Mentioned a few, the square loss is used to measure the distance between the two states in the regression; the Bhattacharyya divergence has employed to medical image segmentation [6]; and the Kullback-Leibler (KL) divergence has been widely used to measure the information difference between two probability distributions [7]. These metrics have achieved good results in many applications. In our previous work [8], we have studied a geometric detection method based on KL divergence. Experiments have shown that its performance outperforms the traditional FFT-CFAR detector.

In this paper, we study the geometric detector based on different metrics. In particular, the Log-Euclidean distance [9], the Bhattacharyya divergence [7], and the Hellinger distance [10] are used as replacements of the Riemannian distance in the geometric detector. Based on the three metrics, the Log-Euclidean median [9], the Bhattacharyya median [7], and the Hellinger median of a finite set of HPD matrices are derived. As a result, a divergence median-based geometric detector is developed. Moreover, we propose a weighted average filter which is combined within the geometric detector. This filter is conceived from the philosophy of the bilateral filtering in image denoising [11]. As this filter acts as a clutter suppression procedure, the detection performance can be improved.

2. Signal model and signal manifold

The radar usually sends several pulses to a moving target, and receives the return data which contains the phase information of this target. A certain model is used to capture the Doppler of target. In this paper, the Doppler is represented as the correlation of data $z = \{z_1, z_2, \ldots, z_n\}$, and model as a multivariate Gaussian process with zero mean, $z \sim CN(\boldsymbol{0}, \boldsymbol{R})$ [2],

$$p(z|\boldsymbol{R}) = \frac{1}{\pi^n \det(\boldsymbol{R})} \exp\{-z^H \boldsymbol{R}^{-1} z\} \tag{1}$$

here, the matrix $\boldsymbol{R}$ is an HPD matrix, and it can be computed as [2],

$$\boldsymbol{R} = E\left[zz^H\right] = \begin{bmatrix} r_0 & \overline{r}_1 & \cdots & \overline{r}_{n-1} \\ r_1 & r_0 & \cdots & \overline{r}_{n-2} \\ \vdots & \ddots & \ddots & \vdots \\ r_{n-1} & \cdots & r_1 & r_0 \end{bmatrix}, \quad r_k = E\left[z_i \overline{z}_{i+k}\right],\ 0 \le k \le n-1, 0 \le i \le n-1 \tag{2}$$

where r_k denotes the correlation coefficient of pulse data, and $\overline{r}_i$ is the complex conjugate of r_i. As there is not enough sample data to compute the statistical expectation $E\left[z_i \overline{z}_{i+k}\right]$, according to the ergodicity , it can be calculated by a finite time serial,

$$\hat{r}_k = \frac{1}{n-k} \sum_{j=0}^{n-1-k} z_j \overline{z}_{j+k}, \quad 0 \le k \le n-1 \tag{3}$$

The pulse data $z = \{z_1, z_2, \ldots, z_n\}$ in each cell in one CPI is modeled by Equation (1) and (2), and represented by an HPD matrix $\boldsymbol{R}$. The pulse data z is staying in the Euclidean space, and the HPD matrix $\boldsymbol{R}$ is viewed as a point in the manifold. Through this parameterization, the data z is transformed into an n dimensional non-liner manifold space,

$$\psi : C(n) \to P(n), \quad z \to \boldsymbol{R} \in P(n) \tag{4}$$

Here $P(n)$ forms a differentiable Riemannian manifold with non-positive curvature. Through this modelling for the radar echo, the target detection should be performed in the manifold. In particular, the structure of matrix space can be considered.

3. Divergence measures and divergence medians

In this Section, a brief description about the divergence measure and the divergence median are given in the following.

3.1. Divergence measures on the Riemannian manifold of HPD matrices

For any two points $\boldsymbol{R}_1, \boldsymbol{R}_2$ on the HPD matrix space $P(n)$, a natural distance which is the Riemannian distance can be given by,

$$d_R^2(\boldsymbol{R}_1, \boldsymbol{R}_2) = \left\| \operatorname{logm}\left(\boldsymbol{R}_1^{-1/2} \boldsymbol{R}_2 \boldsymbol{R}_1^{-1/2}\right) \right\|_F^2 = \sum_{k=1}^{n} \log^2(\lambda_k) \tag{5}$$

where λ_k is the kth eigenvalue of $\boldsymbol{R}_1^{-1/2}\boldsymbol{R}_2\boldsymbol{R}_1^{-1/2}$, and $\text{logm}(\cdot)$ is the logarithmic map on the Riemannian manifold.

In addition to the Riemannian distance, many divergences can be used to measure the dissimilarity between two HPD matrices. In the following, three geometric measures are presented in Table 1.

Table 1 Divergence measures on the Riemannian manifold

Divergence measures	Formulation
Log-Euclidean [9]	$d_{LE}(\boldsymbol{R}_1,\boldsymbol{R}_2)=\left\\|\text{logm}(\boldsymbol{R}_1)-\text{logm}(\boldsymbol{R}_2)\right\\|_F$
Bhattacharyya [7]	$d_{Bh}(\boldsymbol{R}_1,\boldsymbol{R}_2)=2\sqrt{\log\left(\left\lvert\frac{\boldsymbol{R}_1+\boldsymbol{R}_2}{2}\right\rvert\Big/\sqrt{\lvert\boldsymbol{R}_1\rvert\lvert\boldsymbol{R}_2\rvert}\right)},\quad \boldsymbol{R}_1,\boldsymbol{R}_2\in P(n)$
Hellinger [10]	$d_H(\boldsymbol{R}_1,\boldsymbol{R}_2)=\sqrt{2-2\left(\lvert\boldsymbol{R}_1\rvert^{1/4}\lvert\boldsymbol{R}_2\rvert^{1/4}\Big/\left\lvert\frac{1}{2}(\boldsymbol{R}_1+\boldsymbol{R}_2)\right\rvert^{1/2}\right)},\quad \boldsymbol{R}_1,\boldsymbol{R}_2\in P(n)$

3.2. Divergence medians for a finite set of HPD matrices

The divergence median has not received much attention in the research field. Based on the divergence metric, many medians have been derived. Here, four divergence-based medians are listed in Table 2.

Table 2 Divergence medians related to different measures

Divergence median	Iteration formulation
Riemannian [3]	$\hat{\boldsymbol{R}}_{t+1}=\hat{\boldsymbol{R}}_t^{1/2}\,\text{expm}\left\{\varepsilon_t\left(\sum_{k\in G_{\hat{R}_t}}\frac{\text{C}_t}{\left\\|\text{C}_t\right\\|_F}\right)\right\}\hat{\boldsymbol{R}}_t^{1/2}$, $\text{C}_t=\text{logm}\left(\hat{\boldsymbol{R}}_t^{-1/2}\boldsymbol{R}_k\hat{\boldsymbol{R}}_t^{-1/2}\right),\ G_{\hat{R}_t}=\left\{k/\boldsymbol{R}_k\neq\hat{\boldsymbol{R}}_t\right\}$
Log-Euclidean [9]	$\hat{\boldsymbol{R}}_{t+1}=\text{expm}\left(\left(\sum_{i=1}^{m}\frac{1}{d_{LE}(\hat{\boldsymbol{R}}_t,\boldsymbol{R}_i)}\right)\left(\sum_{i=1}^{m}\frac{\text{logm}(\boldsymbol{R}_i)}{d_{LE}(\hat{\boldsymbol{R}}_t,\boldsymbol{R}_i)}\right)\right)$
Bhattacharyya [7]	$\hat{\boldsymbol{R}}_{t+1}=\left(\left(\sum_{i=1}^{m}\frac{\boldsymbol{R}_i^{-1}}{d_{Bh}(\hat{\boldsymbol{R}}_t,\boldsymbol{R}_i)}\right)\Big/\left(\sum_{i=1}^{m}\frac{1}{d_{Bh}(\hat{\boldsymbol{R}}_t,\boldsymbol{R}_i)}\right)\right)^{-1}$
Hellinger	$\hat{\boldsymbol{R}}_{t+1}=\left(\sum_{i=1}^{m}\frac{\lvert\hat{\boldsymbol{R}}_t\rvert^{1/4}\lvert\boldsymbol{R}_i\rvert^{1/4}\left\lvert\frac{1}{2}(\hat{\boldsymbol{R}}_t+\boldsymbol{R}_i)\right\rvert^{-1/2}}{\sqrt{2-2\frac{\lvert\hat{\boldsymbol{R}}_t\rvert^{1/4}\lvert\boldsymbol{R}_i\rvert^{1/4}}{\left\lvert\frac{1}{2}(\hat{\boldsymbol{R}}_t+\boldsymbol{R}_i)\right\rvert^{1/2}}}}\right)\left(\sum_{i=1}^{m}\frac{\lvert\hat{\boldsymbol{R}}_t\rvert^{1/4}\lvert\boldsymbol{R}_i\rvert^{1/4}\left\lvert\frac{1}{2}(\hat{\boldsymbol{R}}_t+\boldsymbol{R}_i)\right\rvert^{-1/2}}{\sqrt{2-2\frac{\lvert\hat{\boldsymbol{R}}_t\rvert^{1/4}\lvert\boldsymbol{R}_i\rvert^{1/4}}{\left\lvert\frac{1}{2}(\hat{\boldsymbol{R}}_t+\boldsymbol{R}_i)\right\rvert^{1/2}}}}\left(\frac{1}{2}(\hat{\boldsymbol{R}}_t+\boldsymbol{R}_i)\right)^{-1}\right)^{-1}$

4. Divergence median-based geometric detector

As mentioned above, the divergence measures and their corresponding medians are presented. The pulse data reflected from a moving target is modeled as an HPD matrix. The element of this matrix denotes the correlation of data. Then, a target detection algorithm is designed similar to the framework of Barbaresco's work [2] according to these HPD matrices in this paper. As the detection algorithm is performed on the space of HPD matrix, and considers the structure of this space which is essentially a manifold. In this sense, our designed detection algorithm can be viewed as a geometric detector. The main differences between our designed detection algorithm and the Riemannian median-based geometric detector are twofold: 1) the different structures are considered owing to different metrics utilized; and 2) a weighted averaging filter which is carried out prior to the target detection is combined within our detection framework. This filter is borrowed from the bilateral filtering [11] in image denoising, and acts as a clutter suppressed procedure.

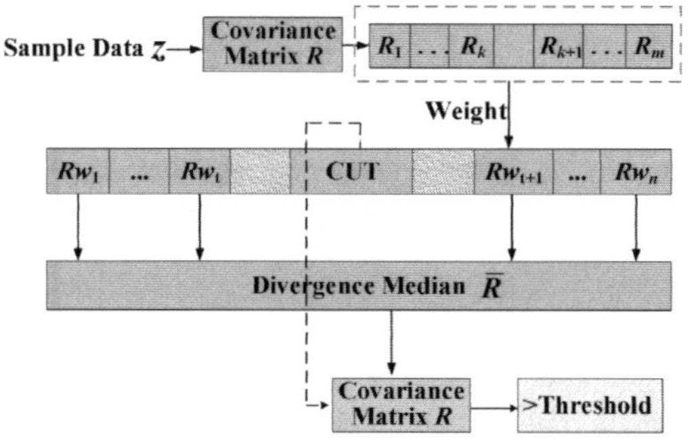

Figure 1. Divergence median-based geometric detector

As illustrated in Figure 1, the correlation of pulse data z returned from a moving target in each cell is modelled as an HPD matrix $\boldsymbol{R}$. All these HPD matrices constitute a Riemannian manifold. A weighted averaging filter is carried on according to the similarity of these matrices on manifold. Specially, each estimated matrix is replaced by a weighted average of its surrounding matrices. Based on these estimated matrices, the detection statistic is defined as the distance between the HPD matrix $\boldsymbol{R}_D$ of cell under test and the mean matrix $\overline{\boldsymbol{R}}$ calculated by the matrices of reference cells. The mean matrix denotes the clutter power level. Finally, the decision is made by comparing the statistic in each cell with an adaptive threshold γ.

For an HPD matrix $\boldsymbol{R}_t$ in the tth cell in one CPI, and its surrounding m matrices are given as $\{\boldsymbol{R}_1, \boldsymbol{R}_2, ..., \boldsymbol{R}_m\}$. A weighted strategy is employed, and the weighted matrix $\boldsymbol{R}_{wt}$ can be estimated as follow,

$$\boldsymbol{R}_{wt} = \sum_{i=1}^{m} w_i \boldsymbol{R}_i, \quad 0 \le w_i \le 1, \ \sum_{i=1}^{m} w_i = 1 \tag{6}$$

where w_i is the weight of ith HPD matrix. Similar to the bilateral filtering in image processing, the weight here is defined according the similarity between the matrix $\boldsymbol{R}_t$ and a surrounding matrix $\boldsymbol{R}_i$. In general, there are many ways to formulate the similarity between two elements. For instance, the reciprocal, the exponent, and the logarithm can be utilized into the definition. Here, we use the exponent of distance between $\boldsymbol{R}_i$ and $\boldsymbol{R}_t$, as,

$$w_i = \frac{1}{W} \exp\left\{-d^2\left(\boldsymbol{R}_i, \boldsymbol{R}_t\right)/h^2\right\} \tag{7}$$

where $d\left(\boldsymbol{R}_i, \boldsymbol{R}_t\right)$ is the Riemannian distance. $W = \sum_{i=1}^{m} \exp\left\{-d^2\left(\boldsymbol{R}_i, \boldsymbol{R}_t\right)/h^2\right\}$ is a normalizing factor, and h is a filtering parameter, which controls the exponential decay. m and h are two free parameters, and they are chosen according to the context.

5. Application of the target detection method

In this section, numerical experiments and real clutter data are performed to evaluate the detection performances of our proposed three divergence medians-based geometric detectors with the weighted averaging filter, namely the Log-Euclidean median-based geometric detector (LogEMGD), the Bhattacharyya median-based geometric detector (BhatMGD), and the Hellinger median-based geometric detector (HelMGD). The performances of these detectors are compared with the Riemannian median-based geometric detector (RiemMGD) [3] and the classical FFT-CFAR detector [1].

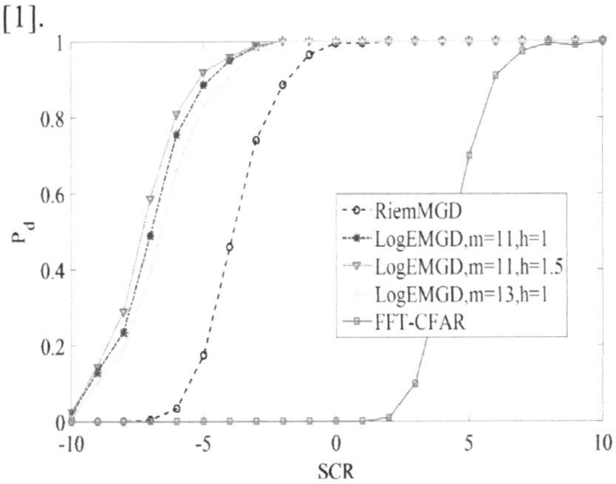

(a) P_d versus SCR, LogEMGD, $Pfa = 10^{-5}$

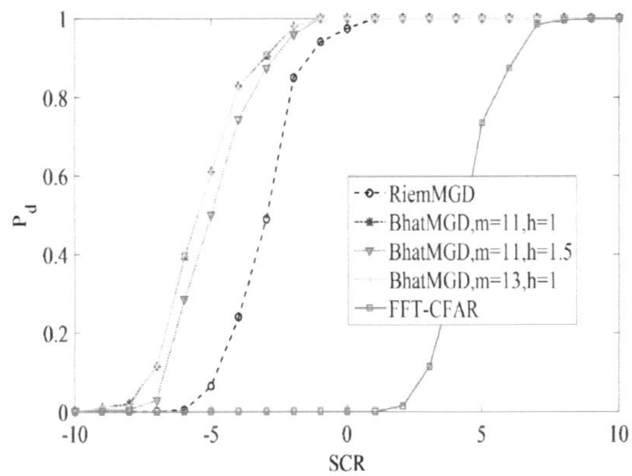

(b) P_d versus SCR, BhatMGD, $Pfa = 10^{-5}$

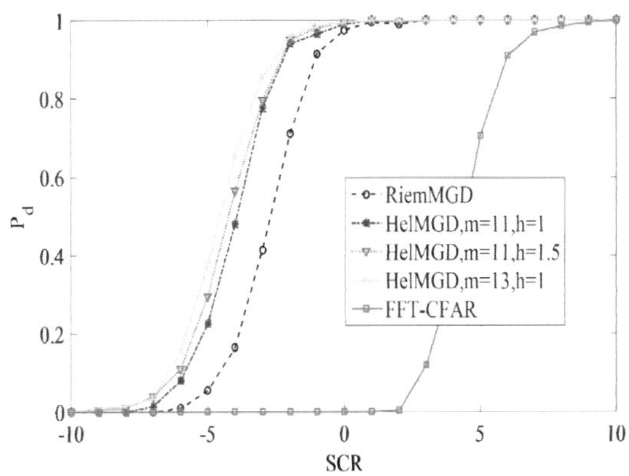

(c) P_d versus SCR, HelMGD, $Pfa = 10^{-5}$

Figure 2. P_d versus SCR in K distribution with different parameter values

The performances of: 1) the LogEMGD detector, the BhatMGD detector, and the HelMGD detector, 2) the RiemMGD detector, (3) the FFT-CFAR detector are compared via Monte Carlo simulations. The data samples are from N = 7 received pulses. In our simulations, the radar received echo contains 7 pulses. The radar central frequency fc is 9 GHz, and the pulse repetition frequency (PRF) is 1000 Hz. The target signal model is $\tilde{a}\boldsymbol{p}$. $\tilde{a}$ is a parameter accounting for the target backscattering and the channel propagation effects, and $\boldsymbol{p}$ is the target steering vector,

$$\boldsymbol{p} = \frac{1}{\sqrt{N}}\left[1, \exp\left(j2\pi f_d T_r\right), \ldots, \exp\left(j2\pi\left(N-1\right) f_d T\right)\right]^T \tag{8}$$

where T_r is the pulse repetition interval, and f_d is the Doppler frequency. A target has an approximate constant velocity v = 5 m/s. As few pulses are considered, $\tilde{a}$ is assumed to be constant. M = 16 range cells are considered as reference cells around cell under test, and used for averaging. In the simulation, the clutter is assumed as the K distribution.

A series of Monte Carlo simulation experiments are used to compare the performances between our proposed detector, the RiemMGD detector and the FFT-CFAR detector. Since there is not enough prior information, an analytical expression for the detection threshold cannot be obtained. In the experiment, the detection threshold is obtained by Monte Carlo experiment. Particularly, the detection statistic in the absence of the target is calculated by the 10^6 experiment, and the threshold is determined according to a given false alarm probability P_{fa}. To ensure the detection probability is accurately estimated, we repeat 200 times simulations to estimate the detection probability under different SCRs. The detection probability is estimated by the relative frequencies.

The Pd vs SCRs under different P_{fa} are given in Figure 2. The SCR varies from -10 to 10 dB, and the P_{fa} is set to 10^{-5}. The parameter values m and h are chosen as m=11, h=1; m=11, h=1.5; m=13, h=1. From Figure 2 we can know that the proposed three detectors have better detection performance than the RiemMGD detector. Moreover, the detection performances of these geometric detectors are better than that of the FFT-CFAR detector. These results prove the advantage of our proposed geometric detection method sufficiently.

6. Conclusion

In this paper, we have proposed a divergence median-based geometric detector. In particular, we have derived the Hellinger median for a set of HPD matrices. The three metrics, namely the log-Euclidean distance, the Bhattacharyya divergence, and the Hellinger distance, are utilized in our detector. These metrics can achieve different structures of Riemannian manifold. In addition, a pre-processing procedure which is borrowed from the bilateral filtering in image processing is carried out before the target detection. As it acts as a clutter suppression function, the performance can be improved. At the analysis stage, we have assessed the detection performance by giving plots of Pd vs SCR in simulated experiments. These results have shown that our proposed geometric detectors have better performance than the Riemannian mean-based geometric detector and the FFT-CFAR detector.

Reference

[1] M.A. Richards, Fundamentals of Radar Signal Processing, *Second Edition*, McGraw-Hill (2014).

[2] J. Lapuyade-Lahorgue, F. Barbaresco, Radar detection using Siegel distance between autoregressive processes, application to HF and X-band radar, in: 2008 *IEEE Radar Conference*, 2008, pp. 1-6.

[3] F. Barbaresco, Robust statistical Radar Processing in Fréchet metric space: OS-HDR-CFAR and OS-STAP Processing in Siegel homogeneous bounded domains, in: 2011 *12th International Radar Symposium (IRS)*, 2011, pp. 639-644.

[4] Z. Liu, F. Barbaresco, Doppler Information Geometry for Wake Turbulence Monitoring, *Springer Berlin Heidelberg*, 2013.

[5] F. Barbaresco, U. Meier, Radar monitoring of a wake vortex: Electromagnetic reflection of wake turbulence in clear air, *Comptes Rendus Physique* 11 (2010) 54-67.

[6] S. Kullback, R.A. Leibler, On Information and Sufficiency, *Annals of Mathematical Statistics* 22 (1951) 79-86.

[7] M. Charfi, Z. Chebbi, M. Moakher, B.C. Vemuri, Bhattacharyya median of symmetric positive-definite matrices and application to the denoising of diffusion-tensor fields, 2013 *IEEE 10th International Symposium on Biomedical Imaging*, 2013, pp. 1227-1230.

[8] Y. Cheng, X. Hua, H. Wang, Y. Qin, X. Li, The Geometry of Signal Detection with Applications to Radar Signal Processing, *Entropy*, 18 (2016).

[9] V. Arsigny, P. Fillard, X. Pennec, N. Ayache, Log-Euclidean metrics for fast and simple calculus on diffusion tensors, *Magnetic Resonance in Medicine* 56 (2006) 411-421.

[10] H.M. Anver, Mean Hellinger distance as an error criterion in univariate and multivariate kernel density estimation, *Dissertations & Theses - Gradworks* (2010).

[11] C. Tomasi, R. Manduchi, Bilateral filtering for gray and color images", *ICCV* (1998) 839-846.

Obstacle Detection in Indoor Environment for Visually Impaired Using Mobile Camera

Sami ur Rahman[1, *], Sana Ullah[1] and Sehat Ullah[1]

[1]Department of Computer Science and IT, University of Malakand, Pakistan

[*]srahman@uom.edu.pk

Abstract. Obstacle detection can improve the mobility as well as the safety of visually impaired people. In this paper, we present a system using mobile camera for visually impaired people. The proposed algorithm works in indoor environment and it uses a very simple technique of using few pre-stored floor images. In indoor environment all unique floor types are considered and a single image is stored for each unique floor type. These floor images are considered as reference images. The algorithm acquires an input image frame and then a region of interest is selected and is scanned for obstacle using pre-stored floor images. The algorithm compares the present frame and the next frame and compute mean square error of the two frames. If mean square error is less than a threshold value α then it means that there is no obstacle in the next frame. If mean square error is greater than α then there are two possibilities; either there is an obstacle or the floor type is changed. In order to check if the floor is changed, the algorithm computes mean square error of next frame and all stored floor types. If minimum of mean square error is less than a threshold value α then flour is changed otherwise there exist an obstacle. The proposed algorithm works in real-time and 96% accuracy has been achieved.

1. Introduction

Obstacle detection is the process of detecting some object or barrier around the autonomous system that can affect movements of a system. Obstacle detection is used to avoid collisions in paths and to ensure safety of vehicles, robots etc. Blind persons have difficulties in navigation. Obstacle detection has a lot of importance in blind navigation. Mostly, blind people use stick cane to detect obstacle and to navigate, but this is not much accurate. Obstacle detector can aid blind peoples in path navigations and can avoid collisions. Different techniques have been developed to detect obstacles. Some of these techniques are using SONAR (Sound Navigation and Ranging) while other using laser camera. These systems are successfully integrated in automatic driving cars, e.g. Google cars, BMW etc. Different mechanisms are used for example in some cases obstacle are known while in some cases environment is known. In this paper we are presenting an algorithm for detecting obstacle in known environment. We are using an android based mobile camera and scan selected area in front of the camera for obstacle detection.

2. Related work

In the available literature, there exist many approaches for obstacle detection. For example, Tapu et al. [1] have proposed an approach for obstacle detection using smartphone. In this system, a set of interest points are extracted from image and is tracked by Lucas Kanade algorithm [2], then camera and the background motion is estimated through homographic transforms. Obstacles are marked as urgent or normal based on their distance to the subject and associated vector orientation. The detected obstacle is then sent to an object classifier. Once the obstacle is identified, the degree of danger is determined. Batavia et al. [3] have proposed a system that uses color segmentation and stereo based color homography. Color segmentation classify image areas as free space or obstacle. This system is based on training algorithm. Hesch et al. [4] have proposed sensors combination (e.g. foot mounted pedometer, 2D laser scanner and 3 axis gyroscope) for helping indoor localization. They have introduced two layered estimators. In first layer, the white cane orientation is tracked while in second layer, the 2D position of the person (holding white cane) is tracked. Another system proposed by Saffoury et al. [5] consist of two hardware components, a laser pointer and an android based smartphone. To measure distance of objects in environment, image processing based techniques are used for obstacle avoidance. This is based on laser light triangulation. In this approach edge detection is used for the enhancement of obstacle detection. User is also warned by recognizing stairs if they are present in field of view of the camera. When obstacle is detected, user is informed by acoustic signal. Saez et al. [6] have embodied full six DOF SLAM solution into a stereo wearable device working in real time. Localization technique is used for visually impaired people. The system presented by Fukasawa et al. [7] consists of map information and navigation system. Sensor is installed on white cane which detects navigation line drawn on the floor. RFID tags are used for map information system. Mahmud et al. [8] have used ultrasonic sensor and microcontroller for range determination and action. User is warned by audio and vibration. Ultrasonic signals are sent in three directions (left, right, front). Vlaminck et al. [9] have used RGB-D camera to perceive environment. They implement three things: obstacle detection, obstacle classification and self-localization. In self-localization, depth is perceived and color information based tracking technique is used. To provide semantic information to the user, obstacle is classified as walls, steps and doors etc. Pradeep et al. [10] have used head mounted stereo vision. Here the authors combine features, based on SLAM and visual odometry to create 3D map in order to detect obstacles. Leung et al. [11] estimate ego motion in dynamic environments. Their system consists of head mounted stereo vision for visually impaired. They find ground planes, and then decompose 6DOF into ground plane motion and planer motion of the ground plane. From the analysis of the disparity array they estimate ground plane and normal to the ground plane using visual data. Gangawane et al. [12] have used camera and ultrasonic sensors to detect obstacles. Obstacle is detected using ultrasonic sensor while from camera image, object size is calculated. In the work of Maidenbaum et al. [13] they have used two infrared emitters and sensors. These sensors emit beams and reflected beam is then received by sensors. One of these two sensors perceive waist height obstacles in 5-meter range and the direction is ahead to user. The second one is towards the ground at 45 degrees to detect ground level obstacles. Lee et al. [14] have used mobile parking sensor consisting of three modules (sensing module, processing module and warning module). Ultrasonic pulses are sent and then received if it is reflected by obstacles. Processing module then decides to warn the user or not, by using buzzer. Peng et al. [15] have histogram and edge cues to detect obstacle on floor regardless of obstacle height. Kumar et al.[16] have proposed a system where free space is computed. The system reduces the 3D point cloud obtained from stereo camera for stochastic representation. The algorithm assigns each cell of values (free or obstacle). In the work of Novotny et al. [17] and Benet et al. [18], IR sensors have been used for measuring distance for various surfaces. In the work of Mocanu [19], obstacle information is acquired from smartphone camera and ultrasonic sensors. Then obstacle recognition module recognizes obstacle and inform the user by acoustic signal. The work of Khan et al. [20] uses depth information to assist user. This system is able to detect dynamic obstacle and transmit information to user by audio feedback. In [21] author proposed a system using Microsoft Kinect sensor to detect 3D objects. Tactile device is used for user feedback. When object is found in front of Kinect sensor, user is informed. Brock et al. [22] have also used Microsoft Kinect sensor and depth information for detecting obstacles. Jayron et al. [23] used RFID readers and RFID cards. Cards are used as checkpoints for visually impaired people. In [24] RFID tags and GPRS have been used to guide blind people. User location is sent using GPRS and server calculates shortest path to destination. [25] Used RFID to self-locate blind in specific locations while [26] Used IR sensor for perceiving environment and four micrometers for haptic display to the user. These motors are adjusted to back vest of the user and provide directional feedback to the user. In [27] a system is proposed where input data is collected from laser data. Obstacle is detected by laser and from images, traversable area is identified using stored knowledge for humanoid robot. Jafari et al. [28] used RGB-D camera for environment perception. In [29] a wearable system is proposed to detect and avoid obstacles based on ultrasonic sensor. Feedback is given to the user by audio and vibration. Cartin et al. [30], presents a system for obstacle detection using multi-sonar system. The system sends vibrotactile feedback to the user.

In our proposed system, we use only a smartphone camera without using any other extra sensors to detect obstacles.

3. Proposed algorithm

The proposed algorithm is developed for indoor navigation. In indoor environment all unique floor types are considered and a single image is stored for each unique floor type. These floor images are considered as reference images.

First the algorithm needs to find the floor type in which the blind person is currently standing. The camera takes an image of the floor and segment a region of interest (ROI) from this image. This ROI is compared with all the stored images for different floor types. Mean Square Error $(MSE)_k$ is computed for every flour type k. $(MSE)_k$ is computed as follows.

$$(MSE)_k = \frac{1}{mn}\sum_{i=1}^{m}\sum_{j=1}^{n}[CF(i,j) - (F)_k(i,j)]^2 \tag{1}$$

Where CF represents ROI of the current floor, F_k represents ROI of floor k, and m and n represent the number of rows and columns in both images.

The algorithm set the current floor type as the one for which minimum of $(MSE)_i$ is less than a threshold value α. The algorithm then compares the present frame and the next frame and compute mean square error MSE of the two frames. If MSE is less than the threshold value α then it means that there is no obstacle in the next frame. If MSE is greater than α then there are two possibilities; either there is an obstacle or the floor type is changed. In order to check if the floor is changed, the algorithm computes $(MSE)_i$ of next frame and all stored floor types k. If minimum $(MSE)_i$ is less than a threshold value α then flour is changed otherwise there exist an obstacle.

The different steps of the algorithm are as follows:

1. CF = Read frame
2. $(MSE)_k = \frac{1}{mn}\sum_{i=1}^{m}\sum_{j=1}^{n}[CF(i,j) - (F)_k(i,j)]^2$
3. $Min\,(MSE)_k$ = minimum of $(MSE)_k$
4. If $Min\,(MSE)_k > \alpha$
   a. Obstacle exist
   b. Handle obstacle
   c. Goto step 1
5. If $Min\,(MSE)_k \leq \alpha$
   a. CF = floor for which $Min\,(MSE)_k \leq \alpha$
   b. NF= read next frame
   c. Compute MSE of current frame and next frame

d. If MSE <= α
e. No obstacle exists
f. Current frame = NF
g. Goto to step 5 (b)
h. If MSE > α
i. Goto step 4 (a)

4. Introduction experimental setup
For experiments we have considered three different building. These buildings are:

Academic Block, University of Malakand: Corridor of the Academic block is made of marble having white color. These marbles have zigzag patterns on it. Floor of the rooms are made of chips having white color.

Education Block, University of Malakand: Floor of the Education block is made of yellowish marble having dot patterns on it. It has one unique floor.

Social Science Block, University of Malakand: Social science block has one unique color floor. All the corridors and rooms floor are made of chips gray color.

Obstacle: Different obstacles that have been detected are tables, chairs, rotated chairs, tubs, bricks, human, stone, flower pot, Jug, cottons and dustbin etc.

Blind People: Tests were conducted by actors who worked as blind people.

Selection of threshold value: Threshold values are selected heuristically.

5. Results
We tested this algorithm in indoor environment having obstacles. Twenty different obstacles were placed in front of the camera and were tested on different floors. Algorithm has been tested by five users holding smartphone with tilt angle of 40 degrees. Details of the tests are as follows:

Test 1: During first test, actor was allowed to walk through in open corridor of academic block of the university. Five obstacle were placed in front of the camera.

Test 2: In second test, actor was allowed to walk through in closed corridor of the academic block of University. This time four obstacle were placed in front of the user.

Test 3: In third test, actor was allowed to walk around in classroom having chairs.

Test 4: Fourth test was conducted in the corridor of the education block. Ten obstacles were placed in front of the user.

Test 5: Fifth test was conducted in Social Sciences block of the University. Actor were allowed to walk through corridor having obstacles. Obstacles contained tables, chairs, bricks, dustbins etc. having various colors and sizes.

The result of these tests is shown in figure 1. The overall accuracy of the algorithm is 96%

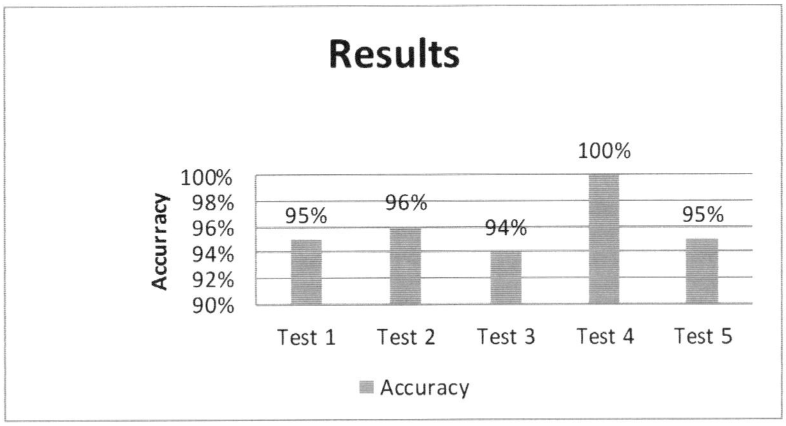

Figure 1. Accuracy of the different tests

6. Conclusion

In this paper, we have presented preliminary results of our algorithm for obstacle detection in indoor navigation using android-based mobile camera. The algorithm produces 96% accuracy. However, this algorithm has some limitations. Currently the threshold value α is computed heuristically which will be addressed in our future work and an automatic method will be developed for computing this value. Similarly, in this paper we have not presented mechanism for obstacle handling which will be presented in our next paper.

References

[1] Tapu R, Mocanu B, Bursuc A, Zaharia T. A smartphone-based obstacle detection and classification system for assisting visually impaired people. *In Proc. of the IEEE International Conf. on Computer Vision Workshops* 2013 (pp. 444-451).

[2] Lucas BD, Kanade T. An iterative technique of image registration and its application to stereo. In Proc. Int. Joint Conf. on Artificial Intelligence 1981 Aug (Vol. 1).

[3] Batavia PH, Singh S. Obstacle detection using adaptive color segmentation and color stereo homography. In Robotics and Automation, 2001. 2001 ICRA. IEEE International Conf. on 2001 (Vol. 1, pp. 705-710). IEEE.

[4] Hesch JA, Roumeliotis SI. Design and analysis of a portable indoor localization aid for the visually impaired. The International Journal of Robotics Research. 2010 Sep;29(11):1400-15.

[5] Saffoury R, Blank P, Sessner J, Groh BH, Martindale CF, Dorschky E, Franke J, Eskofier BM. Blind path obstacle detector using smartphone camera and line laser emitter. InTechnology and Innovation in Sports, Health and Wellbeing (TISHW), International Conf. on 2016 Dec 1 (pp. 1-7). IEEE.

[6] Saez, J. Manuel, Francisco Saez JM, Escolano F, Penalver A. First steps towards stereo-based 6DOF SLAM for the visually impaired. InComputer Vision and Pattern Recognition-Workshops, 2005. CVPR Workshops. IEEE Computer Society Conf. on 2005 Jun 25 (pp. 23-23). IEEE.

[7] Fukasawa AJ, Magatani K. A navigation system for the visually impaired an intelligent white cane. In Engineering in Medicine and Biology Society (EMBC), 2012 Annual International Conf. of the IEEE 2012 Aug 28 (pp. 4760-4763). IEEE.

[8] Mahmud N, Saha RK, Zafar RB, Bhuian MB, Sarwar SS. Vibration and voice operated navigation system for visually impaired person. In Informatics, Electronics & Vision (ICIEV), 2014 International Conf. on 2014 May 23 (pp. 1-5). IEEE.

[9] Vlaminck M, Hiep QL, Vu H, Veelaert P, Philips W. Indoor assistance for visually impaired people using a RGB-D camera. In Image Analysis and Interpretation (SSIAI), 2016 IEEE Southwest Symp. on 2016 Mar 6 (pp. 161-164). IEEE.

[10] Pradeep V, Medioni G, Weiland J. Robot vision for the visually impaired. In Computer Vision and Pattern Recognition Workshops (CVPRW), 2010 IEEE Computer Society Conf. on 2010 Jun 13 (pp. 15-22). IEEE.

[11] Leung, Tung-Sing, and Gerard Medioni. "Visual navigation aid for the blind in dynamic environments." Proc. of the IEEE Conf. on Computer Vision and Pattern Recognition Workshops. 2014.

[12] Shahdib F, Bhuiyan MW, Hasan MK, Mahmud H. Obstacle detection and object size measurement for autonomous mobile robot using sensor. International Journal of Computer Applications. 2013 Jan 1;66(9).

[13] Maidenbaum S, Hanassy S, Abboud S, Buchs G, Chebat DR, Levy-Tzedek S, Amedi A. The "EyeCane", a new electronic travel aid for the blind: Technology, behavior & swift learning. Restorative neurology and neuroscience. 2014 Jan 1;32(6):813-24.

[14] Lee CL, Chen CY, Sung PC, Lu SY. Assessment of a simple obstacle detection device for the visually impaired. Applied ergonomics. 2014 Jul 31;45(4):817-24.

[15] Peng E, Peursum P, Li L, Venkatesh S. A smartphone-based obstacle sensor for the visually impaired. In International Conf. on Ubiquitous Intelligence and Computing 2010 Oct 26 (pp. 590-604). Springer, Berlin, Heidelberg.

[16] Kumar S. Binocular stereo vision based obstacle avoidance algorithm for autonomous mobile robots. In Advance Computing Conf., 2009. IACC 2009. IEEE International 2009 Mar 6 (pp. 254-259). IEEE.

[17] Novotny PM, Ferrier NJ. Using infrared sensors and the Phong illumination model to measure distances. In Robotics and Automation, 1999. Proc. 1999 IEEE International Conf. on 1999 (Vol. 2, pp. 1644-1649). IEEE.

[18] Benet G, Blanes F, Simó JE, Pérez P. Using infrared sensors for distance measurement in mobile robots. Robotics and autonomous systems. 2002 Sep 30;40(4):255-66.

[19] Mocanu B, Tapu R, Zaharia T. When ultrasonic sensors and computer vision join forces for efficient obstacle detection and recognition. Sensors. 2016 Oct 28;16(11):1807.

[20] Khan A, Moideen F, Lopez J, Khoo WL, Zhu Z. KinDectect: kinect detecting objects. In International Conf. on Computers for Handicapped Persons 2012 Jul 11 (pp. 588-595). Springer, Berlin, Heidelberg.

[21] Takizawa H, Yamaguchi S, Aoyagi M, Ezaki N, Mizuno S. Kinect cane: An assistive system for the visually impaired based on three-dimensional object recognition. In System Integration (SII), 2012 IEEE/SICE International Symposium on 2012 Dec 16 (pp. 740-745). IEEE.

[22] Brock M, Kristensson PO. Supporting blind navigation using depth sensing and sonification. In Proc. of the 2013 ACM conf. on Pervasive and ubiquitous computing adjunct publication 2013 Sep 8 (pp. 255-258). ACM.

[23] Sanchez J, Yumang A, Caluyo F. Rfid based indoor navigation with obstacle detection based on a* algorithm for the visually impaired. International Journal of Information and Electronics Engineering. 2015 Nov 1;5(6):428.

[24] Chumkamon S, Tuvaphanthaphiphat P, Keeratiwintakorn P. A blind navigation system using RFID for indoor environments. In Electrical Engineering/Electronics, Computer, Telecommunications and Information Technology, 2008. ECTI-CON 2008. 5th International Conf. on 2008 May 14 (Vol. 2, pp. 765-768). IEEE.

[25] Willis S, Helal S. RFID information grid for blind navigation and wayfinding. In Wearable Computers, 2005. Proc.. Ninth IEEE International Symposium on 2005 Oct 18 (pp. 34-37). IEEE.

[26] Ertan S, Lee C, Willets A, Tan H, Pentland A. A wearable haptic navigation guidance system. In Wearable Computers, 1998. Digest of Papers. Second International Symposium on 1998 Oct 19 (pp. 164-165). IEEE.
[27] Maier D, Stachniss C, Bennewitz M. Vision-based humanoid navigation using self-supervised obstacle detection. International Journal of Humanoid Robotics. 2013 Jun;10(02):1350016.
[28] Jafari OH, Mitzel D, Leibe B. Real-time RGB-D based people detection and tracking for mobile robots and head-worn cameras. In Robotics and Automation (ICRA), 2014 IEEE Inter. Conf. on 2014 May 31 (pp. 5636-5643). IEEE.
[29] Shin BS, Lim CS. Obstacle detection and avoidance system for visually impaired people. Haptic and Audio Interaction Design. 2007:78-85.
[30] Cardin S, Thalmann D, Vexo F. A wearable system for mobility improvement of visually impaired people. The Visual Computer. 2007 Feb 1;23(2):109-18.

A BSS-Based Sum-difference Four-channel Algorithm for Suppressing Multiple-Mainlobe Jammings

ZHOU Bi-lei [1,*], LI Rong-feng[1], SHAO Yin-bo[1], CAI Guan[1] and GUAN Wei[1]

[1]Air Force Early Warning Academy, Wuhan 430019, China

[*]zhoubilei666888@sina.com

Abstract. The target signal loss seriously when jamming enters into mainlobe. The existing studies can only suppress one MLJ. Hence, this paper proposes a BSS-based sum-difference four-channel algorithm for suppressing multiple-mainlobe jammings (Multiple-MLJs). The inputs channels are consisted of sum-difference four-channel. According to the statistical independence, the target and MLJ can be separated by adopting the BSS. A remark feature of the proposed approach is that it does not require priori knowledge about the real target or jammings, and it is easy to implement for engineering applications.

1. Introduction

In recent years, the US military has begun to equip EA-18G aircraft, which gradually replace EA-6B. EA-18G is not only the strongest electronic jammer, but also the strongest attack aircraft. It will be the major airborne electronic attack resources of US Navy in the future. What is more, EA-18G can fly by attack aircraft. Therefore, mainlobe jamming (MLJ) is a severe thereat to radar. In generally, the MLJ has two types. The first is self-defensive jamming, which comes from the target. The second enters into the radar mainbeam, whose direction is different with the target, so we can call it as near-mainlobe jamming. In this paper, near-mainlobe jamming is called mainlobe jamming (MLJ) for short. At present, there are many technologies for suppressing self-defensive jamming [1-6]. However, they are not effective to suppress the MLJ. The sum-difference mainlobe canceller (MLC) [7] can suppress the MLJ effectively. Kai-Bor Yu et al. introduced another adaptive monopulse angle estimation algorithm, composed of ADBF and MLC [8-9]. Two anti-jamming algorithms which combine the sum-difference and auxiliary beam were proposed in [10-11]. However, those methods can only suppress one MLJ.

The BSS approach is recovering a number of unobserved signals from observed mixtures. At present, many researchers have done some works based on the BSS technology for suppressing the MLJ. The Fast ICA (independent component analysis) algorithm based on negative entropy maximization proposed in [12] and the joint approximation diagonalization (JADE) algorithm proposed in [13-15] are more widely used at present. However, these methods can only suppress one MLJ as [7-11]. In this paper, we propose a BSS-based sum-difference four-channel algorithm for suppressing multiple-MLJs.

2. The basic principle of the BSS

We consider the problem where M independent non-stationary signals are received by an array of N ($N \geq M$) channels. The classical narrowband signal model is used:

$$x(t) = Hs(t) + v(t); \quad t \in \mathbb{R} \tag{1}$$

where $x(t) \in \mathbb{C}^{N\times 1}$ is the vector of the channel outputs, $s(t) \in \mathbb{C}^{M\times 1}$ is the signal vector and $v(t) \in \mathbb{C}^{N\times 1}$ is the white Gaussian noise vector. The unknown mixed matrix $H \in \mathbb{C}^{N\times M}$ maps the source signals to the channel outputs, and is assumed to be of full column rank.

Given K snapshots, $\{x(n)\}_0^{K-1}$, of the array output vector, the problem is to estimate the mixed matrix $\boldsymbol{H}$, where $\boldsymbol{HB}$ is essentially equal to the identity matrix, i.e. $\boldsymbol{HB} = \boldsymbol{PD}$ where $\boldsymbol{P}$ is a permutation matrix and $\boldsymbol{D}$ is a non-singular diagonal matrix.

3. Mixed signal model

This paper considers rectangular array which has N_2 columns, and each column has N_1 elements, shown as figure 1. The distance between the elements in the direction of x (and y) is d, the wavelength of the transmitted signal is λ. θ means the azimuth angle of signal, and ϕ means the elevation angle of signal. Assuming there exist M signals $s(t) = [s_1(t), s_2(t), \ldots, s_m(t), \ldots, s_M(t)]^T \in \mathbb{C}^{M\times 1}$, located at (θ_m, ϕ_m) $(m = 0, 1, \ldots, M, M \le 3)$.

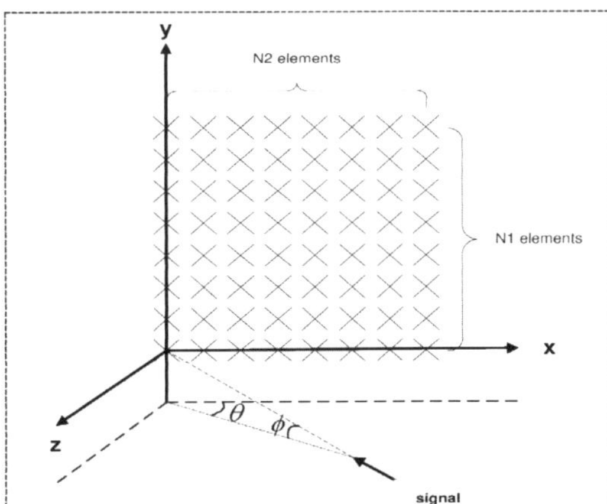

Figure 1. array model

This paper considers sum-difference four beams as input channels, containing of sum beam Σ, azimuth difference beam Δ_A, elevation difference beam Δ_E and double difference beam Δ_Δ. Hence, the total number of channels is four, which must be larger than the number of signals, that is $M \le 4$. And the mixed $\boldsymbol{H}$ is

$$\boldsymbol{H} = \begin{bmatrix} w_\Sigma^H \boldsymbol{a}(\theta_0,\varphi_0), & w_\Sigma^H \boldsymbol{a}(\theta_1,\varphi_1), \ldots, & w_\Sigma^H \boldsymbol{a}(\theta_{M-1},\varphi_{M-1}) \\ w_{\Delta_A}^H \boldsymbol{a}(\theta_0,\varphi_0), & w_{\Delta_A}^H \boldsymbol{a}(\theta_1,\varphi_1), \ldots, & w_{\Delta_A}^H \boldsymbol{a}(\theta_{M-1},\varphi_{M-1}) \\ w_{\Delta_E}^H \boldsymbol{a}(\theta_0,\varphi_0), & w_{\Delta_E}^H \boldsymbol{a}(\theta_1,\varphi_1), \ldots, & w_{\Delta_E}^H \boldsymbol{a}(\theta_{M-1},\varphi_{M-1}) \\ w_{\Delta_\Delta}^H \boldsymbol{a}(\theta_0,\varphi_0), & w_{\Delta_\Delta}^H \boldsymbol{a}(\theta_1,\varphi_1), \ldots, & w_{\Delta_\Delta}^H \boldsymbol{a}(\theta_{M-1},\varphi_{M-1}) \end{bmatrix} = \begin{bmatrix} c_{1,0} & c_{1,1} & \cdots & c_{1,M-1} \\ c_{2,0} & c_{2,1} & \cdots & c_{2,M-1} \\ c_{3,0} & c_{3,1} & \cdots & c_{3,M-1} \\ c_{4,0} & c_{4,1} & \cdots & c_{4,M-1} \end{bmatrix}_{4\times M} \tag{2}$$

where $\boldsymbol{a}(\theta_i,\varphi_i) \in \mathbb{C}^{N_1N_2\times 1}$ is the direction vector of the m th signal $s_m(t)$, an $\boldsymbol{a}(\theta_i,\varphi_i) = \boldsymbol{a}(\varphi_i) \otimes \boldsymbol{a}(\theta_i), i = 0, 1, \ldots, M-1$. The symbol $\otimes$ is Kronecker product. $\boldsymbol{a}(\varphi_i) = [1, e^{-j\alpha_i}, \cdots, e^{-j(N_1-1)\alpha_i}]$, $\boldsymbol{a}(\theta_i) = [1, e^{-j\beta_i}, \cdots, e^{-j(N_2-1)\beta_i}]$, $\alpha_i = 2\pi d \sin\varphi_i / \lambda$, $\beta_i = 2\pi d \cos\theta_i \cos\varphi_i / \lambda$. $\boldsymbol{w}_\Sigma \in \mathbb{C}^{N_1N_2\times 1}$ is the weighting vector of sum beam, $\boldsymbol{w}_{\Delta_E} \in \mathbb{C}^{N_1N_2\times 1}$ is the weighting vector of elevation difference beam, $w_{\Delta_A} \in \mathbb{C}^{N_1N_2\times 1}$ is the weighting vector of azimuth difference beam, and $w_{\Delta_\Delta} \in \mathbb{C}^{N_1N_2\times 1}$ is the weighting vector of double difference beam.

$$\begin{cases} w_\Sigma = \left(w_{Taylor\_y} \odot a(\varphi_i)\right) \otimes \left(w_{Taylor\_x} \odot a(\theta_i)\right) \\ w_{\Delta_A} = \left(w_{Taylor\_y} \odot a(\varphi_i)\right) \otimes \left(w_{Bay\_x} \odot a(\theta_i)\right) \\ w_{\Delta_E} = \left(w_{Bay\_y} \odot a(\varphi_i)\right) \otimes \left(w_{Taylor\_x} \odot a(\theta_i)\right) \\ w_{\Delta_\Delta} = \left(w_{Bay\_y} \odot a(\varphi_i)\right) \otimes \left(w_{Bay\_x} \odot a(\theta_i)\right) \end{cases} \tag{3}$$

where $w_{Taylor\_x} \in \mathbb{C}^{N_2\times 1}$ and $w_{Bay\_x} \in \mathbb{C}^{N_2\times 1}$ is the Taylor vector and Bayliss vector in the direction of x, $w_{Taylor\_y} \in \mathbb{C}^{N_1\times 1}$ and $w_{Bay\_y} \in \mathbb{C}^{N_1\times 1}$ is the Taylor vector and Bayliss vector in the direction of y. The symbol $\odot$ is Hadamard product.

4. Algorithm principle

The algorithm principle of this paper is shown as figure 2.

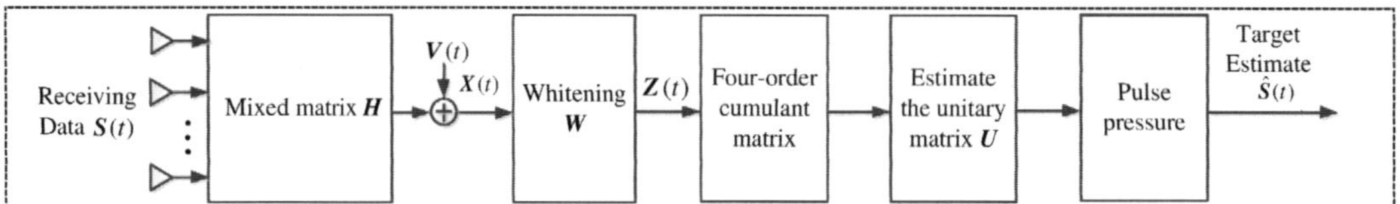

Figure 2. The algorithm procedure

The steps of the new algorithm:

(1) Mixed process: $x(n) = Hx(n) + v(n)$.

where $x(n)$ is the digitized mixed signal of $x(t)$.

(2) Whitening process: $z(n) = Wx(n)$.

where W is the whitening matrix, obtained by the subspace of covariance matrix:

$$R_{XX} = \frac{1}{4}\sum_{n=1}^{4} x(n)x^H(n) \in \mathbb{C}^{4\times 4} \tag{4}$$

$$W = \left[\left(\mu_0 - \sigma^2\right)^{-1/2} g_0, \left(\mu_1 - \sigma^2\right)^{-1/2} g_1, \ldots, \left(\mu_{M-1} - \sigma^2\right)^{-1/2} g_{M-1}\right]^H \tag{5}$$

where $\mu_0, \mu_1, \ldots, \mu_{M-1}$ is the large eigenvalue, $g_0, g_1, \ldots, g_{M-1}$ is the corresponding eigenvector. And σ^2 is the nosie variance estimation, which equals to the average of the remaining eigenvalue.

Whitening signal $z(n)$ can also be expressed as:

$$\begin{aligned} z(n) &= Wx(n) = WHs(n) + Wv(n) \\ &= Us(n) + Wv(n) \end{aligned} \tag{6}$$

Obviously, in order to restore the signal $s(n)$, unitary matrix U should be estimated.

(3) Four-order cumulant matrix: $Q_Z(\Omega) \in \mathbb{C}^{M\times M}$.

$$[Q_Z(\Omega)]_{ij} \overset{def}{=} \sum_{k=1}^{M}\sum_{l=1}^{M} K_{ijkl}(z)^{*}\Omega_{kl};\quad i,j \in [1,2,\ldots,M] \tag{7}$$

where Ω_{kl} is one element of the matrix $\boldsymbol{\Omega} \in \mathbb{C}^{M\times M}$, $\Omega_{kl} = \boldsymbol{\Omega}(k,l)$, $K_{ijkl}(z) = cum\left(z_i(n), z_j^*(n), z_k(n), z_l(n)^*\right)$, $z_i(n)$ is the i-th row of $z(n)$, $cum(\cdot,\cdot,\cdot,\cdot)$ is the four-order cumulant operation.

(4) Estimate the unitary matrix U : the estimated unitary matrix $\hat{U} \in \mathbb{C}^{8\times M}$ is obtained by characteristic decomposition as $Q_z(\boldsymbol{\Omega}) = \hat{U}\Sigma\hat{U}^H$, where $\Sigma \in \mathbb{C}^{M\times M}$ is a diagonal matrix.

(5) Signal separation: $\hat{s}(n) = \hat{U}z(n)$.

(6) Pulse pressure:

$$y(n) = conv\left(\hat{s}(n), s_0^*(-n)\right) \tag{8}$$

where $conv(\ ,\ )$ is convolution operation, $y(n)$ is the outputs. By peak detection, the target can be obtained.

5. Simulation

Consider a rectangular planar array which has 48 columns, and each column has 22 elements, placed half a wavelength apart. The transmission signal is LFM signal, bandwidth is $5MHz$, pulse width is $20\mu s$, and sampling rate is $20MHz$. The sumbeam static weight in the direction of elevation and azimuth is sampled for -35dB of Chebyshev. The difference beam static weight in the direction of azimuth is sampled for -35dB of Bayliss. The boresight of sumbeam is (90°, 30°). The 3dB sumbeam width along azimuth and elevation are 3.2° and 7.1° respectively. The incoming angle of the target with -11dB is located at (90°, 30°) and the 3000th sampling point. The noise in each channel follows additive white Gauss noise.

Simulation 1: one target and three MLJs.

Simulation conditions: The incoming angle of the three MLJs with 50dB is located at (90.8°, 30°), (90°, 31.775°), (89.2°, 30°), which is located at the one quarter of 3dB beam width (along the azimuth and elevation direction).

Figure 3 shows the outputs of the sum-different four channels by pulse pressure, where we cannot find the target.

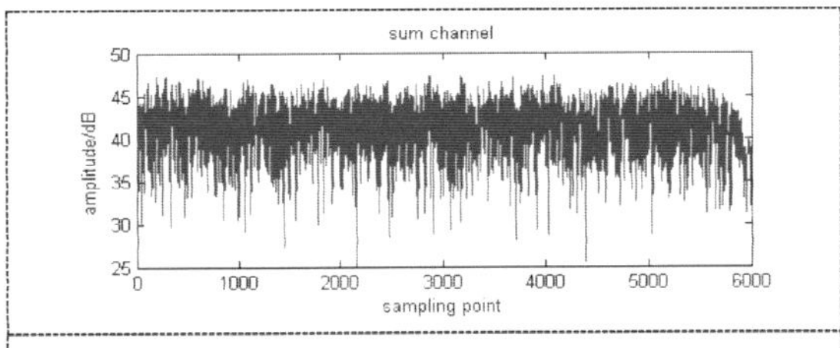

(a) sum channel

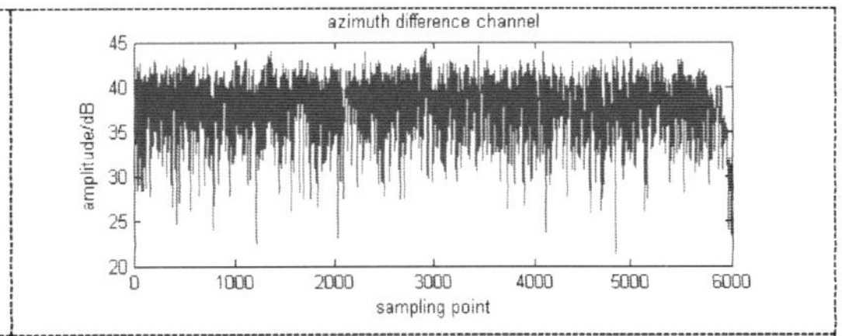

(b) azimuth difference channel

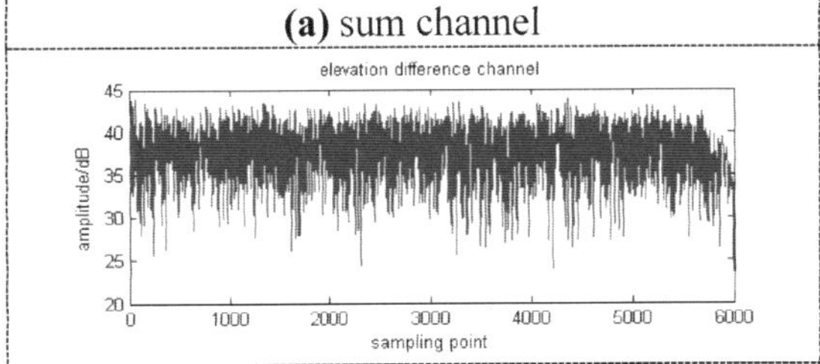

(c) elevation channel

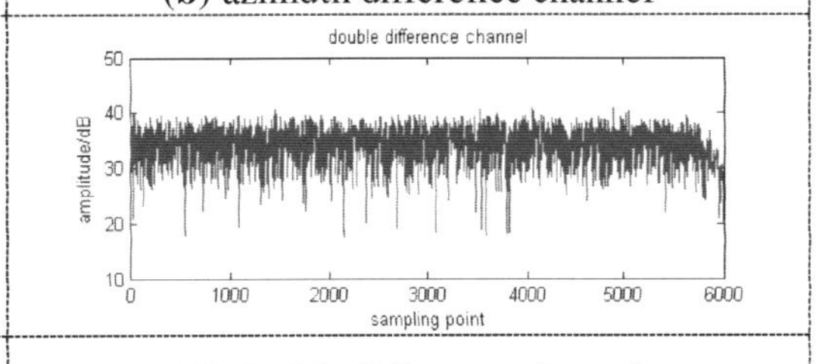

(d) double difference channel

Figure 3. mixed signal with pulse pressure

Figure 4 shows the signal separation result by the new algorithm, where the target is obtained.

When the MLJs are closer with the target, coming from (89.6°, 30°), (90.4°, 30°), (90.4°, 31°), located at one eighth of 3dB beam width, the target still can be obtained in the figure 5. Hence, the new algorithm can be applied to more severe mainlobe jamming environment.

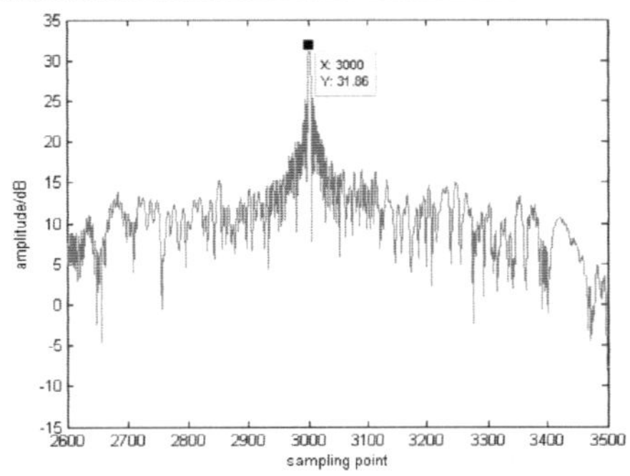

Figure 4. Separation result by the new algorithm

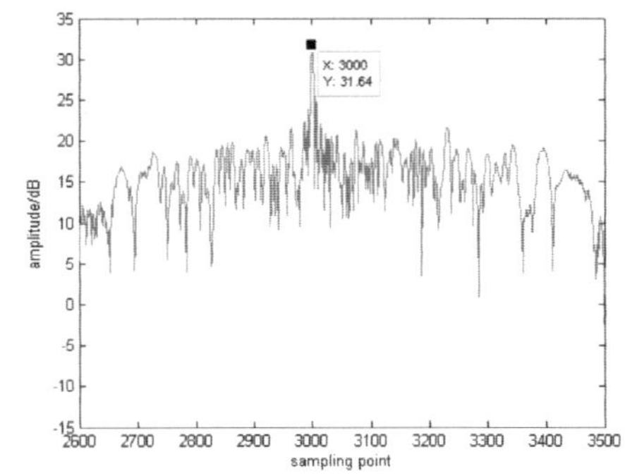

Figure 5. Separation result by the new algorithm

Simulation 2: one target, one mainlobe dense false-target (MDFJ) and two MLJs.

The simulation conditions: The incoming angle of the two MLJs with 50dB is located at (90.8°, 30°), (90°, 31.775°), and the MDFJ with -5dB is located at (89.2°, 30°). These jammings are all located at the one quarter of 3dB beam width. And the MDFJ located at the 2100th to 4000th sampling point.

Figure 6 shows the separation result by the new algorithm. Obviously, the target and MDFJ have been separated, and it is clearly that the channel 1 has the target signal.

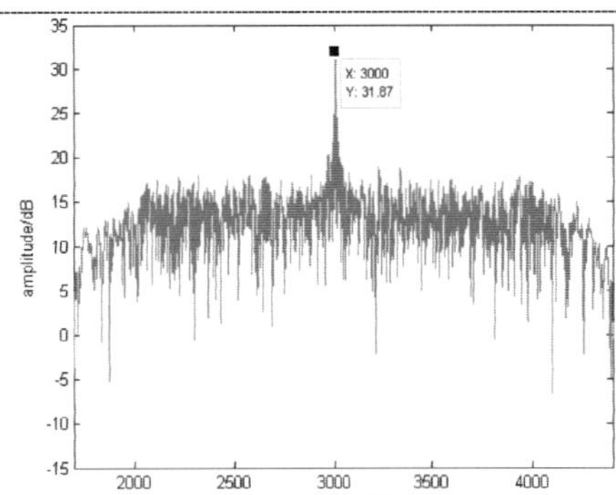

(a) channel 1

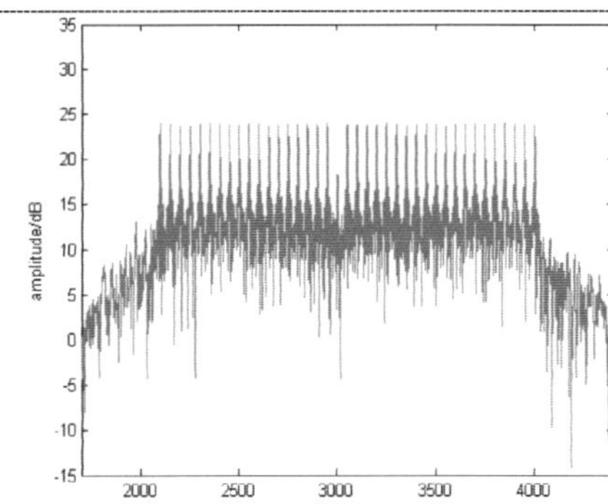

(b) channel 2

Figure 6. Separation result by the new algorithm

6. Conclusions

Mainlobe jamming is still a severe problem in the anti-jamming field currently. And the existing studies can only suppress one MLJ. In order to solve this problem, this paper proposes a BSS-based four-channel sum-difference algorithm for suppressing multiple-mainlobe jammings. According to the simulation result, the new algorithm can suppress multiple-MLJs and MDFJ effectively. Moreover, the new algorithm can be applied to more severe mainlobe jamming environment. And it is easy for implement in engineering. Hence, the algorithm of this paper has an important engineering application value.

References

[1] SHI Yanbin, ZHANG An, GAO Xianjun. Research on the influence of SSJ on the survivability of aerial attack aircraft [J]. Electronics Optics & Control, 2008, 15(6): 7-9.
[2] LI Yongzhen, WANG Guoyu, WANG Liandong, WANG Xuesong. Study on Polarization Cancellation Algorithm of Self-defensive Blanket Jamming Based on Radar Sidelobe—canceller Antenna [J]. SIGNAL PROCESSING, 2008, 24(5): 775-779.
[3] Liming Ding, Rongfeng Li, Yongliang Wang, Lingyan Dai, Fengbo Chen. Discrimination and identification between mainlobe repeater jamming and target echo by basis pursuit. IET radar, sonar & navigation, 2017, 11(1):11-20.
[4] Liming Ding, Rongfeng Li, Lingyan Dai, Fengbo Chen, Yongliang Wang. Discrimination and identification between mainlobe repeater jamming and target echo via sparse recovery. IET radar, sonar & navigation, 2017, 11(2):235-242.
[5] Liming Ding, Lingyan Dai, Rongfeng Li, Fengbo Chen, Yongliang Wang. Discrimination and identification of time-delay repeater jamming and target echo by basis pursuit. 2015 IET International Radar Conference, Hangzhou, China.
[6] LUO Shuangcai, TANG Bin. An Algorithm of Deception Jamming Suppression Based on Blind Signal Separation [J]. Journal of Electronics & Information Technology, 2011, 33(12): 2801-2806.
[7] S.P. Applebaum, R. Wasiewicz. Main beam jammer cancellation for monopulse sensors. Final technical report DTIC RADC-TR-86-267, 1984.
[8] Yu K B, Murrow D J. Adaptive digital beamforming for angle estimation in jamming [J]. IEEE Transactions on Aerospace Electronic Systems, 2001, 37(2): 508-522.
[9] Yu K B, Murrow D J. Combining sidelobe canceller and mainlobe canceller for adaptive monopulse radar processing. US Patent 6867726, B1 2005.
[10] LI Rongfeng, RAO Can, DAI Lingyan , WANG Yongliang. Combining sum-difference and auxiliary beam for adaptive monopulse in jamming. Journal of Systems Engineering and Electronics, 2013, 24(3): 372-381.
[11] ZHOU Bilei, LI Rongfeng, DAI Linyan, WANG Yongliang. Adaptive Monopulse Algorithm Based On Combining Sum-Difference And Auxiliary Beam For Anti-Jamming At Subarray Level [J]. Radar Science And Technology, 2014, 12(4): 379-388.
[12] WANG Wentao, ZHOU Qingsong, LIU Xinghua, LI Lei. A Study on Radar Mainlobe Jamming Suppression Based on Blind Source Separation of Fast ICA [J]. Modern Radar, 2015, 37(12): 40-48.
[13] WANG Jianming, WU Guangxin, ZHOU Weiguang. A Study on Radar Mainlobe Jamming Suppression Based on Blind Source Separation Algorithm [J]. Modern Radar, 2010, 32(10): 46-49.
[14] WANG Wentao, ZHANG Jianyun, LIU Xinghua, LI Lei. Radar Anti-mainlobe-Jamming Based on Blind Source Separation Algorithm of JADE [J]. Fire Control & Command Control, 2015, 40(9): 104-108.
[15] Cardoso J. F. Blind beamforming for non-Gaussian signals [J]. IEEE Proceedings -F. 1993, 140(6): 224-230.

Study on Data Clustering and Intelligent Decision Algorithm of Indoor Localization

Zexi Liu[1]

[1]College of Computer Science, Chongqing University, Chongqing, 400044, China

Email: *345442314@qq.com

Abstract. Indoor positioning technology enables the human beings to have the ability of positional perception in architectural space, and there is a shortage of single network coverage and the problem of location data redundancy. So this article puts forward the indoor positioning data clustering algorithm and intelligent decision-making research, design the basic ideas of multi-source indoor positioning technology, analyzes the fingerprint localization algorithm based on distance measurement, position and orientation of inertial device integration. By optimizing the clustering processing of massive indoor location data, the data normalization pretreatment, multi-dimensional controllable clustering center and multi-factor clustering are realized, and the redundancy of locating data is reduced. In addition, the path is proposed based on neural network inference and decision, design the sparse data input layer, the dynamic feedback hidden layer and output layer, low dimensional results improve the intelligent navigation path planning.

1. Introduction

Indoor positioning technology extends the relation of human perception space geography from outdoor to indoor, so that human beings have the ability of positional perception in architectural space.

Unlike satellite localization technology, due to the complexity of indoor environment, signal transmission attenuation, indoor location technology to conquer signal real-time tracking, multipath signals more effective compensation, and the key technology of network coordination [1].

United States proposed the E911 program to develop the indoor localization technology. Apple has introduced beacon's indoor localization technology to raise indoor positioning accuracy from 30 meters to about 10 meters. Ericsson used 4G base station technology to optimize the positioning accuracy of wide-area to about 30. Japan has also proposed more advanced indoor targeting technology. EU has proposed a law to promote the development of indoor localization technology. In 2009, China developed a technology called "XiHe". It used TC - OFDM to improve the ability of indoor and outdoor seamless positioning. According to the survey statistics, indoor localization technology will usher in nearly one trillion in the market space by 2020, and by using base stations, terminals, platforms and services, it will form a new industrial chain. Indoor positioning technology will be applied to fire rescue, emergency management and other social services.

The current indoor localization technology is mainly based on distance measurement, fingerprint and inertial navigation method. These three methods have different advantages and disadvantages in the range, precision and complexity of indoor positioning. But it still has the following problems:

(1) The coverage of single network is insufficient, and multi-source network coordination is required. The localization algorithm based on distance measurement depends on the transmission interval of pilot frequency, and it is difficult to achieve high precision in one room [2]. However, the indoor localization technology based on fingerprint needs dense nodes to realize the high precision of local area. Therefore, multi-network association has become an important idea of indoor localization.

(2) Localization data is redundant, data processing is very complex and location services need precise planning for the navigation path. In the condition of multi-network co-variation, the location data involved is huge [2]. At the same time, the indoor scene is complex and needs to be effectively perceived, only in the way can navigation be more reliable.

To solve the problems of lacking of precision and data redundancy in indoor positioning, this paper put forward an algorithm based on the data clustering and intelligent decision research of indoor localization. By optimizing the clustering process of massive indoor localization data, the data normalization pretreatment, multi-dimensional controllable clustering center and multi-factor clustering are realized, and the redundancy of locating data is reduced. In the end, to make the path navigation more intelligent, we used neural network inference and decision, designed the sparse data input layer, the dynamic feedback hidden layer and low dimensional output layer.

2. Basic knowledge of multisource indoor localization

2.1 Localization based on distance measurement

Using the principle of geometry to calculate the location of the target, it can determine the location of the target by the channel arrival time or the arrival angle.

Wang Qin from University of Science and Technology Beijing proposed a TOA localization algorithm of ranging error classification. Based on the RSSI and RITEM values in the TOA ranging process, we can estimate the error level. By using the maximum likelihood method, we may choose the label position of maximum probability as positioning results to improve the positioning precision of the algorithm[3].

Jing Hao from University of Nottingham put forward a wireless signal distance weighted indoor co-localization method. He used particle filter integrating with course-calculated data, WiFi data, distance between the users, indoor maps and other information to eliminate the error when signal is not stable or database dose not update on time.

Xu Peipei from Southeast University put forward a scheme of TOA without time synchronization process. By using bi-directional ranging method to realize positioning devices. She loaded the transmission time and receiving time to the data frame, to complete the process of ranging between two points. This scheme reduced the consumption of the time synchronization module [4], at the same time, this improved localization algorithm made the whole positioning system more accurate and robust.

2.2 Localization based on fingerprint

By using the distribution characteristics in different spaces, comparing the characteristics of the actual current environment to the measured feature, we can determine the current relative position of the target. Fingerprint localization generally uses signal strength as the characteristic, the localization accuracy depends on the distribution density of the node.

Tao Zheng from Dalian University of technology put forward a WLAN location fingerprinting positioning algorithm based on the chi-square distance. In this method, we can use fast K-means clustering algorithm to cluster reference point of location area. This method enhanced the robustness of data processing. By using the method of sensitivity and chi-square distance as the distance measure function in position fingerprint localization algorithm [5], we can reduce the effects of environmental noise on the positioning accuracy.

Dan Yu from Taiyuan University of science and technology proposed an algorithm based on technology of SLAM to optimize the fingerprint data quality and to enhance the uniqueness of fingerprint. This algorithm ensured the quality of fingerprint data and essentially improved the performance of the fingerprint orientation.

Li Wenhao from Liaoning University proposed that by using Matrix Completion (MC) theory we can establish the off-line position fingerprint effectively. He also put forward a new type of algorithm which is called BSA – SVT. In the online positioning stage he put forward a positioning algorithm based on the DE - BCS, further improving the positioning accuracy.

2.3 Localization based on inertial device

Inertial navigation measures the acceleration of the carrier in the inertial reference system, by integrating it with time and converting it into the navigation system, we can get information in the navigation system such as speed, yaw angle and position.

Gong-min Yan of Northwestern Polytechnic University put forward the compass initial alignment algorithm based on position error [6]. He combined it with the strap down inertial navigation attitude algorithm and get the Positioning damping strap down inertial navigation algorithm. This algorithm is easy to calculate and it's very stable, but compared with Kalman filter for integrated navigation, there're still lots of shortcomings in the application conditions.

Xi Wen et al. from Nanchang University obtained the inertial navigation device data on the basis of the Kinect visual range. Using Kalman filter, he optimized the error of position and posture as observed quantity to fix the error. This method improve the positioning accuracy and stability of indoor mobile robot.

Lei Wang from Tongji University used UWB module installed on the mobile node to produce the estimated value of position. This method calibrated the position of moving nodes obtained by inertial navigation algorithm. It has higher indoor positioning accuracy.

3. Clustering processing of large amount of indoor location data

Because in the multi-source indoor positioning there are base station, Wi-Fi, inertial navigation and infrared and other infrastructure, data collected including images, voice, text, etc. are heterogeneous. There are great difference between size of different expression data, and at different time and location, there exists great information redundancy. For this reason, we introduce clustering process of large amount of indoor positioning data. In this process, we set the Euclidean distance as the criteria for the classification of elements, set the simplified location information as the aim of clustering processing. We iterate this classification process until the location information has no obvious change.

3.1 Euclidean similarity measure

Assume a given location data set $X = \{x_m | m = 1,2, \dots, n\}$, we use d successive attributes $A_1, A_2 \dots A_d$ to describe samples in X. $x_i = (x_{i1}, x_{i2}, \dots, x_{id})$ describes all attributes of ith sample.

We usually use $d(x_i, x_j)$ as distance to describe the similarity between sample x_i and x_j, the greater the distance, the greater the difference. According to Euclidean distance formula, we can calculate the similarity:

$$d(x_i, x_j) = \sqrt{\sum_{k=1}^{d} (x_{ik} - x_{jk})^2} \quad (1)$$

3.2 Error sum of squares criterion and function clustering criterion

In this paper, we use error sum of squares function as clustering performance criteria function. Assuming data set X includes k cluster subsets $X_1, X_2, \dots X_k$ and the sample number of this data set are $n_1, n_2 \dots, n_k$; the center of clustering subsets are $m_1, m_2, \dots m_k$, the formula of error sum of squares function can be presented as:

$$E = \sum_{i=1}^{k} \sum_{p \epsilon X_i} ||p - m_i||^2 \quad (2)$$

Cluster average value of similarity

There are four main steps:

1. Allocate all of the objects to k clusters which are not empty.
2. Calculate the average of each cluster and use that average to represent the cluster.
3. Allocate each object according to the distance between the object and the cluster center.
4. Turn to step2, recalculate each cluster's average value. Iterate this process until it satisfies some criteria function.

In the process of clustering, assume that the location data meets the characteristics of the checkerboard data set. 486 positive data is used, and the data is transformed to [-1, 1]. The distribution is shown in figure 1.

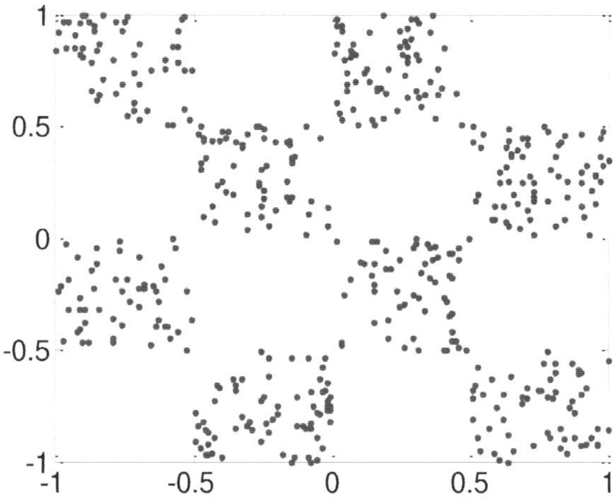

Figure 1. initial distribution of the data

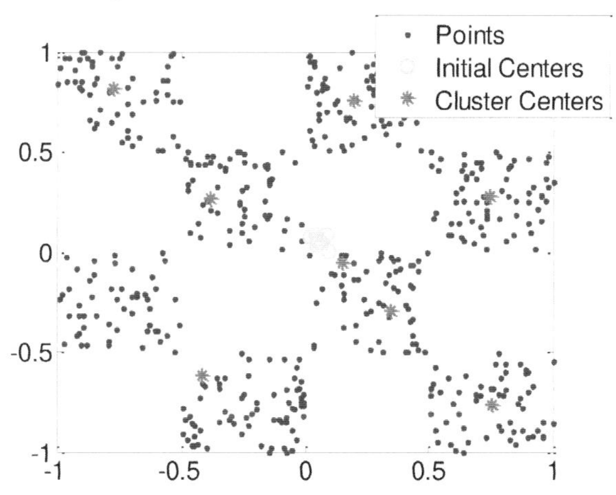

Figure 2. clustering distribution of data

Through the iterative process, the location data set is divided into different categories, which makes the estimation of the location information of the clustering performance to be optimal. The results are shown in figure 2.

4. Path reasoning and decision making based on neural network

Each unit in artificial neural network has the extensive connection, parallel-distributed information storage and processing ability, adaptive learning ability, etc. Thus, it can provide better fault tolerance, strong adaptability and fusion processing ability for the path reasoning and decision making of indoor positioning [7]. We will design sparse data input layer, dynamic feedback implicit layer and low-dimensional result output layer for the path reasoning and decision making based on neural network.

4.1 Data input layer

Indoor positioning data are mass and redundant. Through clustering algorithm, the data category will be simplified. But the number of same type of data is still large and its correlation with path planning is still low, which makes it difficult to process fast.

Therefore, at first, we should input the clustered data concurrently. Scale transformation and pretreatment are required after the most important input data are determined. The input vector of n dimensions are defined as M. After the scale transformation, data will be transformed into $\frac{M}{|\sum M|}$. We must check if there are any exceptions before scaling.

Then the data transformed will be controlled at input layer. Any neuron of a certain layer j's input is $net_j(\frac{M}{|\sum M|})$, output is y_j; The output of any neuron in an adjacent lower layer i is y_i. At this time, the interface between the input layer and the implicit layer can be obtained as follows:

$$net_j\left(\frac{M}{|\sum M|}\right) = \sum_i w_{ij} y_i \quad (3)$$

$$y_i = f(net_j) \quad (4)$$

w_{ij} refers to the weight of the neuron input unit. $f()$ refers to the output function of the neuron network. The sparse data input layer is shown in figure 3.

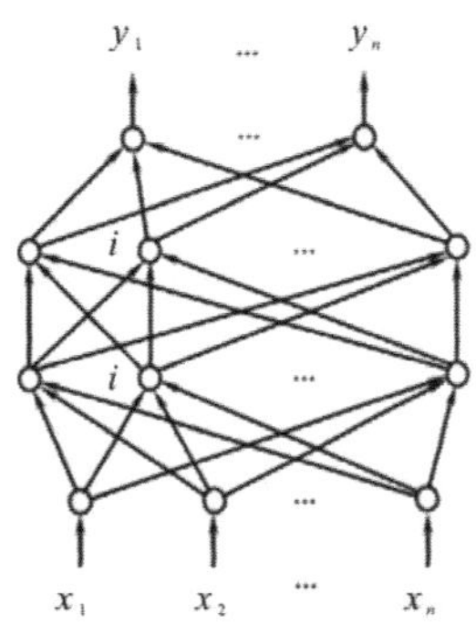

Figure 3 .Sparse data input layer

4.2 Dynamic feedback implicit layer

Learning is the most important characteristics of neural network. Sample input and output mode of the same training set are acted on the network repeatedly [8]. The network automatically adjusts the connection strength or topology structure between neurons according to certain training rules, so that the actual output meets the expected requirements or tends to be stable.

We can set threshold function as below:

$$y = sgn(\sum_{i=1}^{n} w_{ij} y_i - \theta) \quad (5)$$

$\theta = -w_{n+1}$, The formula (5) can be further written as the dot product relationship, $y = sgn(W^T X)$, η is the learning factor and represents the proportionality constant of learning speed. Assume that $y_i(t)$and $y_j(t)$ represent the state (output) of ith and jth neuron. We can get the formula:

$$w_{ij}(t+1) = w_{ij}(t) + \eta[y_j(t)x_j(t)] \quad (6)$$

The learning process can be designed as below:

(1) Select a set of initial weights $w_{ij}(1)$

(2) Calculate the error between the actual output and the expected output result.

(3) Refresh the weights. The threshold value can be considered as a input constant (-1);

(4) Return to step (2), until all output of the network under all training mode satisfy the requirement.

4.3 Low dimensional result output layer

Through the dynamic feedback implicit layer, we can output the post-processing data into the neural network. Output unit sums up all input values, and the output function of threshold model produces a set of output modes.

In this paper, we used Matlab 7.0 to build and test the experiment environment. Neural Networks Toolbox for Matlab was mainly used.

(1) Select the number of neurons: There is no direct expression control of the number of hidden layer neurons. When the number of hidden layer is 20 x 10 and 8 x 4, the training result of the navigation path is shown in figure 4. When there are too few neurons in the hidden layer, it is not possible to deduce the correct navigation path from the location data; when the number of hidden layer neurons is too much [9], the location data study time will be too long, and it can hardly identify the samples that were not directly received before.

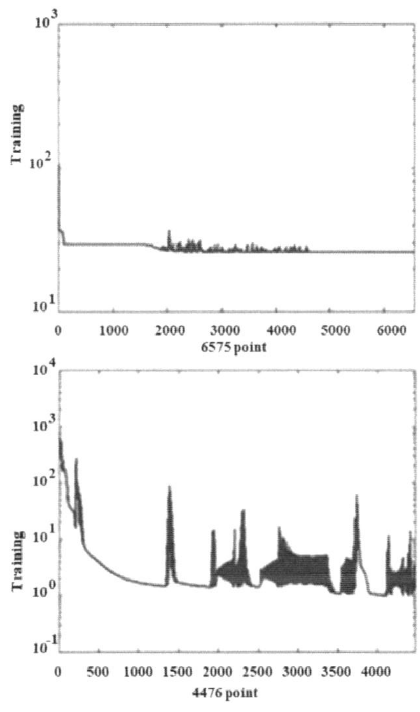

Figure 4. Training error curve of the same hidden layer number of different nodes

(2) Because the learning process of neural network uses gradient descent method, we use the first derivative information of the error to guide the direction of next step to get the minimum error. In order to guarantee the convergence of the navigation path algorithm, learning rate η has to be less than a certain limit. We usually set $0<\eta<1$. The closer it gets to the minimum, the convergence of the algorithm gets slower as the gradient change is going to zero. Accuracy of path prediction under different learning factors is shown in figure 5.

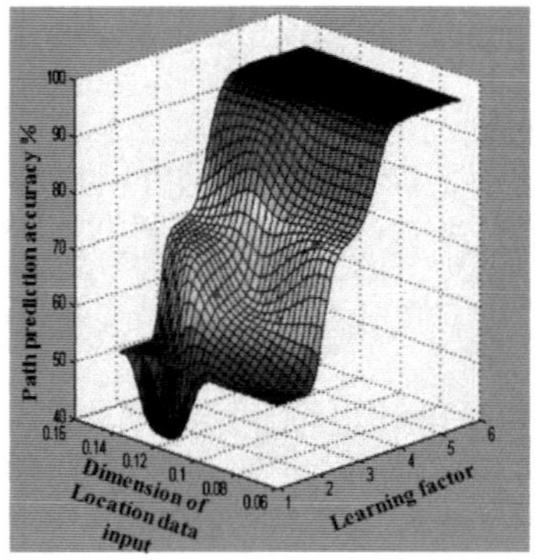

Figure 5. Accuracy of path prediction under different learning factors

5. Conclusion

In this article, through indoor positioning data clustering algorithm and intelligent decision-making research study. We designed massive indoor location data clustering process and path reasoning and decision making based on neural network. The problem of insufficient accuracy and data redundancy in indoor positioning is solved effectively. The intelligence of navigation path planning was improved.

Reference

[1] Hatamlou. Black hole: A new heuristic optimization approach for data clustering. Information Sciences, 2013, 222(3):175-184.

[2] W. Lin, Z. Deng. Dimensional functional differential convergence for Cramer-Rao lower bound. Journal of difference equations and applications. 2017, 23(1-2):249-257.

[3] Niknam T, Fard E T, Pourjafarian N, et al. An efficient hybrid algorithm based on modified imperialist competitive algorithm and K-means for data clustering. Engineering Applications of Artificial Intelligence, 2011, 24(2):306-317.

[4] Yan X, Zhu Y, Zou W, et al. A new approach for data clustering using hybrid artificial bee colony algorithm. Neurocomputing, 2012, 97(1):241-250.

[5] Haykin S S. Kalman Filtering and Neural Networks. Adaptive & Learning Systems for Signal Processing Communications & Control, 2014:170 - 174.

[6] Schmidhuber J. Deep Learning in neural networks: An overview.[J]. Neural Networks the Official Journal of the International Neural Network Society, 2014, 61:85–117.

[7] Hinton G E, Srivastava N, Krizhevsky A, et al. Improving neural networks by preventing co-adaptation of feature detectors. Computer Science, 2012, 3(4): 212-223.

[8] Graves A, Mohamed A R, Hinton G. Speech recognition with deep recurrent neural networks. 2013, 38(2003):6645-6649.

[9] Lin W, Deng Z, Fang Q, et al. A new satellite communication bandwidth allocation combined services model and network performance optimization. International Journal of Satellite Communications & Networking, 2016,35(3): 263-277

Research on Radar Micro-Doppler Feature Parameter Estimation of Propeller Aircraft

Zhihua He[1,*], Feixiang Tao[1], Jia Duan[1] and Jingsheng Luo[1]

[1]AVIC Leihua Electronic Technology Research Institute, Wu'xi 214063, China

Email: *shimenguest@163.com

Abstract. The micro-motion modulation effect of the rotated propellers to radar echo can be a steady feature for aircraft target recognition. Thus, micro-Doppler feature parameter estimation is a key to accurate target recognition. In this paper, the radar echo of rotated propellers is modelled and simulated. Based on which, the distribution characteristics of the micro-motion modulation energy in time, frequency and time-frequency domain are analyzed. The micro-motion modulation energy produced by the scattering points of rotating propellers is accumulated using the Inverse-Radon (I-Radon) transform, which can be used to accomplish the estimation of micro-modulation parameter. Finally, it is proved that the proposed parameter estimation method is effective with measured data. The micro-motion parameters of aircraft can be used as the features of radar target recognition.

1. Introduction

Radar is an important tool for un-cooperated air targets recognition, based on which military schemes can be made, such as the judgment of military motives, the establishment of the battle scheme, and the fire control [1] etc. Air targets mainly consist of jet plane, propeller aircraft and helicopter. Among them, the propeller aircraft has the advantages in low velocity and taking off-landing properties, which make it an important role in air transportation, reconnaissance and surveillance, logistics etc. In general, air targets include obvious powerful rotating components, such as the compressor blades of engine in jet planes, the propeller in propeller aircrafts, the rotor and empennage in helicopters. These rotating components work periodically, which induce a frequency modulation on the returned radar echo, namely micro-Doppler effect. Therefore, the air target recognition can be realized by estimating the micro-Doppler parameters of powerful components from the radar echo of air targets. For example, the number and length of blades, the rotating velocity can be deducted from the estimated micro-Doppler parameters. Resultantly, the target recognition can be realized.

Most available references have studied the micro-Doppler effect and parameter estimation of rotating components in air targets deeply. The mathematical model of main rotor in helicopters has been given by Chen, in [2] and [3]. The radar echo of rotor is modelled as sinc-function behaved periodical pulse in time domain, while extends in frequency domain. By analyzing the parameters of available helicopter rotors from different countries, C. E. Rotander realizes that the ratio of length to number of blades can be used as a critical feature to identify different kinds of helicopters in [4]. Fortunately, the ratio of length to number of blades can be obtained via the Doppler spectrum of rotors. However, the type of available helicopter is too much to identify the exact type by using the ratio of length to number of blades, which limits its application in reality. In [5], by applying the time-to-frequency transform to the radar echo of helicopters, the number and length of rotor blades have been estimated via inverse-Radon (I-Radon) transform by A. Cilliers. However, the estimation method of the blade length cannot be extended to other scenarios, due to the neglect of the incident angel between radar echo and the rotate plane [6].

Different from the micro-Doppler feature of rotors in helicopter, the blades of propeller aircrafts have such properties as short length, high rotating speed, large rotate angle, the low energy from the wingtip [7] etc. Most available researches focus on the micro-Doppler feature of rotors in helicopter. While, there are few researches on the micro-Doppler feature of propeller aircrafts. In this paper, the model of the radar echo of propeller aircrafts has been built based on both simulated experiments and real measured data. Through the analysis of micro-Doppler feature and spatial transformation, the parameters of micro-motion can be deducted. Finally, a method for parameter estimation of propeller aircrafts rotating components is proposed, which is helpful for air target classification and recognition.

2. Parametric model of radar echo of radar echo from propeller in aircrafts

Due to the rotation of propeller in propeller aircrafts, their echo behaves differently from normal radar echoes. In time domain, the echo from propeller is behaved as sinc-function pulse. While in frequency domain, its spectrum extends. All these features can be used to model the echo of propellers blades. First, the echo of a single propeller blade is modelled by assuming the blade as an ideal line source with same length and velocity of real one.

Under the assumption of far field and long wavelength working condition, a blade of propeller can be equally set as a line source with length, which rotates around the cicatricle center with constant speed. Let the range from the cicatricle center of propeller blade to radar be R, as show in Figure. 1.

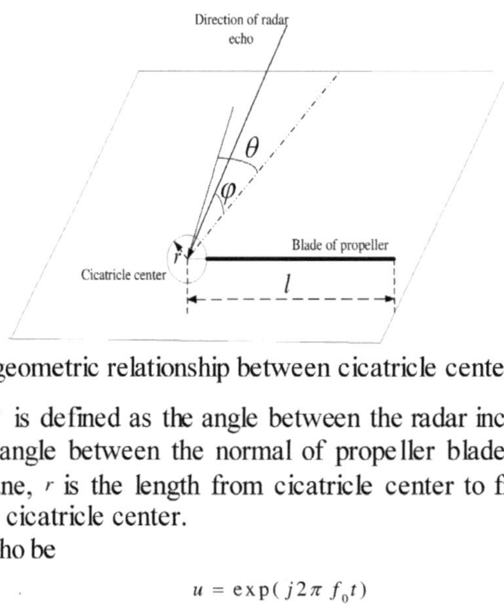

Figure.1 The geometric relationship between cicatricle center and radar echo

As shown in Figure.1, φ is defined as the angle between the radar incident beam and the propeller plane, θ is defined as the angle between the normal of propeller blade and the projection of radar beam into the propeller plane, r is the length from cicatricle center to front of blade, l is the length from the end of blade to the cicatricle center.

Let the incident radar echo be

$$u = \exp(j2\pi f_0 t) \tag{1}$$

in which, f_0 stands for the frequency of the incident beam. Thus, the radar echo of a point in the blade at time t can be expressed as

$$v(t) = \exp(j2\pi f_0 (t - \tau(t))) \tag{2}$$

in which, $\tau(t) = 2(R - Vt - x\sin(\theta_0 + 2\pi F_{rot} t)\cos(\varphi)) / c$ denotes the delay time, V is the velocity of plane, x means the distance from the point to cicatricle center, θ_0 is the initial angle of the blade at time $t = 0$, F_{rot} is denoted by the rotate frequency of such propeller blade. Thus, the radar echo of such blade can be regarded as the sum of the echo of all the points in that blade as

$$v_s = \int_r^l v(t)dx = \int_r^l \exp(j2\pi f_0(t - \frac{2(R - Vt - x\sin(\theta_0 + 2\pi F_{rot}t)\cos(\varphi))}{c}))dx$$
$$= (l-r)\sin c(\frac{2(l-r)\sin(\theta_0 + 2\pi F_{rot}t)\cos(\varphi)}{\lambda}) \times \tag{3}$$
$$\exp(j2\pi(f_0 t - \frac{2(R - Vt)}{\lambda} + \frac{(l+r)\sin(\theta_0 + 2\pi F_{rot}t)\cos(\varphi)}{\lambda}))$$

in which, λ is denoted by the wavelength of the radar echo. As can be seen from (3), the ideal maximum amplitude of propeller blade is decided by its length. The shape of the amplitude depends on such parameters as propeller blades length, rotating frequency, the angle between radar beam and propeller plane, and wavelength. Meanwhile, the carried frequency of echo from propeller relies on the frequency of incident beam. The Doppler shift is introduced by both the vertical motion of blade and the rotation of blade around the cicatricle center.

Let the propeller is consisted of N blades, the radar echo can be represented by

$$v_w = (l-r)\sum_{n=0}^{N-1} A_n B_n \tag{4}$$

in which,

$$A_n = \exp(j2\pi(f_0 t - \frac{2(R - Vt)}{\lambda} + \frac{(l+r)\sin(\theta_n + 2\pi F_{rot}t)\cos(\varphi)}{\lambda})) \tag{5}$$

$$B_n = \sin c(\frac{2(l-r)\sin(\theta_n + 2\pi F_{rot}t)\cos(\varphi)}{\lambda}) \tag{6}$$

$$\theta_n = \theta_0 + \frac{(n-1)2\pi}{N} \tag{7}$$

in which, θ_n means the initial angle of the nth blade. Since the blade distributes equally in propeller plane, the initial angle of other blades can be deducted from that of certain blade.

Based on above analysis, the echo of propeller aircraft is affected by the structure, motion, working conditions of the aircraft, and the relative geometric position to aircraft propeller etc.

From (4) to (7), the term A_n determines the phase of propeller echo, while B_n affects its amplitude. Therefore, under the assumption that long illuminated time is promised by radar, the receiver obtains sinc-function modulated pulses cluster. When the direction of radar beam is consistent with the normal direction of blade, the amplitude maximizes. When the direction of radar beam departures the normal direction of blade, the amplitude attenuates. In that way, the radar echo of propeller behaves specifically in time-to-frequency domain, which is concerned with the certain rotation of the propeller blade. Generally, it behaves as sine modulated function, which can be utilized to estimate the micro-Doppler parameter of propeller blade.

3. Micro-Doppler parameter estimation of propeller

According to the domain, available researches on radar echo analysis of micro-motion components can be classified into three types, namely analysis based on time domain, Doppler spectrum domain and time-frequency domain. For time domain, the radar echo of micro-motion components behave as shining pulses, which the energy of radar echo concerns mainly in a short time but with high level of clutter power. For Doppler spectrum, the energy of clutter concerns around 0, while the energy of airframe concerns on a short spectrum around the peak frequency. Resultantly, the energy of clutter remain in micro-Doppler units is rather low. However, the micro-Doppler of rotation components extends in spectrum, leading to low energy of certain micro-Doppler units yet. For time-frequency domain, the variation of Doppler modulation with time caused by the rotation of micro-motion components can be clearly represented by the means of short time Fourier transforms (STFT), Wigner-Ville transforms and so on. In reality, the radar echo is consisted of noise, clutter and interference etc. Due to the fact that the cross term after second order time-to-frequency transform is too obvious to pollute the micro-Doppler units with low energy, STFT is adopted in this paper to estimate the micro-Doppler parameters.

3.1. Micro-Doppler Spectrum Analysis Based on I-Radon Transform

Suppose $f(x,y)$ is a two-dimensional function in time-frequency plane (x,y). The radon transform is used to project a single line $u = x\cos\theta + y\sin\theta$ in the plane (x,y) into another plane defined by (u,θ). It can be expressed mathematically as

$$g(u,\theta) = \iint_{-\infty}^{+\infty} f(x,y)\delta(u - x\cos\theta - y\sin\theta)dxdy \quad (8)$$

According to the image reconstruction theory, the projection of a sine behaved curve into parameters domain can be realized by the I-Radon transform. Here, suppose there is a point (x_0, y_0) in the time-frequency plane (x,y), a line across this point with $u = x_0\cos\theta + y_0\sin\theta$ can be transformed by triangle transform such as

$$u = B\sin(\theta + \varphi_0) \quad (9)$$

in which, $B = \sqrt{x_0^2 + y_0^2}$; $\sin\varphi_0 = x_0 / \sqrt{x_0^2 + y_0^2}$; $\cos\varphi_0 = y_0 / \sqrt{x_0^2 + y_0^2}$.

As can be illustrated, after applying the radon transform with different θ, u varies as a sine modulated function in plane (u,θ). Therefore, a sine behaved curve in the plane (u,θ) can be projected as a single point (x_0, y_0) in the plane (x,y). Fortunately, the point (x_0, y_0) can be obtained by I-Radon transform. Namely, the I-Radon transform can realize the projection of a sine behaved curve into parameter space.

The projection angle can be decided by the rotation velocity of the propeller blade. For example, an I-Radon transform image of the radar micro-Doppler spectrum is illustrated as Figure.2(a).The real measured data is recorded, in which the propeller is composed of 4 blades. As can be seen from the image, the energy of each blade is accumulated, from which the rotation velocity, the number of blades, the radius of cicatricle and the length of blade can be estimated.

3.2. Micro-Doppler parameter estimation of propeller

Since different types of propeller aircrafts are composed of different propellers and powerful engines, air target classification can be realized by estimating the micro-Doppler parameters of propellers. The feature extraction method is accomplished by following steps.

The time-frequency distribution image of target can be achieved by applying the STFT to the signal of target in time domain.

- According to the range of the rotation velocity, I-radon transform is applied with different rotation velocity during this range. The rotation velocity is estimated which focuses the time-frequency distribution image best.
- By applying image segmentation to the best focused image, the image of the propeller can be obtained to estimate the rotational velocity of blades in the propeller.
- Let the I-Radon transform center be the center of a circle, a circle with radius R (here let R=1m) is plotted to estimate the number of the blades. The number of the blades is equal to the number of intersections of the circle and the image of the propeller.
- Let the I-Radon transform center be the benchmark. The projected radius of the cicatricle is estimated by the length from the intersection of a line equally between two blades to the benchmark. According to the given velocity from tracking progress, the incident angle can be obtained, thus the real radius of the cicatricle can be deducted.
- Let the I-Radon transform center be the benchmark as well. The projected length of the blades can be estimated by applying detection progress to the energy of the blades. Since the incident angle can be easily got, the real length of the blade can be deducted. It should be noticed that

the estimated length may be shorter than real ones due to the low energy of the blade end by I-Radon accumulation.

4. Experimental results and analysis based on real measured data

The real measured radar echo of a certain kind of propeller aircraft is utilized in this section. The carried frequency is 10GHz. The number of blades is 4 with length 1.7m and rotation velocity around 1300r/min. After applying STFT to the radar echo, figure.2 (b) can be obtained, in which micro-Doppler modulation is obvious in certain time with sine distribution. The average energy reaches its maximum when the velocity is 1250r/min. I-Radon transform is applied with the velocity to obtain such image as Figure. 2 (d). The segmented image can be obtained by the proposed method in the above section, as shown in Figure. 2 (c). In Figure.2 (d), the top and end of blade are marked, in which there are 4 blades as known. The extracted length is 1.08m, which is smaller than real values. This is due to the fact that the energy of blade ends can hardly be accumulated by I-Radon transform. Although there is relatively high error of blade length estimation, the micro-Doppler parameters extracted can still be utilized to realize air target recognition.

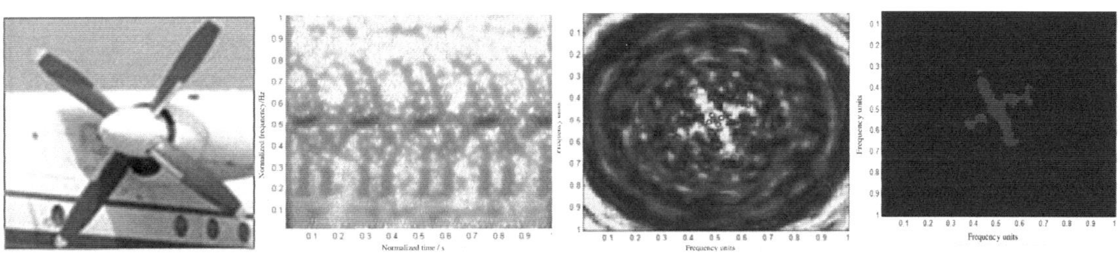

(a) Real image of propeller (b) Time-frequency distribution of propeller (c) Segment image after I-Radon (d) Parameter extraction on I-Radon images

Figure. 2 The parameter extraction of real measured radar echo of propeller aircraft

5. Conclusion

In this paper, the micro-Doppler features of propeller aircraft are analyzed by real measured data. The parameters of propeller blades can be estimated by accumulating the energy of each blade with time-frequency transform and I-Radon Transform. The proposed micro-Doppler parameter estimation method is illustrated in this paper. Real measured data experimental results validate the effectiveness of the proposal. The micro-motion parameters of aircraft can be used as the features of radar target recognition.

References

[1] Tait 2005 Introduction to Radar Target Recognition. United Kingdom: The Institution of Electrical Engineers.

[2] Chen V C *2001 Radar signatures of rotor blades* (*Proceedings of SPIE vol 4391*) pp 63~70.

[3] Chen V C Li F Ho S and Wechsler H 2006 Micro-doppler effect in radar: phenomenon, model, and simulation study (IEEE Transactions on Aerospace and Electronic Systems vol 42 no 1) pp 2–21.

[4] Rotander C E and Von H Sydow 1997 Classification of helicopters by the L/N quotient (Proc of IEE Radar vol 449) pp 629~632.

[5] Cilliers A and WAJ Nel 2008 Helicopter parameter extraction using joint Time-Frequency and Tomographic Techniques (Record of the IEEE International Radar Conference) pp 598–603.

[6] S. Qureshi, S. Mirza, and M. Arif 2005 "Inverse radon transform-based image reconstruction using various frequency domain filters in parallel beam transmission tomography," Engineering Sciences and Technology, 2005. SCONEST 2005. Student Conference on, pp. 1–8.

[7] Martin J and Mulgrew B 1990 Analysis of the theoretical radar return signal form aircraft propeller blades (Record of the IEEE International Radar Conference) pp 569–572.

Application Research of Horn Array Multi-Beam Antenna in Reference Source System for Satellite Interference Location

Ping Zhou[1], Hui Lin[1,*] and Qi Zhang[1]

[1]ShenZhen Station, State Radio Monitoring Center, Shenzhen, Guangdong, China

[*]linhui@srrc.org.cn

Abstract. The reference source system is a key factor to ensure the successful location of the satellite interference source. Currently, the traditional system used a mechanical rotating antenna which leaded to the disadvantages of slow rotation and high failure-rate, which seriously restricted the system's positioning-timeliness and became its obvious weaknesses. In this paper, a multi-beam antenna scheme based on the horn array was proposed as a reference source for the satellite interference location, which was used as an alternative to the traditional reference source antenna. The new scheme has designed a small circularly polarized horn antenna as an element and proposed a multi-beamforming algorithm based on planar array. Moreover, the simulation analysis of horn antenna pattern, multi-beam forming algorithm and simulated satellite link cross-ambiguity calculation have been carried out respectively. Finally, cross-ambiguity calculation of the traditional reference source system has also been tested. The comparison between the results of computer simulation and the actual test results shows that the scheme is scientific and feasible, obviously superior to the traditional reference source system.

1. Introduction

Dual-satellite geolocation technology is most widely used in finding the interference source of satellite up-link station. In order to get more accurate positioning results, several known ground position of the uplink reference source signals should be transmitted for calibration ephemeris [1][2]. The reference source system generally uses a large caliber parabolic antenna, due to frequent use, that will inevitably lead to high mechanical failure, and if there are not enough number of normal working reference antennas ,which can lead to positioning failure.

The reference source system is an important short board which plays an important role in restricting the performance of satellite geolocation system. At present, there is no open literature at home and abroad to put forward new ideas and new schemes for the next generation reference source system. The new reference source system proposed in this paper tries to overcome the defects of the traditional system and greatly improve the success rate and timeliness of satellite interference source localization.

2. How the reference source system works in dual-geolocation

The accuracy of the dual-satellite geolocation depends on the accuracy of the ephemeris of both satellites (main satellite and adjacent satellite), which is usually from forecast ephemeris in the foreign public website, in the case of actual interference geolocation[3]. Due to the insufficient accuracy, the direct use of public ephemeris will result in geolocation error of thousands of kilometers, leaving the location results meaningless. Five different signals need to be transmitted up-to the satellite through five geographically dispersed reference source systems, which will be used to calibrate the ephemeris.

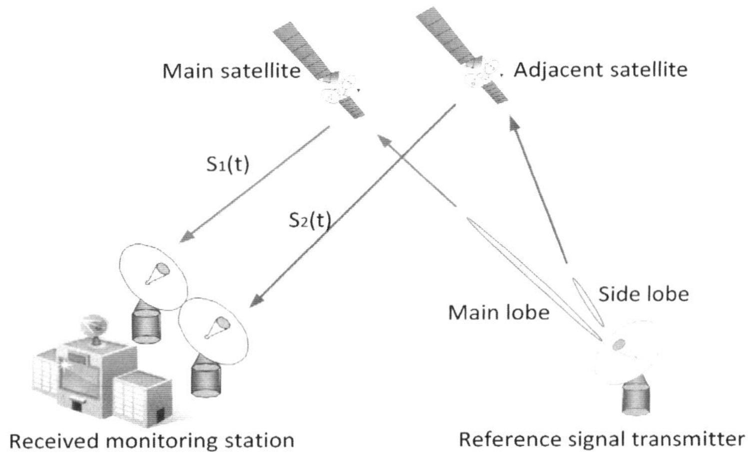

Figure 1. Workflow diagram of reference source system based on parabolic antenna

The reference signal is transmitted to the main satellite and "leaked" to the adjacent satellite respectively by the main lobe and side lobe through the transmitting antenna, as shown in figure 1. Then the same two signals are respectively received by different ground receiving antennas after transparent forwarding. By calculating the cross ambiguity function (CAF) of the two signals (S1 (t) and S2(t)), it is able to estimate the arrival time difference (DTO) of the two signals, and also their relative doppler frequency difference (DFO) due to the relative movement of the two satellites[4]. After successfully estimating parameters of the five reference signals, the ephemeris calibration algorithm can be used to precisely correct the ephemeris of the satellites, then get high-precise results of interference geolocation.

3. Shortcomings in the traditional reference source system

The current reference source system uses a parabolic reflector antenna, which has the advantage of high gain and affordable price. However it has some serious drawbacks as below:

- Mechanical rotation speed is slow and time-consuming, which causes that short-term interference signals can't be located in time by the geolocation system.
- Long working time and frequent use of the reference source system can easily lead to mechanical failure.
- For the side lobe from transmitting antenna points to an adjacent satellite randomly, the side lobe pointing to the adjacent satellite may be just the zero point between the side lobes. In this case, it cannot work.

4. Horn array antenna

The beam forming of the array antenna is achieved by the phase control of each element without mechanical movement, which keeps the horn antennas pointing to the satellite quickly and accurately. The use of array antenna for transmitting reference signals can completely avoid the disadvantages of traditional reference source due to mechanical rotation, as shown in figure 2. As mentioned above, the reference signals are invalid when the side lobe signal pointing to the adjacent satellite is too low or zero. However, the reference signal based on the horn antenna array can directly form two main beams with high gain when transmitting.

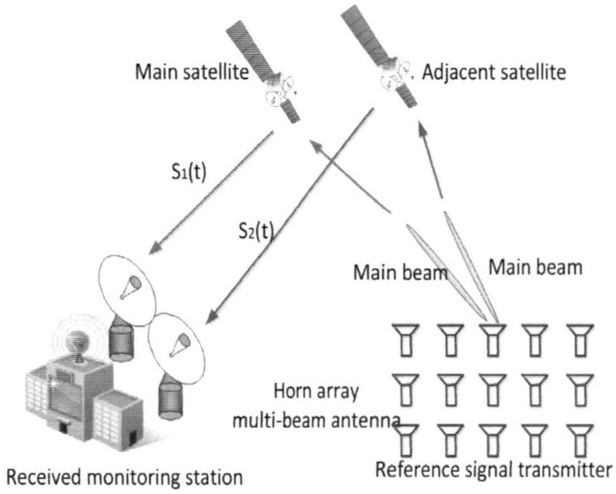

Figure 2. Workflow diagram of reference

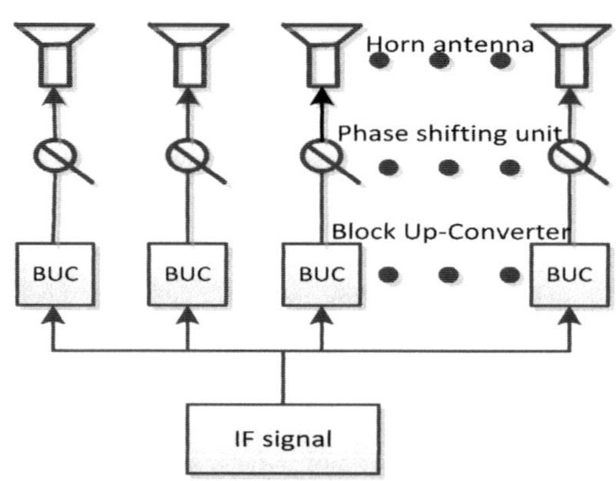

Figure 3. Structure of horn antenna array element source system based on horn array

The array antenna with the same gain as the parabolic antenna requires thousands of antenna elements. And each element of phased-array antenna is equipped with a separate power amplifier unit and phase shifter, a large number of amplifiers and phase shifters will make the cost of the antenna array extremely expensive, thus lost its practical applicability[5][6]. If the gain of the antenna element is high, the number of antenna elements can be greatly reduced, thus reducing the cost of the antenna array. The gain of the horn antenna is proportional to the caliber, in order to avoid the phase ambiguity effect, the antenna element spacing should not exceed half a wavelength. This condition determines that the horn antenna diameter needs to be controlled in a certain range [7].

The horn array antenna is a rectangular array consisting of circular horn antennas. The array unit consists of an IF signal source, block up-converters (BUCs), phase shifters, and horn antennas. The phase shifter is a phase control unit, and the system configures the corresponding phase for each antenna element according to a specific multi-beam forming algorithm to realize multi-beam transmission of the antenna array. BUCs is the block up-converters which converts the IF signal to the RF signal and then amplifies the power, as shown in figure 3.

5. Beamforming algorithm

The beamforming algorithm at the receiver side is widely studied[8][9], unlike beamforming at the receiver side, the beamforming algorithm at the transmitter side does not amplify the noise, and the phase shift vector can be solved by using the least square method with some constraint conditions[10][11]. As shown in figure 4, the horn array antenna is a uniform planar array antenna and the spacing of the antenna array elements are $\lambda/2$, the origin of the axis is the reference point. Since the geostationary orbit satellites are mainly distributed along the azimuth plane, if the azimuth beam of the antenna array beam is too wide, it may interfere with the adjacent satellite, and the beam does not have the problem on the pitch surface. Thus the array is designed to have a number of elements in the azimuth direction greater than the number of elements in the pitch direction. Here, we assume the X-axis (in the azimuth plane) equidistant arrangement of 40 horn antennas, and the Y-axis (in the pitch plane) has 10 horn antennas.

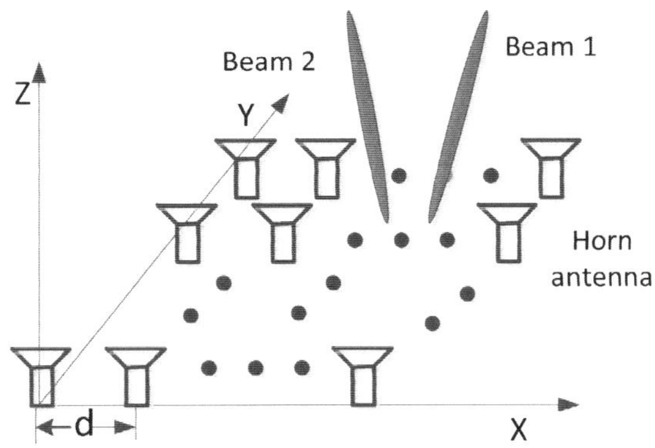

Figure 4. Structure of horn array

Assuming that the array has N beams, then the i-th beam (0 <i <= N, i is an integer), the angle between the beam and the Z axis is θ_i, and the angle between the projection of the beam on the X-Y plane and the X axis is φ_i.

The vector of the i-th beam along the X-axis array is then expressed as:

$$a_x(\theta_i,\varphi_i) = [1 \ \ e^{j*2*pi*d*\sin(\theta_i)*\sin(\varphi_i)/\lambda} \ \cdots \ e^{j*2*pi*d*(M_1-1)\sin(\theta_i)*\sin(\varphi_i)/\lambda}]^T \tag{1}$$

Similarly, the vector of the i-th beam along the Y-axis array is:

$$a_y(\theta_i,\varphi_i) = [1 \ \ e^{j*2*pi*d*\sin(\theta_i)*\cos(\varphi_i)/\lambda} \ \cdots \ e^{j*2*pi*d*(M_2-1)\sin(\theta_i)*\cos(\varphi_i)/\lambda}]^T \tag{2}$$

The symbol T means transpose. Where M1=40, M2=10. Then, the array steering vector of the i-th beam based on the rectangular array can be expressed as :

$$a(\theta_i,\varphi_i) = a_x(\theta_i,\varphi_i) \otimes a_y(\theta_i,\varphi_i) \tag{3}$$

Where ⊗ denotes the Kronecker product and equation (3) is equivalent to the straightening operation of the steering matrix of the array.

Each array element in the array has a corresponding phase shift for a certain wave direction, and the corresponding phase shift matrix is

$$W' = \begin{bmatrix} \beta_1 & \beta_{M+1} & \beta_{2M+1} & \cdots & \beta_{(3M+1)M+1} \\ \beta_2 & \beta_{M+2} & \beta_{2M+2} & \cdots & \beta_{(3M+1)M+2} \\ \cdot & \cdot & \cdot & & \cdot \\ \cdot & \cdot & \cdot & & \cdot \\ \beta_M & \beta_{2M} & \beta_{3M} & \cdots & \beta_{4MM} \end{bmatrix} \tag{4}$$

By the vectorizen calculation of matrix, then we have

$$W = vec(W') \tag{5}$$

When the product of the conjugate transpose of the steering vector a (θ_i,φ_i) and the array phase shift vector is 1, the array forms a main beam in direction a (θ_i,φ_i), otherwise, when the product is 0, the array forms null steering beams at the direction of a (θ_i,φ_i). H denotes conjugate transpose.

$$a(\theta_i,\varphi_i)^H * W = \begin{cases} 1 \\ 0 \end{cases} \tag{6}$$

C and f denote the constraint matrix and the constraint response vector, respectively. And we can express the terms C and f as following:

$$C = [a\ (\theta_1,\varphi_1)\ \ a\ (\theta_2,\varphi_2)\ \ a\ (\theta_3,\varphi_3)\ \ \cdots\ a\ (\theta_P,\varphi_P)]$$
$$f = [1\ 1\ \cdots\ 1\ 0\ 0\ \cdots\ 0] \tag{7}$$

In equation(7), the number of 1 in the constraint response vector f is the number of array beams and the number of 0 is the number of null steering beams formed by the array. Unlike the beamforming method at receiver side, the expression of W in the array that satisfies all the constraints is

$$C^H * W = f \tag{8}$$

Therefore, the least squares solution for W in restricted all conditions is:

$$W = (C * C^H)^{-1} * C * f \tag{9}$$

6. Computer simulation experiment

The simulation results are carried out in three steps: the first step is to calculate the gain pattern of the horn antenna. The second step is to make two beamforming of the antenna array according to the beamforming algorithm proposed above. The EIRP (Equivalent Isotropic Radiated Power) value range of two beams is calculated by combining the information obtained in the first two steps. In the third step, based on the EIRP values of the two beams, the two signals are transmitted through the satellite link, and the two signals are calculated by CAF calculation to verify the feasibility of the scheme.

6.1. Simulation experiment of the gain of horn antenna

Based on the avoidance of phase blurring, the aperture of the horn antenna is λ/2. The frequency is 6GHz, λ is 5cm, right circular circular polarization. The HFSS simulation shows that the gain of the antenna is shown in figure 5 and figure 6. Figure 7 shows the horn antenna model. The maximum gain of the antenna at 6GHz is 7dB on both E and H plane, gain range of 3dB beam is 4-7 dBi.

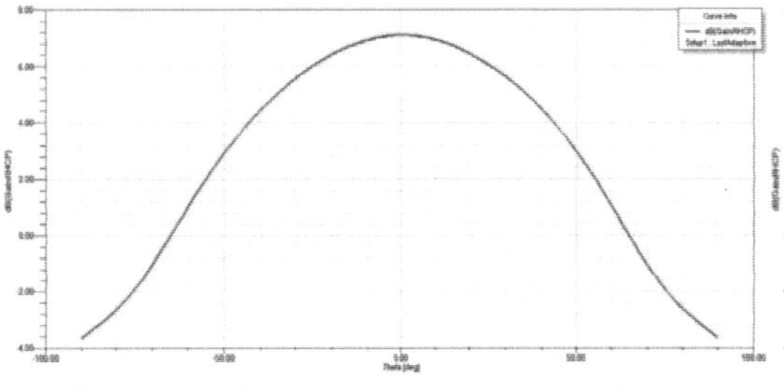

Figure 5. Horn antenna pattern on E plane

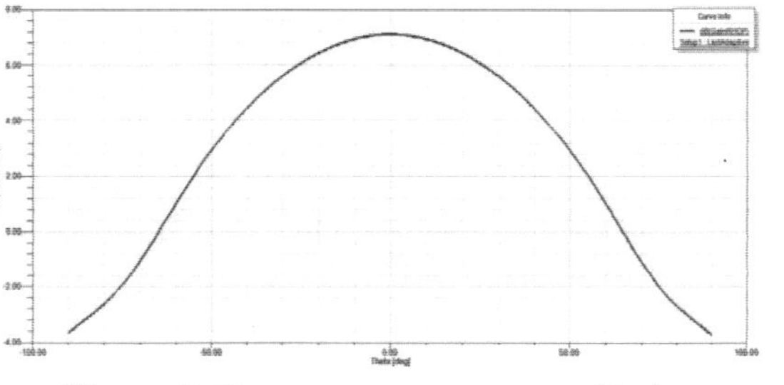

Figure 6. Horn antenna pattern on H plane

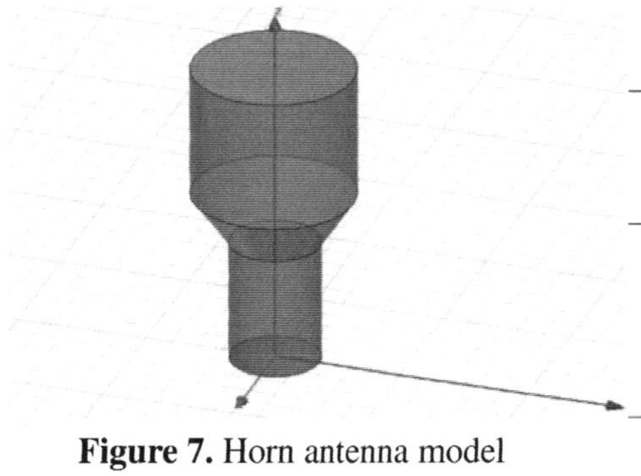

Figure 7. Horn antenna model

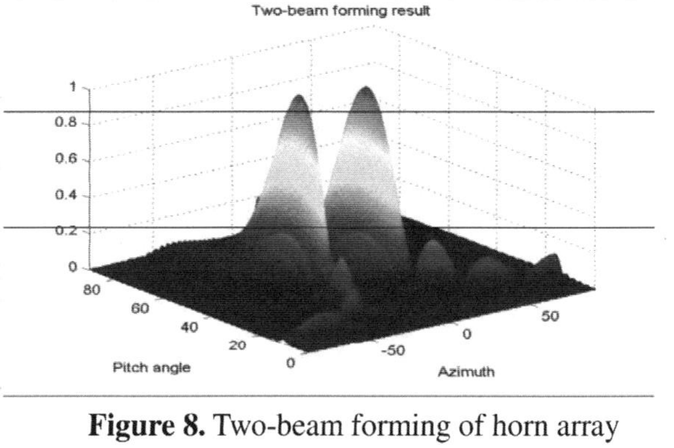

Figure 8. Two-beam forming of horn array

6.2. Simulation experiment of array antenna beamforming

Assume that the antenna is installed in Shenzhen, its pointing angles to the main satellite (China Sat 6B) and the adjacent satellite (Asian Sat 5) are respectively (22°, 49.2°) and (-13°, 53.6°) (the left side is the angle of azimuth, the right side is the angle of pitch).

According to the algorithm provided by formula (9), the array phase shift vector is calculated by Matlab, and corresponding phase configuration is carried out for each array unit. The two beam simulation results of the rectangular array are shown in figure 8, figure 9, figure 10.

The simulation result shows that each array processing gain of the two beams of the antenna is 26dB, the beam-steering angle to the adjacent satellite is (-13 °, 53.6 °), and the beam-steering angle to the main satellite is (22 ° , 49.2 °). It meets the design requirements. Two beam azimuth 3dB beam width of the azimuth of the two beams are respectively 2.8 ° and 2.6 °. The span between the satellites which cover china is generally greater than 3 °. So the two main beams don't interfere with neighbor satellites.

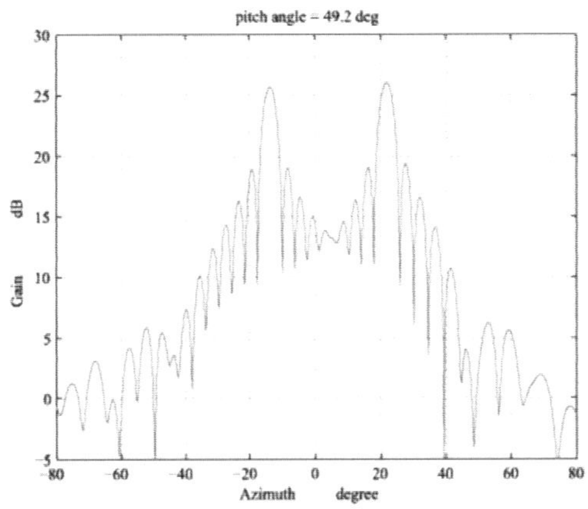

Figure 9. Beam-forming on azimuth Plane (pitch angle=49.2 deg)

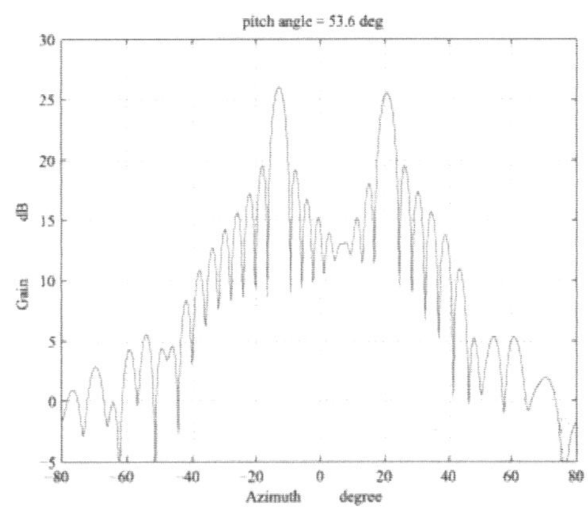

Figure 10. Beam-forming on azimuth plane (pitch angle=53.6 deg)

The EIRP of the array antenna is shown in equation (10):

$$EIRP = G_{array} + G_{horn} + P_{BUC} \quad (dBW) \tag{10}$$

Ghorn is the gain of horn antenna, PBUC is the transmitting power of BUC, and Garray is the gain of the synthetic beam relative to the single horn array. Garray = 26dB, Ghorn values are from 4dB to 7dB. The output power of PBUC ranges from 0dBw to 10dBw. The EIRP is from 30dBw to 43dBw.

6.3. Simulation of cross ambiguity function computing

The estimation of DTO and DFO of reference source is calculated by using CAF[12]:

$$A(\tau, f) = \int_{-T/2}^{T/2} s_1(t) s_2(t+\tau)^* e^{-j2\pi ft} dt \tag{11}$$

S1(t) and S2(t) are the complex envelops of the two signals, τ and f represent the time difference and frequency difference between the two signals, and T is the signal sampling time.

Simulation conditions: the signal transmission frequency is 6GHz, the bandwidth is 128kHz, and the EIRP of signals of the two channels transmitted through satellite link is from 30dBw to 43dBw. The Sampling bandwidth is 128KHz, the sampling time is 16s, the processing gain is 63dB.

Obtain the ephemeris of the main satellite and the adjacent satellite at a certain moment through STK software. The two main beam signals are transmitted up-to the two satellites and received by the ground station , then the related peak of two signals are computed by Matlab.

The actual test is known: when the EIRP value of the antenna is 35dBw with bandwidth 128kHz, then the carrier-to-noise ratio (CNR) of the signal after forwarding on the transponder is 0dB.

The EIRP of the satellite beam is 31dBw, 35dBw, 39dBw, 43Bw (the bandwidth is 128KHz, the same below), the range of the EIRP of the adjacent satellite beam is from 30dBw to 43dBw, and the step is 1dBW. The simulation results are shown in figure 11. It shows that when the two signals take the minimum value, the EIRP value of the main satellite is 31dBw (CNR = -4dB), the EIRP value of the adjacent satellite is 30dBw (CNR = -5dB), the CAF correlation peak is 49 dB, and it beyond the available threshold(20dB) of 29dB. When the two signals take the max value, the EIRP value of the main satellite is 43dBw (CNR = 8dB), the EIRP value of the adjacent satellite is 43dBw (CNR = 8dB), the cross ambiguity function correlation peak is 57dB, and it beyond the available threshold of 37dB. Within the range of the EIRP value, the CAF correlation peak of the two signals can be significantly beyond the threshold. And it fully meets the positioning requirements.

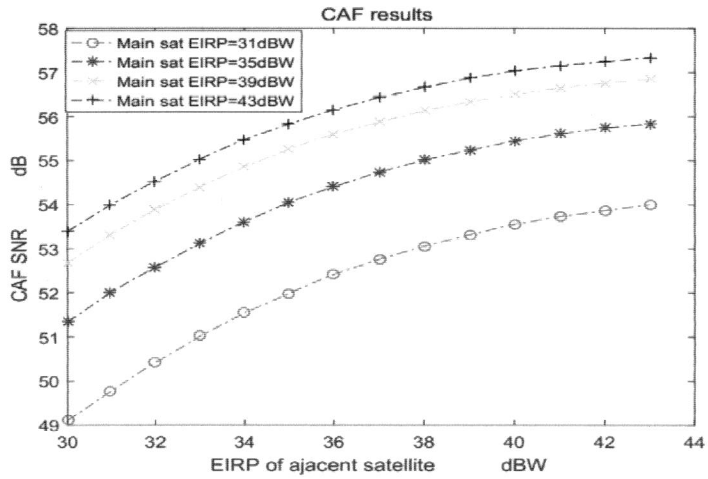

Figure 11. Cross ambiguity function computing result

6.4. Actual test of conventional reference source

Here are the actual test conditions of conventional parabolic reference source antenna. The gain of antenna is 40dB. The range of the transmitter power amplifier output is between 1 and 13 dBi, and the EIRP output of the main satellite ranges from 40dBW to 53dBW. The transmitting bandwidth is 128kHz. The test results are shown in table 1.

Table 1. The experimental results of traditional reference source antenna

Processing Gain (dB)	EIRP of Main Sat(dBw)	The Peak of CAF (dB)
63	40	0
63	43	0
63	46	20.8
63	49	22.1
63	53	26.8

In the test, the CAF confidence threshold is set to 20 dB. The peak values will be displayed if they are higher than 20 dB, while those lower than the threshold will be displayed as 0, which are determined not trustable. As shown as the results, only when the radiated power of the main beam from the parabolic reference source antenna reaches 46 dBW(CNR=9dB), which is 16 dBW higher than the minimum EIRP of the array antenna, the correlation peak above the threshold can be extracted.

6.5. Comparison of the experiment results

It can be concluded by comparing the test results of two kinds of the reference source antennas:

- Under the premise that the EIRP value is much smaller than that of the parabolic reference source, the CAF SNR of the horn antenna array reference source is significantly larger than that of the traditional reference source, that means the confidence of the parameter unbiased estimator is higher.
- The horn array antenna still has a large CAF SNR which is much larger than the threshold, when the two beam EIRP values are less than 35 dBW(CNR<0 dB, and in this case, the reference signal does not adversely affect the normal operation of the transponder). The parabolic reference source however requires the CNR of the main satellite larger than 9 dB to get a trusted correlation peak.

7. Conclusion

When the horn array antenna is used to the reference source system of satellite interference source, it has the advantages of high reliability of parameter estimation and can reduce the time of satellite interference source localization by more than 5 times, which can improve the positioning timeliness. And it can greatly improve the satellite interference system on the short-term interference source. It is the feasible option for next generation of reference source system, and has a huge application prospects.

Due to the limitation of the wavelength which is less than half the wavelength, the horn antenna gain is small, that the array number still has a large scale, the next work will consider the use of non-uniform horn array, increase the horn spacing. Use a larger caliber horn antenna to reduce the number of array elements and their associated BUCs to further reduce array costs.

Reference

[1] Qu W Z, Ye S F and Sun Z B 2005 Algorithm of position calibrator for satellite interference location *Chinese Journal of Radio and Science.* **20** 342-46

[2] Xia C X, Ye S F and Wang J H 2008 Ephemeris Determination in Satellite interference location system Journal of Astronautics. 29 342-46

[3] Sun Z B and Ye S F 2004 Satellite interference location using cross ambiguity function Chinese Journal of Radio Science. 19 525-29

[4] Haworth D P, Smith N G, Bardelli R, et al 1997 Interference localization for EUTELSAT satellites-the first European transmitter location system International Journal of Satellite Communications. 15 155-83

[5] Jun Goto, Satoshi Yamaguchi, Miyuki Tanaka, et al 2013 An injection-molded circular polarized horn array antenna with low axial ratio Conf.on Antennas and Propagation. 1590-94

[6] Elena Pucci, Eva Rajo-Iglesias, Jose-Luis and Vazquez-Roy 2014 Planar dual-mode horn array with corporate-feed network in inverted microstrip gap waveguide IEEE Transactions on Antennas and Propagation. 62 3534-42

[7] Sekiguchi S, Sugimoto M, Shu S, et al 2017 Broadband corrugated horn array with direct machined fabrication IEEE Transactions on Terahertz Science and Technology. 736-41

[8] Liu B and He Y 2016 Improved DBF algorithm for multichannel high-resolution wide-Swath SAR IEEE Transactions on Geoscience and Remote Sensing. 54 1209-25

[9] Zhao X F, Zhang W, Yang T L, et al 2012 A DBF method of localizing accuracy improvement and speech enhancement based on a new microphone arrays International Conf. on Signal Processing. 303-06

[10] Liang G, Gong W, Liu H, et al 2009 61-channel digital beamforming (DBF) transmitter array Asia Pacific Microwave Conference. 739-42

[11] Liu C 2012 Robust Adaptive Beamfoeming Algorithms (XiAn: Xidian University Press) pp 85-96

[12] Ye S F, Sun Z B and Xia C X 2013 Satellite interference location technology and engineering application (BeiJing: National Defence Industry Press) pp 92-99

SVM-based Partial Discharge Pattern Classification for GIS

Yin Ling[1,*], Demeng Bai[1], Menglin Wang[2], Xiaojin Gong[2] and Chao Gu[1]

[1] State Grid Shandong Electric Power Research Institute, Jinan, 250012, China.
[2] The College of Information Science and Electronic Engineering, Zhejiang University, Hangzhou, 310027, China.

[*]lysgwork@163.com

Abstract. Partial discharges (PD) occur when there are localized dielectric breakdowns in small regions of gas insulated substations (GIS). It is of high importance to recognize the PD patterns, through which we can diagnose the defects caused by different sources so that predictive maintenance can be conducted to prevent from unplanned power outage. In this paper, we propose an approach to perform partial discharge pattern classification. It first recovers the PRPD matrices from the PRPD2D images; then statistical features are extracted from the recovered PRPD matrix and fed into SVM for classification. Experiments conducted on a dataset containing thousands of images demonstrates the high effectiveness of the method.

1. Introduction

Gas insulated substation (GIS) is extensively used in electrical industry [1] due to its compactness and insensitivity to environment. To ensure the safety of electric power systems, predictive maintenance of GIS is routinely performed by experienced personnel. They measure partial discharges (PD), which are the localized dielectric breakdowns in small regions of insulating systems [2], to diagnose the condition of GIS. The partial discharge patterns can reveal the defects caused by different sources.

Traditional PD pattern classification was conducted manually by experts. With the increase of sensing data, it becomes more and more desirable to make classification and diagnosis automatic. For this purpose, a bunch of methods have been developed [3-7]. For instance, James and Phung calculated the IEC-270 integrated quantities, statistical moments and other fingerprints to recognize the PD patterns within the power frequency cycle [4]. Gao et al. investigated frequency characteristics of PD for classification [6]. As summarized in [7], the common framework shared in most of the methods is composed of two stages, which are feature extraction and pattern classification. Besides these two stages, there is another problem we need to face. That is, the partial discharges collected by PD sensors are stored in the form of graph. The corresponding raw data is unavailable due to the lack of standard protocols from manufacturers.

This paper proposes an approach to classify the PD patterns captured by a PD-Detector. Our method first parse the raw information from PD graphs via image processing techniques. Then, statistical features are calculated and fed into a support vector machine (SVM) [8] for classification. Our approach is validated in a dataset containing thousands of PD graphs and demonstrates promising results.

IOP Conf. Series: Journal of Physics: Conf. Series **960** (2018) 012051 doi:10.1088/1742-6596/960/1/012051

2. The proposed method

There are two ways to perform PD pattern classification. One is extracting features from PD graphs directly, and the other recovers raw information first and then extract features. Our approach takes the latter way. In this paper, we first design an approach to parse raw data from PD graphs collected by a PD-Detector. Then, we extract statistical features from the recovered information and use SVM for classification. Each step is introduced in the following subsections.

2.1. Graph parsing

A PD-Detector gathers the phase-resolved partial discharge (PRPD) spectrum. The PRPD spectrum consists of three-dimensional discharge patterns (φ, q, n), in which φ stands for the phase angle, q is the discharge magnitude and n is the discharge rate [7]. The patterns are stored in either 2D or 3D forms as shown in Figure 1 (a). Graph parsing in this work aims to recover the raw PRPD matrix when a PRPD2D image is given.

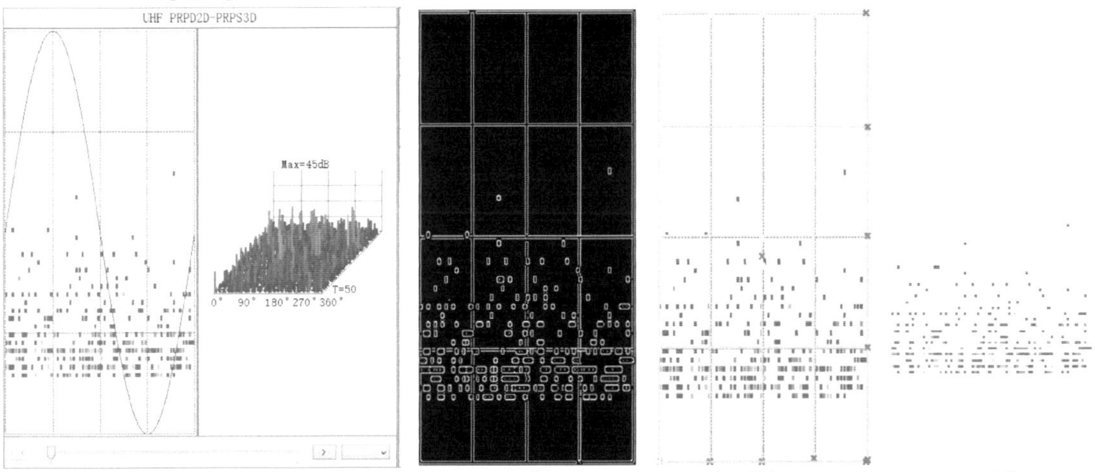

(a) (b) (c) (d)

Figure 1. An example of the partial discharge graph. (a) is PRPD2D and PRPS-3D images; (b) is edge detection result; (c) is Hough transform result; (d) is the visualization of the recovered PRPD matrix.

To this end, we design a parsing procedure based on some image processing techniques. The procedure consists of a couple of steps as presented in Figure 2.

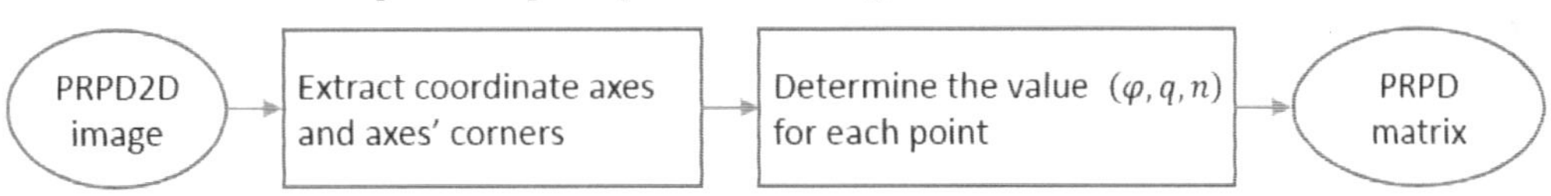

Figure 2. The procedure of graph parsing.

2.1.1 Extract coordinate axes and axes' corners. When a PRPD2D image is input, we first filter out the blue phase line and convert the remaining color image into gray level. Then, the Canny operator is applied to detect edges, by which we obtain the edge detection result as shown in Figure 1(b). Based on this, Hough transform is performed to extract the horizontal axis (φ), the vertical axis (q) and grid lines, as presented in Figure 1(c). Finally, the bottom-left corner (bl_x, bl_y) and top-right corner (tr_x, tr_y) of axes are determined.

2.1.2 Determine the value (φ, q, n) *for each point.* Based on the axes' corners detected in the previous step, we can estimate the size of each bin on both the phase and the magnitude axis according to the following equations:

$$\begin{cases} X_{bin} = (tr_x - bl_x)/72 \\ Y_{bin} = (tr_y - bl_y)/100 \end{cases} \quad (1)$$

where the discharge magnitude axis Y is divided into 100 bins and the phase axis X is divided into 72 bins, each of which stands for 5o.

Once the scale of axes is determined, we recover the PRPD matrix. We first initialize a 2D matrix of size 100×72 and all entries are set be to 0. Each entry (i, j) corresponds a grid (x, y) on the PRPD2D image, in which the grid (x, y) is estimated by

$$\begin{cases} x = round(bl_x + i \cdot X_{bin} - 0.5 \cdot X_{bin}) \\ y = round(bl_y - j \cdot Y_{bin} + 0.5 \cdot Y_{bin}) \end{cases} \quad (2)$$

The RGB value of each grid on the PRPD2D image represents the discharge rate. According to the color table which is prior constructed, we can roughly get the discharge rate and therefore reconstruct the PRPD matrix. Figure 1(d) visualizes the recovered PRPD matrix, which validated the effectiveness of our graph parsing method.

2.2. Feature extraction

In this work, the features used for PRPD pattern classification is extracted from the reconstructed PRPD matrix. 18 statistical features are extracted in total. They are skewness, kurtosis, cross correlation factor (cc) and asymmetry calculated from three distributions $q_a \sim \varphi$, $q_m \sim \varphi$, and $n \sim \varphi$, in which $q_a \sim \varphi$ refers to the distribution of the average discharge magnitude with respect to the phase angle, $q_m \sim \varphi$ is the distribution of the maximum discharge magnitude w.r.t the phase angle, and $n \sim \varphi$ is the distribution of the discharge rate and the phase.

The procedure of feature extraction is introduced in the following:

Algorithm 1. Feature extraction
1. Calculate the three distributions from the PRPD matrix;
2. Separate each distribution into two parts, one is of positive phase and the other has negative phase;
3. Calculate the mean μ and the standard deviation σ for both positive and negative parts of all distributions;
4. Calculate the skewness and kurtosis for each part of three distributions, by which we obtain 12 features in total;
5. Calculate the cross correlation factor and asymmetry for each distribution, by which we get 6 features in total.

The above mentioned features are widely used in PRPD classification. For the purpose of self-containedness, we briefly introduce them as follows.

2.2.1. Skewness. The skewness feature measures the degree of titling of a distribution with respect to normal distribution. It is defined by

$$S = \frac{\sum_{i=1}^{N} (x_i - \mu)^3 \cdot f(x_i)}{\sigma^3 \sum_{i=1}^{N} f(x_i)} \tag{3}$$

Here, x_i is the i -th phase and $f(x_i)$ is the probability at x_i. N is the number of phase bins in half cycle (either positive or negative).

2.2.2. Kurtosis. Kurtosis measures the peakness of a distribution. It is defined as

$$K = \frac{\sum_{i=1}^{N} (x_i - \mu)^4 \cdot f(x_i)}{\sigma^4 \sum_{i=1}^{N} f(x_i)} - 3 \tag{4}$$

When $K = 0$, it indicates that the distribution has the same sharpness as a normal distribution. If $K < 0$, it is flatter; otherwise, it is sharper.

2.2.3. Cross correlation factor (CC). CC measures the similarity between the positive half cycle and the negative half cycle of a distribution. It is defined by

$$CC = \frac{\sum_{i=1}^{N} x_i^+ x_i^- - \sum_{i=1}^{N} x_i^+ \sum_{i=1}^{N} x_i^- / N}{\sqrt{\left[\sum_{i=1}^{N} (x_i^+)^2 - \left(\sum_{i=1}^{N} x_i^+\right)^2 / N\right]\left[\sum_{i=1}^{N} (x_i^-)^2 - \left(\sum_{i=1}^{N} x_i^-\right)^2 / N\right]}} \tag{5}$$

When CC equals 1, it indicates that the two halves are highly similar; otherwise, they are not.

2.2.4. Asymmetry. Discharge asymmetry measures the difference of discharge level between positive and negative half cycles. It is

$$Q = \frac{Q_s^+ / N^+}{Q_s^- / N^-} \tag{6}$$

in which Q_s^+ and Q_s^- are, respectively, the sum of discharges of a distribution in the positive and negative half cycles. N^+ and N^- are the number of discharges in the corresponding half cycles. When $Q = 1$, it indicates equal discharge levels.

2.3. PD pattern classification

When electrical insulations work normally, there should be no discharge detected. Otherwise, PD detectors will capture partial discharges. According to the sources that lead to defects, we can categorize the PD patterns into four major classes: 1) corona discharge, which is an electrical discharge brought on by air surrounding an electrically charged conductor; 2) surface discharge, which occurs along the surface of solid insulations when the surface tangential electric field is high enough to cause a breakdown; 3) floating electrode partial discharge, which happens when there is an ungrounded conductor within the electric field between conductor and ground; and 4) particle discharge, which occurs when conductive particles contaminate insulation medium. Typical examples of each category are presented in Figure 3.

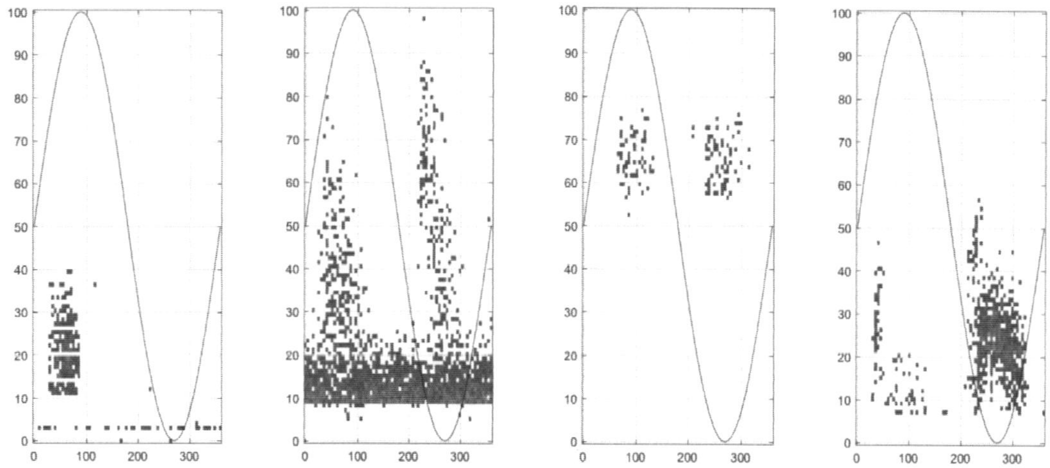

(a) Corona (b) Particle (c) Floating (d) Surface

Figure 3. Typical examples of PD graph for each category.

In this work, we employ the support vector machine (SVM) [8] for classification. SVM constructs a set of hyperplanes in the high-dimensional feature space that have the largest distances to the nearest training-data point of each class. It is extensively used in supervised machine learning for classification when the dimensionality of features are low. Therefore, SVM fits our task well.

When a set of training data (x_i, y_i) is given, SVM determines the hyperplane (w, b) by minimizing the following objective function:

$$\min \frac{1}{n}\sum_{i=1}^{n} \zeta_i + \lambda \|w\|^2 \quad s.t. \quad y_i(w \cdot x_i + b) \geq 1 - \zeta_i \text{ and } \zeta_i \geq 0 \tag{7}$$

Here, x_i is the i-th feature vector, y_i is the labelled class, and λ is a scaling factor. This function can be optimized via quadratic programming algorithms.

3. Experiments

To validate the proposed method, we build a dataset that contains 4600 PRPD2D images collected by PD-Detectors when performing routinely maintaince. Each category is of 1150 images. In our experiment, we randomly take 60% images for training and the remaining for test.

Table 1 lists the classification accuracy for each class and the mean accuracy reaches to 0.9158. The experiment validates the effectiveness of our proposed method.

Table 1. The classification accuracy of each class.

Category	Corona	Particle	Floating	Surface
Accuracy	0.9457	0.9043	0.8804	0.9326

4. Conclusions

In this paper, we have presented an approach to perform partial discharge pattern classification. Our approach first reconstructs the PRPD matrix from the PRPD2D images collected by PD-Detectors; then 18 statistical features are extracted from the recovered PRPD matrix and fed into SVM for classification. We have validated our approach in a dataset containing 4600 images. Experiments have demonstrated the high effectiveness of the method.

References

[1] Bolin P and Koch H 2005 Introduction and applications of gas insulated substation (gis) *Power Engineering Society General Meeting IEEE San Francisco*, CA, USA
[2] Aaradhi V and Gaidhani K 2013 Partial discharge in gas insulated substations (gis): effects, mitigation & analysis International Journal of Scientific & Engineering Research 4(3) pp 1-6
[3] Guo Z H, Wang H, Xiu-Ming D U, et al 2016 GIS Partial Discharge Pattern Recognition Based on Dimension Reduction Based on Rough Set Theory 2016 International Conference on Power, Energy Engineering and Management
[4] James R E and Phung B T 1995 Development of computer-based measurements and their application to pd pattern analysis IEEE Trans. Dielectr. Electr. Insul. 2(5) pp 838-856
[5] Fangcheng L, Hu J, Wang Z, et al 2015 GIS Partial Discharge Pattern Recognition Based on Principal Component Analysis and Milticlass Relevance Vector Machine Transactions of China Electrotechnical Society 30(6) pp 225-231
[6] Gao W, Zhao D, Ding D, Yao S, Zhao Y and Liu W 2015 Investigation of frequency characteristics of typical pd and the propagation properties in gis IEEE Trans. Dielectr. Electr. Insul. 22(3) pp 1654-1662
[7] Sahoo N and Salama M 2005 Trends in partial discharge pattern classification: a survey IEEE Trans. Dielectr. Electr. Insul. 12(2) pp 248-264
[8] Chang C and Lin C 2011 Libsvm: a library for support vector machines ACM Transactions on Intelligent Systems and Technology 2(3) pp 1-27

Edge Detection Based On the Characteristic of Primary Visual Cortex Cells

M M Zhu[1,*], Y L Xu[1] and H Q Ma[1]

[1]Air Force Engineering University, Xi'an, China

[*]ming_paper@163.com

Abstract. Aiming at the problem that it is difficult to balance the accuracy of edge detection and anti-noise performance, and referring to the dynamic and static perceptions of the primary visual cortex (V1) cells, a V1 cell model is established to perform edge detection. A spatiotemporal filter is adopted to simulate the receptive field of V1 simple cells, the model V1 cell is obtained after integrating the responses of simple cells by half-wave rectification and normalization, Then the natural image edge is detected by using static perception of V1 cells. The simulation results show that, the V1 model can basically fit the biological data and has the universality of biology. What's more, compared with other edge detection operators, the proposed model is more effective and has better robustness

1. Introduction

Biological vision system in the image processing has the superiority that the current computer vision system can't match. Therefore, it is an inevitable trend to use the research results of biological vision to promote the development of computer vision. With the development of brain science and physiological science, scholars have been studying the mechanism of biological vision system, which is applied to the research of image retrieval and image recognition, and has achieved many practical applications [1]. In the field of image processing, edge detection as one of the most basic technology has an important role in the subsequent image processing process. At present, the classical edge detection algorithm with image gradient or second derivative uses different differential operators to get the edge of image, such as Sobel operator, Log operator, Canny operator [2]. Some new edge detection methods have been proposed, such as wavelet multi-scale, fractal theory, mathematical morphology and so on, as scholars continue to explore in the field of image edge detection. However, most of the existing algorithms have a question about detection accuracy and anti-noise performance balance. Although there are some ways to solve this problem, those methods greatly increase the amount of computation. Therefore, finding an edge detection algorithm, which holds simple algorithm, good detection performance and strong anti-noise performance, has been an important research direction of image processing [3].

A large number of studies have shown that the biological vision system consists of dorsal pathway and ventral pathway whose structures and functions are relatively independent [4]. As a common starting point for both pathways, V1 has both dynamic and static feature-aware capabilities. Hubel and Wiesel [5] pointed out that V1 cells are the starting point of visual cortex information processing, generally divided into simple cells and complex cells. Simple cells have a small receptive field and are rectangular in shape, with a central region elongated and a parallel antagonist region on one or both sides of the central region [6]. Simple cells have orientation selectivity, which is almost unresponsive to diffuse light stimulation, whereas for specific strip stimuli, the response of simple cells varies with the direction of stimulation [7]. The experimental results obtained by many scholars through models show that the cells with orientation selection characteristics are not only responsible for dealing with the orientation information of the edge contours, but also play an important role in the perception and processing of motion signals [8]. This orientation selectivity is basic biological characteristic of V1 cells and plays an important role in edge detection [9]. Complex cells are integrated by simple cells, so the receptive field is larger than the simple cells and the structure is also more complex, as shown in Figure 1.

Based on the dynamic and static characteristics of V1 cells, this paper establishes a V1 cell model with orientation selection for edge detection, and compares with two methods of commonly used edge detection operator Canny and Log. The simulation results show that the cell model has better detection performance and robustness, and the fusion of biological vision and computer vision has been explored.

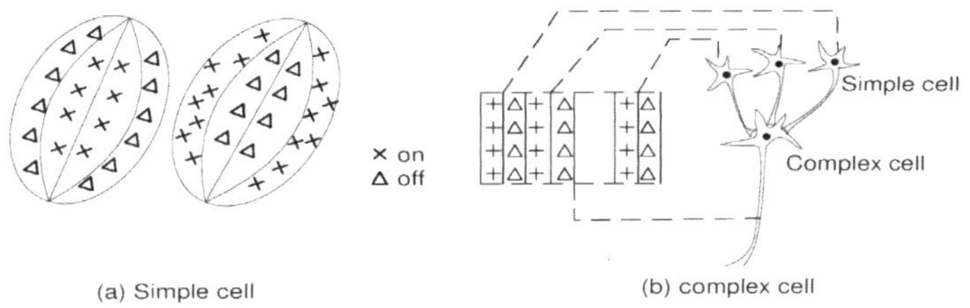

Figure 1. The receptive field.

2. The model

2.1. Modeling

Some scholars use Gabor function model, wavelet model or derivative model to simulate the receptive fields of simple cell. [10] Many linear systems and pulse response techniques show that the V1 simple cell is similar to the linear filtering phase of the spatiotemporal energy model [11]. Therefore, this paper uses the spatiotemporal filters of the spatiotemporal energy model to simulate the receptive fields of simple cell, and the spatiotemporal filters are decomposed into separable filters in space and time in order to achieve low computational complexity. In fact, the receptive field of simple cell is the same as the Gabor kernel, and the exponential decay function is a good way to simulate the orientation of cells in time. So spatial component of the filter is described by Gabor filters s and temporal component by an exponential decay function T .Given the spatial and temporal frequencies f_s and f_t of a receptive field, we define the spatial filter and the time filter by:

$$S(X,Y) = e^{\left(-\left(X^2+Y^2\right)/2\sigma^2\right)} e^{j2\pi\left(f_s\cos(\theta)X + f_s\sin(\theta)Y\right)} \quad (1)$$

$$T(t) = e^{(-t/\tau)} e^{j2\pi f_t t} \quad (2)$$

Where σ and τ define the spatial and temporal scales. And θ define the preferred orientation of a simple cell. A preferred velocity V_c related to the frequencies by the relation:

$$V_c = f_t / f_s \quad (3)$$

Through the Euler formula, equation (1) and (2) can be rewritten as:

$$S(X,Y) = e^{\left(-\left(X^2+Y^2\right)/2\sigma^2\right)}\left(\cos 2\pi\left(f_s\cos(\theta)X + f_s\sin(\theta)Y\right) + j\sin 2\pi\left(f_s\cos(\theta)X + f_s\sin(\theta)Y\right)\right) \quad (4)$$

$$T(t) = e^{(-t/\tau)}\left[\cos(2\pi f_t t) + j\sin(2\pi f_t t)\right] \tag{5}$$

Denoting the real and imaginary components of the filters S and T as S_o, T_o and S_e, T_e respectively:

$$S(X,Y) = S_o(X,Y) + jS_e(X,Y) \tag{6}$$

$$T(t) = T_o(t) + jT_e(t) \tag{7}$$

We introduce the odd and even spatiotemporal filters defined as follows:

$$G_o(X,Y,t) = S_o(X,Y)T_e(t) + S_e(X,Y)T_o(t) \tag{8}$$

$$G_e(X,Y,t) = S_o(X,Y)T_o(t) - S_e(X,Y)T_e(t) \tag{9}$$

The luminance and contrast normalization [10] are implemented as a input $P(x,y,t)$ pre-processing stage:

$$P'(x,y,t) = \frac{P(x,y,t) - Lum(t)}{Con(t)} \tag{10}$$

$Lum(t)$ and $Con(t)$ are defined as:

$$Lum(t) = \sum_x \sum_y P(x,y,t) \tag{11}$$

$$Con(t) = \sqrt{\sum_x \sum_y (P(x,y,t) - Lum(t))^2} \tag{12}$$

We define the response of simple cells with a preferred orientation of contrast sensitivity in the spatial domain, with a preferred velocity:

$$R_{o/e}(x,y,t) = \left(G_{o/e}(X,Y,t) \overset{(x,y,t)}{*} P'\right)(x,y,t) \tag{13}$$

The complex cells are not a simple superposition of simple cells. it exhibits a strong non-linear property[11]. Many scholars have studied the integration of simple cells. The normalized model [12] can explain the saturation characteristics of cell responses. Energy model [13] can achieve the phase invariance of cell responses. Max model [14] can achieve the scale and local translation invariance. However, the above mechanisms only simulate the partial characteristics of complex cells, without the effective integration of simple cell responses. Based on the existing biological experiment data and models, this paper aims to achieve edge detection by integrating simple cell responses by the following ways:

The half-wave rectification (HR) [10] is used for nonlinear processing:

$$R_{o/e}'(x,y,t) = HR\left[R_{o/e}(x,y,t)\right] \tag{14}$$

The integration of the above responses is as follows:

$$E(x,y,t) = \sqrt{R_o(x,y,t)^2 + R_e(x,y,t)^2} \tag{15}$$

Followed by normalization: the final V1response is defined by:

$$E_{V1}(x,y,t) = \frac{E^n(x,y,t)}{k_{tuned} I_i^n + \sum_{i=1}^{N} w_i E^n(x,y,t,\theta_i) + \varepsilon} \tag{16}$$

Where $0 \le \varepsilon \le 1$ is constant. w are weights, k_{tuned} and n are the key parameters. When $k_{tuned} = 0, n = 1$, this equation represents a standard normalization.

According to the experimental simulation results, when $k_{tuned} = 0, n = 2, w = 1$, the cell model can best fit the biological characteristics.

$$E_{V1}(x,y,t) = \frac{E^2(x,y,t)}{\sum_{i=1}^{N} E^2(x,y,t,\theta_i) + \varepsilon} \tag{17}$$

2.2. Model analysis

In order to verify the biological characteristics of the V1 cell model, the response of the model to the input stimulus is compared with the biological data. Figure 2(a) and Figure 2(b) show that stimuli are drifting gratings, and plaid patterns composed of two gratings.

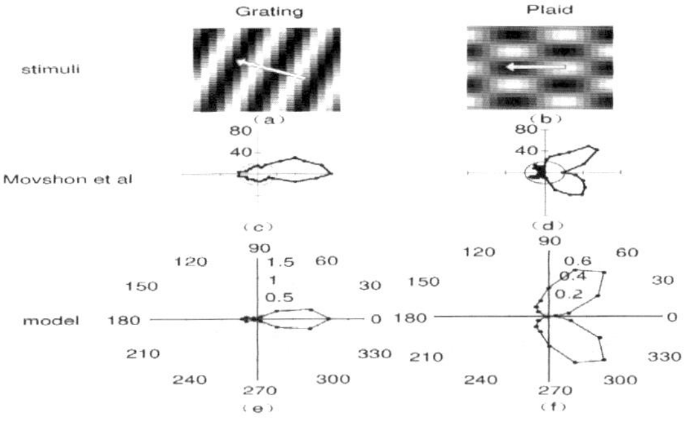

Figure 2. Direction tuning curves.

From Figure 2(c)-2(f), the responses of model are basically consistent with the biological experiment results. The direction tuning curve for gratings is unimodal. but the direction tuning curve for plaids shows two distinct lobes. Each lobe corresponds to one of the plaid's component gratings. The above experimental results show that the V1 cell model can sense the direction of the stimulus and has selectivity and the universality of biology. So it has laid a good biological foundation for the edge detection experiment.

3. Experiment

In order to verify the effectiveness of the V1 cell model, the edge detection experiment is performed using the MATLAB R2014b software. All images in the experiment are from the USF (University of South Florida) image edge detection database.

3.1. Parameters analysis

It is found that the size and quantity of spatial filters have a great influence on the edge detection of natural images. The experiment result is shown in figure 3(c)-(h) with only changing the size or number of filters (the other parameters constant). Figure 3(a) and 3(b) are the original image and the ground truth.

From the Figure 3, it can be seen that as the size of the spatial filters increases, the edge becomes blurred and even some fine edges overlapping or disappear. When the size of filters is $3*3$, the edge detection performance is best. When the number of filters is 4, some edges can't be detected. When the number of filters is 6 and 12, the edge detection performance is consistent with the Ground truth. When the number of filters exceeds 6, continuing to increase the number of filters can't further enhance the edge detection performance, but it will increase the amount of computation. Based on the above analysis, we can see that the edge detection effect of the image is optimal when the size of the spatial filter is $3*3$ and the number is 6.

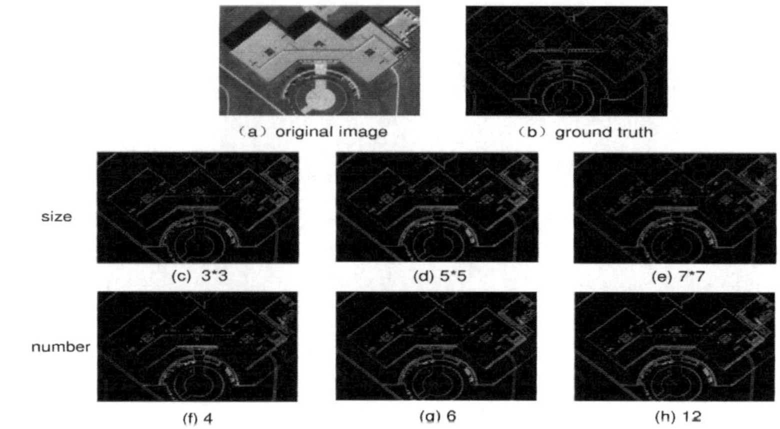

Figure 3. Edge detection with different Parameters.

3.2. Results and analysis

In this paper, the edge detection of all the images in the USF image edge detection database is carried out and compared with the classical edge detection operator Canny and Log operator in order to demonstrate the performance of the model. Three images are selected for demonstration analysis. As shown in Figure 4, The first row is the original image and the second is the ground truth in the USF database, and the third row to the fifth row are the Canny operator, the Log operator and the edge detection result of the model, respectively.

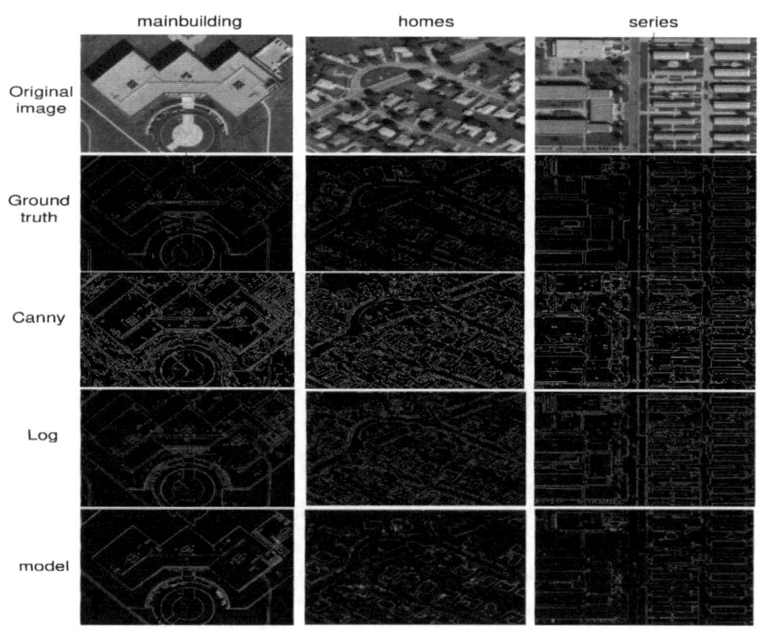

Figure 4. Edge detection without noise.

It can be seen from Figure 4, compared with the Ground truth, the results of Canny operator have been more pseudo-edge, so it can't be used to identify the target. Compared with the results of our model, Log operator has the same detection performance. And both of them can detect edge accurately. But our model contains fewer background pseudo-edges and has a better detection performance. The reason can be concluded that the operator represented by Canny and Log depends on the local calculation of the linear filter and uses the bottom feature (the image pixel gray value) to detect the edge of the image, which amplifies some small and unimportant edges. In this paper, the two-dimensional Gabor function and the exponential decay function are combined effectively so that our model has the direction selection characteristics, And the direction (primary feature) is not only higher than the pixel and gray (underlying features), but also lower than the edge (high-level semantic features), which is the most useful form to understand the high-level semantic features (edge). Therefore, the model can detect the edge of the image by using the direction feature, and can achieve better results.

In order to further demonstrate the anti-noise and robustness of our model, after adding Gaussian noise (mean 0, variance 0.01), Poisson noise and Spot noise (density 0.05), the results shown in the Figure 5.

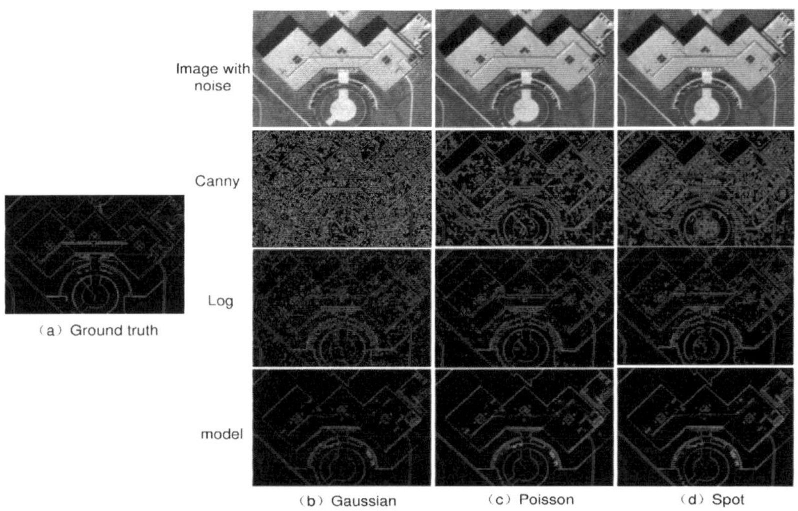

Figure 5. Edge detection with noise.

As shown in Figure 5, after adding three different kinds of noise, the original image becomes slightly blurred. Canny operator is very sensitive to three kinds of noise, so a large number of noise edges appear in its results. After adding Gaussian noise, Log operator will detect a lot of noise edge and can't extract the target edge accurately. Under the Poisson and speckle noise conditions, although the target shape can be identified, more noise is introduced. For the images with three kinds of noise respectively, the results of our model are almost no noise. It can detect edge accurately and has stronger the anti-noise performance and robustness. Noise can significantly affect the gray value of the image, as a result that the Canny and Log operators can't effectively distinguish the edge and noise. Our edge detection model is based on the biometric visual mechanism, which uses the direction of the characteristics to detect the edge of the image. It can be seen from the equation (15) and (16) that the model output is an effective integration of a simple cell, which can combine local information to enhance its ability to distinguish the main body edge and background noise in the image.

For quantitative analysis, this paper calculates the peak signal-to-noise ratio (PSNR) corresponding to each edge algorithm, as shown in Table 1.

Table 1.PSNR(dB) corresponding to each algorithm.

algorithm	Gaussian	*Poison*	Spot
Canny	17.16	19.98	18.86
Log	18.85	21.39	22.10
ours	**22.48**	**25.07**	**27.25**

As can be seen from Table 1, compared with Canny operator and Log operator, the proposed model is more effective and has a better robustness.

4. Conclusion

Based on the dynamic and static characteristics of V1 cells, this paper establishes a V1 cell model to detect the natural images, which makes a helpful exploration to the fusion of biological vision and computer vision. Simulation results show that the V1 model can basically fit the biological data and has the universality of biology. The influence of the size and number of the spatial filter on the edge detection of the natural image is analyzed by the control variable method, and the optimal parameters are finally determined. What's more, compared with the traditional edge detection operators, the proposed model is more effective and has a better robustness. This paper constructs a biological vision model and applies it to image processing. The next stage work will continue to establish a more perfect visual cortex perception model in order to achieve better information representation ability, which will provide more new methods and ideas for the fusion of biological vision and computer vision.

Acknowledgments

This work was supported by NSFC (6137217, 61379104).

References

[1] Namboodiri V M K, Huertas M A and Monk K J 2015.Visually cued action timing in the primary visual cortex. *J. Neuron.* **86(1)** 319-330

[2] Gao S, Peng Y and Guo H 2014.Texture analysis and classification of ultrasound liver images. J. Bio-medical materials and engineering. 24(1) 1209-16

[3] Alberto C, Pietro C and Kiran M K 2004. Calculation of tensile and flexural strength of disordered materials using fractional calculus. J. Chaos, Solitions and Fractals. 37(1) 17-25

[4] Hinton G E, Salakhutdinov R R 2006.Reducing the dimensionality of data with neural networks. J. Science. 313(5786) 504-507.

[5] Hubel D H and Wiesel T N 1959.Receptive fields of single neurons in the cat's striate cortex. J. Journal of Physiology. 148(3) 574-591

[6] Zou H Z, Xu Y L, Ma S P, Li S and Zhang W D 2016. Neuron-filtering feature extraction based on LNP model. J. Journal of Image and Graphics. 21(10) 1376-82

[7] Gao S, Peng Y and Guo H 2014.Texture analysis and classification of ultrasound liver images. J. Bio-medical materials and engineering. 24(1) 1209-16

[8] Burr D C and Thompson P 2011. Motion psychophysics: 1985-2010. J. Vision Research. 51(13) 1431-56

[9] Poo M, Du J and Ip N Y 2016. China Brain Project: Basic Neuroscience, Brain Diseases, and Brain-Inspired Computing. J, Neuron 92(3) 591-96

[10] Xu Y L, Lv C and Ma S P 2017.Motion Feature Extraction of Random-Dot Video Sequences with the Visual Cortex Mechanism, J. Journal of University of Electronic Science and Technology of China. 46(4) 630-635

[11] Yi Z and Bauer P H 2016.Spatiotemporal energy demand models for electric vehicles. J. IEEE Transactions on Vehicular Technology 65(3) 1030-1042.

[12] Kasabov N 2015.Integrative computational neurogenetic modelling. J. Brain Mapping. 667-674

[13] Rust N C, Schwartz O and Movshon J A 2005.Spatiotemporal elements of macaque v1 receptive fields. J. Neuron. 46(6) 945-956

[14] Bradley D and Goyal M S 2008.Velocity computation in the primate visual system. J. Nature Reviews Neuroscience. 9(9) 686-695

Institute of Physics
Dirac House, Temple Back
Bristol BS1 6BE UK

ISSN: 1742-6588
ISBN 978-1-5108-5851-0